Springer Series in Reliability Engineering

Today's modern systems have become increasingly complex to design and build, while the demand for reliability and cost effective development continues. Reliability is one of the most important attributes in all these systems, including aerospace applications, real-time control, medical applications, defense systems, human decision-making, and home-security products. Growing international competition has increased the need for all designers, managers, practitioners, scientists and engineers to ensure a level of reliability of their product before release at the lowest cost. The interest in reliability has been growing in recent years and this trend will continue during the next decade and beyond.

The Springer Series in Reliability Engineering publishes books, monographs and edited volumes in important areas of current theoretical research development in reliability and in areas that attempt to bridge the gap between theory and application in areas of interest to practitioners in industry, laboratories, business, and government.

Now with 100 volumes!

Indexed in Scopus and EI Compendex

Interested authors should contact the series editor, Hoang Pham, Department of Industrial and Systems Engineering, Rutgers University, Piscataway, NJ 08854, USA. Email: **hopham@soe.rutgers.edu, or Anthony Doyle, Executive Editor, Springer, London. Email:** anthony.doyle@springer.com.

Seongwoo Woo

Design of Mechanical Systems

Accelerated Lifecycle Testing and Reliability

Springer

Seongwoo Woo
Department of Mechanical Engineering
Ethiopian Technical University
Addis Ababa, Ethiopia

ISSN 1614-7839 ISSN 2196-999X (electronic)
Springer Series in Reliability Engineering
ISBN 978-3-031-28937-8 ISBN 978-3-031-28938-5 (eBook)
https://doi.org/10.1007/978-3-031-28938-5

This Springer imprint is published by the registered company Springer Nature Switzerland AG
The registered company address is: Gewerbestrasse 11, 6330 Cham, Switzerland

Preface

Customers in modern life utilize a mechanical product because of their good performance, reliability, ease of use, cost reduction, etc. In the beginning of the twentieth century, new advanced mechanical systems such as airplanes, automobiles, refrigerators, machine tools, appliances were designed for people to guide a good life. It will be generally implemented in the product development process as follows: (1) define the situations and their specifications; (2) develop the prototype, detailed product design, and testing; and (3) produce. A mechanical system transmits (produced) power to fulfill its purpose that requires forces and movement. It thus produces mechanical advantages by adapting proper mechanisms. For instance, an automobile is a wheeled motor vehicle used for transit. Its power is produced by the engine and transmitted to each wheel through mechanisms such as gear transmission and drive train. In the process, the mechanical system is subjected to repetitive stresses so that it might be designed in a sturdy and strong way.

Due to the increase in (intended) functions and parts, complexity, and growing competition in the marketplace, new distinctive attributes in mechanical products, such as refrigerators, are quickly developed and transported to the marketplace. With incomplete testing or no knowledge of how they are employed by the consumer, introduction will develop product failure from the field and influence the manufacturer's brand in a negative way. The new qualities of technological complex-designed systems should be assessed in the developing stage before being released into the field to avoid accidents that suggest design flaws. Employing a traditional mechanical product therefore must be required to have some methodologies with reliability quantitative (RQ) specifications.

To stop recalls of a mechanical system from the field that has design defects, it should be designed to survive typical circumstances operated by the person who actually acquires and makes use of the product. For instance, the Boeing 737 MAX passenger aircraft from March 2019 to December 2020 was grounded after 346 persons lost their lives in pair crashes. The airplanes adopted the CFM International LEAP-1B engines using the optimized 68-inch fan design; these engines consumed 12% less fuel and were 7% lighter than other engines. Investigators, including the Ethiopian Civil Aviation Authority, tentatively conjectured that the crash was created by the aircraft's design engine. The reliability of a product thus is one of its achievable attributes, regardless of both the type of product and the number of components combined into it. To ensure that a product is not unsuccessful in the market, problematic parts need to be verified by proper reliability testing, such as ALT, which may generate reliability quantitative (RQ) statements.

Devices, working structures, and their elements in multimodule mechanical products are subjected to the influence of various loads. These can be static, cyclic, or dynamic loads. As a result, the accumulation of damage and the development of mechanical failure, such as fatigue cracks, under the influence of loads is a common phenomenon that occurs in metals. Fracture, fatigue, creep, and environmentally assisted cracking, among other critical and subcritical processes, are still open issues in the design of products. To slow crack growth and ensure an adequate level of safety and optimal durability of structural elements, experimental tests and simulations are required to determine the influence of various factors. Such factors include the impact of microstructure, voids, notches, the environment, etc. Attention is being paid from the very basic aspects explaining the material response at the atomic and microstructural scales to the development of engineering procedures defining the structural integrity conditions of a given structural component.

Fatigue is often the principal origin of failure in metallic components, accounting for approximately 80–95% of all structural failures. It exhibits itself in the configuration of cracks that originate from high stress concentrations—grooves, thin surfaces, holes, etc., in mechanical products and propagate it. Fatigue thus is the weakening of a material created by cyclic loading. A considerable amount of consideration is being focused on low-cycle fatigue, such as superalloys, particularly in the field of turbine engines made of nickel-based polycrystalline materials. Fatigue also has some quantifiable elements, such as the stress ratio, R ($=\sigma_{min}/\sigma_{max}$), or mean stress, which could be interpreted as the relationship of the maximum cyclic stress to the minimum cyclic stress. Employing a stress ratio, which is manifested as an accelerated factor (AF) in parametric ALT, pinpoints the structural imperfections—stress raisers—in the mechanical system.

Engineers often have used the strength of materials as one of the traditional design aspects. Recent fracture mechanics manifests that the critical factor could be fracture toughness as a different property of material strength. With the application of quantum mechanics, designers recognize that product failure occurs from microscale or nanoscale voids observed in engineering plastics or metallic alloys. Because they judge limited parts and testing time, contemporary (test) skills cannot reproduce the structural imperfections of parts in a multimodule structure and do not pinpoint the fatigue(s) that happen to the parts by end-users in the field.

As an alternative that can be used along with the FEM, engineers believe that field failures could be judged by (1) strict mathematical representation such as Lagrangian or Newtonian skills; (2) attaining its time response for (dynamic) loading, which is acquiring the stress/strain in a product; (3) utilizing the established method of rainflow counts with von Mises stress; and (4) assessing system usefulness by Palmgren–Miner's principle. Nonetheless, employing this structured method, which shall supply closed formation, correct answers might necessitate some presumptions that did not identify fatigue failures in a multimodule system caused by matter imperfections such as microvoids, thin surfaces, contacts, etc.

Product reliability can be defined as the ability of a system or module to function under specified conditions for a stated period of time. To develop a new reliability methodology, we have to combine the design concepts of the mechanical systems with statistical concepts, including probability. This book suggests parametric ALT as a general way to pinpoint design imperfections of a new mechanical product and alter them. This established process to enhance reliability and identify the reasons for failure mode is elucidated to explore the right design answers. It is made up of: (1) an ALT strategy produced on BX life, (2) load inspection, (3) parametric ALTs with the structural modifications, and (4) an estimation of whether the product design(s) secures the desired BX life. We also propose a quantum-transported time-to-failure prototype and sample size derivation in this book.

The main purpose of writing this book, therefore, is to explain a parametric ALT embedded in the product development process. It consists of eight chapters. Chapter 1 explains the definition and design of the mechanical system. It generally moves power from one place to another by choosing a suitable mechanism. For example, after briefly reviewing the design principles of mechanical products, especially refrigerators, we explain the historical reliability concepts for evaluating the product lifetime. Finally, we stress the necessity of a new reliability methodology for assessing product design because of the incompleteness of current concepts.

Chapter 2 will discuss the basic concepts of design mechanical systems, such as load analysis. We also deal with modeling techniques such as Newtonian, D'Alembert's principle, Lagrangian, and bond graph. Chapter 3, from the standpoint of probability/statistics and reliability engineering, will discuss what (reliability) data in the mechanical system are. As customers start to use products, failure data start to emerge. Based on the field data, we can estimate the product lifetime and describe it on bathtub curves. However, because it is very difficult to obtain complete data, companies independently try to obtain it from reliability testing. However, there is no reliability methodology based on the time-to-failure model and sample size equation. This chapter also addresses probability and its distributions, such as binomial, Poisson, exponential, and Weibull distributions, which provides a clear understanding of reliability testing and lifetime from product testing (or field data). It requires numerous concepts, such as bathtub, Weibull distribution, failure rate, and lifetime measures, such as MTTF (or BX life) in reliability engineering. BX life is defined as the time at which X percent of the items in a population will have failed. On the other hand, the mean time to failure (MTTF) equals a B60 life. BX life is popularly used because MTTF takes a long time to reach it. To do this, we need to merge the design concepts of mechanical products, such as the life stress model and sample size equation.

Chapter 4 will address mechanical structures, including mechanisms such as belts, gears, bearings in a mechanical product. The engineer should understand how to design the type of mechanism and its analysis, such as displacement, velocity, and acceleration, that will be a basis for load (or stress) analysis. Chapter 5 will discuss the basic concepts of design mechanical systems such as strength and stiffness, based on load analysis. Chapter 6 will address the mechanical system. However, each mechanical product has different design principles—thermodynamics, mechanism, dynamics, fluid mechanics, strength of mechanics, statistics, etc.—we can find that their failure mechanisms—fatigue, fracture, and erosion—due to the design problems are similar. Chapter 7 addresses parametric ALT as a powerful tool for future engineering development and its case studies. Chapter 8 addresses the case studies of parametric ALT.

Addis Ababa, Ethiopia

Seongwoo Woo

Contents

About the Author

Seongwoo Woo has a B.S. and an M.S. in Mechanical Engineering, and he has obtained his Ph.D. in Mechanical Engineering from Texas A&M. His major interests are in energy systems such as HVAC and its heat transfer, optimal design and control of refrigerators, reliability design of mechanical components, and failure analysis of thermal components in marketplace using non-destructive methods such as scanning electron microscopy and X-ray. In particular, he has developed parametric accelerating life testing (ALT) as a new reliability methodology, which would determine if there is a design fault in a mechanical system that is subjected to repetitive stress. Using this parametric ALT, engineers can find the design faults before the product is launched and thus can avoid failures.

From 1992 to 1997, he worked in the Agency for Defense Development, Chinhae, South Korea, where he was the researcher in charge of the Development of Naval Weapon System. From 2000 to 2010, he worked as a senior reliability engineer in Side-by-Side Refrigerator Division, Digital Appliance, SAMSUNG Electronics, where he focused on enhancing the lifetime of refrigerators using a parametric ALT. Now, he is working as a professor at the Mechanical Department, Ethiopian Technical University.

Chapter 1
Introduction to the Necessity of Design Methodology

1.1 Introduction

As people want to spend their modern human life diversified and wealthy, mechanical products equipped with new technology, such as refrigerators, automobiles, airplanes, etc., not used before the 20 centuries have been developed. To survive the competitive market situation, it should fulfill market requirements such as high response, high control of performance, low noise, precision control over a wide frequency range, long lifetime, compact and highly portable weight, high reliability, low price, advanced hardware design, contamination resistance, and energy efficiency. Moreover, every year, companies have to develop new features through the product development process. Well-developed designs can offer tremendous advantages in the market. On the other hand, they can also expose potential problems under customer usage or environmental conditions because most products have inherent design imperfections, such as cracks, notches, and local thin areas.

As the development period of products such as automobiles has shortened from 65 to 24 months, the reliability grows from 0.9 to 0.99. Moreover, because of the successful possession of a strong desire in the global marketplace, manufacturers should innovate their products with new technology and swiftly deliver them to the field. With either finite testing or no understandable comprehension of how a current design can be utilized by the customer, the product will launch. That is why product recall frequently occurs in the field. Only a few global companies embedded in high technologies and their organizational structures might survive in the competitive marketplace. To meet market needs such as quality/reliability and developing time, manufacturers need to be a structured reliability method tightly bound to the product developing process. However, there are no systematic reliability methods that can demonstrate the robustness of product design. That is, after engineers find inherent design faults by new proper reliability methodology such as parametric ALT and correct them, manufacturers could safely implement new designs into products because such an approach reduces quality costs due to the frequent failures in the marketplace.

S. Woo, *Design of Mechanical Systems*, Springer Series in Reliability Engineering,
https://doi.org/10.1007/978-3-031-28938-5_1

Devices, working structures, and their elements in multimodule mechanical products are subjected to the effect of numerous loads. They may be static, cyclic or dynamic loads. As a result, the accumulation of damage and the evolution of mechanical failure, such as fatigue, is a common phenomenon. Fracture, fatigue, creep or environmentally assisted cracking, among other critical and subcritical processes, still open issues in the design of products. To decrease crack growth and ensure a sufficient level of safety and optimal durability of structural parts, testing and simulations are necessary to decide the effect of some factors. They comprise the impact of microstructure, voids, notches, the environment, etc. Attention is being paid from the very basic aspects explaining the material response at the atomic and microstructural scales to the development of engineering procedures defining the structural integrity conditions of a given structural component.

The application of all this newly developed knowledge affects a wide range of sectors where structural safety is a major concern: nuclear power plants, civil engineering structures, oil and gas, pressurized equipment, aircraft, naval structures, etc. Structural failures in any of these sectors may have evident serious consequences in terms of human lives, environmental disasters or economic losses. Therefore, to avoid structural failures, it is necessary to understand the different mechanisms generating critical and subcritical processes in the structural materials and to develop assessment techniques and management procedures for the corresponding structures.

This book supplies the fundamental ideas and tools for the design of mechanical systems, such as parametric ALT, as a structured reliability method that can assess the design of mechanical products in the developing process. This established process for improving reliability and identifying the causes of failure mode is clarified to examine the correct design responses to address failures in the product. It includes the following: (1) an ALT strategy produced on BX life, (2) load inspection, (3) parametric ALTs with the structural alternations, and (4) an assessment of whether the current design(s) secures the desired BX life. A mechanical system transfers power, which implements mechanical advantages by adapting product mechanisms. In this process, a mechanical product is subjected to repetitive stress. If there are design faults that can cause an inadequacy of strength (or stiffness), the product will fail in its lifetime. To design the mechanical system, it is necessary to comprise the definition for mechanical systems, probability/reliability concepts, numerous mechanisms, load analysis, strength design, failures of mechanical system design, etc.

1.2 Mechanical Product—Refrigerator (or Air Conditioner)

Mechanical products transfer power to fulfill a job which necessitates forces and motion, which generates mechanical advantages by adapting system mechanisms. The term 'mechanical' originates from the Latin 'machina', which may be explained as connecting to machinery. A mechanical product usually is made up to (1) a power source (electric motor) and actuators which produce forces and motion, (2) structured mechanisms which form the actuator input to attain a particular implementation of output forces and motion, and (3) a controller with sensors.

1.2.1 Refrigerator (or Air Conditioner)

To store fresh (or frozen) food, chilled air in a refrigerator is supplied from the heat exchanger (evaporator) to both the freezer and the refrigerator. To fulfill this, a vapor-compression refrigeration cycle is employed in refrigerators. Refrigeration is a procedure of lowering the temperature of a matter less than that of its environment. The capacity is defined as the tone of refrigeration, identical to the heat necessitated for melting one ton of ice per day. There are two types as follows: (1) vapor-compression refrigerating systems, (2) vapor-absorption refrigerating systems, etc. There are main applications, such as (1) central air conditioning, (2) food storage in refrigerators, (3) removal of heat in the areas of chemical processes, and (4) cryonics projects.

To produce a continuous cooling effect, it operates at two pressure levels (high and low). Major components of a refrigerator are (1) compressor, (2) condenser, (3) capillary tube (or expansion valve), and (4) evaporator (Fig. 1.1).

On the other hand, there is another mechanical product, namely, room air conditioner (RAC), that is operated by a vapor-compression refrigerating cycle. Air conditioning is the procedure of eliminating heat and moisture from the inside of a filled area to enhance the comfort of residents. There are some examples as follows: (1)

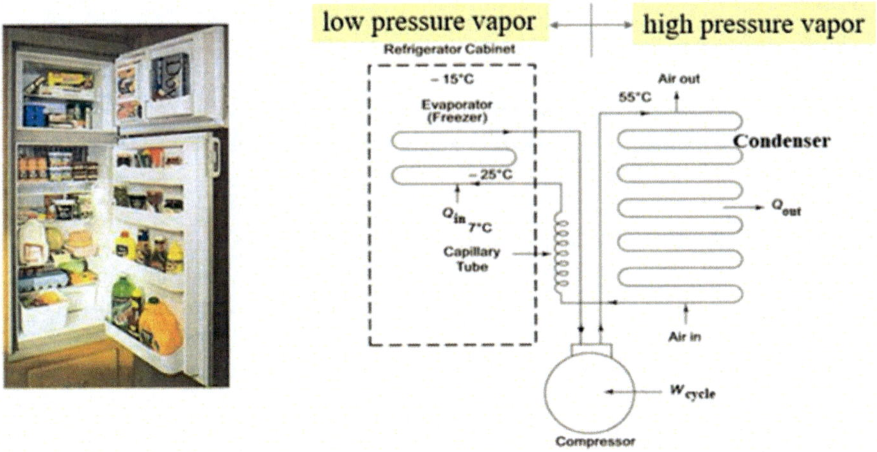

Fig. 1.1 Refrigerator and its diagram

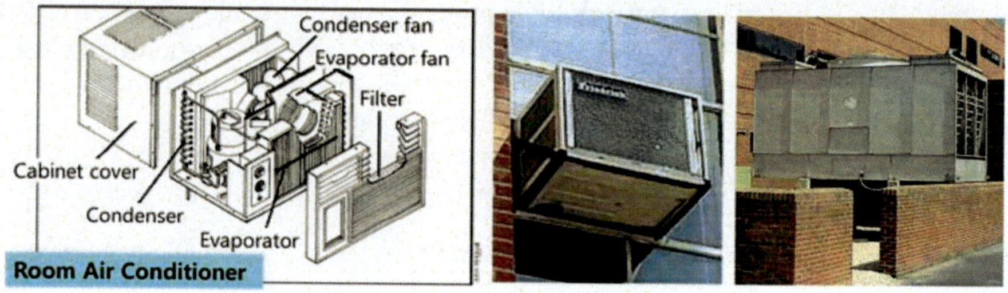

Fig. 1.2 A typical air conditioner

upgrade the human comfort in offices, hotels, buses, cars, etc., (2) utilized in industries—food processing, printing, chemicals, pharmaceuticals and machine tools, etc.

For example, assume that a room is continually kept at a temperature of 25 °C. In the air conditioner, the air in the room is pulled by a fan and is passed through a *cooling coil*, the surface of which is continued, say, at a temperature of 10 °C. By going through the coil, the air is cooled down (e.g., to 15 °C) before being released into the room. As taking up the room heat, the air is passed again through the cooling coil at 25 °C (Fig. 1.2).

1.2.2 Brief History of Refrigeration

Approximately BC 2000 years ago, residence in Egypt are known to make ice by maintaining water in the porous pots outside the home at night. Approximately 1000 BC, China utilizes ice for chilling beverages. In the fourth century, as liquefying salt in water, East Indians made ice. In 1834, Jacob Perkins (American) evolved a hand worked refrigeration system utilizing ether as the working fluid (Fig. 1.3).

Dr. John Gorrie of Florida in 1851 acquired the first American patent of a cold air machine to make ice to heal people hurting from high fever. Refrigeration with compressed ammonia (NH_3) was introduced by David Boyle in 1872. Four years later, Carl Linde invented the first reliable and efficient ammonia refrigerator. In 1904 on the New York Stock Exchange, an approximately 450-ton cooling machine was placed for use. In 1915, the first two-stage contemporary compressor was evolved. Ferdinand Carre, French engineer, used this possession of ammonia's affinity for water to make the earliest absorption refrigeration product in 1859 (Fig. 1.4).

Until the 1920s, the evolution of refrigeration systems was limited to improvements in vapor-compression systems and cold-air apparatus. After the 1920s, there has been large diversifying in the extension of refrigeration systems, conducting to new evolutions such as vortex tubes, thermo-electrics, pulse tubes, steam jets, and centrifugal compression systems. In the 1930s, the most critical thing was the

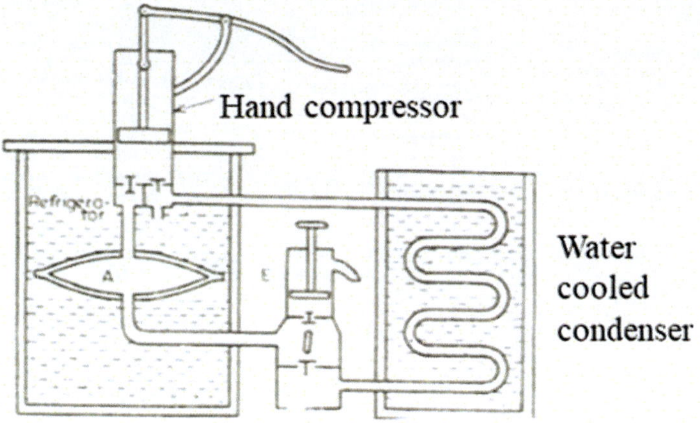

Fig. 1.3 The hand-operated refrigeration machine (Jacob Perkins)

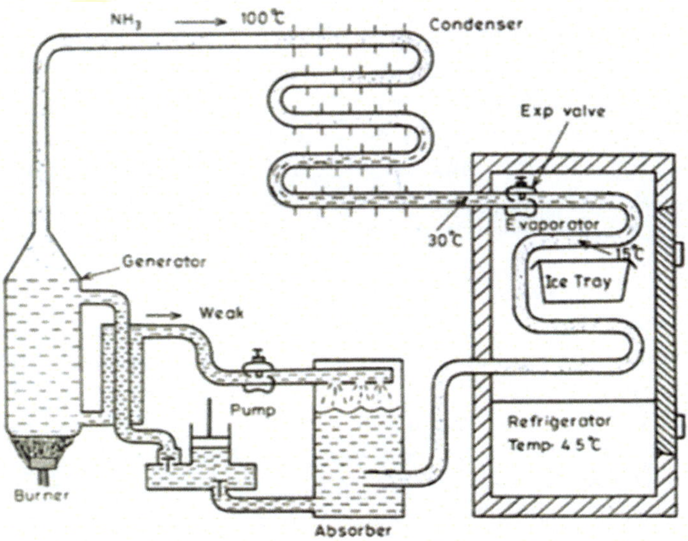

Fig. 1.4 Vapor-absorption machine (Ferdinand Carre)

creation of refrigerants by Midgley (American), which were chlorofluorocarbons (CFCs)—Freons (Fig. 1.5).

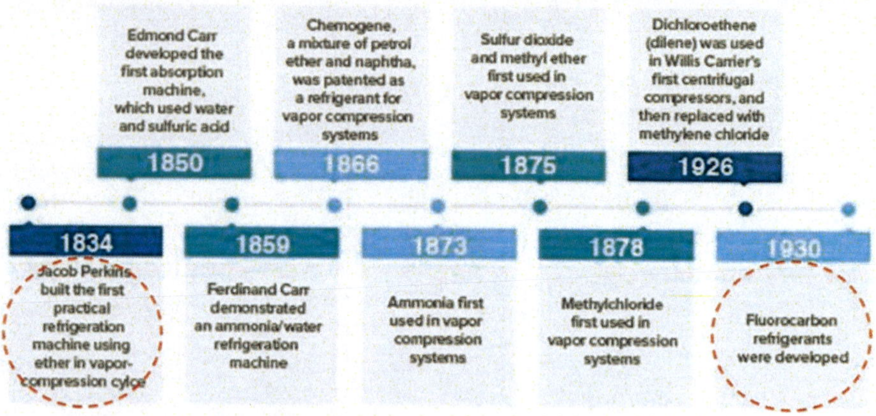

Fig. 1.5 Timeline of refrigeration history

1.2.3 Vapor-Compression Refrigeration (VCR) System

The VCR cycle is a procedure of dropping the temperature of a substance to less than that of its surroundings. That is, the VCR cycle is a general way to move heat from a low temperature to a high temperature. Figure 1.6 manifests the purposes of refrigerators and heat pumps. The objective of a refrigerator is the elimination of heat, defined as the cooling load, from a low-temperature medium. On the other hand, the objective of a heat pump is the move of heat to a high-temperature medium, defined as the heating load.

As the reversed Carnot cycle is used, the cyclic refrigeration implement is functioning between two constant temperature spaces and the T-s diagram for the working fluid. That is, you should remember that heat transfers in the Carnot cycle occur at constant temperature. If our concern is the cooling load, the cycle is specified as the Carnot refrigerator. On the other hand, if our concern is the heat load, the cycle is defined as the Carnot heat pump (Fig. 1.7).

The principle of contrast for refrigeration cycles is the reversed Carnot cycle. A refrigerator or heat pump which works on the reversed Carnot cycle is defined as a Carnot refrigerator or a Carnot heat pump. Their COPs are expressed as follows:

$$\mathrm{COP}_{R,Carnot} = \frac{1}{T_H/T_L - 1} = \frac{T_L}{T_H - T_L} \tag{1.1a}$$

$$\mathrm{COP}_{HP,Carnot} = \frac{1}{1 - T_H/T_L} = \frac{T_H}{T_H - T_L} \tag{1.1b}$$

The VCR cycle consists of an evaporator, compressor, condenser, and expansion (or throttle) valve. In a perfect VCR cycle, the refrigerant comes into the compressor as a saturated vapor and comes it out as superheat. It is chilled down to the saturated

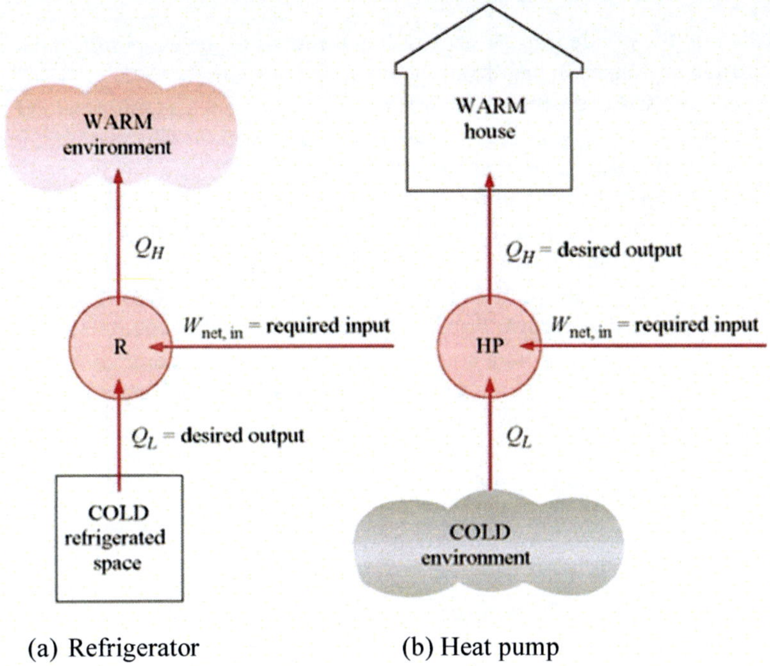

Fig. 1.6 Vapor-compression refrigerating cycle

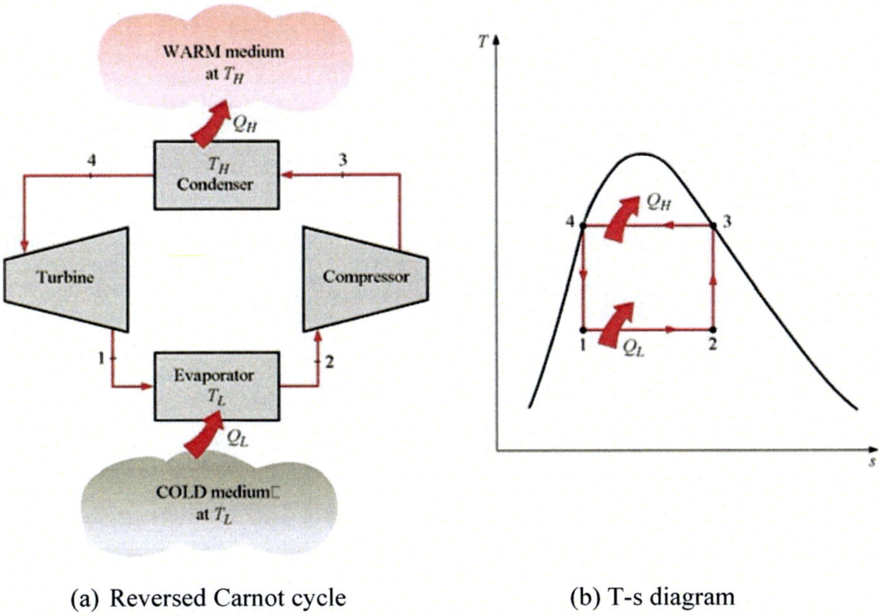

Fig. 1.7 Ideal refrigeration cycle—reversed Carnot cycle

liquid state in the condenser. After it is thus throttled to the evaporator pressure, it will vaporize as it takes in heat from the refrigerated room.

The perfect VCR cycle thus is made up to four procedures as follows: (1) Process 1–2: isentropic compression in compressor, (2) Process 2–3: constant pressure heat refusal in the condenser, (3) Process 3–4: throttling in an expansion valve, and Process 4–1: constant pressure heat addition in the evaporator (Fig. 1.8). That is, the refrigerant comes into the compressor at a low pressure (state 1). It thus runs away from the compressor and arrives at the condenser at a raised pressure (state 2). As heat is moved to the surroundings, the refrigerant is condensed (state 3). The pressure of the liquid is lessened as it runs through the expansion valves (state 3). The residual liquid at a low pressure and temperature is vaporized in the evaporator (state 4).

The performance of refrigerators and heat pumps is defined as the coefficient of performance (COP),

$$\mathrm{COP}_R = \frac{Desired\ output}{Required\ input} = \frac{Cooling\ effect}{Worked\ input} = \frac{\dot{Q}_{\mathrm{L}}}{W_{\mathrm{net,\ in}}} \tag{1.2a}$$

$$\mathrm{COP}_{HP} = \frac{Desired\ output}{Required\ input} = \frac{Heating\ effect}{Worked\ input} = \frac{\dot{Q}_{\mathrm{H}}}{W_{\mathrm{net,\ in}}} \tag{1.2b}$$

Both $\mathrm{COP_R}$ and $\mathrm{COP_{HP}}$ shall be bigger than 1. Under the identical working states, the COPs are expressed as:

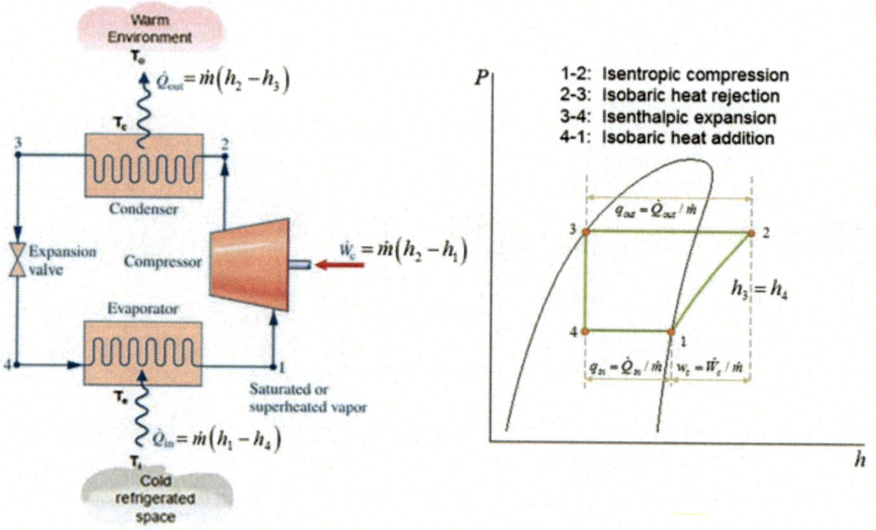

(a) Refrigeration cycle (b) P-h diagram

Fig. 1.8 Vapor-compression refrigeration cycle

$$\mathrm{COP}_{HP} = \mathrm{COP}_R + 1 \tag{1.2}$$

The cooling coefficient of performance ($\mathrm{COP_C}$), the refrigeration cycle efficiency, is expressed as

$$\eta_E = \mathrm{COP}_C = \frac{\dot{Q}_{\text{in}}}{W_{\text{cycle}}} = \frac{\dot{m}(h_1 - h_4)}{\dot{m}(h_2 - h_1)} = \frac{(h_1 - h_4)}{(h_2 - h_1)} \tag{1.3}$$

where $\dot{m}$ = mass flow rate, compressor work: $\mathrm{W}_{cycle} = \dot{m}(h_2 - h_1)$, refrigeration effect: $\dot{q}_{in} = h_1 - h_4$ and refrigeration capacity: $\dot{Q}_{in} = \dot{m}(h_1 - h_4)$.

We observe that η_E may be > 1. Therefore, a refrigeration system for cooling is named as a refrigerator.

Generally, cooling capacity is defined as the amount of heat lowered relative to the surrounding environment, which is expressed in tons. This is identical to the heat necessitated for melting one ton of ice in one day. One ton of refrigeration is the steady-state heat transfer rate necessitated to melt 1 ton of ice at 0 °C (32 °F) in 24 h. That is, it can be expressed as

$$1\ \text{ton} = 144\ \text{Btu/lb} \times 2000 = 288{,}000\ \text{Btu/day} = 12{,}000\ \text{Btu/h}\ (= 3.516\ \text{kW})$$

For example, if a system has a capacity of 1,200,000 Btu/h, we know that it has 100 tons (= 1,200,000/12,000).

Example 1.1 Calculate the COP and refrigerating efficiency for an ideal single-stage cycle working with a condenser pressure of 1253 kPa and an evaporator pressure of 201 kPa utilizing R-134a as the refrigerant. The cycle work is 31 kJ/kg of refrigerant.

Solution

The COP may be expressed as follows:

$$\eta_E = \mathrm{COP}_C = \frac{\dot{Q}_{\text{in}}}{W_{\text{cycle}}} = \frac{\dot{m}(h_1 - h_4)}{\dot{m}(h_2 - h_1)} = \frac{(h_1 - h_4)}{(h_2 - h_1)}$$

From the thermodynamic table, because $h_1 = 392.8$ kJ/kg, $h_4 = 268.5$ kJ/kg, $W_{\text{cycle}} = 31$ kJ/kg, COP_C is

$$\eta_E = \mathrm{COP} = \frac{\dot{Q}_{\text{in}}}{W_{\text{cycle}}} = \frac{(392.8\ \text{kJ/kg} - 268.5\ \text{kJ/kg})}{31\ \text{kJ/kg}} = 4.0$$

A Carnot cycle functioning between the saturation temperatures in accordance with the stated condenser and evaporator pressures has a COP as follows:

$T_e = -10 + 273 = 263$ K.

$T_c = 48 + 273 = 321$ K.

Therefore, we have

$$\mathrm{COP}_C = \frac{263}{321 - 263} = 4.5$$

The refrigerating efficiency is thus

$$\eta_R = \frac{COP}{COP_C} = \frac{4.0}{4.5} = 0.89 = 89\%$$

Example 1.2 Refrigerant-134a is the operating fluid in a theoretical VCR cycle. The refrigerant enters the evaporator at – 20 °C and has a condenser pressure of 0.9 MPa. The mass flow rate is 3 kg/min. Discover $\mathrm{COP_R}$ and $\mathrm{COP_{R,\,Carnot}}$ for the same $T_{\max}$ and $T_{\min}$, and the tons of refrigeration.

Solution

Utilizing the Refrigerant-134a Tables, we have

$$\left.\begin{array}{l} State\ 1 \\ Compressor\ inlet \\ T_1 = -20\ ^\circ\mathrm{C} \\ x_1 = 1.0 \end{array}\right\}\left\{\begin{array}{l} h_1 = 238.41\frac{\mathrm{kJ}}{\mathrm{kg}} \\ s_1 = 0.9456\frac{\mathrm{kJ}}{\mathrm{kg{\cdot}K}} \end{array}\right. \quad \left.\begin{array}{l} State\ 2 \\ Compressor\ exit \\ P_{2s} = P_2 = 900\ \mathrm{kPa} \\ s_{2s} = s_1 = 0.9456\frac{\mathrm{kJ}}{\mathrm{kg{\cdot}K}} \end{array}\right\}\left\{\begin{array}{l} h_{2s} = 278.23\frac{\mathrm{kJ}}{\mathrm{kg}} \\ T_{2s} = 43.79\ ^\circ\mathrm{C} \end{array}\right.$$

$$\left.\begin{array}{l} State\ 3 \\ Condenser\ exit \\ P_3 = 900\ \mathrm{kPa} \\ x_3 = 0.0 \end{array}\right\}\left\{\begin{array}{l} h_3 = 101.61\frac{\mathrm{kJ}}{\mathrm{kg}} \\ s_3 = 0.3738\frac{\mathrm{kJ}}{\mathrm{kg{\cdot}K}} \end{array}\right. \quad \left.\begin{array}{l} State\ 4 \\ Throttle\ exit \\ T_4 = T_1 = -20\ ^\circ\mathrm{C} \\ h_4 = h_3 \end{array}\right\}\left\{\begin{array}{l} x_4 = 0.358 \\ s_4 = 0.4053\frac{\mathrm{kJ}}{\mathrm{kg{\cdot}K}} \end{array}\right.$$

$$\begin{aligned} COP_R &= \frac{\dot{Q}_L}{\dot{W}_{net,im}} = \frac{\dot{m}(h_1 - h_4)}{\dot{m}(h_2 - h_1)} = \frac{h_1 - h_4}{h_2 - h_1} \\ &= \frac{(238.41 - 101.61)\frac{\mathrm{kJ}}{\mathrm{kg}}}{(278.23 - 238.41)\frac{\mathrm{kJ}}{\mathrm{kg}}} \\ &= 3.44 \end{aligned}$$

The tons of refrigeration, expressed as the cooling load or refrigeration effect, are

$$\dot{Q}_L = \dot{m}(h_1 - h_4) = 3\frac{\mathrm{kg}}{\mathrm{min}}(238.41 - 101.61)\frac{\mathrm{kJ}}{\mathrm{kg}}\frac{1\ \mathrm{Ton}}{211\frac{\mathrm{kJ}}{\mathrm{min}}} = 1.94\ \mathrm{Ton}$$

$$\begin{aligned} COP_{R,\ \mathrm{Carnot}} &= \frac{T_L}{T_H - T_L} \\ &= \frac{(-20 + 273)K}{(43.79 - (-20))K} \\ &= 3.97 \end{aligned}$$

Another measure of the efficacy of the refrigeration cycle is how much input power to the compressor, in horsepower, is necessitated for each ton of cooling.

The unit conversion is 4.715 hp/ton of cooling.

$$\frac{\dot{W}_{net,in}}{\dot{Q}_L} = \frac{4.715}{COP_R} = \frac{4.715}{3.44}\frac{\text{hp}}{\text{Ton}} = 1.37\frac{\text{hp}}{\text{Ton}}$$

1.2.4 Air-Conditioning Systems and Equipment

Air conditioning is the procedure of eliminating heat and moisture from the inside of an occupied room to enhance the comfort of occupants. Before it passes through the main air conditioning devices, outside air is drawn in, filtered and heated. The colored lines in the lower portion of the diagram show the changes in temperature and water vapor concentration (not RH) as the air streams through the system. In summer, air conditioning will cool and dehumidify. On the other hand, in winter, it will heat and humidify (Fig. 1.9).

In 1902, Willis Carrier in Buffalo, New York, invented the earliest modern electrical air conditioning unit. There are four types of air conditioning as follows: (1) Window air-conditioning system, (2) split air-conditioning system, (3) centralized air-conditioning system, and (4) unitary air-conditioning system (Fig. 1.10).

The unitary air conditioning systems work in accordance with the principle of the vapor compression refrigeration cycle. In this system, factory-assembled air conditioners can be mounted in or adjacent to the room to be conditioned. The unitary air conditioning systems are of the following two types: (1) window units and (2) vertical packed units.

A typical window-type air conditioner works as follows: (1) consider that a space is kept at a constant temperature of 25 °C, (2) in the air conditioner, the air from the room is pulled by a fan and is made to go over a *cooling coil*, the surface of which is kept, i.e., at a temperature of 10 °C, (3) after going over the coil, the air is cooled down (for instance, to 15 °C) before being distributed to the space, and (4) after taking up the room heat, the air is again returned to the cooling coil at 25 °C (Fig. 1.11).

Centralized air-conditioning systems with air handling units (AHUs) are utilized in big buildings, theatres, hotels, airports, and shopping malls to be completely conditioned. It also uses a vapor-compression refrigeration system that consists of compressor—increases the temperature and pressure of the refrigerant, condenser—eliminates heat which was put in the system by the evaporator and compressor,

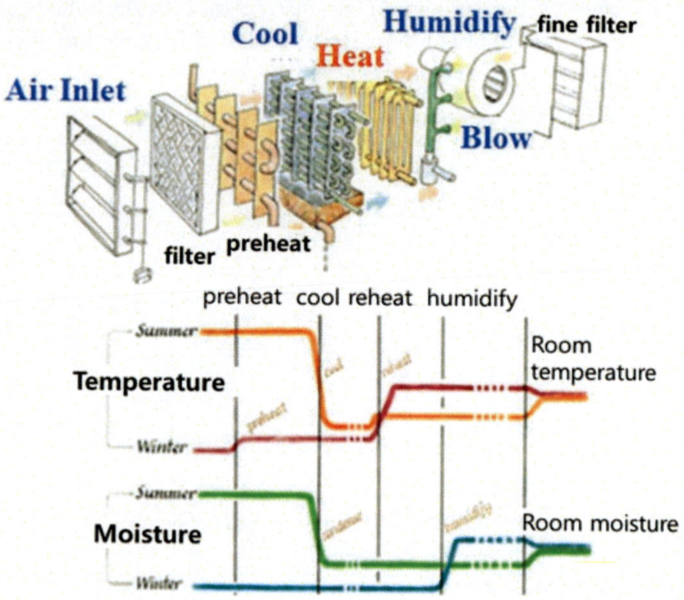

Fig. 1.9 Total air conditioning

metering device—controls refrigerant flow to the evaporator, and evaporator—heat is sucked up from the space.

An AHU is a device utilized to circulate, distribute, and control the conditioned air for HVAC buildings. It is a big metal box having a blower, heating or cooling elements, filter racks or chambers, humidity (45–55%) and temperature (19–23 °C) control loops. There are components for AHU as follows: (1) heating unit—heat the air to the controlled temperature, (2) cooling unit/dehumidifier—chill the air to the necessitated temperature or to eliminate moisture from the air, (3) humidifier—bring the air to the contrilled humidity, if too low, (4) filters—remove particles of predetermined dimensions and/or microorganisms, (5) ducts—transport the air, (6) control damper—fastened adjustment of volume of air, (7) silencer—reduce noise caused by air circulation, and (8) weather louvre—prevent insects, leaves, dirt and rain from entering (Fig. 1.12).

There are several types of centralized air-conditioning systems as follows: (1) all air systems and (2) air-and-water systems.

An all-air system can be expressed as a system supplying the whole sensible/latent cooling capacity in the chilled air provided by the system. It requires no extra cooling and is classified as follows: (1) single duct systems supply either cooling or heating utilizing the same duct, but not both heating and cooling simultaneously. Therefore, they have a constant volume with single zone systems, a constant volume with multiple zone systems, and variable volume systems. (2) Dual duct systems can

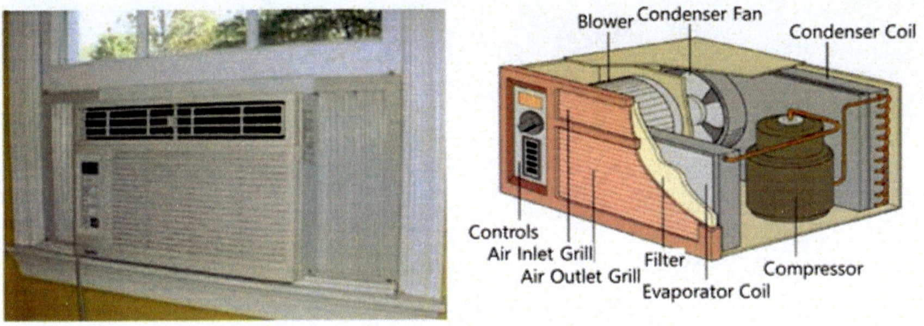

(a) Window air-conditioning system

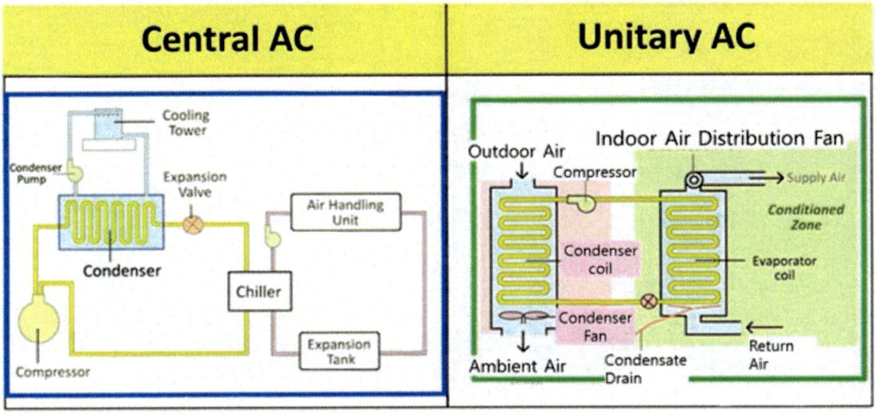

(b) Centralized air conditioning system and Unitary Air conditioning System.

Fig. 1.10 Kinds of air conditioning

supply both cooling and heating at the same time. Therefore, they have a dual duct with constant volume systems and a dual duct with variable volume systems.

A single zone system is a straightforward form of an all-air system that has a single conditioner being in the service of a single temperature zone. It is applicable to small department stores/individual shops in shopping centers, individual class rooms, etc. (Fig. 1.13).

On the other hand, because it is expensive to supply a separate system for each zone of a large building, multiple zone systems are used. For such occasions, the fundamental control system idea is enlarged to satisfy the cooling and heating needs of the multi zone. There are two types as follows: (1) constant-air volume systems with terminal reheat zones and (2) variable-air volume systems with dual ducts (Fig. 1.14).

As air is the medium utilized to stabilize the load, engineers can select to vary either the supply air temperature constant volume or changing the volume as the room load alters. That is, a variable air volume (VAV) system can be applicable to inside or perimeter zones with common or separate fan systems, common or separate

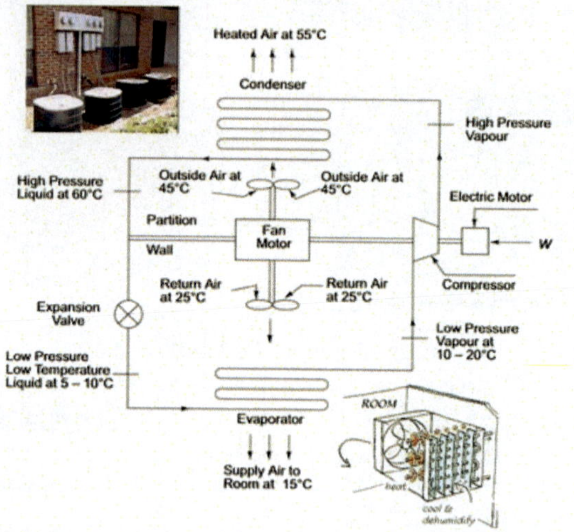

(a) Schematic diagram of a normal window-type air conditioner

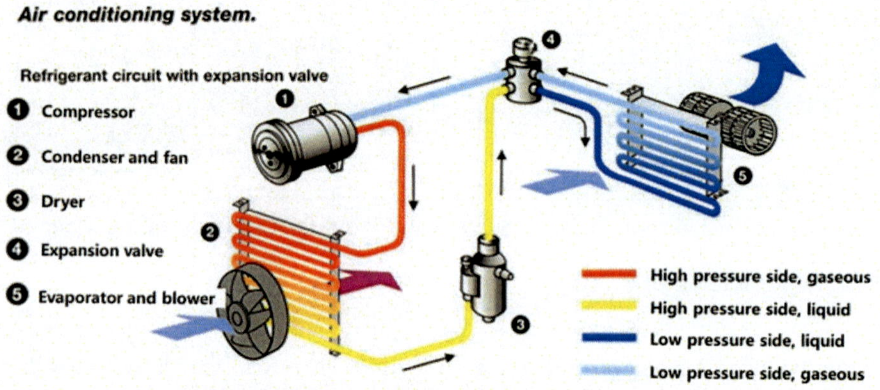

(b) Vapor-compression refrigeration system in an air conditioner

Fig. 1.11 A typical (window-type) air conditioner

air temperature control, and with or without auxiliary heating devices. The variable volume idea can be applied to volume variation in the main system. Variation of flow under control of a space thermostat can be fulfilled by placing a straightforward damper or a volume regulating device in a duct, a pressure reducing device, or at the terminal diffuser or grill.

To regulate cooling or heating for the VAV system, a single stream of cool air fulfills all the zones, and a thermostat in each zone adjusts a damper to control the flow rate of chilled air into the zone. The implementation is inside an office building with no heating loads and where only cooling loads exist (Fig. 1.15).

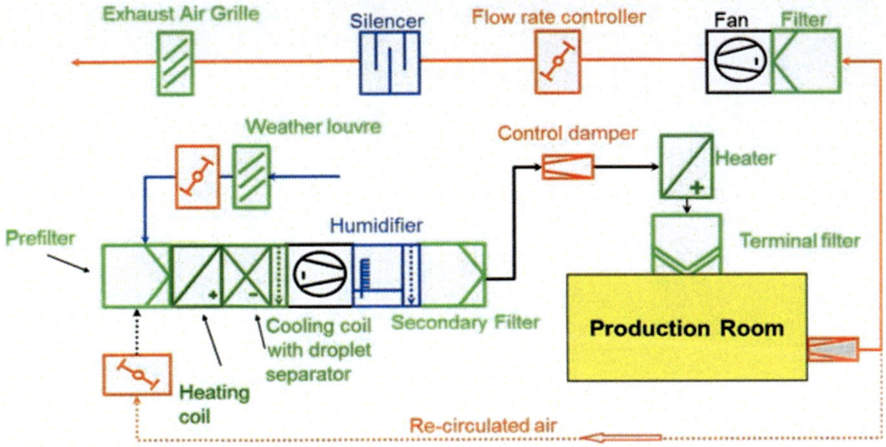

Fig. 1.12 Air handling unit in air conditioner

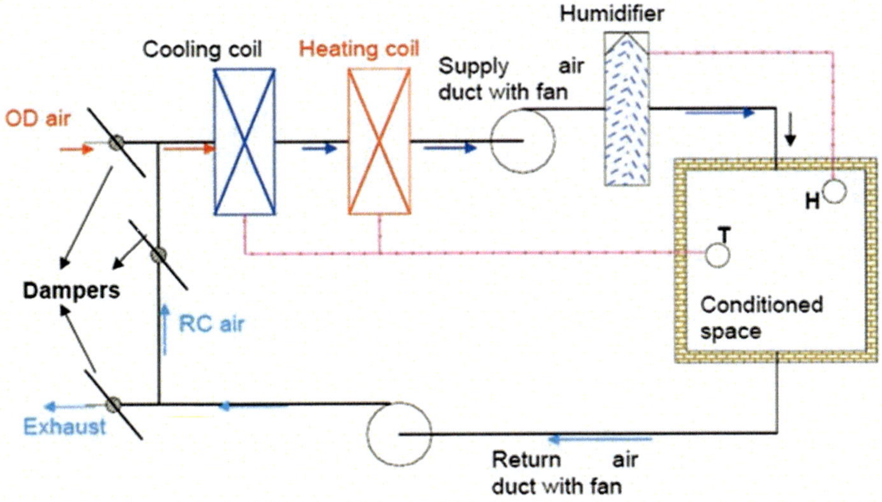

Fig. 1.13 A constant volume, single zone system

On the other hand, the dual duct system conditions all air in a central equipment and supplies it to the regulated rooms through two ducts. One duct moves chilled air and the other moves warm air, therefore supplying air sources for both cooling and heating. In the conditioned zone, in response to a room thermostat, warm/cold air is blended in proportions to fulfill the existing heat load of the space. The dual duct system reacts quickly to changes in zone load and can house heating in some zones and cooling in others (Fig. 1.16).

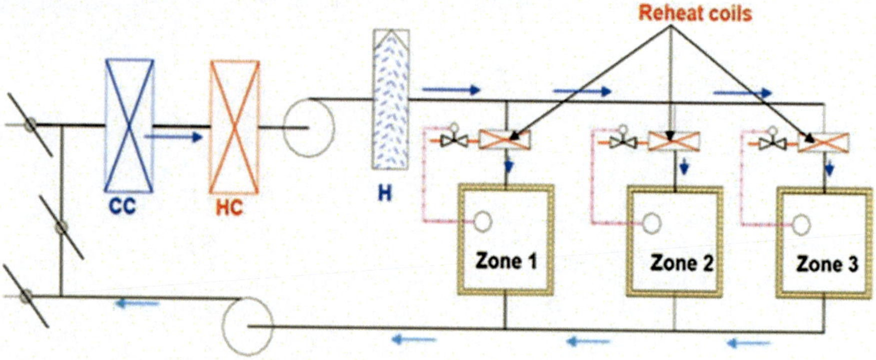

Fig. 1.14 A multizone system with terminal reheat zone

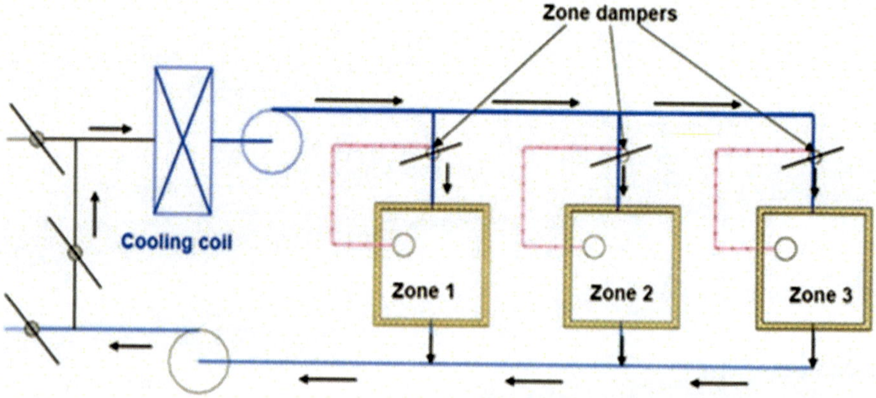

Fig. 1.15 A multizone system with a single duct and a variable volume system

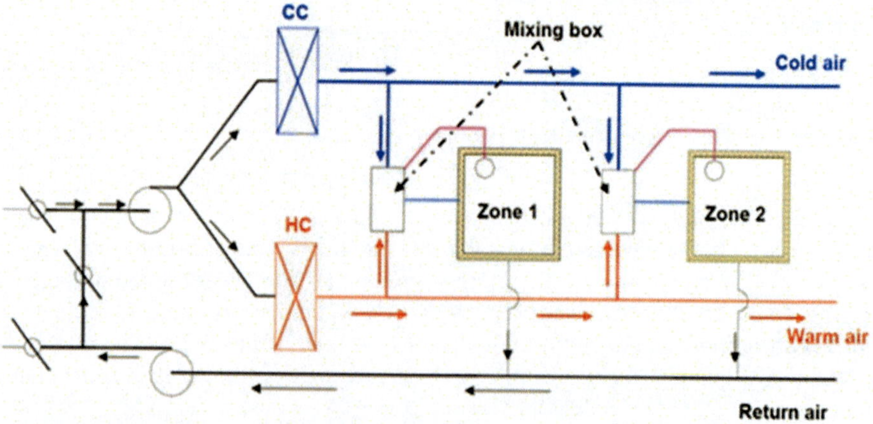

Fig. 1.16 A multi zone system with dual duct and constant volume system

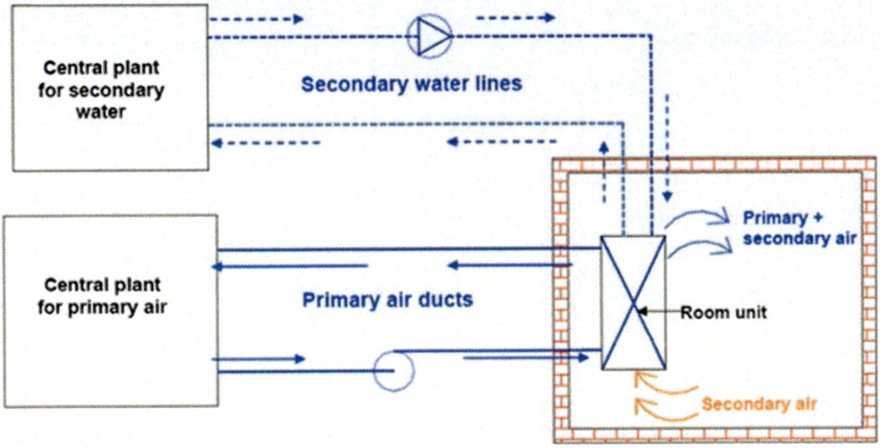

Fig. 1.17 A basic air–water system

Air-and-water systems fulfill conditioned space through the utilize of both air and water sources, which are given out to terminal units mounted in the room. The air and water are cooled or heated in central mechanical apparatus spaces and supplied to the room in which comfort conditions are being kept. The most common type of terminal utilized with air and water systems is the high-pressure induction unit. The air provided to the induction unit is specified as primary air. It is dispersed from the central mechanical equipment room at high velocities between 10 and 24 m/s. The water provided to the induction unit is specified as secondary water.

Air-and-water systems are essentially relevant to multizone type outer spaces of buildings where a wide span of sensible loads prevails and where careful control of humidity is not necessary. Hospitals, schools, apartment houses, research laboratories, etc., are applicable. The main subsystems are central air conditioning equipment, duct distribution and water distribution systems and a room terminal. The air supply is usually a constant volume and supplies outside clean air for ventilation (Fig. 1.17).

Finally, all-water systems fulfill both sensible and latent space cooling by flowing cooled water from the AC central refrigeration system through cooling coils in terminal units found in building inhabited rooms. The most usual terminals are fan coil units, unit ventilators, and valence units. The all-water system supplies individual room control, with no cross contamination of recomputed air from one space to another (Fig. 1.18).

1.2.5 Refrigerant

A refrigerant is a kind of working fluid in refrigeration. It is operated at boiling (evaporator) and condensing (condenser) as a form of two phases (change) and has

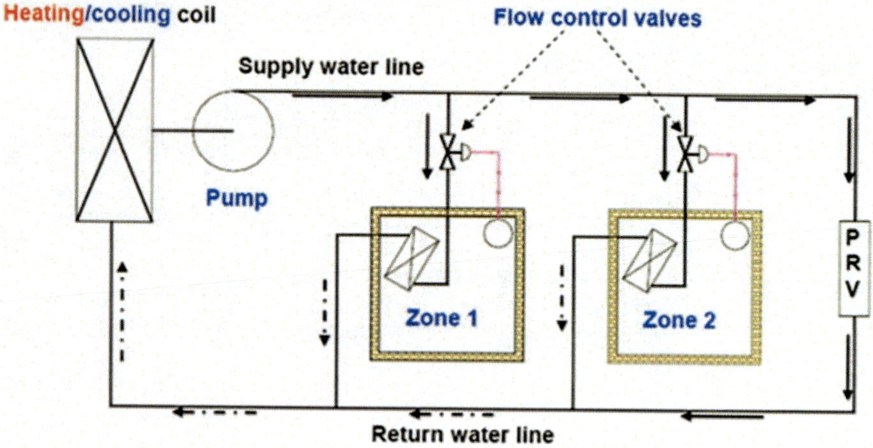

Fig. 1.18 A two-pipe, all water system

thermodynamic characteristics of low-temperature boiling fluids. Refrigerant choice relies on the capability to eliminate heat, toxicity, density, availability, cost, etc. It is classified as follows: (1) primary refrigerants and (2) secondary refrigerants.

The main refrigerants are those which go through the operations of compression, condensation, expansion and evaporation in cyclic processes. This class of refrigerants includes ammonia, R12, R22, and carbon dioxide. On the other hand, the medium that does not pass through the cyclic processes in a refrigeration system and is only utilized as a medium for heat transfer. This category is water, brine solutions of sodium chloride and calcium chloride (Fig. 1.19).

The American Society of Refrigerating Engineers (ASRE) has evolved certain practices for utilizing different types of refrigerants. A three-digit nomenclature can constitute these. Therefore, the first digit stands for the number of carbon atoms in the

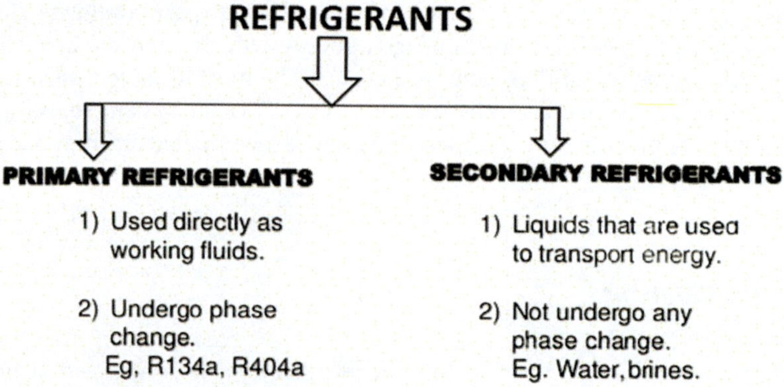

Fig. 1.19 Primary and secondary refrigerants

compound minus one, the second digit represents the number of hydrogen atoms plus one, and the third digit represents the number of fluorine atoms. The existing atoms are still chlorine. For example, there are refrigerants such as ammonia, R12, R22, carbon dioxide, etc. Most refrigerants are halogenated hydrocarbons. The naming convention embraced by ASHRAE is expressed as follows:

$$\mathbf{R}(\mathrm{a}-1)(\mathrm{b}+1)\mathrm{d}=\mathbf{C}_{\mathrm{a}}\mathbf{H}_{\mathrm{b}}\mathbf{Cl}_{\mathrm{c}}\mathbf{F}_{\mathrm{d}} \tag{1.4}$$

where $\mathrm{c}=2(a+1)-\mathrm{b}-\mathrm{d}$

Example 1.3 Calculate the chemical formulation of refrigerant R22 that has three-digit nomenclature.

Solution

According to the convention.

R22 (R022) is

No. of C atoms in R22: $a-1=0\rightarrow a=1$,

No. of H atoms in R22: $b+1=2\rightarrow b=2$,

No. of F atoms in R22: $d=2$, $\mathrm{c}=2(a+1)-\mathrm{b}-\mathrm{d}=2\,(1+1)-1-2=1$.

Therefore, we can calculate the chemical formulation of R22 as follows:

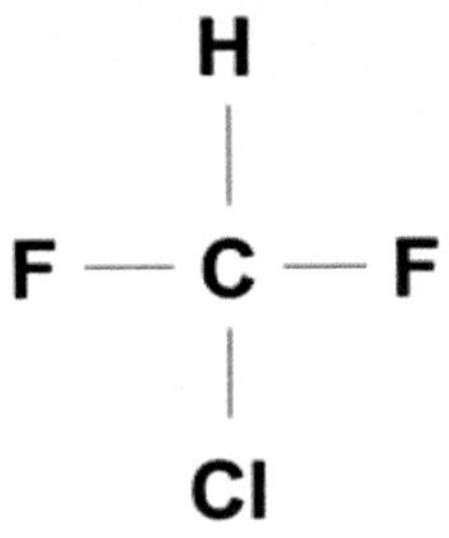

chlorodifluoromethane

That is, since there is only one carbon atom in the compound, this compound has come from the methane series (CH_4). From the computation, we can see that there is one hydrogen atom and two fluorine atoms. The existing valence bond of carbon will still be balanced by chlorine.

1.2.6 Vapor Absorption Refrigeration System

The crucial disadvantage of the vapor-compression refrigeration system is that it requires big mechanical power, such as compressor work. Another form of refrigeration which has merit is absorption refrigeration, where the refrigerant is absorbed by a transport medium and compressed in liquid form. The absorption refrigeration system popularly utilized is the ammonia-water system, where ammonia serves as the refrigerant and water as the transport medium. In the absorption system, the compressor is substituted by an absorber, generator and pump (Fig. 1.20).

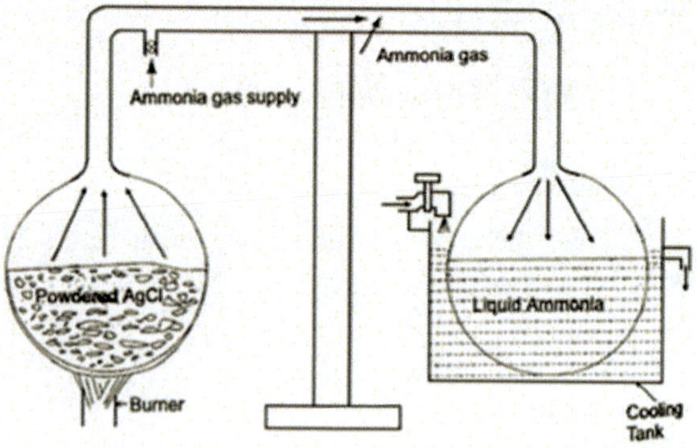

(a) Intermittent vapor absorption refrigeration system

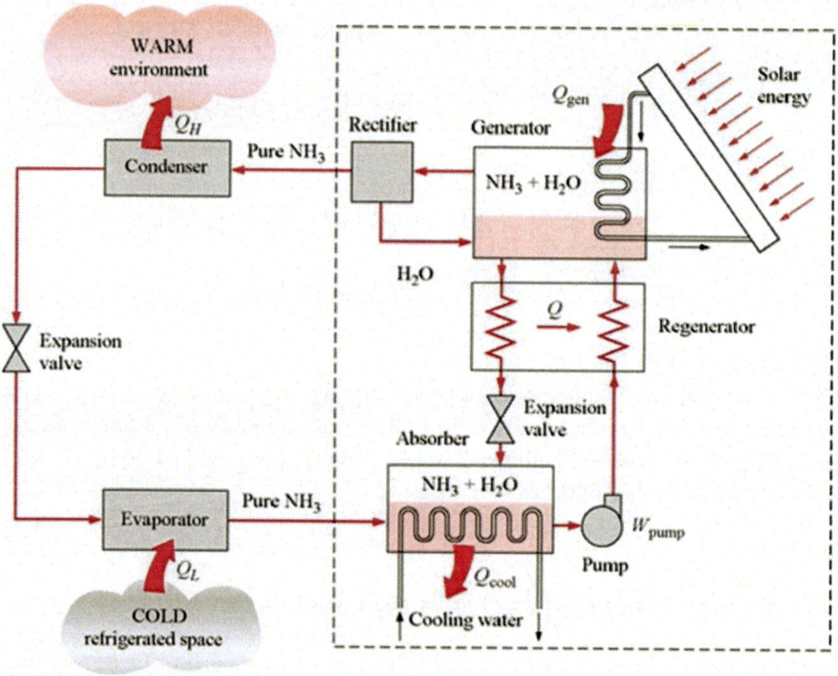

(b) Vapor absorption refrigeration system using solar energy (example)

Fig. 1.20 Vapor absorption refrigeration system

The concept for the operating concept of a vapor absorption system was produced by Michael Faraday in 1824. Two chambers are united with the help of a tube. The white powder was maintained inside the first chamber to which ammonia gas was provided and sealed. The powder was heated, while the other end was cooled utilizing circulating water. Liquid ammonia was attained in the cool end of the equipment. As ceasing heat, the liquid ammonia instead of sitting there began boiling (bubbles generated), and vapor was reabsorbed by the white powder.

On touching the boiling end, it was amazed to recognize that the container was very cold. He repeated the experiments, and cooling was noticed again. This led to the invention of an intermittent vapor absorption system with a solid as an absorber (Fig. 1.20b).

The refrigerant comes into the evaporator in the shape of a cool, low-pressure mixture of liquid and vapor (4). Heat is moved from the comparatively warm water to the refrigerant, creating the liquid refrigerant to boil. The absorber pulls in the refrigerant vapor (1) to blend with the absorbent. The pump pushes the blend of refrigerant and absorbent up to the high-pressure side of the system.

The generator supplies the refrigerant vapor (2) to the rest of the system. The refrigerant vapor (2) leaving the generator enters the condenser, where heat is moved to water at a lower temperature, bringing the refrigerant vapor to condense into a liquid. This liquid refrigerant (3) thus flows to the expansion device, which produces a pressure drop which drops the pressure of the refrigerant to that of the evaporator. The following mixture of liquid and vapor refrigerant (4) travels to the evaporator to reiterate the cycle (Fig. 1.21).

The absorption systems utilize heat energy in the form of steam, direct fuel firing or waste heat to attain the refrigerant effect. The absorption cycle utilizes a liquid pump and a compressor to produce the pressure rise between the evaporator and

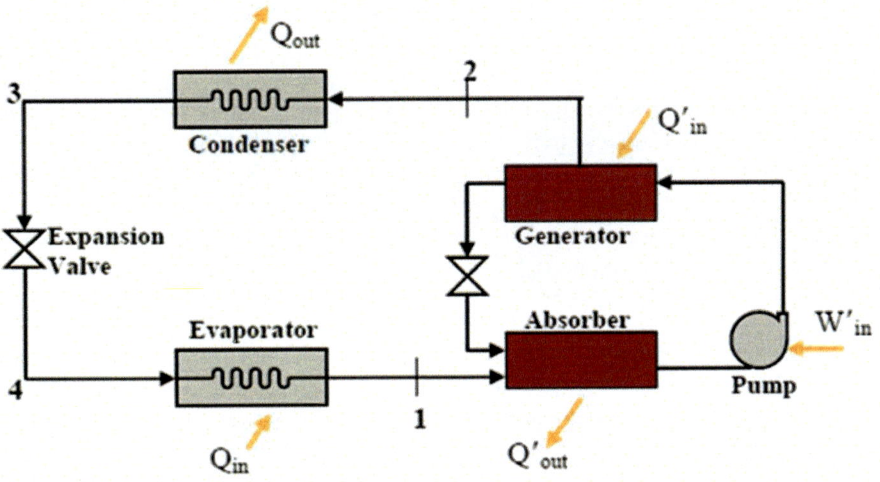

Fig. 1.21 A vapor-absorption refrigeration cycle

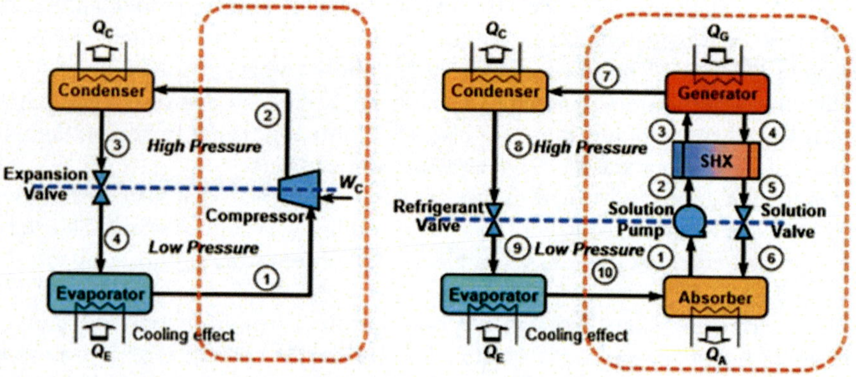

(a) Vapor compression refrigeration (VCR) (b) vapor absorption refrigeration (VAR)

Fig. 1.22 Differences between VCR and VAR

condenser. Pumping a liquid is much easier and cheaper than compressing a gas, so the system takes less work input. However, there is a big heat input in the generator. Therefore, the system primarily substitutes the work input of a vapor-compression cycle with a heat input (Fig. 1.22).

Therefore, although the VCR cycle utilizes a halocarbon (such as HCFC-123, HCFC-22, HFC-134a, etc.) as the refrigerant, the absorption cycle utilizes different refrigerants which have no related environmental hazard, ozone depletion or global warming potential. For instance, the lithium bromide absorption system uses distilled water as the refrigerant. Compared to compression chillers in VCR, absorption systems have very few moving components, provide less noise and vibration, are dense for large capacities and necessitate little maintenance. The performance of absorption systems is insensitive to load variations and does not rely very much on evaporator superheat. Compared with mechanical chillers, absorption systems have a low coefficient of performance. However, absorption chillers may considerably lessen working costs because they are powered by low-grade waste heat.

The renewal of the compressor by the straightforward arrangement of VARS is not very cheap in application. To improve it, certain extra auxiliary items are supplied in the system. They cover analyzer, a rectifier, and two heat exchangers. The feasible absorption cycles as developed after incorporating these auxiliaries are shown in Fig. 1.23.

(a) **Analyzer:** The ammonia vapors going away from the generator may have certain moisture, and it might be eliminated from any indication of water vapor before going on to the condenser and thus to the expansion valve; otherwise, the water vapor will freeze in the little valve passage and choke the flow.

The analyzer is to eliminate moisture as much as possible. It is an open type of cooler and forms an essential portion of the generator, installed on its top.

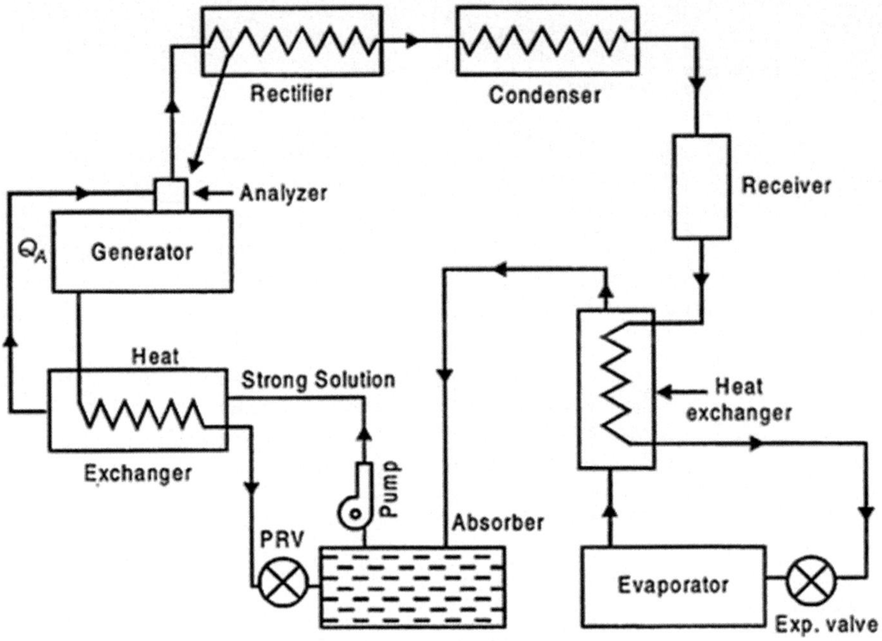

Fig. 1.23 Practical absorption system

Both the strong aqua ammonia solution from the absorber and the condensate eliminated in the rectifier are introduced from the top and flow downward.

The hot rising vapor of ammonia thus comes in contact with the same and is cooled. Therefore, most of the water vapor is condensed and drips back into the generator. It helps in salvaging a certain portion of heat in outgoing vapor, which would otherwise have been rejected out through the condenser.

(b) **Rectifier:** It is a closed kind of cooler and is a miniature condenser where any signs of water vapor left in the ammonia vapor are eliminated by condensation. The cooling is fulfilled by flowing water as is done in a condenser. The condensed water is flowed back to the generator through the analyzer.

(c) **Heat exchangers:** Two heat exchangers are supplied to change heat from the higher temperature fluid to the lower temperature fluid so that one is cooled and the other is heated. One heat exchanger is supplied between the liquid receiver and evaporator so that the liquid is subcooled and vapor is heated up. Another heat exchanger is situated between the generator and absorber so that the strong aqua is heated before going on to the analyzer and the weak aqua is cooled before flowing the absorber.

1.2.7 Major Equipment of Vapor-Compression Refrigeration Systems

The main equipment of the VCR system is as follows: (1) Compressors raise the temperature and pressure of the refrigerant. It suctions low-temperature and low-pressure refrigerant vapors from the cooling coil through the suction line and compresses it. It discharges the refrigerant vapors of high temperatures and pressure to the condenser through the discharge line. (2) Condenser—eliminates heat which was attached to the system by the evaporator and compressor. It is a heat exchanger that removes heat from the hot vapor refrigerant discharged from the compressor. (3) The metering device controls the refrigerant flow to the evaporator. It can be classified as follows: capillary tube, thermostatic expansion valve, constant pressure expansion valve, high side float, and orifice plates, and (4) evaporator—absorbs heat from the space. Its function is to take in heat from the surroundings or medium (Fig. 1.24).

Compressor is classified as positive displacement and dynamic type. For positive displacement, there are rotary and reciprocating. On the other hand, for the dynamic type, there are centrifugal and axial types (Fig. 1.25).

Reciprocating compressors are equipped with pistons and cylinders and suction and discharge valves. The valve position is regulated by the pressure difference across it. They can be classified as open, semi-hermetic or fully hermetic. Rotary compressors use circular motion to obtain compression. After refrigerant is confined by the rotating vanes, refrigerant compresses as volume lessens (Fig. 1.26).

Helical rotary (screw type) compressors utilize two rotors, a male and a female. The volume of the refrigerant decreases, and a continuous, flowing output is produced as it flows through the compressor. Compressor capacity is also regulated by a slide valve. Scroll compressors use two machined scrolls in exactly the same way. That is, one scroll is stationary and the other is orbits. As the nesting of the scroll confines vapor, gas is established from the outer edge, and refrigerant is released from the center. Centrifugal compressors depend on centrifugal force. There are no pistons,

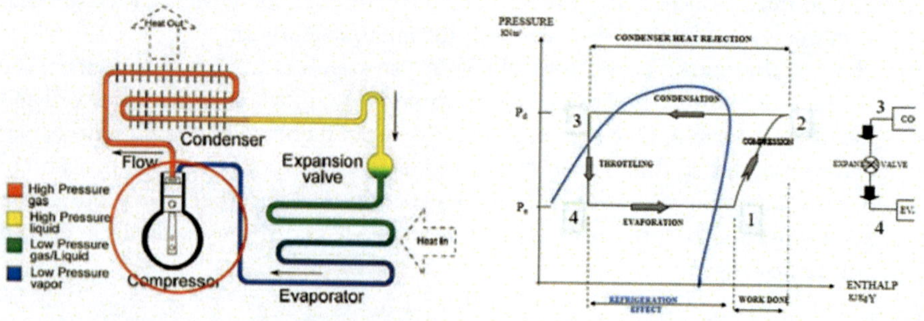

(a) Major equipment of refrigeration

(b) P-h diagram

Fig. 1.24 Major equipment of refrigeration

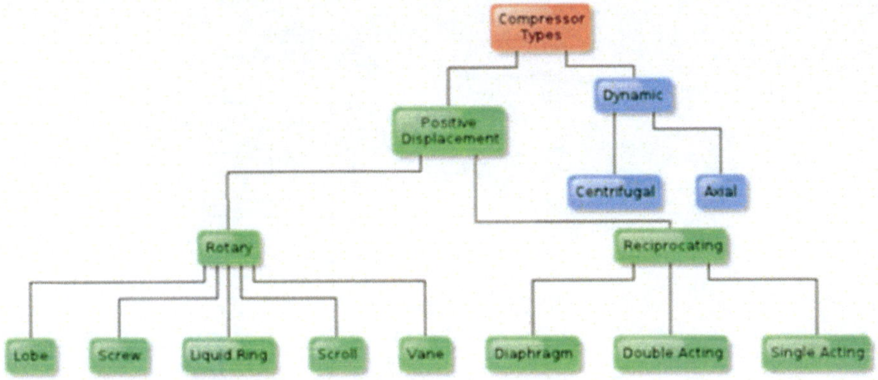

Fig. 1.25 Compressor types

(a) Reciprocating compressors (b) Rotary compressors

Fig. 1.26 Compressor types: reciprocating and rotary

valves or cylinders. By utilizing an impeller, the capacity is controlled by inlet vanes. It is used for very large applications (Fig. 1.27).

Condensers are categorized as air cooled (tube and fin), water cooled (shell and tube), and evaporative condenser. Air cooling (tube and fin) occurs as follows: (1) de-superheating: removes sensible heat (superheat) from the vapor refrigerant, (2) condensing: vapor condenses into a liquid (latent heat), and (3) subcooling: addition sensible heat is removed from the liquid refrigerant. Water cooled (shell and tube) is a heat exchanger. That is, as water circulates through the tubes and makes several passes before exiting, hot gas enters the shell and condenses. It can be mechanically cleanable. The evaporative condenser utilizes air and water to heat off the cooling load. Air flows over the coil. Water can also flow over the coil if needed. Latent and sensible heat transfers take place as water is heated and evaporated (Fig. 1.28).

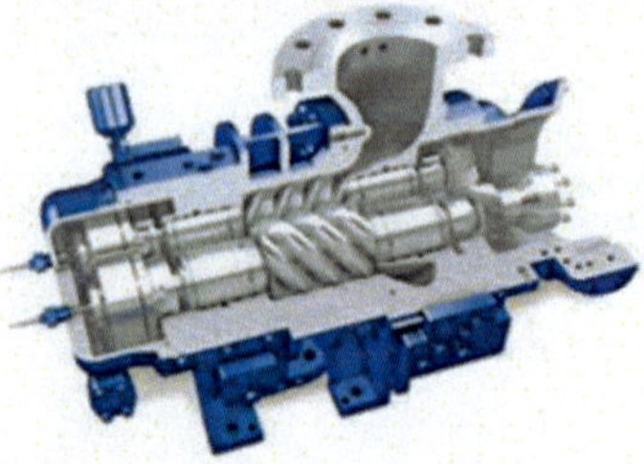

(a) helical rotary (screw type)

(b) scroll type

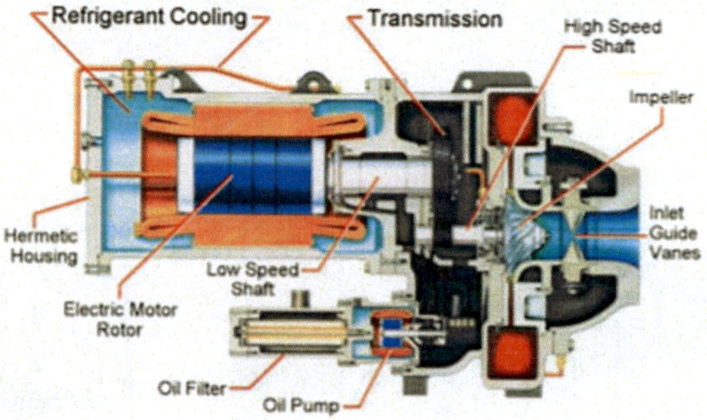

(c) centrifugal type

Fig. 1.27 Compressor types: helical rotary, scroll and centrifugal

The evaporator is classified as a direct expansion chilled water evaporator and flooded shell-and-tube. A direct expansion chilled water evaporator occurs as follows. That is, as refrigerant flows in the tubes, water fills the shell. On the other hand, for flooded shell-and-tube, as refrigerant is in the shell, water flows in the tubes (Fig. 1.29).

The metering device is classified as follows: (1) thermostatic expansion valve, (2) constant pressure expansion valve, (3) capillary tube, (4) high side float, and (5) orifice plates. The thermostatic expansion valve maintains superheat in the evaporator. Three pressures control the valve position (Bulb, spring and evaporator). That is, as the load increases, the valve opens. On the other hand, as the load decreases, the valve closes. The constant pressure expansion valve maintains a constant evaporator pressure. A growth in evaporator pressure will bring the valve to close. A reduce in evaporator pressure shall cause the valve to open. It is commonly used on systems with constant loads. Capillary tubes are seamless tubing with a small inside diameter and no moving parts. A high side float valve introduces refrigerant to the evaporator at the same rate as it leaves.

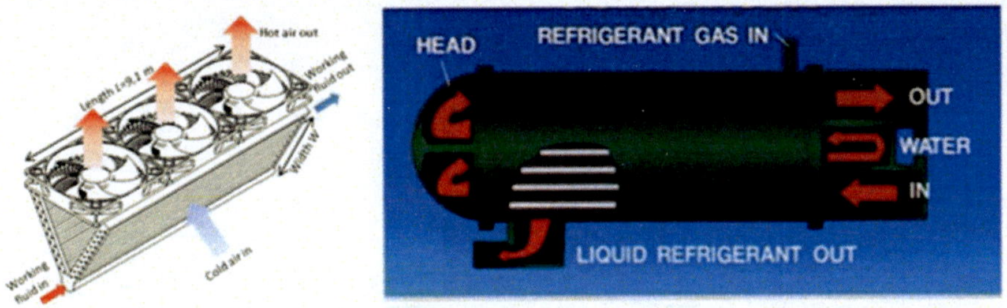

(a) Air cooled (tube and fin) (b) Water cooled (shell and tube)

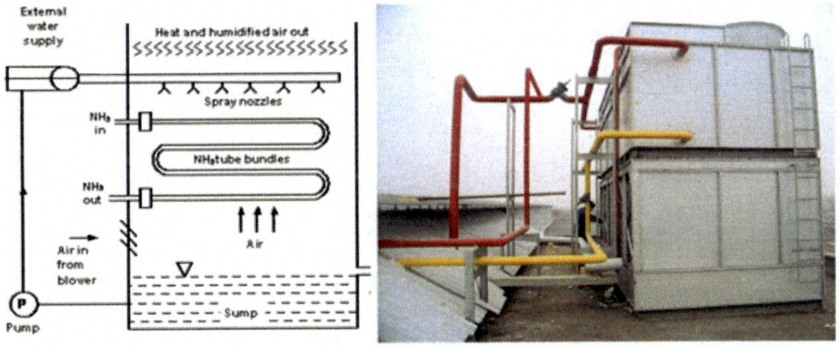

(b) Evaporative condenser

Fig. 1.28 Condenser types

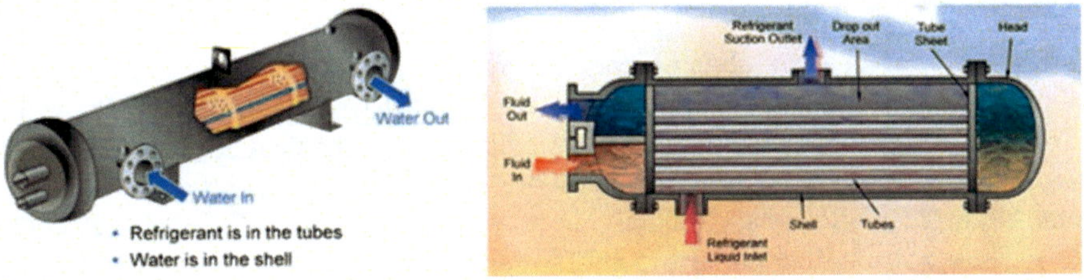

(a) Direct expansion chilled water evaporator (b) Flooded shell and tube

Fig. 1.29 Evaporator types

A float on the high side maintains proper flow. Orifice plates are used primarily on centrifugal chillers. Two plates have a series of holes in each. The pressure drop across the plates changes with the load. For a high load, liquid passes the plates. On the other hand, for low loads, some liquid flashes to a vapor (Fig. 1.30).

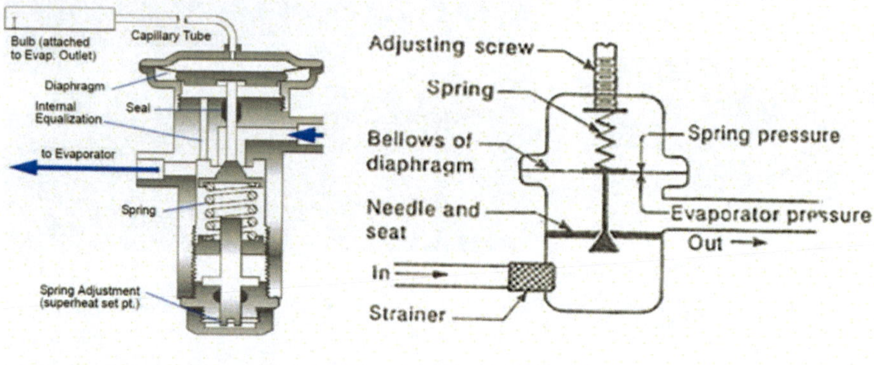

(a) Thermostatic expansion valve (b) Constant pressure expansion valve

Fig. 1.30 Evaporator types

1.3 Other Mechanical Products—Automobile, Airplane, Etc.

1.3.1 Automobile

An automobile used for transportation is a wheeled motor vehicle. The term 'automobile' originates from the Ancient Greek term autós, meaning 'self', and the Latin term mobiliz, signifying 'movable'. Over the past decades, extra features and controls have been attached to vehicles, thus making them steadily more complicated but also more reliable and easier to run. These include air conditioning, rear reversing cameras, and navigation systems.

The power of an automobile is generated by an internal combustion engine which functions in the Carnot cycle, a theoretical ideal thermodynamic cycle. Then, the power is transmitted to each wheel, thus obtaining a mechanical advantage through mechanisms such as transmission and drive systems. An automobile is made up of some different modules, such as engine, transmission, drive, electrical, and body parts. The whole number of parts in an automobile can be as many as 20,000. Hence, the whole failure rate of an automobile in its life is the total of the failure rates of each module. A car's life is anticipated to have a B40 life of at least 12 years (Figs. 1.31 and 1.32).

1.3.2 Airplane

Airliners are designed to move cargo payloads or passengers. It is propelled forward by a thrust force from an engine. In particular, as the aircraft moves forward, the wings deflect the air downward, thus generating a lifting force to carry it in flight

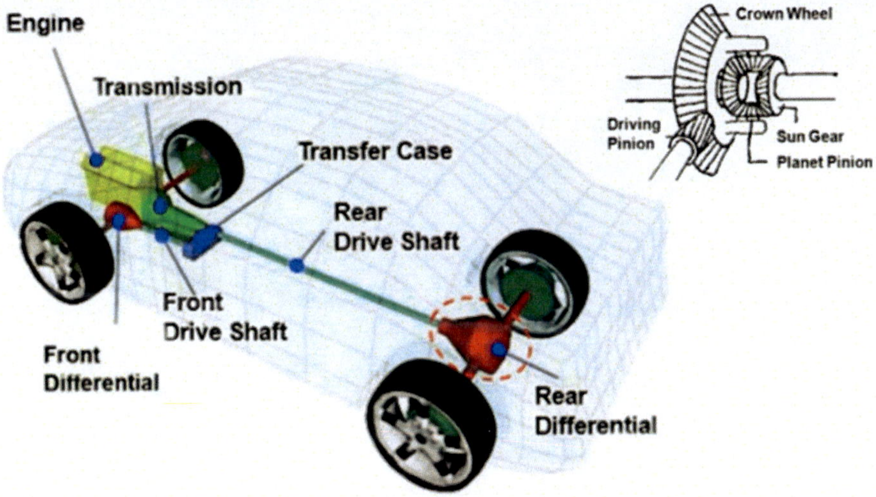

Fig. 1.31 Automobile structure

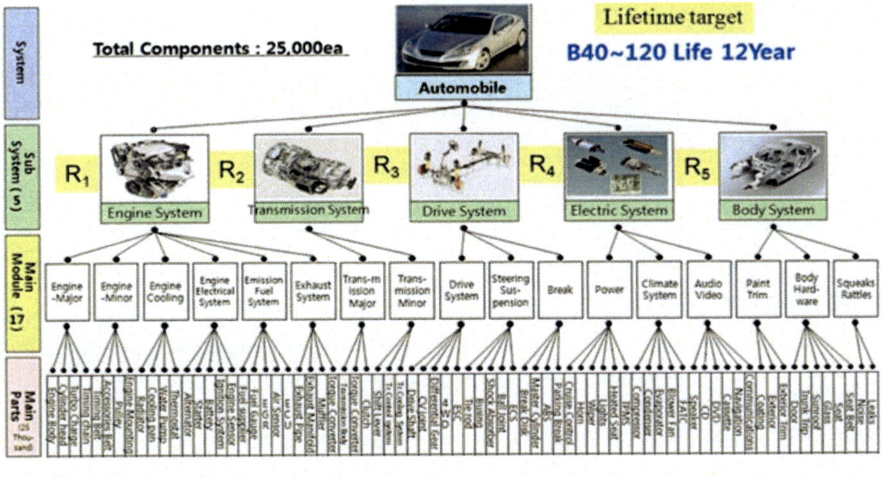

Fig. 1.32 Breakdown of automobiles with multiple modules

that is dependent on a variety of wing configurations. The term 'airplane' originates from the French aéroplane, which arises from the Greek word ἀήρ (air), and either the Latin word 'planus', which means 'level', or the Greek word πλάνος (planos), which means 'wandering'. It also means the wing moves through the air.

An aircraft utilizes an onboard propulsion from the mechanical power produced by an aircraft engine, which is either a propeller or jet propulsion. In particular, jet engines supply airplane thrust by sucking in the air, compressing the air, and introducing fuel into the hot-compressed air mixture in a combustion chamber, and

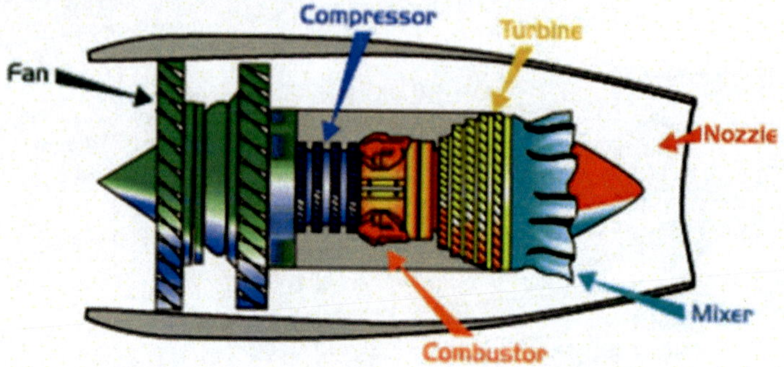

Fig. 1.33 Jet-engine structure

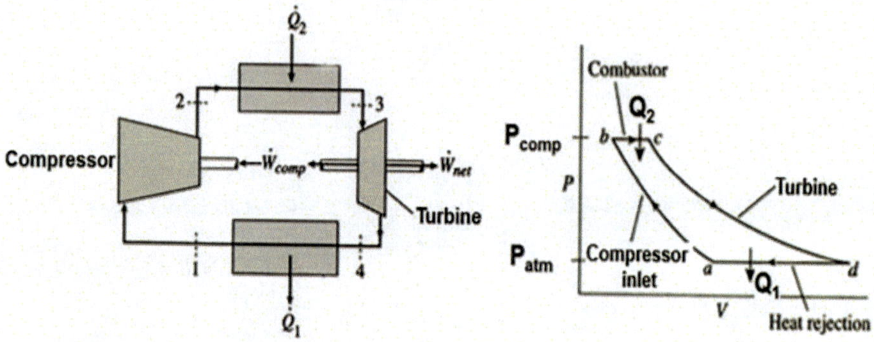

Fig. 1.34 Brayton cycle

the elevated exhaust emits rearwards through a turbine. Modern gas turbine engines are operated by the Brayton cycle, which comprises a gas compressor, a combustion chamber, and an expansion (Figs. 1.33 and 1.34).

The mechanical structure of an aircraft is composed of approximately 200,000 components, which include engine systems, aviation control systems, power delivery systems, door systems, and machine systems. Airplanes are built with multiple modules that have their own structures and mechanisms (Fig. 1.35).

1.3.3 Heavy Machinery

Heavy machinery is designed for performing construction tasks and is operated by hydraulic power, which functions through the mechanical advantage of a straightforward machine. By properly using hydraulic circuits, the machine transmits a hydraulic fluid throughout some hydraulic motors and hydraulic cylinders and becomes pressurized in hoses and tubes of circuits. The hierarchical layout of a

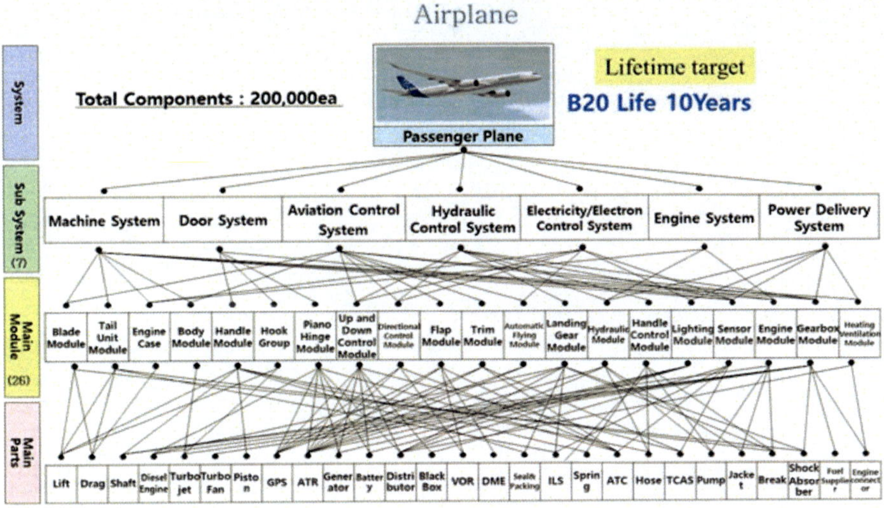

Fig. 1.35 Structure of an airplane with multiple modules

construction machine, such as an excavator, is made up to an engine device, a track system, an electric device, an upper appearance system, a driving system, a chief control valve unit, a hydraulic operation machine system, a cooling system, and other numerous parts. Multiple modules in a product have their own mechanisms and structures. A typical appliance contains over 5000 parts (Fig. 1.36).

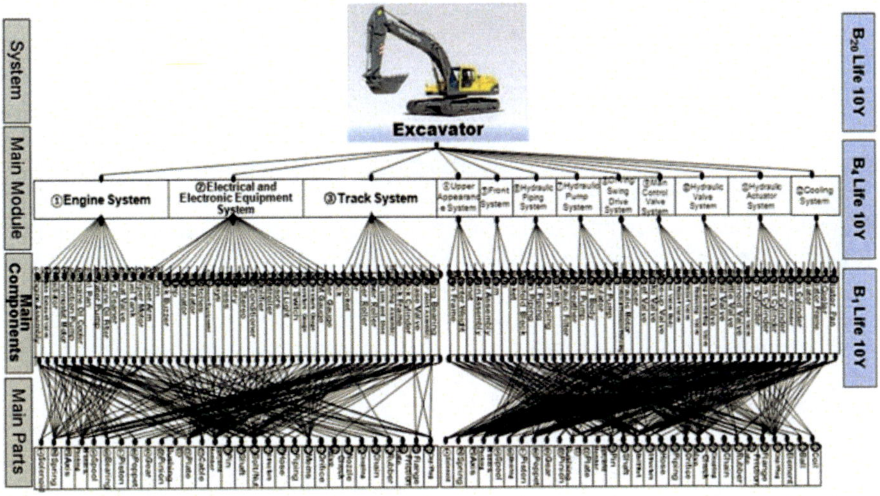

Fig. 1.36 Structure of excavator with multiple modules

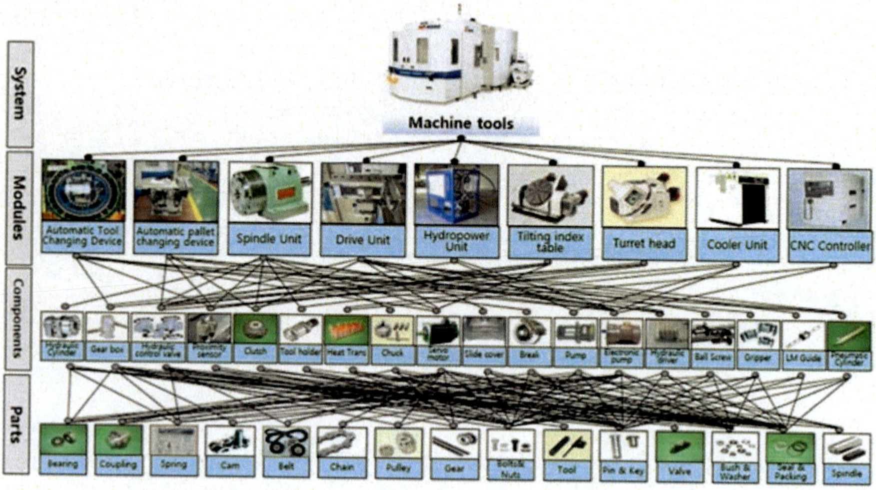

Fig. 1.37 Structure of machine tools with multiple modules

1.3.4 Machine Tools

A machine tool is usually a machine for forming or machining metal or other rigid materials by boring, cutting, shearing, grinding, etc. It utilizes some sorts of cutting or shaping tools. All machine tools have some ways of restricting the work piece and supply a guided motion of the components of the machine. Therefore, the relative motion between the cutting tool (which is specified as the 'tool path') and the work piece is constrained or regulated by the machine to at least some extent, rather than being totally 'freehand' or 'offhand'.

The hierarchical layout of machine tools comprises an automatic tool or pallet altering device, a drive unit, a spindle unit, a tilting index table, a hydropower unit, a turret head, a computer numerical control (CNC) controller, a cooler unit, etc. The machine tools possess over 1000 parts. Its reliability design will concentrate on the modules. The machine tools can effortlessly compute the module reliability because of the serial connection system (Fig. 1.37).

1.4 Reliability Disasters and Its Assessment Significance

1.4.1 Quality and Reliability

Customers choose a product because of its better performance, ease of usability, lack of problems, and technological simplicity. From the standpoint of a customer, the reliability of the product becomes one of the acceptable attributes, regardless of

Table 1.1 Quality flaws and failures

	Quality faults	Failure
Notion	Out of common specifications	Physical failure (e.g., fatigue)
Indicator	Defect rate, ppm	Lifetime, failure rate
Measure	Percent, ppm	Percent/year, year
Field	Manufacturing	Design
Probability distribution	Normal distribution $f(x) = \frac{1}{\sigma\sqrt{2\pi}} e^{-\frac{(x-\mu)^2}{2\sigma^2}}$	Exponential/Weibull $F(t) = 1 - R(t) = 1 - e^{-\lambda t}$

both the kind of product and the number of components connected to the product. When a global company manufactures a product with good quality—performance and reliability—the brand image will remain in the marketplace. The term 'quality of product' in daily usage is vaguely defined to signify the inherent level of the product's excellence. That is, it can be expressed as 'conformance to specification at the beginning of product use'.

On the other hand, recalls emerge in the marketplace when customers start to use a new product. From the investigation of the root causes of product recalls, we recognize that they result from the design faults missed in the developing process. Unless the company ensures the quality of product lifetime for new (intended) functions, it shall be dismissed from the worldwide marketplace. We can determine the concept of 'reliability' from the following question: how many of these samples satisfy their specifications at the end of the one-year assurance period? (Table 1.1).

1.4.2 Reliability Disasters

A disaster—such as an automobile recall, a nuclear plant accident, and an oil spill—is a failure of a product's intended function due to design flaws, which causes catastrophic people and environmental or economic impacts. This disaster also has no forecasting power of the society or community to control its own resources because people will recognize its significance after revelation. Thus, people often learn lessons from their past mistakes so that they can be prevented by recognizing their root causes early. For example, in 1912, we can think the RMS Titanic that was sank in the ocean injured more than 2200 people onboard. As the crew saw the approaching icebergs very late, they couldn't turn the ship quickly, which led the ship to hit on a floating iceberg at its right-hand side. It only took 2 h and 40 min to sink, which cost the lives of more than 1500 people. The reason behind this incident is that the ship metal iron was too fragile to bear the iceberg impact due to the cold seawater of the Atlantic Ocean. There were no alternative plans for rescue, and the ship sank quickly within a few hours (Fig. 1.38).

Fig. 1.38 RMS Titanic of the White Star Line sinking approximately 2.20 AM, Monday morning, April 15, 1912 (Wikipedia)

The Titanic sinking, from the standpoint of material, was primarily caused by the increased brittleness of the steel utilized to build the hull of the ship due to the impact of cold seawater of the North Atlantic. Although the Collison of the iceberg on the ship is small, it brought a huge amount of damage. Consequently, the bolts holding on the steel plates fractured, which led to the breakdown of the ship hull. Immediately after the incident of RMS Titanic, every ship was mandated to have a proper emergency evacuation plan.

After WWII, the product continued to evolve as a hybrid system with multiple functions, composed of many parts, such as airplanes, automobiles, refrigerators, television sets, and electronic computers. Moreover, as elaborated control and devising for system safety also became more necessitated and added to their system, the chance for reliability disaster increased in the market due to their complexities. As the life of a mechanical product was decided by its flawed parts, it was critical to assess them in the product developing process before the release of the final product.

For example, a Boeing 747 jumbo jet airplane commonly consists of roughly 4.5 million parts covering fasteners, multiple modules, and subsystems. An automobile consists of more than 22,000 components, multiple modules, and subsystems. In 1935, a farm tractor consisted of 1200 components, but in 1990, the number expanded to approximately 2900. Even for comparatively simpler mechanical products such as bikes, there has been a notable increase in complexity with respect to parts. As a consequence, the design of mechanical products such as automobiles depends on these parts. If one of them is problematic (or wrongly designed) in the field, there will be a massive recall of all mechanical systems (Fig. 1.39).

As product complexity increases when a new product structure is continually adapted to satisfy product performance requirements and reduce costs, there are

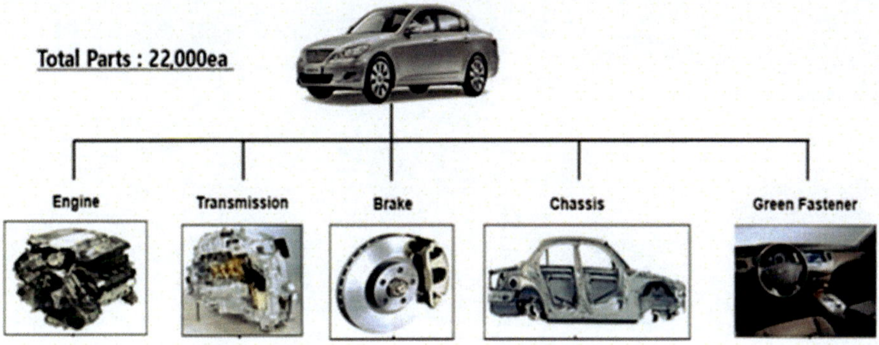

Fig. 1.39 Structure of an automobile composed of multiple modules

other opportunities for the problematic design of parts. According to an investigation carried out by the U.S. Navy concerning components failure, 43% of the product failures were attributed to design, 30% to operation and maintenance, 20% to manufacturing, and 7% to various factors. As the design cost holds only 5% of the actual cost, the cost effect due to a recall takes 70% of the actual cost.

The basic cause of most structural failures is generally negligence during the application of a new design or material. As an 'improved' design is released in the marketplace, there are some factors that the product engineer does not expect. That is, new materials (or a new design) can not only supply large merits but also bring potential issues. As a consequence, a new design might be changed into customer service only after proper testing, such as parametric ALT. Such an approach will decrease the chance of failures and prevent them.

Thus, we list a few typical examples of reliability disasters in the following:

- The Versailles rail crash occurred in 1842 on the railway between Paris and the Versailles. Following King Louis Philippe I's celebrations at the Versailles Palace, a train coming back to Paris caught train to leave its tracks accidentally at Meudon. As the main train broke an axle, the carriages behind piled into it and caught fire. With roughly 200 deaths, including that of the explorer Jules Dumont d'Urville, it was recorded as one of the earliest and deadliest railway accidents in the world. Because most of the passengers who died were having on the seat belt, the disaster led to the abandonment of the practice of locking passengers in their carriages. It began the investigation of metal fatigue subjected to repeated loads such as S–N curves (Fig. 1.40).
- As the first commercial jet engine airplane, the de Havilland DH 106 Comet was replaced with the propeller airplane. People can have a transatlantic flight with this plane. The earliest original Comet plane had an aerodynamic design with four turbojet engines in two wings, an aerodynamic fuselage, and large square-type windows. It flew on July 27, 1949. Since 1952, it has supplied a silent and pleasant passenger cabin. After a year in 1953, the Comets started to experience design issues, and three airplanes broke up while flying. Due to metal fatigue in

Fig. 1.40 Versailles rail accident (1842) from Wikipedia

its airframe, a design defect in the corners of the square windows, which was subjected to repeated stresses, was discovered in the Comet. Consequently, the Comet was designed again with oval windows, structural reinforcement, and other changes (Fig. 1.41).

- Point Pleasant Bridge Collapse: Situated on the West Virginia/Ohio border, the bridge fell in 1967. This disaster aroused the loss of 46 lives, and the root cause was the metal fatigue of a critical eye bar in the suspension bridge (Fig. 1.42).
- Because welded designs are faster and cheaper than riveted designs, earlier Liberty ships were fabricated by all-welded hulls. As measured by the Charpy impact tests, the steel employed to construct the Liberty ships also had poor toughness. The Liberty ships hurted hull and deck cracks. During WWII, there were almost 1500 examples of notable brittle fractures. Covering three of the 2710 Liberties,

Fig. 1.41 De Havilland DH 106 Comet (1954) from Wikipedia

Fig. 1.42 Point pleasant bridge disaster (Wikipedia)

twelve ships suddenly broke in half, including the SS John P. Gaines that sank on November 24, 1943, costing the lives of ten people (Fig. 1.43).

- Space Shuttle Challenger: This disaster happened in 1986, in which all the crew were dead. The principal cause of this disaster was design flaws—rubber O-rings under a chilled winter environment in southern Texas. At that time, engineers from the booster manufacturer detected a possible issue with the O-ring

Fig. 1.43 Early liberty ships disaster (Wikipedia)

seals and advised that the launch time should be detained. However, they had no evidence data to convince National Aeronautics and Space Administration (NASA) officials. Many Americans are crumpled in pride because of Space Shuttle Challenger's failure.

- Nuclear Reactor Explosion in Chernobyl: This disaster happened in 1986 in the former Soviet Union, in which 31 people died, and this area is yet polluted due to the presence of radioactive materials. The disaster was the consequence of design flaws, such as a problematic switch in the reactor design.
- A Leak at McKee Refinery in 2007: Propane gas escaped from the McKee Refinery's Propane Deasphalting Unit in Sunray, Texas. Three laborers were hurt with critical burns, and the refinery was closed for 2 months. Gas prices soared 9 cents per gallon in the west.

1.5 Historical Review of Development of Reliability Methodologies

The modern idea of reliability started in 1816. Reliability in statistics at that time was defined as the measurement consistency to define a test. That is, a test is reliable if the identical consequence is repetitive. As an attribute of a product, reliability was a general concept that had been recognized. We summarize the historical milestones in the early twentieth century in Table 1.2.

In 1842, rail accidents often occurred in Versailles, France. August Wöhler studied the root causes of fracture in railroad axles and began the earliest structured investigation of the S–N curve (or Wöhler's curve) [1, 2]. To stop railroad accidents, S–N curves of matters shall be utilized to reduce fatigue incidents by dropping the critical stress in a part.

During World War I (WWI), Griffith studied fracture mechanics to determine the failure of brittle materials. He proposed the first law of thermodynamics to express a fracture theory entrenched on a straightforward energy balance. As a flaw in material is endurable for (dynamic) loading, fracture starts to occur when the strain-energy change is sufficient to endure the surface energy of the matter. He also proposed that the low fracture strength found in tests was due to the existence of microscopic defects in the bulk matter, which may be effective (Fig. 1.44) [3].

Failure happens when the free energy pertains to a peak value at a critical crack length.

$$C = \sqrt{\frac{2E\gamma}{\pi}} \tag{1.5}$$

where E is the Young's modulus of the material and γ is the surface energy density of the material.

Vacuum tubes, originating in 1904 by John Ambrose Fleming, were a critical part of electronic instruments such as television, diffusion of radio, radar, sound recording

Table 1.2 History abridge of reliability technology

~ 1950	–	Wilhelm Albert publishes the first article on fatigue (1837) A. Wöhler summarized fatigue test results on rail-road axles (1870) O.H. Basquin proposes a log–log relationship for S–N curves (1901) John Ambrose Fleming invented vacuum tubes in 1904 Griffith'st theory of fracture (1921) A.M. Miner introduces a linear damage hypothesis (1945)
WW II	Germany	V-I, V-II Rocket development (R. Lusser's Law)
WW II	US	Reliability of the Electron Power Tube (Aircraft electronic devices failure in the WW II)
1954	Japan	Surveys and studies on the Electron Power Tube Reliability in the Vacuum Committee of the Institute of Electrical Engineers
1952–1957	US	US DOD formed the Advisory Group on the Reliability of Electronic Equipment (AGREE) AGREE suggest vacuum tube follows the bathtub curve
1954	US	First National Symposium on Reliability and Quality Control, New York
1950s	US	Several conferences began to focus on various reliability topics (e.g., 1955 Holm Conference on Electrical Contacts)
1961	Italy	The Rome Air Development Center (RADC) introduced a PoF program
1962	US	Launched the Apollo program (FMEA & FTA), First Reliability and Maintainability Conference
1962	US	First Symposium on Physics of Failure in Electronics, Chicago
1965	IEC	Reliability and Maintainability Technical Committee, TC 56, Tokyo
1968	–	Tatsuo Endo introduces the rain-flow cycle count algorithm
1971	Japan	First Reliability and Maintainability Symposium

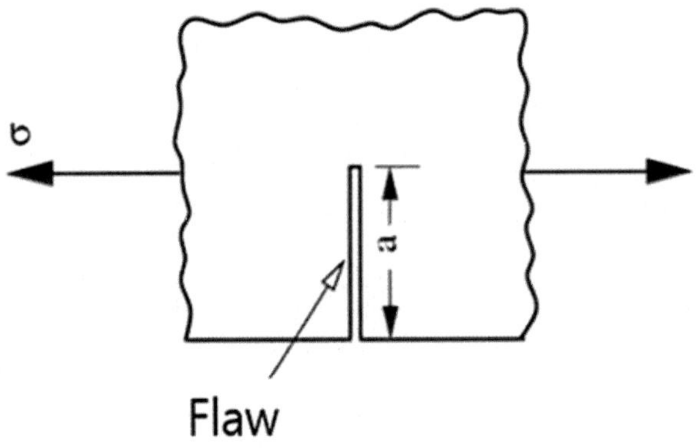

Fig. 1.44 An edge crack (flaw) of length in matter

Fig. 1.45 British engineer John Ambrose Fleming and his vacuum tubes patents [4]

and reproduction, sound reinforcement, analog and digital computers, large telephone networks, and industrial process control. As the vacuum tube was invented, it made modern technologies applicable to the product. The use of radio with vacuum tubes began in the public in 1916, and the reliability concept of troublesome vacuum tubes began to develop (Fig. 1.45).

In 1895, Karl Pearson stated the concept of a 'negative exponential distribution'. His exponential distribution had many amazing things which were applicable in the 1950s and 1960s. In other words, one property of the serial system is the capacity to add on failure rates of different parts in products. Simply adding it was more applicable at the time when utilizing mechanical and later electric systems:

$$R(t) = R_1(t) \cdot R_2(t) \cdots R_n(t) \tag{1.6}$$

$$R(t) = \mathrm{e}^{-\lambda_1 t} \cdot \mathrm{e}^{-\lambda_2 t} \cdots \mathrm{e}^{-\lambda_n t} \tag{1.7}$$

$$R(t) = \mathrm{e}^{-(\lambda_1 + \lambda_2 + \cdots + \lambda_n)t} \tag{1.8}$$

where R is the reliability function, λ is the failure rate, and t is the use time.

As automobiles started to utilize in the early 1920s, Walter A. Shewhart at Bell Laboratories introduced product advance by statistical quality control. In 1924, he presented the control chart and the idea of statistical control. As a measurement tool, statistics were combined with the evolution of reliability concepts. As engineers were responsible for product quality and reliability, technicians took care of the failures. In the 1930s, quality and process measures in automobiles were still developing.

In a military short lecture, W. Edwards Deming in the 1940s mentioned the management's responsibility for quality. He stated that quality issues occur due to design flaws, not due to workers' mistakes [5]. For example, an incentive reliability idea was applicable to the spark transmitter telegraph due to the simple design. It was a battery-powered system connected with uncomplicated transmitters by wire. The leading failure mode was a broken wire or inadequate voltage. After WWI, this system was changed with considerable ungraded transmitters based on vacuum tubes.

Before WWII, there were still no ideas in reliability engineering. However, numerous new electronic systems, such as vacuum tube portable radios, electronic switches, electronic detonators, and radar, were initiated into the army during WWII. As the war started, it was found that half of the airborne electronic apparatuses in store failed in life and were unable to satisfy the army needs (the Air Core and Navy). Reliability effort for this time had to do with new (metal) material testing. The root causes for the failures were only the products' fatigue or fracture. For example, Miner issued a seminal manuscript entitled 'Cumulative Damage in Fatigue' in 1945 in *ASME Journal*. B [6]. In 1948, Epstein published a paper titled 'Statistical Aspects of Fracture Problems' in the *Journal of Applied Physics* [7].

During WWII, Wernher von Braun, developing the V-1 missile, one of the Germany research groups, was collaborating for better designs. After the end of WWII, he announced that the first V-1 missiles were completely non-successful. Although they attempt to supply high-quality components and pay cautious attention in detail, all earliest missiles either detonated on the pad or landed 'too soon'. By developing reliability theory, they had to find solutions.

During WWII, Germany applied the fundamental reliability idea to upgrade the reliability of its V1 and V2 rockets with multiple modules. To finish the assignment of rockets VI and VII, German engineers had to improve the reliability of rockets VI and VII. Robert Lusser, a mathematician, was called in as one of the consultants. His work was to examine the missile system, and he briefly elaborated the product probability law for series parts. This proposition is to define systems functioning only if all parts are working and is logical under exceptional presumptions. That is, he defines that the reliability of a series system is equal to the system of the part reliability. In other words, a series system is 'weaker than that of its weakest link', as the product reliability of a series of parts shall be less than the lowest-reliability part. If the system consists of a large number of parts, the system reliability may thus be rather low, even though each part has high reliability. The product lifetime, thus, is decided by the weakest chain (or module) and continued from its design modification. After WWII, the Department of Defense in the United States severely acknowledged the inevitability for reliability of its military apparatus. It turned out to be the conceptual starting point of MIL-HDBK-217 and MIL-STD-756. However, reliability engineering has yet to be completely undeveloped because of the restrictions on understanding the design, failure mechanism, and reliability testing.

In the beginning of the 1950s, the principal military application areas of reliability design were the vacuum tube in radar systems, military equipment, or other electronics because these products were troublesome, as in WWII. Vacuum tube computers, which have a 1024-bit memory, were devised to take a huge space and use kilowatts of power, although they are very ineffective in modern computers. During WWII, many airplanes were wrecked during bombarding. The vacuum tubes installed in these airplanes have been demonstrated to be troublesome components, although the part price is not expensive. After the warfare, half of the electronic equipment for shipboard was unsuccessful in achieving its life. The basic cause of most failures was also the vacuum tubes. Failure modes of vacuum tubes in sockets were irregular working issues. The action plans for a problematic electronic system were to breakdown the system, eliminate the tubes, and remounting them at a right period.

In the modernization process, because the army should think about the cost matters, the operation and logistics costs for the vacuum tubes would become enormous. To address the issue of vacuum tubes, the Institute of Electrical and Electronic Engineers (IEEE) organized the Reliability Society in 1948. In 1948, Birnbaum established the Laboratory of Statistical Research at the University of Washington, which sufficed to utilize the idea of statistics. In 1951, the Air Development Center (RADC) was found in New York and Rome to investigate reliability problems with Air Force Rome.

In 1950, a study faction in the army was commenced, which was thus named the Advisory Group on the Reliability of Electronic Equipment (AGREE). By 1959, this faction in its result report proposed three suggestions for reliable systems such as vacuum tubes: (1) there was a necessity to evolve reliable parts for suppliers, (2) the army might initiate quality and reliability specifications for part suppliers, and (3) real field data might be gathered on parts to determine the basic causes of issues.

A close meaning of modern system life originated in 1957 AGREE Commission Report. Task Group 1 in AGREE has evolved minimum-acceptable units for the reliability of various types of military electronic apparatus, demonstrated in terms of Mean Time Between Failures, although describing life for electronic parts was presently improper. The last report of the AGREE committee proposed that the reliability of products such as vacuum tubes followed the bathtub curve. As a consequence, the reliability of parts is usually expressed in 'the bathtub curve', which shows trends such as early failure, useful lifetime, and wear-out failure. At the moment, the accumulative distribution function corresponding to a bathtub curve is substituted with a Weibull chart in reliability engineering (Fig. 1.46).

The AGREE committee also suggested officially testing systems with a statistical confidence level. It might perform the environmental tests under extreme temperature and vibration conditions, which changed to Military Standard 781. The AGREE report first defined reliability as 'the probability of a product functioning with no failure under designated circumstances for a specified period of time'.

At the start of the 1950s, a conference on electrical contacts and connectors was commenced to investigate the reliability physics—failure mechanisms and reliability topics. In 1955, Naresky in RADC published 'Reliability Factors for Ground Electronic Equipment' [8]. The proceeding, entitled 'Transaction on Reliability and Quality Control in Electronics', was presented in a conference, which was integrated

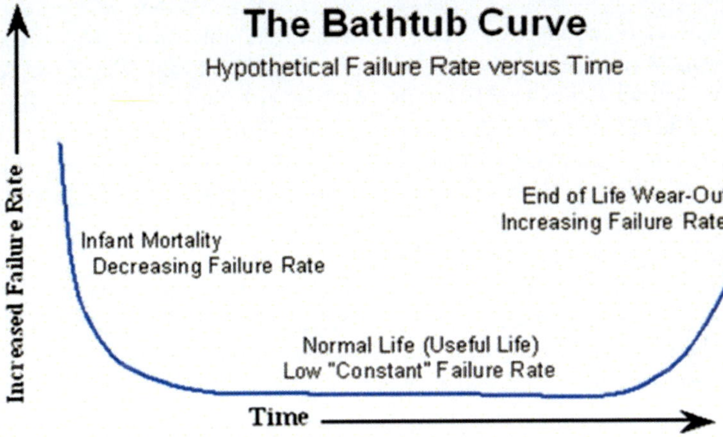

Fig. 1.46 Bathtub curve for vacuum tube radio systems

with an IEEE Reliability conference and grew the Reliability and Maintainability Symposium.

In the 1950s, as television usage was started at American homes, more vacuum tubes were already used. Due to the failure of one or more vacuum tubes, repair problems usually occur. One of the key switching devices in TV was a vacuum tube, which controls electric current through a vacuum in a sealed container—cathode ray tube. Representative reliability issues developed in the tubes with oxide cathodes are as follows: (1) decreased ability to release electrons, (2) a stress-connected fracture of the tungsten wire, (3) air leakage into the tube, and (4) glowing plate—a sign of an overloaded tube. Most vacuum tubes used in radio systems that follow a bathtub-type curve are easy to evolve, replaceable electronic modules—standard electronic modules (SEMs)—and thus reinstate an unsuccessful system.

In a 1957 report titled 'Predicting Reliability', Robert Lusser in Redstone Arsenal stated that 60% of the failures of one army missile system were due to its parts. He also emphasized that contemporary quality methods for electronic parts were improper and new ideas for electric parts should be executed. Aeronautical Radio INC. planned to upgrade the suppliers of vacuum tubes and lessen early mortality removals by a factor of four. This period of 10 years ended up with Radio Corporation of America publishing information in TR1100 on the failure rates of some army parts. RADC utilized these ideas as the fundamentals for Military Handbook 217. Over the next several decades, Birnbaum suggested Chebychev's inequalities, reliability of complex systems, nonparametric statistics, competing risk, cumulative damage models, survival distributions, and mortality rates.

Weibull in Sweden studied the fatigue of matters and immediately suggested a Weibull distribution. Weibull in 1939 proposed an easy mathematical probability distribution, which could show a wide span of failure attributes by altering two parameters. Although his failure distribution would not be applicable to every failure mechanism, it is a functional tool to examine many reliability issues. With seven case studies, his most famous papers in 1951 were suggested to the American Society of Mechanical Engineers (ASME) on the Weibull distribution. Between 1955 and 1963, he conducted studies on the creep and fatigue mechanisms of materials. He also suggested the Weibull distribution based on the weakest chain model of failures in materials. In a Wright Air Development Center Report 59-400 for the US military, he produced 'Statistical Evaluation of Data from Fatigue and Creep Rupture Tests: Fundamental Concepts and General Methods' in 1959 [9].

While working as an adviser for the US Air Force Materials Laboratory in 1961, he issued one book on materials and fatigue testing [10]. In 1972, the ASME granted its gold medal to Weibull. King Carl XVI Gustaf of Sweden personally awarded the Great Gold Medal from the Royal Swedish Academy of Engineering Sciences to him in 1978.

As his analysis methods and applications were spreading worldwide, many people started to utilize the Weibull plot. In the late 1950s, Dorian Shainin wrote down the earliest booklet on Weibull, while Leonard Johnson in General Motors in 1964 could make better the charting techniques by proposing median ranks and beta binomial confidence bounds. Professor Gumbel stated that the Weibull distribution is a Type III smallest extreme value distribution, such as Eqs. (1.9) and (1.10). For the Weibull function, Dr. Abernethy suggested many implementations, investigation methods, and rectifications [11].

$$F(x) = \exp\left(-\left(\frac{a-x}{b}\right)^c\right) \tag{1.9}$$

where $x \le a,\ b > 0,\ c > 0,$

$$K = A\exp\left(-\frac{E_a}{RT}\right) \tag{1.10}$$

where k is the rate constant of a chemical reaction, T is the absolute temperature, A is the pre-factor, E_a is the activation energy, and R is the universal gas constant.

While visiting an Institute in Columbia as a professor for the Study of Fatigue and Reliability in 1963, Weibull collaborated with professors Gumbel and Freudenthal. As a consultant for the US Air Force Materials Laboratory, he issued the associated reports and a book on materials and fatigue testing until 1970.

After WWII, manufacturers produced more complex products created of an ever-increasing number of parts (radio, television, computers, etc.), the development continued throughout the world. As the introduction of automatic equipment in manufacturing is increasing, the need for complex control and safety systems is also becoming more crucial.

In the 1960s, as the transistor in 1947 and transistor radio in 1954 were invented, many failures occurred in the field. To improve it, physics of failure (PoF) for electronic parts began in the 1960–1970s. People enjoyed music everywhere using

Fig. 1.47 A transistor radio with multiple parts (Wikipedia)

pocket-size radios. These tools had some difficulties, such as transistor failure, capacitor problems, and electromechanical faults. To prevent these failures, PoF was used as a structured method for the design and development of reliable products. By understanding the basic causes of the failures, the products may be made with better performance (Fig. 1.47).

In one electronics conference financed by Illinois Institute of Technology (IIT), RADC earnestly functioned the PoF. In the 1960s, NASA was established to achieve America's forceful dedication to space exploration. Their crucial attempts were to make the reliability of parts and systems better, which might function satisfactorily to finish the space commissions. RADC issued the document 'Quality and Reliability Assurance Procedures for Monolithic Microcircuits'. Semiconductors are favored in compact movable transistor radios. Next, inexpensive price germanium and silicon diodes could satisfy the requirements. Although the Institute of Radio Engineers (IRE), Dr. Frank M. Gryna issued a Reliability Training Book.

At this time, as the nuclear power industry and airplanes, helicopters, missiles, and submarines in the army were utilized, the reliability issues of various technologies in their product started PoF. The effects of the electromagnetic compatibility system were studied at RADC in the 1960s (Fig. 1.48).

As one of the milestones, the success of the Arrhenius model for semiconductors was proven in the Proceedings of the Seventh National Symposium of Reliability and Quality Control in 1962. G.A. Dodson and B.T. Howard in Bell Labs issued the paper, titled as 'High Stress Aging to Failure of Semiconductor Devices' [12]. This conference also published many other papers and improved the reliability of other electronic parts. It was then given a new name to the Reliability Physics Symposium

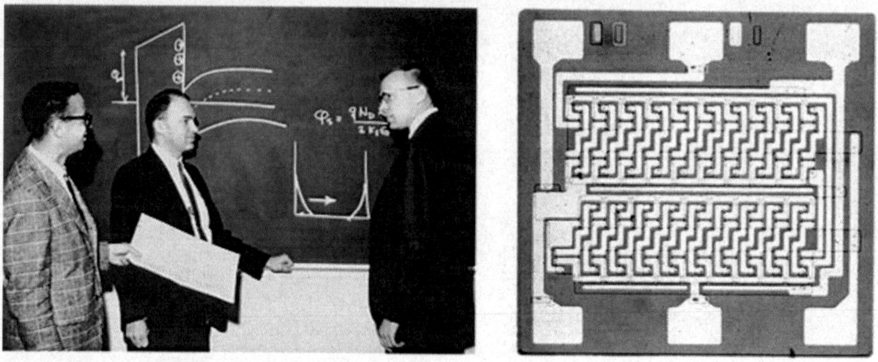

Fig. 1.48 Andy Grove, Bruce Deal, and Ed Snow at the Fairchild Palo Alto R & D laboratory and first commercial Metal Oxide Semiconductor (MOS) IC in 1964 (Wikipedia)

(RPS) in 1967. When applied to integrated circuits (ICs), Shurtleff and Workman in the late 1960s published a paper on step stress testing which sets up its limits.

As metals are pressurized at high current densities, electromigration in the electronic part is one of the critical failure mechanisms that can be applicable to the transportation of mass in metals. In the physics of electromigration, J.R. Black issued his work in 1967. As the quantity of free-charge carriers is growing with temperature, silicon in semiconductors has started to dominate reliability activities in a variety of industries. The U.S. Army Material Command published a Reliability Handbook (AMCP 702-3) in 1968. Shooman's Probabilistic Reliability was also published to describe statistical methods.

Automotive business issued a FMEA handbook for the technical development of suppliers to study the failure mode of electronic parts, not yet issued as a military standard. As many commercial satellites were released, the International Telecommunications Satellite Organization (INTELSAT), which was supplying international broadcast services between the United States and Europe in 1965, also intensified the reliability study for communications. World experts participated in reliability conferences. When Apollo was landing on a moon, people acknowledged how far reliability had progressed in the new decade.

$$f(t) = \frac{1}{2\mu^2\gamma^2\sqrt{\pi}}\left(\frac{t^2-\mu^2}{\sqrt{\frac{t}{u}}-\sqrt{\frac{u}{t}}}\right)\exp\left[-\frac{1}{\gamma^2}\left(\frac{t}{\mu}+\frac{\mu}{t}-2\right)\right] \tag{1.11}$$

where γ is a shape parameter and μ is a scale parameter.

As seen in Eq. (1.11), Birnbaum and Saunders in 1969 provided a life distribution model which might be obtained from a physical fatigue procedure where crack growth leads to failure. Since one of the leading methods to select a life distribution prototype is to obtain it from a physical/statistical argument which is compatible with the failure mechanism, the Birnbaum-Saunders fatigue life distribution is worth taking into consideration.

After the microcomputer was designed in the 1970s, RAM dimensions were rapidly growing. ICs were substituted by vacuum tubes. The variations of ICs such as NMOS, bipolar, and CMOS rapidly increased. In the middle of the 1970s, Electrostatic Discharge (ESD) and Electrical Over Stress (EOS) were issued by some papers and finally turned the debate issues of a conference at the end of the decade.

In the same way, International Reliability Physics Symposium (IRPS) studies for passive parts such as resistor, capacitor, and inductor passed to a Capacitor and Resistor Technology Symposium (CARTS). The forward-thinking papers on gold aluminum intermetallic, elevated testing, and the usage of scanning electron microscopy (SEM) were in this decade.

Based on field data, Hakim and Reich in the middle of the 1970s issued a paper on the assessment of plastic-encapsulated transistors and ICs. In addition, the two most significant reliability papers were published: one on soft errors from alpha particles first issued by Woods and May and the other on accelerated testing of ICs with activation energies computed for a variety of failure mechanisms by D.S. Peck. Bellcore at the end of the decade gathered commercial market data and became the basis of the Bellcore reliability prediction methodology utilized extensively with MIL-STD-217F.

Toward the end of the 1950s and the beginning of the 1960s, interest in the United States was focused on space research and intercontinental ballistic missiles, chiefly joined to the Gemini and Mercury programs. In the competition with Russia, the launching of a manned spacecraft was critical. During the Apollo space program, the spacecraft and its parts reliably functioned the entire road to the moon. In returning to the Navy, all agreements should hold specifications for reliability instead of exactly performance requirements. As Military Standard 1629 on FMEA was published in 1974, NASA could jump at designing and developing spacecrafts such as the space shuttle. Their accent was on risk management through the reliability, utilization of statistics, system safety, QA, human factors, and software assurance. As technology rapidly advanced, reliability had enlarged into a number of new areas.

Highlighting random vibration and temperature cycling became environmental stress screening testing, which was finally published as a Navy document P-9492 in 1979 and issued on Random Vibration with Tustin in 1984. The bygone quality procedures were changed with the Navy Best Manufacturing Practice program. An engineer association working with reliability questions was soon found. IEEE Transactions on reliability came out in 1963, and many books on the subject were issued in the 1960s.

In the 1970s, concerns on safety and risk features of the construction and operation of nuclear power plants expanded in the other areas of the world as well as in the United States. In the United States, a huge research commission financed by the multimillion-dollar project, the so-called Rasmussen report, was established to examine the problem. Despite its ineffectiveness, this report points out the first safety analysis of the complex system as a nuclear power plant.

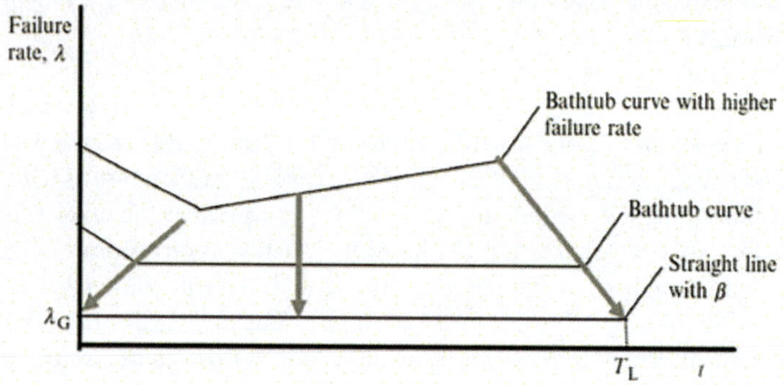

Fig. 1.49 Bathtub curve and straight line with slope β

Comparable efforts in the business of Asia and Europe have also been made in the analysis of risk and reliability issues. In particular, the offshore oil industry of Norway occurs. As in the North Sea, offshore gas and oil development is moving forward into deeper waters, and a growing number of faraway-operated subsea production systems are working. In many regards comparable to the reliability of spacecrafts, the reliability of subsea systems is critical because moderate reliability shall not be compensated by large maintenance.

During this decade, as the failure rates of numerous electronic components, including mechanical components, fell by a factor of 10, engineers doubted the bathtub curve. For such a circumstance, the conventional failure rate attributed by the bathtub curve shall be lessened to resemble the failure rate expressed by a flat, straight line with the shape parameter β (Fig. 1.49).

References

1. Wöhler A (1855) Theorie rechteckiger eiserner Brückenbalken mit Gitterwänden und mit Blechwänden. Zeitschrift für Bauwesen 5:121–166
2. Wöhler A (1870) Über die Festigkeitsversuche mit Eisen und Stahl. Zeitschrift für Bauwesen 20:73–106
3. Griffith AA (1921) The phenomena of rupture and flow in solids. Philos Trans Roy Soc London A 221:163–198
4. Fleming JA (1815) US Patent 803,684, 17 Nov 1815
5. Deming WE, Stephan F (1940) On a least squares adjustment of a sampled frequency table when the expected marginal totals are known. Ann Math Stat 11(4):427–444
6. Miner MA (1945) Cumulative damage in fatigue. J Appl Mech 12(3):59–64
7. Epstein B (1948) Statistical aspects of fracture problems. J Appl Phys 19
8. Naresky JJ (1962) Foreword. In: Proceedings of first annual symposium on the physics of failure in electronics, September 26–27
9. Weibull W (1959) Statistical evaluation of data from fatigue and creep rupture tests, part I: fundamental concepts and general methods. Wright Air Development Center Technical Report 59-400, Sweden

10. Weibull W (1961) Fatigue testing and analysis of results. Pergamon Press, London
11. Abernethy R (2002) The new Weibull handbook, 4th edn self-published ISBN 0-9653062-1-6
12. Lloyd D, Lipow M (1962) Reliability: management, methods and mathematics. Prentice Hall, Englewood Cliffs

Chapter 2
Engineering (Dynamic) Load Analysis

2.1 Introduction

As mentioned, mechanical systems convert (generated) power into forces (effort) and velocity (flow) that eventually provide mechanical advantages to accomplish a task by adapting product mechanisms. In power transmission, a product will be subjected to a variety of stresses due to repeated loads. If there are design defects in a product which produces an insufficiency in the product's stiffness (or strength) when subjected to loads, it may cause a structural failure of parts in a mechanical system, i.e., the (low/high) cycle fatigue that is characterized by repeated plastic deformation. While designing a mechanical system, an engineer should recognize the structural loading of the mechanical system. Basic analysis will be carried out as follows: (1) state the problem, (2) draw figures, (3) describe motion: number of degrees of freedom ⇒ number of coordinates ⇒ assign coordinate ⇒ number of EOM (equation of motion) ⇒ do the kinematics (calculate velocity and acceleration).

There are two kinds of loadings: statics and dynamics. Static loads—tension, compression, shear, or twist—cause displacement, strain, and stress. The mechanical system is devised to provide sufficient strength to a product in its lifetime to bear such repeated (static) loads as follows: (1) structural load and deflection versus matter stress and strain, (2) tension and compression loads, and (3) torsion and bending loads. A dynamic load is a force applied by a body in motion, generally in a comparatively short time period. Because such loads are generally changeable, we can define the dynamic load. It causes motions such as vibration, cumulative damage, fatigue, and fracture. Thus, a product should be designed to have sufficient stiffness in its structure during its lifetime to endure repeated (dynamic) loads as follows: (1) impact, vibration, and shock loads and (2) unbalanced inertia loads.

S. Woo, *Design of Mechanical Systems*, Springer Series in Reliability Engineering,
https://doi.org/10.1007/978-3-031-28938-5_2

2.2 Statics Principles

2.2.1 (Static) Force as a Vector

Forces always occur in pairs acting on two different objects. Force pairs act in opposite orientations and have the same magnitude. A single force can be substituted by component forces if their combination generates the same results as the original force. The resultant force is a single force which has the same result as two coincident forces.

Example 2.1 Based on the following table, we find the result force as follows (Fig. 2.1).

Force	X-components	Y-components
F1 = 5 k	$5(1)\surd 2 = 3.5355$	$-5(1)\surd 2 =- 3.5355$
F2 = 30 k	$30(1)\surd 5 = 13.4164$	$30(2)\surd 5 = 26.8328$
F3 = 20 k	$20(2)\surd 5 = 17.8885$	$-20(1)\surd 5 =- 8.94427$
F4 = 10 k	−10	0
Resultant, R	24.8404	14.35301

$$\text{Resultant Force, R} = \sqrt{(24.8404^2 + 14.35301^2)} = 28.689\ \text{k}$$

$$\text{Resultant Angle, } \theta = \tan^{-1}(14.35301/24.8404) = 30.02^\circ$$

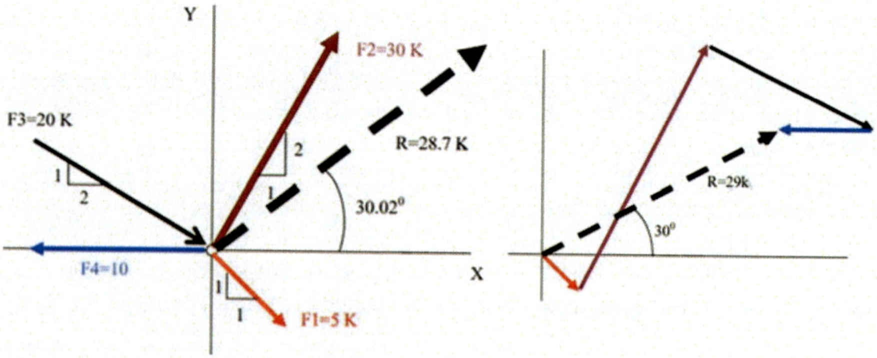

Fig. 2.1 Force sum as a vector (example)

2.2.2 *Moment (or Couple) of a Force*

As seen in Fig. 2.2, the moment of a force in respect of a reference point is equal to the multiplication of the force and the perpendicular distance of the force from the reference point. Orientations for moments about the reference point are either clockwise or counterclockwise rotations. A frequently utilized sign convention is counterclockwise rotation as positive (+) and clockwise rotation as negative (−).

A couple is expressed as two forces possessing equal magnitude, parallel in lines of action, but opposite in directions. Couples produce rotational results on a body without capacity of movement of the body in any direction. The moment of a couple, M, is calculated as the multiplication of the force (F) times the perpendicular distance (d) between the two equal and opposite forces (Figure 2.3).

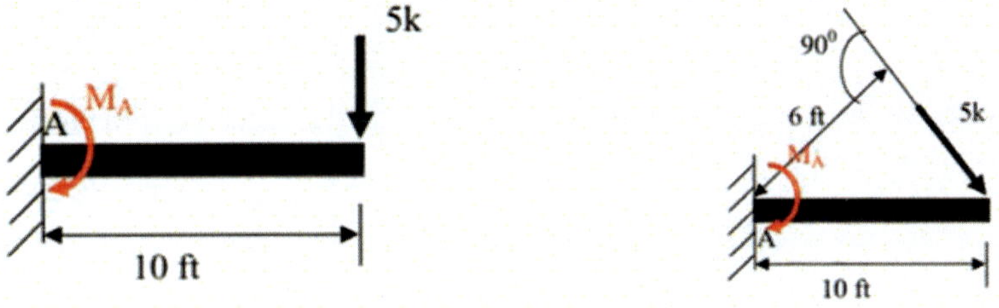

Moment at A, M_A = -5 x 10 = -50 k-ft (clockwise)

Moment at A, M_A = -5 x 6 = -30 k-ft (clockwise)

Fig. 2.2 Moment (or couple) of a force (example)

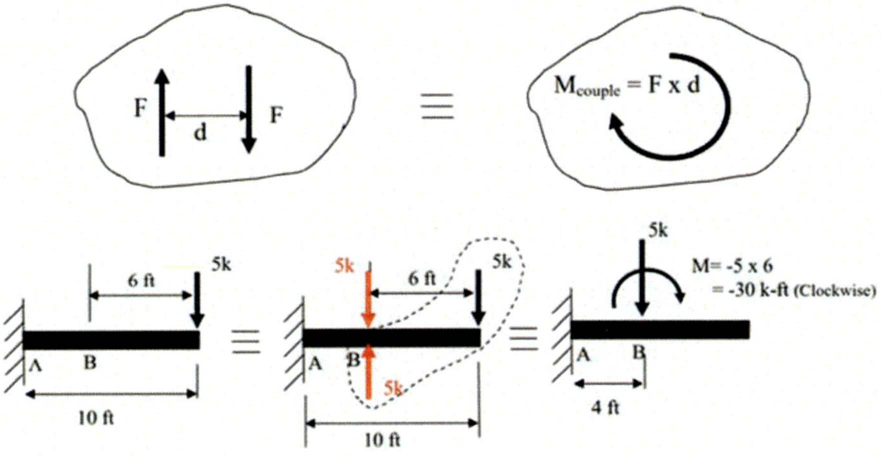

Fig. 2.3 Couple of a force

2.2.3 (Mass) Moment of Inertia

The moment of inertia is a standard of an object's resistance to changes in its rotation and is also expressed as the potential of a cross-section to withstand bending. Think about the moment of inertia I_x of an area A in respect of an axis AA′. Designate by y, the interval from an element of area dA to AA′ (Table 2.1).

For example, discover the moment of inertia of rectangular based on Fig. 2.4. That is,

$$I_x = \int_{Area} y^2 dA = \int_0^h \int_0^b y^2 dx dy$$
$$= \left[b\frac{y^3}{3} \right]_0^h = \frac{bh^3}{3} \tag{2.1}$$

Therefore, we can compute the moment of inertia at the centroid from the parallel axis theorem such as follows:

$$I_x = I_{x'} + Ad^2 \tag{2.2}$$

where A is area, d is distance

For example, the moment of inertia of rectangular is

$$\frac{bh^3}{3} = \frac{bh^3}{12} + bh\left(\frac{h}{2}\right)^2 = \frac{bh^3}{12} + \frac{bh^3}{2} \tag{2.3}$$

On the other hand, mass moment of inertia also known as rotational inertia is a quantity that is used in measuring a body's resistance to a change in its rotation direction or the angular momentum. It basically characterizes the acceleration undergone by an object or solid when torque is applied.

That is, mass moment of inertia is defined as follows:

$$I_{zz} = \int r^2 dm \tag{2.4}$$

Parallel axis theorem for mass moment of inertia is defined as follows:

$$I_{zz/A} = I_{zz/G} + Md^2 \tag{2.5}$$

where M is mass

For example, find mass moment of inertia in circular plate, as shown in Fig. 2.5.

Table 2.1 Moment of inertia for various cross sections

Rectangle	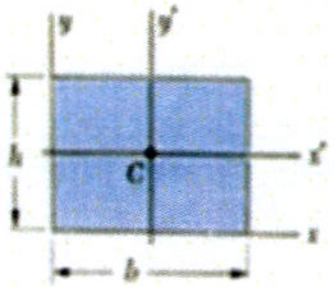	$\bar{I}_{x'} = \frac{1}{12}bh^3$ $\bar{I}_{y'} = \frac{1}{12}b^3h$ $I_x = \frac{1}{3}bh^3$ $I_y = \frac{1}{3}b^3h$ $J_C = \frac{1}{12}bh\left(b^2 + h^2\right)$	Semicircle	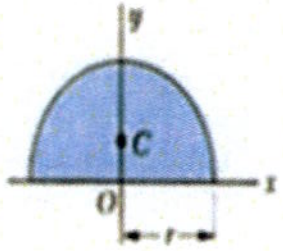	$I_x = I_y = \frac{1}{8}\pi r^4$ $J_O = \frac{1}{4}\pi r^4$
Triangle	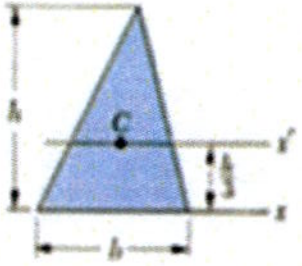	$\bar{I}_{x'} = \frac{1}{36}bh^3$ $I_x = \frac{1}{12}bh^3$	Quarter circle	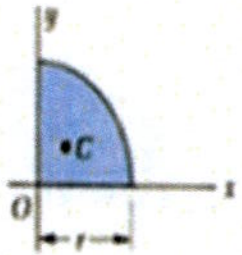	$I_x = I_y = \frac{1}{16}\pi r^4$ $J_O = \frac{1}{8}\pi r^4$
Circle	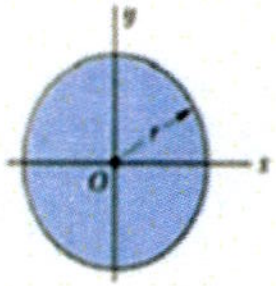	$\bar{I}_x = \bar{I}_y = \frac{1}{4}\pi r^4$ $J_O = \frac{1}{2}\pi r^4$	Ellipse	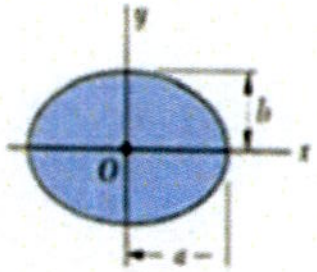	$\bar{I}_x = \frac{1}{4}\pi ab^3$ $\bar{I}_y = \frac{1}{4}\pi a^3b$ $J_O = \frac{1}{4}\pi ab\left(a^2 + b^2\right)$

$$I_x = \int y^2 dA$$

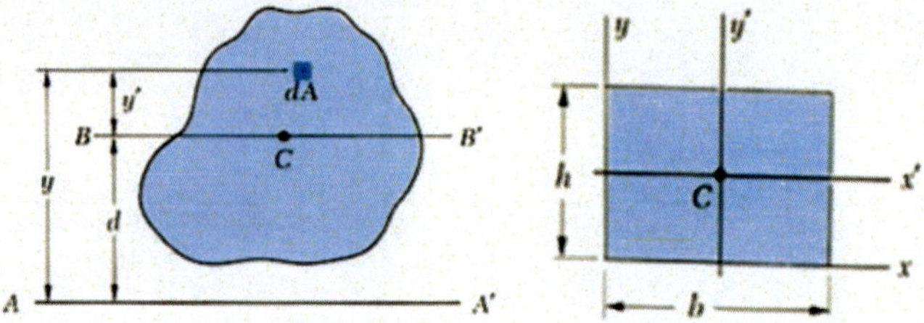

Fig. 2.4 Moment of inertia

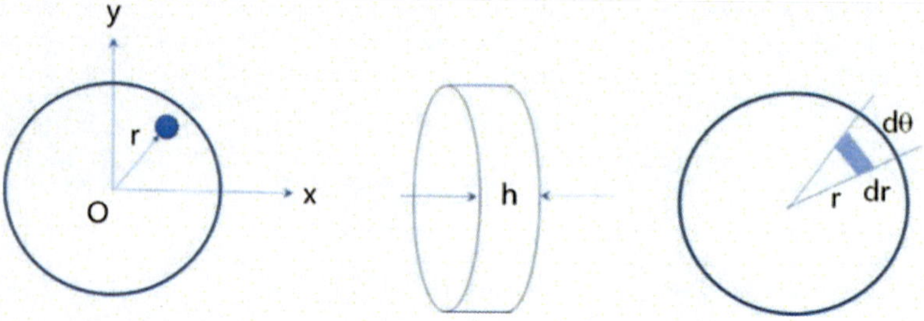

Fig. 2.5 Circular plate

$$I_{zz} = \int r^2 dm = \int (x^2 + y^2) dm = \int_0^R \int_0^{2\pi} r^2 \rho h r dr d\theta = 2\pi \rho h \int_0^R r^3 dr$$

$$= \frac{2\pi R^4}{4} \rho h = \frac{MR^2}{2}$$

where $dm = \rho dV = \rho hrdrd\theta$, $V = hrdrd\theta$

2.2.4 *Equilibrium*

For a rigid object that is not going at all, we have the next circumstances:

(1) Translational Equilibrium: The state in which there are balanced forces acting on a body.

$$\sum F_x = 0 \tag{2.6}$$

(2) Rotational Equilibrium: The condition in which the total of all the clockwise moments is equivalent to the total of all the counterclockwise moments about a pivot point.

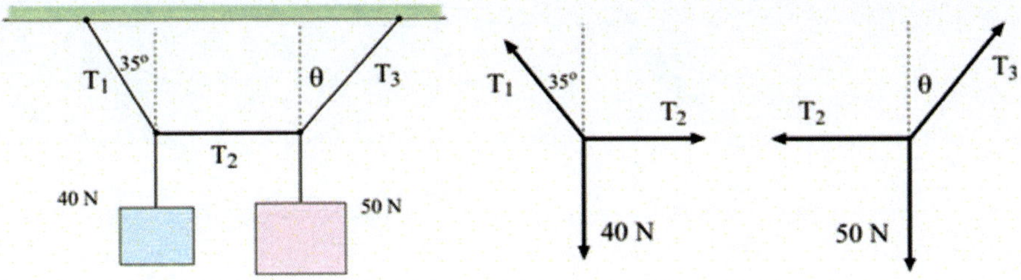

Fig. 2.6 Forces at the left/right junction of the strings

$$\sum M = 0 \tag{2.7}$$

Example 2.2 The system in Fig. 2.6 is in equilibrium with the string in the center, which is precisely horizontal. Discover (a) tension T_1, (b) tension T_2, (c) tension T_3 and (d) angle θ.

Solution

Four unknowns (T_1, T_2, T_3 and θ) to resolve it. This junction in the strings is in static equilibrium. Thus, the total of the y components of the forces in the left junction of the strings is zero:

$$\sum F_y = 0 : +T_1 \cos 35° - 40\ N = 0 \rightarrow T_1 = 48.8\ N$$

and the total of the x components of the forces is zero:

$$\sum F_x = 0 : -T_1 \sin 35° + T_2 = 0 \rightarrow T_2 = 28.0\,N$$

Now, the total of the x or y components of the forces is zero. We look at the right junction of the strings.

$\sum F_x = 0 : -T_2 + T_3 \sin\theta = 0$ or $\sum F_y = 0 : +T_3 \cos\theta - 50.0\ N = 0$
$T_3 \sin\theta = T_2 = 28.0\ N$ or $T_3 \cos\theta = 50.0\ N$
$\tan\theta = 28.0\ N/50.0\ N = 0.560 \Leftrightarrow \theta = \tan^{-1}(0.560) = 29.3°$. Finally, we attain T_3,
$T_3 = (50.0\ N)/(\cos 29.3°) = 57.3\ N$
$T_1 = 48.8\ N\ T_2 = 28.0\ N\ T_3 = 57.3\ N \theta = 29.3°$

Example 2.3 The system in Fig. 2.7 is in equilibrium. A mass of 225 kg attaches from the ending of the uniform strut, whose mass is 45.0 kg. Discover (a) the tension T in the cable and the (b) horizontal and (c) vertical force components applied on the strut by the hinge.

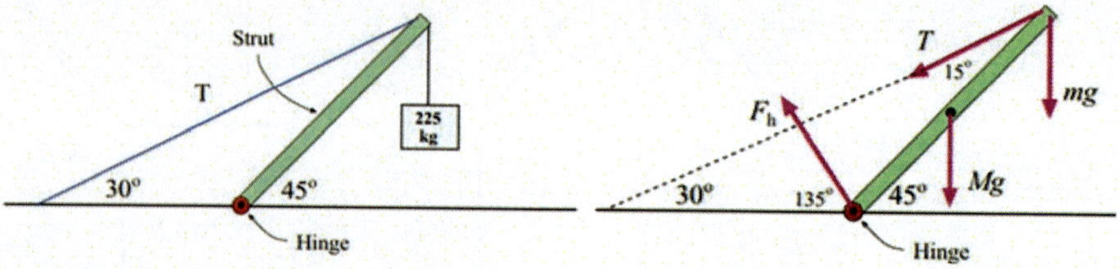

Fig. 2.7 System forces are in equilibrium

Solution

If the dangling mass is m = 225 kg, the string that bears it applies a downward force of magnitude mg at the peak ending of the strut.

$\sum F_x = 0$ (the sum of the F_x's must give zero). That is, $F_{h,x} - T\cos 30° = 0$
$F_{h,x} = T\cos 30° = (6630\ N)\cos 30° = 5.74 \times 10^3\ N$
$\sum F_y = 0$ (the sum of the F_y's must give zero). That is, $F_{h,y} - T\sin 30° - Mg - mg = 0$
$F_{h,y} = T\sin 30° + Mg + mg = (6630\ N)\sin 30° + (45\ kg)(9.80\ m/s^2) + (225\ kg)(9.80\ m/s^2) = 5.96 \times 10^3\ N$
$\sum M_B = 0$ (the condition for zero net torque). $-Mg(L/2)\sin 45° - mg(L)\sin 45° + T(L)\sin 15° = 0$
$T\sin 15° = (Mg/2)\sin 45° + mg\sin 45° = (45\ kg/2)(9.80\ m/s^2)\sin 45° + (225\ kg)(9.80\ m/s^2)\sin 45°$
$T = (1715\ N)/\sin 15° = 6.63 \times 10^3\ N$

Example 2.4 As seen in Fig. 2.8, a ladder with a constant density and a mass m pauses against a frictionless upright wall at an angle of 60°. The bottom ending pauses on a horizontal surface where the coefficient of static friction is $\mu_s = 0.40$. A student with a mass M = 2 m tries to mount the ladder. What fraction of the length L of the ladder will the student have extended when the ladder starts to slip?

Solution

We apply the conditions for static equilibrium. That is,

$\sum F_x = 0$ (the total of the F_x's must give zero). $f_s - N_w = 0$
$\sum F_y = 0$ (the total of the F_y's must give zero). $N_f - Mg(=2\ mg) - mg = 0$
$\sum M_B = 0$ (the condition for zero net torque)
$-(xL)(2\ mg)\sin 30° - (L/2)(mg)\sin 30° + (L)(N_w)\sin 60° = 0$
$-xmg - mg/4 + \sqrt{3}/2\ N_w = 0$
If the ladder is on the brink of slipping, we have the equality that $f_s = \mu_s N_f \Leftrightarrow \mu_s N_f - N_w = 0 \Leftrightarrow \mu_s N_f\ (= 3\ mg) - N_w = 0 \Leftrightarrow N_w = 3\mu_s mg$
$-xmg - mg/4 + \sqrt{3}/2\ N_w\ (= 3\mu_s mg) = 0 \Leftrightarrow -x - 1/4 + 3\sqrt{\ }3/2\ \mu_s = 0$
$x = -1/4 + 3\sqrt{3}/2\ \mu_s(= 0.4) = 0.798$

The student can mount a fraction of nearly 80% of the length of the ladder before it begins to slide.

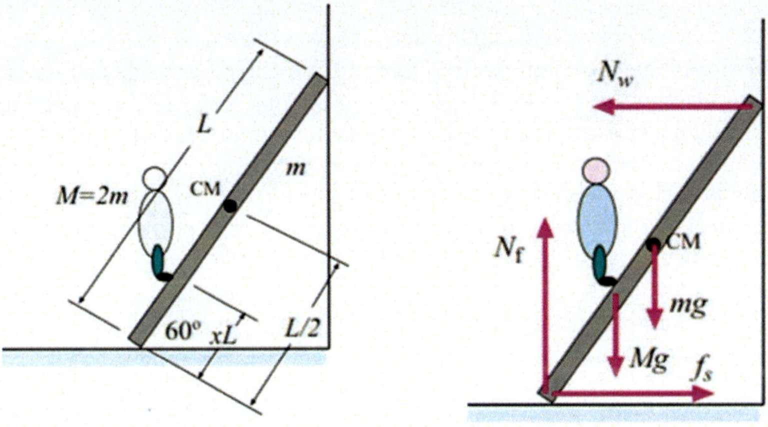

Fig. 2.8 Ladder forces are in equilibrium

2.3 (Dynamic) Modeling of Mechanical System (Power System)—Direct Method'

2.3.1 Introduction

Modeling of a mechanical system, including the structure and mechanism, is a mathematical description of the dynamic structures to understand its attributes under the work of forces. General modeling ways utilized in dynamic systems are Newton, Lagrange, D'Alembert's law, and Hamiltonian mechanics. As a product, models must report the system characteristics, which shall be depicted as random variables in state space. The state space is shown as vectors (displacement, velocity, and acceleration) in time. It also supplies a suitable and dense way to examine systems with numerous inputs/outputs.

When detected in most mechanical parts, field loads fit a random curve in which load amplitudes fluctuate over time. For instance, automobiles have totally random stochastic load curves due to the road roughness, vehicle velocity, and environmental states. In an airplane, a mean load change such as drag due to wind repeatedly occurs on the wings when it takes off, taxi or landing. On the other hand, although the load sequence is still variable, the load of the gas turbine blade in an airplane is to a big extent deterministic such that there is no randomness in the system conditions.

With uncomplicated algorithms and fast processors, an online load measurement for components shall be straightly calculated in operation. However, a measurement in operation is completely time-consuming and literally unfeasible to determine the entire transferred loads in product life. To effectively understand the load background of a system in time, engineers rely on mathematical modeling, such as Newtonian modeling and response analysis.

Mechanics is a physical discipline that discusses the condition of pause or movement of bodies under the influence of forces. No other subjects play a more crucial role in engineering analysis than does mechanics. Thus, the start of engineering originated from the research of mechanics. Modern research and evolution in the area of system reliability might depend on the fundamental scientific theorem of mechanics, including probability and statistical ways.

Mechanics is the elderly of the physical sciences. The premature investigation was that of Archimedes (287–212 B.C.) which is related to the scientific theorem of the buoyancy and lever. The earliest study of a dynamic issue was ascribed to Galileo (1564–1642) regarding his tests with dropping stones. The precise configuration of the law of gravitation and the laws of motion was established by Newton (1642–1727). Basic benefaction to the evolution of mechanics were also achieved by D'Alembert, Laplace, Lagrange, etc.

The concepts of mechanics as a science are dependent on strict mathematics. On the other hand, the aim of engineering mechanics is the implementation of its truths to system design. The fundamental principle of mechanics is comparatively few in number, but they have broad fields such as vibrations, automatic control, engine performance, fluid flow stability and strength of structures and machines, and rocket and spacecraft design.

For example, only by utilizing a few equations can scientists outline the movement of a projectile flying through the air and the haul of a magnet and predict the eclipses of the moon. The mathematical investigation of the movement is defined as Newtonian mechanics because almost the whole investigation is constructed on the achievement of Isaac Newton. Some mathematical principles and laws at the central of Newtonian mechanics involve the following:

- Newton's first law of movement: In the lack of forces, a particle goes with constant velocity.
- Newton's Second Law of Movement: The acceleration of a body is immediately related to the net force and inversely related to its mass.

$$F = ma = \frac{dp}{dt} \tag{2.8}$$

- Newton's Third Law of Movement: As two objects interact, they apply forces to one another which are equivalent to magnitude and opposite in orientation. That is, $f_{21} = -f_{12}$.

The equation of motion (EOM) thus is found as follows: (1) choosing independent coordinate systems, (2) finding number of freedom, (3) selecting EOM, (4) drawing free body diagram, (5) applying force $\left(\sum_{external} F = ma\right)$ or torque $\left(\sum_{external} \tau = \frac{d^A h}{dt} + {}^O v_A \times {}^O P\right)$.

2.3.2 *Reference Frames and Vectors—Displacement, Velocity, and Acceleration*

If the displacement vector is given with respect to the reference frame, velocity and acceleration at Cartesian coordinates can be defined as Fig. 2.9.

Displacement at point B with respect to reference frame O is

$$r_{B/O} = r_{A/O} + r_{B/A} \tag{2.9}$$

where $r_{B/O} = r_{Bx}\hat{i} + r_{By}\hat{j} + r_{Bz}\hat{k}$

If Eq. (2.9) is differentiated, velocity at point B is

$$\frac{dr_{B/O}}{dt} = v_{B/O} = v_{A/O} + v_{B/A} \tag{2.10}$$

where $v_{B/O} = \frac{dr_{B/O}}{dt} = \dot{r}_{Bx}\hat{i} + \dot{r}_{By}\hat{j} + \dot{r}_{Bz}\hat{k}$

If Eq. (2.10) is differentiated, acceleration at point B is

$$\frac{d^2 r_{B/O}}{dt^2} = a_{B/O} = a_{A/O} + a_{B/A} \tag{2.11}$$

where $a_{B/O} = \ddot{r}_{Bx}\hat{i} + \ddot{r}_{By}\hat{j} + \ddot{r}_{Bz}\hat{k}$

If $\Delta r_{B/O}$ moves in Δt, distance $r'_{B/O}$ can be defined as follows (see Fig. 2.9b):

$$r'_{B/O} = r_{B/O} + \Delta r_{B/O} \tag{2.12}$$

As Δt approaches 0, velocity $v_{B/O}$ can be obtained as follows:

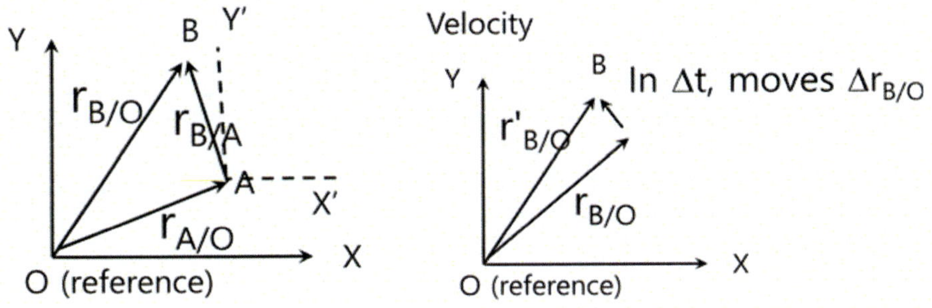

(a) Displacement (b) Velocity

Fig. 2.9 Displacement or velocity vector at the reference frame

$$v_{B/O} = \lim_{\Delta t \to 0} \frac{r_{B/O}}{\Delta t} = \frac{dr_{B/O}}{dt} \tag{2.13}$$

The time derivative of a vector can be defined as follows:

$$\left(\frac{dA}{dt}\right)_{Oxyz} = \left(\frac{dA}{dt}\right)_{Axyz} + \omega_{/o} \times \mathrm{A} \tag{2.14}$$

Therefore, from Eq. (2.9), the velocity with respect to a rotating frame, A, at point O is

$$v_{B/O} = \frac{dr_{B/O}}{dt} = \left(\frac{dr_{A/O}}{dt}\right)_{/O} + \left(\frac{dr_{B/A}}{dt}\right)_{/O} = v_{A/O} + \left(\frac{dr_{B/A}}{dt} = v_{B/A}\right)_{Axyz'} + \left(v_{B/A}\right)_A \tag{2.15}$$

where $\left(\frac{dr_{B/A}}{dt}\right)_{Axyz'}$ is a dog running with $\omega_{/o} = 0$. That is, $\left(\frac{dr_{B/A}}{dt}\right)_{\omega_{/o}=0}$

If $\omega = \omega\hat{k}$, $r_{B/A} = r_{B/A}\hat{i}$, $\left(v_{B/A}\right)_A$ is defined as follows:

$$\left(v_{B/A}\right)_A = \omega_{/O} \times r_{B/A} = \omega_{/O}\hat{k} \times r_{B/A}\hat{i} = \omega_{/O} r_{B/A}\hat{j} \tag{2.16}$$

Finally, Eq. (2.16) can be redefined as follows:

$$v_{B/O} = \frac{dr_{B/O}}{dt} = \left(\frac{dr_{A/O}}{dt}\right)_{/O} + \left(\frac{dr_{B/A}}{dt}\right)_{/O} = v_{A/O} + \left(\frac{dr_{B/A}}{dt} = v_{B/A}\right)_{Axyz'} + \omega_{/O} \times r_{B/A} \tag{2.17}$$

The acceleration from Eq. (2.17) can be defined as follows:

$$a_{B/O} = \left(\frac{dv_{B/O}}{dt}\right)_{Oxyz} = a_{A/O} + \left(a_{B/A}\right)_{Axyz'} + 2\,\omega_{/O} \times \left(v_{B/A}\right)_{Axyz'} + \dot{\omega}_{/O}\left(v_{B/A}\right)_A + \omega_{/O} \times \left(\omega_{/O} \times r_{B/A}\right) \tag{2.18}$$

For cylindrical coordinate system, the velocity and acceleration from Eqs. (2.17) and (2.18) can be expressed as follows:

$$v_{B/O} = v_{A/O} + \left(\frac{dr_{B/A}}{dt}\right)_{Axyz'} + \omega_{/O} \times r_{B/A} = \frac{d\left(r\hat{r} + z\hat{k}\right)}{dt} = v_{A/O} + \dot{r}\hat{r} + r\dot{\theta}\hat{\theta} + \dot{z}\hat{k} \tag{2.19}$$

$$a_{B/O} = a_{A/O} + \left(\ddot{r} - r\dot{\theta}^2\right)\hat{r} + \ddot{z}\hat{k} + \left(\mathrm{r}\ddot{\theta} + 2\dot{r}\dot{\theta}\right)\hat{\theta} \tag{2.20}$$

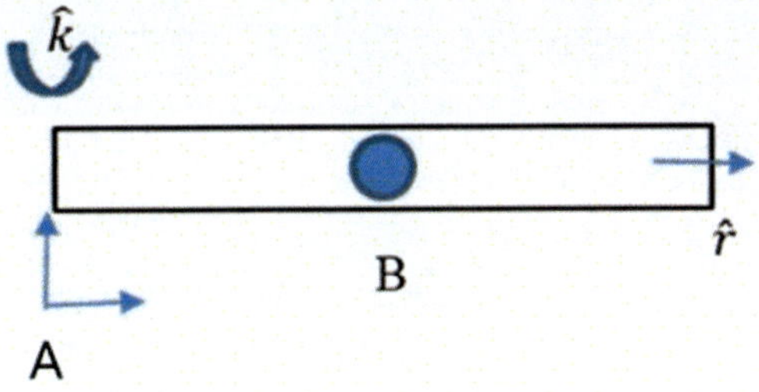

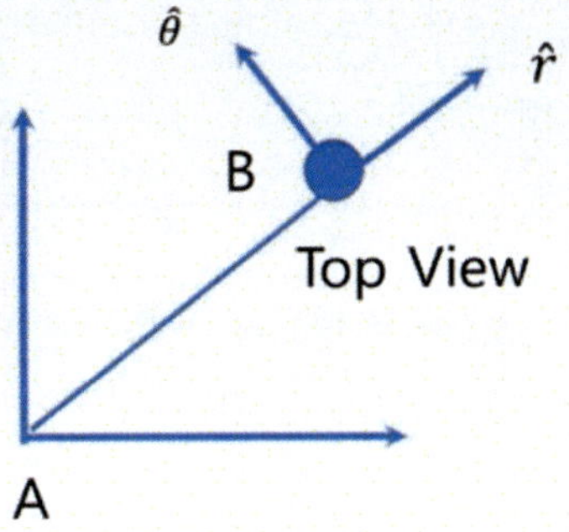

Fig. 2.10 Two rotating frames

For example, seen in the Fig. 2.10, there is a frictionless ball, B, in the hollow cylinder, rotating with constant speed, $\dot{\theta}$. In this case, we can find the velocity and acceleration of a ball based on the definition of Eqs. (2.19) and (2.20).

$$v_{B/O} = 0 + \dot{r}\hat{r} + r\dot{\theta}\hat{\theta} \tag{2.21}$$

$$a_{B/O} = 0 + \left(\ddot{r} - r\dot{\theta}^2\right)\hat{r} + \ddot{z}\hat{k} + \left(r\ddot{\theta} + 2\dot{r}\dot{\theta}\right)\hat{\theta} \tag{2.22}$$

2.3.3 *Velocity at Two Rotating Frames*

There are two rotating frames, seen in Fig. 2.11. We specify velocity at B point with respect to A point.

$$^{O}v_B = {}^{o}v_B + {}^{o}v_{B|\omega=0} + {}^{o}\omega_B \times {}^{A}r_B \tag{2.23}$$

where ${}^{o}\omega_B = {}^{o}\omega_1 + {}^{o}\omega_2\left(arm\widehat{AB}\right) = (\omega_1 + \omega_2)\hat{k}$

If arm length is l, Eq. (2.23) can be specified as follows:

$$^{O}v_B = R\omega_1\hat{j}_1 + (\omega_1 + \omega_2)\hat{k} \times l\hat{i}_2 = R\omega_1\hat{j}_1 + l(\omega_1 + \omega_2)\hat{j}_2 \tag{2.24}$$

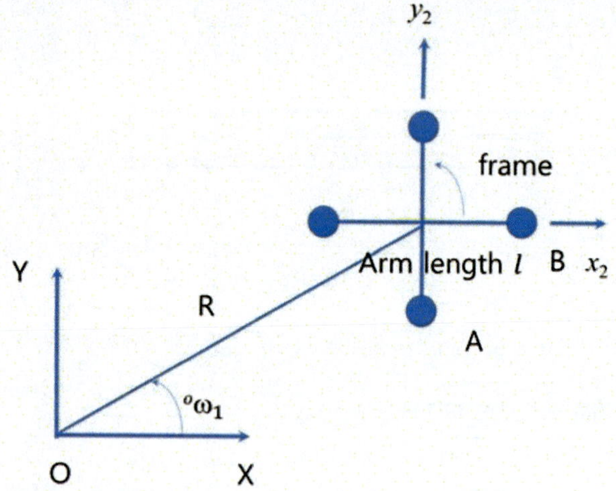

Fig. 2.11 Two rotating frames

2.3.4 *Linear Impulse Momentum*

As an object with mass, *m*, is subjected to external force, F_{ext}, motion of equation is given as follows:

$$\sum F_{ext} = ma = m\frac{dv}{dt} \tag{2.25}$$

If Eq. (2.25) is integrated for interval $[t_1, t_2]$, it can be expressed as follows:

$$\int_{t_1}^{t_2} \sum F_{ext}\mathrm{dt} = \int_{v_1}^{v_2} mdv \text{ or } \sum \int_{t_1}^{t_2} F_{ext}\mathrm{dt} = \mathrm{m}v_2 - \mathrm{m}v_1 \tag{2.26}$$

Finally, Eq. (2.26) can be summarized as follows:

$$\mathrm{m}v_1 + \sum \int_{t_1}^{t_2} F_{ext}\mathrm{dt} = \mathrm{m}v_2 \tag{2.27}$$

For example, as seen in Fig. 2.12, there is a sliding object in the slope plane. In this case, we can find the final velocity after time elapses 3 s, based on Eq. (2.27).

$$\mathrm{m}v_{1x} + \int_0^t [mg\sin\theta - \mu mg\cos\theta]dt = \mathrm{m}v_{2x} \text{ where } \sum F_x = mg\sin\theta - \mu mg\cos\theta \tag{2.28}$$

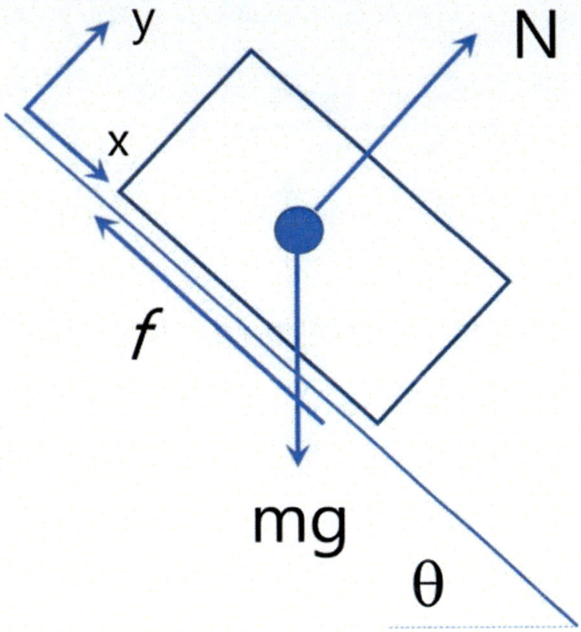

Fig. 2.12 A sliding object in the slope plane

Assume initial velocity $v_{1x} = 0$ in Eq. (2.28), we can final velocity at t = 3 s.

$$[mg\sin\theta - \mu mg\cos\theta]\mathrm{t} = \mathrm{m}v_{2x} \; or \; v_{2x} = 3[g\sin\theta - \mu g\cos\theta] \tag{2.29}$$

2.3.5 Angular Momentum of a Particle (or Rigid Body) for Motion of Equation (MOE)

There is a particle with some velocity (Fig. 2.13). Its angular momentum is specified as follows:

$$h_{B/O} = r_{B/O} \times P_{B/O} \tag{2.30}$$

If Eq. (2.30) is differentiated, it can be found as follows:

$$\frac{dh_{B/O}}{dt} = \sum \tau_{B/O} \tag{2.31}$$

Because $h_{B/A} = r_{B/A} \times P_{B/O}$, torque $\tau_{B/A}$ is

$$\tau_{B/A} = \frac{dh_{B/A}}{dt} + v_{B/A} \times P_{B/O} \tag{2.32}$$

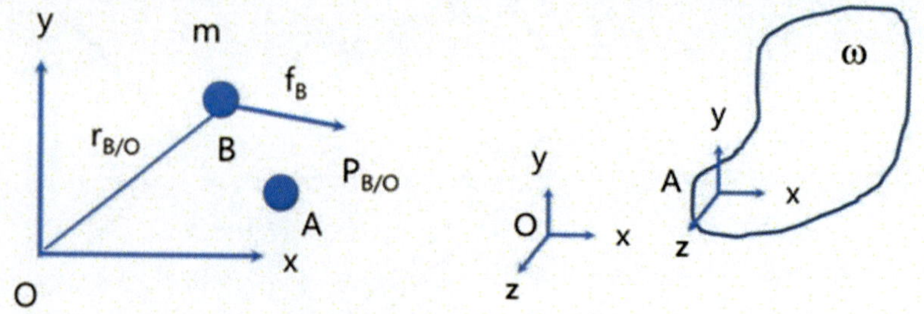

Fig. 2.13 A particle (or rigid body) in the space

The motion of equation at B is

$$\sum_{ext} f_B = \frac{dP_{B/O}}{dt} = m\frac{dv_{B/O}}{dt} \tag{2.33}$$

Torque B with respect to A is

$$\tau_{B/A} = r_{B/A} \times f_B = r_{B/A} \times \frac{dP_{B/O}}{dt} = \frac{d(r_{B/A} \times P_{B/O})}{dt} - \frac{dr_{B/A}}{dt} \times P_{B/O} \tag{2.34}$$

Because $r_{B/A} = r_{B/O} - r_{A/O}$ and $\frac{dr_{B/A}}{dt} = v_{B/O} - v_{A/O}$, Eq. (2.34) is

$$\tau_{B/A} = \frac{dh_{B/A}}{dt} - (v_{B/O} - v_{A/O}) \times P_{B/O} \tag{2.35}$$

where $P_{B/O} = mv_{B/O}$

Case I: If $v_{A/O} = 0$, $\tau_{B/A} = \frac{dh_{B/A}}{dt}$ from Eq. (2.35)

Case II: If $v_{A/O} \|$ to $P_{B/O}$ (true when A = center of mass), $\tau_{/A} = \frac{dh_{/A}}{dt}$ from Eq. (2.35)

For example, if there is an object in the rotating plane with $\dot{\theta}$ = constant and $\dot{r}$ = constant, we can find $h_{B/o}$ and $\tau_{B/o}$ as follows:

$$h_{B/O} = r_{B/O} \times P_{B/O} = r\hat{r} \times m(\dot{r}\hat{r} + r\dot{\theta}\hat{\theta}) = mr^2\dot{\theta}\hat{k} \tag{2.36}$$

$$\tau_{B/o} = \frac{dh_{B/O}}{dt} = \frac{d}{dt}(mr^2\dot{\theta}\hat{k}) = 2mr\,\dot{r}\,\dot{\theta}\hat{k} + mr^2\ddot{\theta}\hat{k} \tag{2.37}$$

In the same manner, if there is a rotating object with $\dot{\theta}$ = constant and $\dot{r} = 0$, we can find torque T from motion of equation.

$$v_{A/O} = r_1\dot{\theta}\hat{\theta} \tag{2.38}$$

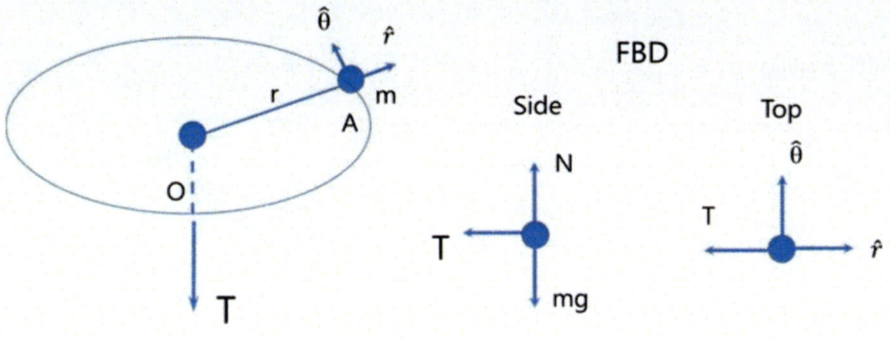

(a) A rotating object (b) Free body diagram

Fig. 2.14 A rotating object with constant speed and radius

$$P_{A/O} = mr_1\dot{\theta}\hat{\theta} \tag{2.39}$$

$$\sum_{ext} F = \frac{dP_{A/O}}{dt} = -mr_1\dot{\theta}^2\hat{r} = -T\hat{r} \tag{2.40}$$

$$T = mr_1\dot{\theta}^2 \tag{2.41}$$

Example 2.5 As seen in Fig. 2.14, there is a rotating object with constant speed and radius. If $\dot{r}$ = constant, find motion of equation.

$$v_{A/O} = \dot{r}\hat{r} + r\dot{\theta}\hat{\theta} \tag{2.42}$$

$$P_{A/O} = m\left(\dot{r}\hat{r} + r\dot{\theta}\hat{\theta}\right) \tag{2.43}$$

$$\frac{dP_{A/O}}{dt} = m\left[\left(-r\dot{\theta}^2\right)\hat{r} + \left(r\ddot{\theta} + 2\dot{r}\dot{\theta}\right)\hat{\theta}\right] = \sum_{ext} F \tag{2.44}$$

$$\hat{\theta}\ \text{direction}: \frac{dP_{A/O}}{dt} = 0 = m\left(r\ddot{\theta} + 2\dot{r}\dot{\theta}\right) \tag{2.45}$$

$$r\ddot{\theta} = -2\dot{r}\dot{\theta} \tag{2.46}$$

$$\hat{r}\ \text{direction}: \sum_{ext} F = -mr\dot{\theta}^2 = -T \text{ or } T = mr\dot{\theta}^2 \tag{2.47}$$

$$h_{A/O} = r_{A/O} \times P_{A/O} = r\hat{r} \times m\left(\dot{r}\hat{r} + r\dot{\theta}\hat{\theta}\right) = mr^2\dot{\theta}\hat{k} \tag{2.48}$$

Fig. 2.15 A rotating object with coordinate (r, z)

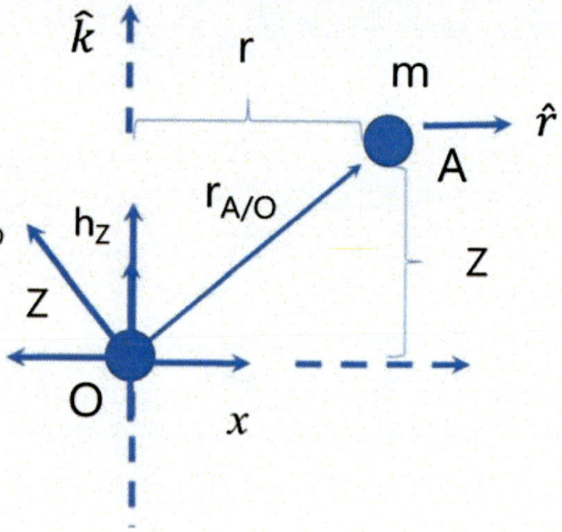

$$\hat{k}\text{ direction}: \sum_{ext} \tau_{A/O} = \frac{dh_{A/O}}{dt} = m\left(2r\dot{r}\dot{\theta} + r^2\ddot{\theta}\right) = 0 \text{ or } 2\dot{r}\dot{\theta} = -r\ddot{\theta} \quad (2.49)$$

As $r = r_1$ at $t_1, h_{A/O} = mr_1^2\dot{\theta}_1$. On the other hand, as $r = r_2$ at $t_2, h_{A/O} = mr_2^2\dot{\theta}_2$. Because $h_{A/O} =$ constant, we know that $mr_1^2\dot{\theta}_1 = mr_2^2\dot{\theta}_2$.

Example 2.6 As seen in Fig. 2.15, find motion of equation.

$$h_{A/O} = r_{A/O} \times P_{A/O} = \left(r\hat{r} + z\hat{k}\right) \times mr\dot{\theta}\hat{\theta} = mr^2\dot{\theta}\hat{k} - mrz\dot{\theta}\hat{r} \quad (2.50)$$

$$\begin{aligned}\sum_{ext} \tau_{A/O} = \frac{dh_{A/O}}{dt} &= mr^2\ddot{\theta}\hat{k} - mrz\ddot{\theta}\hat{r} - mrz\dot{\theta}\dot{\theta}\hat{\theta} \\ &= mr^2\ddot{\theta}\hat{k} - mrz\ddot{\theta}\hat{r} - mrz\dot{\theta}^2\hat{\theta}\end{aligned} \quad (2.51)$$

Consider bodies that rotate about center of mass or fixed points in case that reference frames attached to bodies is utilized in Fig. 2.16.

$$h_{i/A} = r_{i/A} \times P_{1/O} = \left(x_1\hat{i} + z_1\hat{k}\right) \times m_1x_1\omega_z\hat{j} = m_1x_1^2\omega_z\hat{k} - m_1x_1z_1\omega_z\hat{i} \quad (2.52)$$

In this case, external torque, τ^{ext}, is

$$\frac{dh_{i/A}}{dt} = \tau^{ext} \quad (2.53)$$

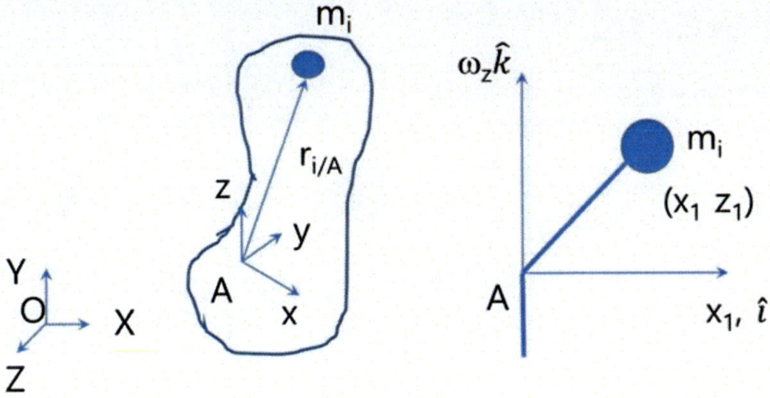

Fig. 2.16 Bodies that rotate about center of mass or fixed points

On the other hand, angular momentum of rigid body is specified as follows:

$$H_{/A} = \sum_i r_{i/A} \times P_{i/A} = \sum_i m_i r_{i/A} \times (\omega_{/o} \times r_{i/A}) = H_x\hat{i} + H_y\hat{j} + H_z\hat{k} \quad (2.54)$$

For matrix form, Eq. (2.53) is expressed as follows:

$$H_{/A} = \begin{Bmatrix} H_x \\ H_y \\ H_z \end{Bmatrix} = \begin{bmatrix} I_{xx} & I_{xy} & I_{xz} \\ I_{yx} & I_{yy} & I_{yz} \\ I_{zx} & I_{yz} & I_{zz} \end{bmatrix} \begin{Bmatrix} \omega_x \\ \omega_y \\ \omega_z \end{Bmatrix} \quad (2.55)$$

where $I_{xz} = -\sum_i m_i x_i z_i$, $I_{zz} = \sum_i m_i (x_i^2 + y_i^2)$

In case $\omega = \omega_z \hat{k}$

$$H_{/A} = \begin{Bmatrix} H_x \\ H_y \\ H_z \end{Bmatrix} = \begin{bmatrix} I_{xx} & I_{xy} & I_{xz} \\ I_{yx} & I_{yy} & I_{yz} \\ I_{zx} & I_{yz} & I_{zz} \end{bmatrix} \begin{Bmatrix} 0 \\ 0 \\ \omega_z \end{Bmatrix} = \begin{Bmatrix} I_{xz}\omega_z \\ I_{yz}\omega_z \\ I_{zz}\omega_z \end{Bmatrix} \quad (2.56)$$

In other word,

$$H_{/A} = H_x\hat{i} + H_y\hat{j} + H_k\hat{k} = I_{xz}\omega_z\hat{i} + I_{yz}\omega_z\hat{j} + I_{zz}\omega_z\hat{k} \quad (2.57)$$

Example 2.7 There is two rotating ball, as seen in Fig. 2.17. Find angular momentum ${}^A H$ at point A.

$${}^o P_1 = m_1 {}^o v_1 = m_1 \Omega x_1 \hat{j} \quad (2.58)$$

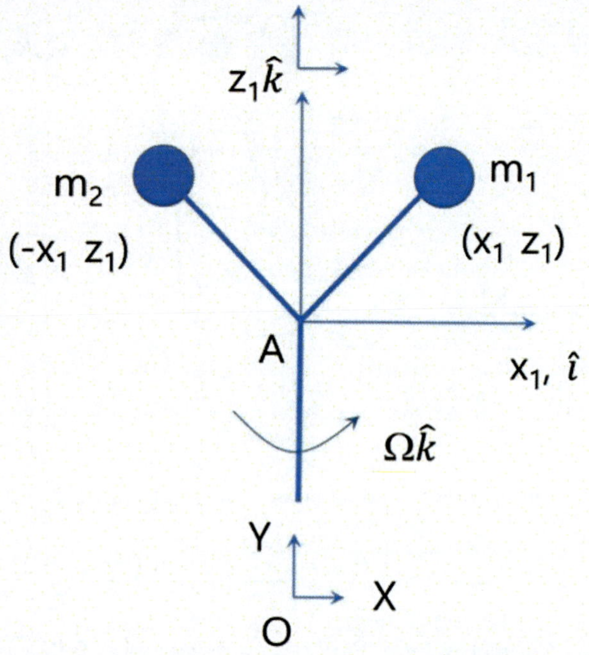

Fig. 2.17 Two rotating ball

$$^{o}P_2 = m_2{}^{o}v_1 = -m_2\Omega x_1\hat{j} \tag{2.59}$$

$$^{A}h_1 = {}^{A}r_1 \times {}^{o}P_1 = \left(x_1\hat{i} + z_1\hat{k}\right) \times m_1\Omega x_1\hat{j} = m_1x_1^2\Omega\hat{k} - m_1x_1z_1\Omega\hat{i} \tag{2.60}$$

$$^{A}h_2 = {}^{A}r_2 \times {}^{o}P_2 = \left(-x_1\hat{i} + z_1\hat{k}\right) \times -m_2\Omega x_1\hat{j} = m_2x_1^2\Omega\hat{k} + m_2x_1z_1\Omega\hat{i} \tag{2.61}$$

If $m_1 = m_2 = m$, $\omega = \begin{Bmatrix} 0 & 0 & \Omega \end{Bmatrix}^T$, angular momentum is

$$^{A}H = {}^{A}h_1 + {}^{A}h_2 = 2mx_1^2\Omega\hat{k} \tag{2.62}$$

In this case, external torque, τ^{ext}, is

$$\frac{d^{A}H}{dt} = \tau^{ext} = 2mx_1^2\,\dot{\Omega}\,\hat{k} \tag{2.63}$$

Principal axis can be defined as a set of orthogonal axes such that $[I] = \begin{bmatrix} I_{xx} & 0 & 0 \\ 0 & I_{yy} & 0 \\ 0 & 0 & I_{zz} \end{bmatrix}$.

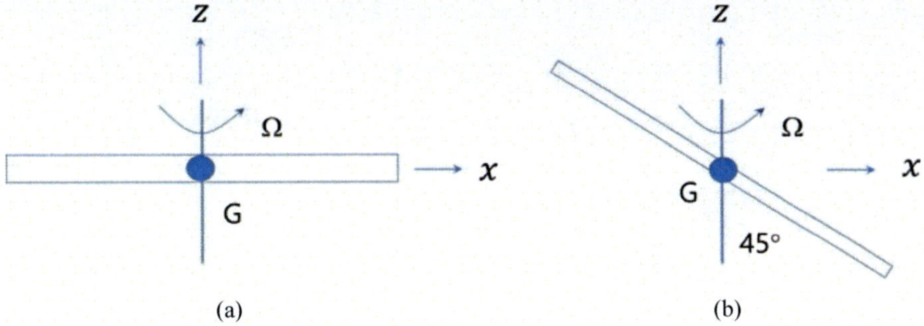

Fig. 2.18 A rotating rod

Rules of symmetry for rigid bodies are as follows: (1) If there is an axis of symmetry, the axis is a principal axis, (2) If there is one plane of symmetry, a principal axis perpendicular to it pass through center of mass, (3) If there are two orthogonal planes of symmetry, their intersection is a principal axis.

Fundamental laws for rigid bodies are defined as follows:

$$\sum_{ext} F = \frac{dP_{/O}}{dt} = Ma \tag{2.64}$$

$$\sum_{ext} \tau = \frac{dH_{/A}}{dt} + v_{A/O} \times P_{/O} \quad \text{where} \quad H_{/A} = \left[{}^{G}I\right]\{\omega\} \tag{2.65}$$

Example 2.8 As seen in Fig. 2.18, find angular momentum, H, of a rotating rod.

For case (a), because L ≫ R, angular momentum, H, is

$$H = \left[{}^{G}I\right]\{\omega\} = \begin{bmatrix} I_{xx} & 0 & 0 \\ 0 & I_{yy} & 0 \\ 0 & 0 & I_{zz} \end{bmatrix} \begin{Bmatrix} \omega_x \\ \omega_y \\ \omega_z \end{Bmatrix} = \begin{bmatrix} \frac{MR^2}{2} \approx 0 & 0 & 0 \\ 0 & \frac{ML^2}{12} & 0 \\ 0 & 0 & \frac{ML^2}{12} \end{bmatrix} \begin{Bmatrix} 0 \\ 0 \\ \Omega \end{Bmatrix}$$
$$= \begin{Bmatrix} 0 \\ 0 \\ \frac{ML^2}{12}\Omega \end{Bmatrix} \tag{2.66}$$

If Eq. (2.66) is differentiated,

$$\frac{dH}{dt} = \frac{ML^2}{12}\dot{\Omega}\hat{k} \tag{2.67}$$

For case (b), angular momentum, angular momentum, H, is

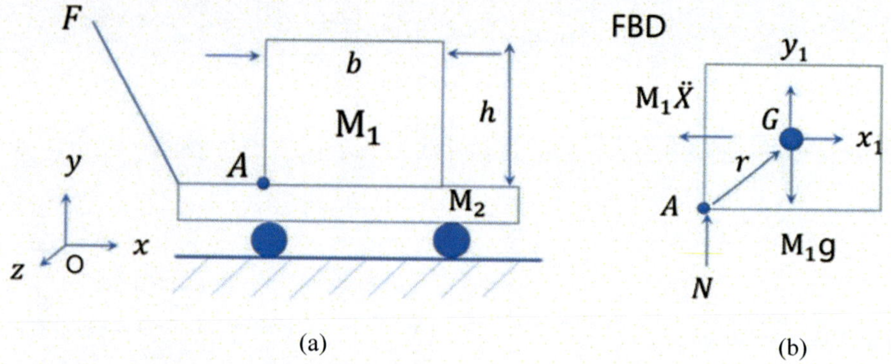

Fig. 2.19 An object on cart

$$H = \begin{bmatrix} 0 & 0 & 0 \\ 0 & \frac{ML^2}{12} & 0 \\ 0 & 0 & \frac{ML^2}{12} \end{bmatrix} \begin{Bmatrix} -\frac{\sqrt{2}}{2}\Omega \\ 0 \\ \frac{\sqrt{2}}{2}\Omega \end{Bmatrix} = \begin{Bmatrix} 0 \\ 0 \\ \frac{ML^2}{12}\frac{\sqrt{2}}{2}\Omega \end{Bmatrix} \tag{2.68}$$

If Eq. (2.68) is differentiated,

$$\frac{dH}{dt} = \frac{ML^2}{12}\frac{\sqrt{2}}{2}\dot{\Omega}\hat{k} \tag{2.69}$$

Example 2.9 As seen in Fig. 2.19, find maximum force, F, of an object on cart.

$$\sum_{ext} F_{x,system(M_1+M_2)} = (M_1 + M_2)\ddot{x} \tag{2.70}$$

$$\sum_{ext} \tau_{A/O} = -M_1 g \frac{b}{2}\hat{k} = \frac{dH_{/A}}{dt} + v_{A/O} \times \left(M_1 v_{G/O}\right) \text{ where } v_{A/O} = \dot{X}\hat{i},\ v_{G/O} = v_{A/O} \tag{2.71}$$

Only rotation considered is

$$\omega = \{0\ 0\ \omega_z\}^T \tag{2.72}$$

$$H_{G/O} = \left[I_{G/O}\right]\begin{Bmatrix} 0 \\ 0 \\ \omega_z \end{Bmatrix} = I_{zz/G}\omega_z\hat{k} \text{ where } H_x = H_y = 0,\ H_z = I_{zz/G}\omega_z\hat{k} \tag{2.73}$$

$$H_{A/O} = H_{G/O} + r_{G/A} \times P_{/O} = I_{zz/G}\omega_z\hat{k} + \left(\frac{b}{2}\hat{i} + \frac{h}{2}\hat{j}\right) \times M_1\dot{X}\hat{i}$$

$$= \left(I_{zz/G}\omega_z - M_1 \frac{h}{2}\dot{X} \right)\hat{k} \tag{2.74}$$

$$\sum_{ext} \tau_{/A} = -M_1 g \frac{b}{2}\hat{k} = \frac{dH_{/A}}{dt} = \left(I_{zz/G}\dot{\omega}_z - M_1 \frac{h}{2}\ddot{X} \right)\hat{k} \tag{2.75}$$

If $\dot{\omega}_z = 0$, we know as follows:

$$M_1 g \frac{b}{2} = M_1 \frac{h}{2}\ddot{X} \tag{2.76}$$

$$\ddot{X}_{\max} = g\frac{b}{h} \tag{2.77}$$

$$F_{\max} = (M_1 + M_2)\ddot{X}_{\max} = (M_1 + M_2) g \frac{b}{h} \tag{2.78}$$

$$\sum_{ext} \tau_{A/O} = M_1 \ddot{X}\frac{h}{2} - M_1 g \frac{b}{2} \tag{2.79}$$

If $r_{G/A} = \left(a\hat{i} + c\hat{k}\right)$, $H_{A/O}$ is,

$$\begin{aligned} H_{\frac{A}{O}} &= H_{\frac{G}{O}} + r_{\frac{G}{A}} \times P_{\bar{O}} = \left[I_{/G}\right]\begin{Bmatrix} 0 \\ 0 \\ \omega_z \end{Bmatrix} + \left(a\hat{i} + c\hat{k}\right) \times Ma\omega_z\hat{j} \\ &= \left[I_{zz/G} + Ma^2\right]\omega_z\hat{k} - Mac\omega_z\hat{i} \end{aligned} \tag{2.80}$$

2.3.6 *Four Classes for Rigid Rotational Motion*

There are four classes for rigid rotational motion as follows: (1) Class 1: Pure rotation about a fixed axis through the center of mass G, (2) Class 2: Pure rotation about a fixed axis @ A that is not identical to the center of mass G (A ≠ G), (3) Class 3: No external constraint, (4) Class 4: Bodies with moving points of constraint.

Class 1 (Pure rotation):

Rotors of all kinds

$$H_{G/O} = \left[I_{G/O}\right]\begin{Bmatrix} \omega_x \\ \omega_y \\ \omega_z \end{Bmatrix} \tag{2.81}$$

For 2D planar motion, Eq. (2.81) is defined as follows:

$$H_{G/O} = I_{zzG}\omega_z \text{ where } \omega = \left\{0\ 0\ \omega_z\right\}^T \tag{2.82}$$

If Eq. (2.82) is differentiated, it is

$$\frac{dH_{G/O}}{dt} = I_{zzG}\dot{\omega}_z = I_{zzG}\ddot{\theta} \tag{2.83}$$

Class 2 (Pure rotation about a fixed axis @ A ≠ G):

$$\sum_{ext}\tau_{A/O} = \frac{dH_{A/O}}{dt} = \frac{d}{dt}\left(\left[I_{/A}\right]\begin{Bmatrix} \omega_x \\ \omega_y \\ \omega_z \end{Bmatrix}\right) \tag{2.84}$$

For 2D planar motion $\omega = \begin{Bmatrix} 0 \\ 0 \\ \omega_z \end{Bmatrix}$, if $\left[I_{G/O}\right] = \left[\ddots\right]$ (principle axis is chosen), use parallel axis theorem to obtain $I_{A/O}$. That is, because $I_{A/O} = I_{G/O} + Md^2$, it is

$$H_{G/O} = \left[I_{G/O}\right]\begin{Bmatrix} \omega_x \\ \omega_y \\ \omega_z \end{Bmatrix} \tag{2.85}$$

Class 3 problem (no external constraint):

To explain the case with no external constraint, an example is a wheel with mass = 75 kg, subjected to end force F = 150 N, as seen in Fig. 2.20.

The degree of freedom (DOF) is

$$DOF = 6 - 3 = 3$$

K = 0.15 m (≡ radius of gyration)

$$I_{zz_G} = MK^2 \text{ or } K = \sqrt{\frac{I_{zzG}}{M}} \tag{2.86}$$

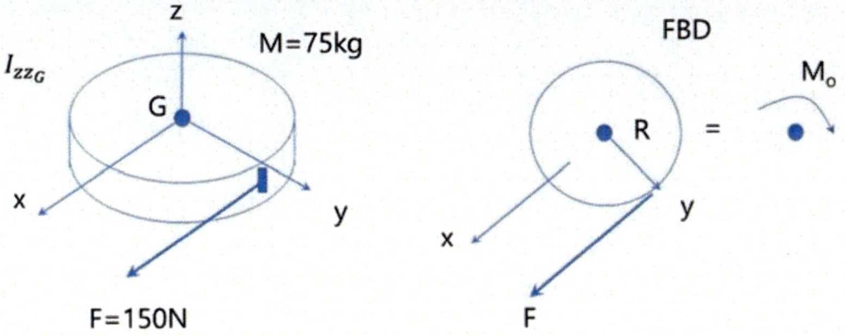

Fig. 2.20 Class 3 problem (no external constraint)

Because two trivial equations have, motion of equation at x direction is

$$\sum F_x = F = M\ddot{x} \text{ or } \ddot{x} = \frac{F}{M} = \frac{150\text{ N}}{75\text{ kg}} = 2\text{m/s}^2$$

Class 4 problem (bodies with moving points of constraint):

Angular momentum of a rigid body (Fig. 2.21) is

$$H_{A/O} \equiv \sum r_{i/A} \times P_{i/O} = \sum_i \left(r_{G/A} + r_{i/G}\right) \times P_{i/O} = r_{G/A} \times \sum_i m_i v_{i/O} + \sum_i r_{i/G} \times m_i v_{i/O} \tag{2.87}$$

$$x_2 = x_1 - R\theta \tag{2.88}$$

$$\ddot{x}_2 = \ddot{x}_1 - R\ddot{\theta} \tag{2.89}$$

Because $\tau_A = \frac{dH_{A/O}}{dt} + v_{A/O} \times P_{/O}$, angular momentum is

$$H_{A/O} = I_{zz_G}\theta + r_{G/A} \times P_{/O} = I_{zz_G}\dot{\theta}\hat{k} + R\hat{j} \times M\dot{x}_2\hat{i} = I_{zz_G}\dot{\theta}\hat{k} - RM\dot{x}_2\hat{k} = I_{zz_G}\dot{\theta}\hat{k} - RM\dot{x}_2\hat{k} \tag{2.90}$$

$$\frac{dH_{A/O}}{dt} = I_{zz_G}\ddot{\theta}\hat{k} - RM\ddot{x}_2\hat{k} = I_{zz_G}\ddot{\theta}\hat{k} - RM\left(\ddot{x}_1 - R\ddot{\theta}\right)\hat{k} = 0 \tag{2.91}$$

$$\ddot{\theta} = \left[MR/\left(I_{zz_G} + MR^2\right)\right]\ddot{x}_1 \tag{2.92}$$

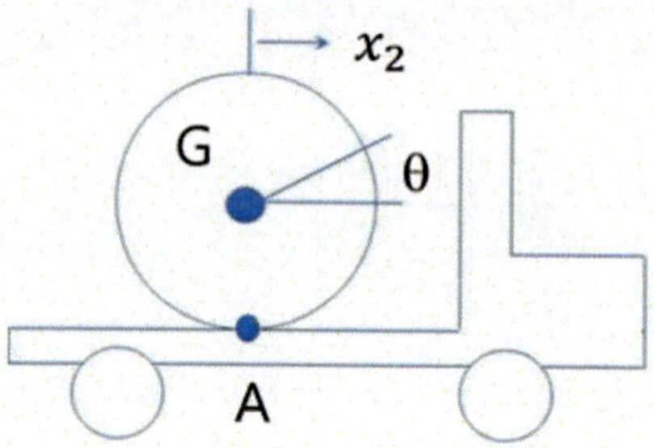

(a) An object moving points of contact

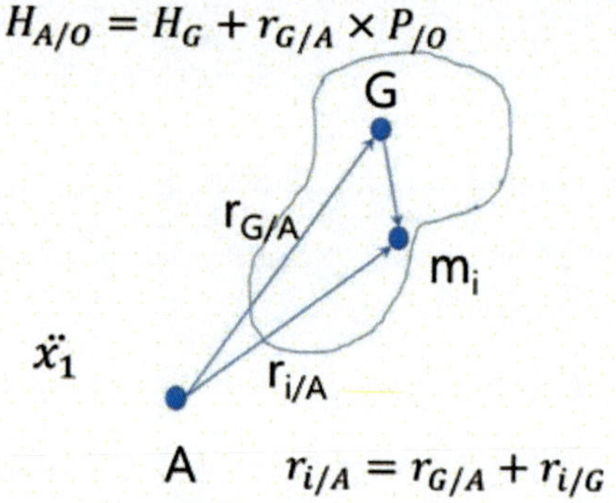

(b) Angular momentum of a rigid body

Fig. 2.21 Class 4 problem (moving points of contact)

Example 2.10 There is a rotating rigid object with rectangular shape, seen in Fig. 2.22. Find motion of equation. Here, L = 32.1 cm, $a = 32.1$ cm, $b = 1.25$ cm, $d = 10.2$ cm.

We know that $I_{zz/G} = M\left(\frac{L^2+a^2}{12}\right)$, $I_{yy/G} = M\left(\frac{L^2+b^2}{12}\right)$, $I_{xx/G} = M\left(\frac{a^2+b^2}{12}\right)$. Because

$$(a/L)^2 = 0.022 \text{ and } (b/L)^2 = 0.002,\ I_{zz/G} = \frac{ML^2}{12} \tag{2.93}$$

If $d = L/2$,

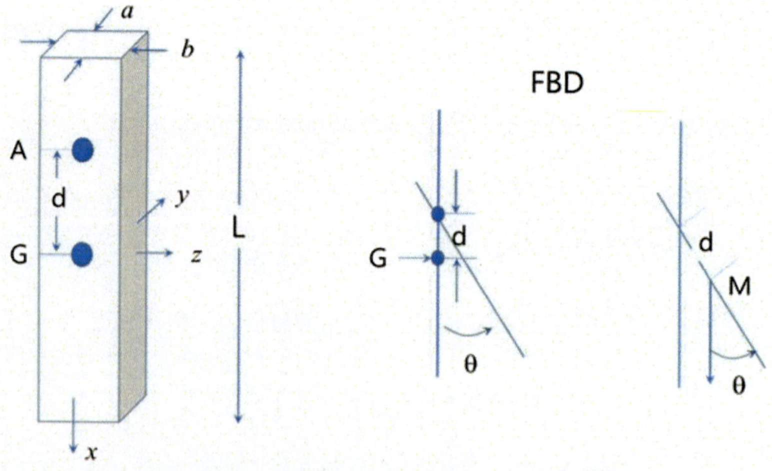

Fig. 2.22 A rotating object with constant speed and radius

$$I_{zz/A} = I_{zz/G} + Md^2 = \frac{ML^2}{12} + \frac{ML^2}{4} = \frac{ML^2}{3} \tag{2.94}$$

$$\sum_{ext} \tau_{A/O} = \frac{dH_{A/O}}{dt} = \frac{dI_{zz/A}\omega_z}{dt} = I_{zz/A}\ddot{\theta}_z$$

$$= -Mgd\sin\theta \text{ or } I_{zz/A}\ddot{\theta}_z + Mgd\sin\theta = 0 \tag{2.95}$$

If small θ, $\sin\theta \approx \theta$. Equation (2.95) is linearized. That is,

$$I_{zz/A}\ddot{\theta}_z + Mgd\theta = 0 \tag{2.96}$$

Assume $\theta(t) = \theta_0\sin\omega t$, we can find ω_n as follows:

$$\left(-\omega^2 I_{zz/A} + +Mgd\right)\theta_0\sin\omega t = 0 \tag{2.97}$$

$$\omega = \sqrt{\frac{Mgd}{I_{zz/A}}} \text{ (generic solution for pendulum with one degree of freedom)} \tag{2.98}$$

2.3.7 *Degree of Freedom (DOF), Coordinate System, Free Body Diagram, and Equation of Motion*

As seen in Fig. 2.23, there is an object in a slope plane. To describe it as motion of equation (MOE), first of all, we have to define degree of freedom (DOF), based on the constraints. DOF is a number of independent coordinates that is necessary to describe the motion. That is,

$$DOF = 6n + 3m - C \tag{2.99}$$

where $n = number\ of\ rigid$, $m = number\ of\ particle$, $C = constraints$

C is 5 as follows:

(1) $y = \dot{y} = \ddot{y} = 0$ is 1
(2) Assume $z = \dot{z} = \ddot{z} = 0$ is 1. That is, $\sum F_z = 0 = ma_z$
(3) Assume no rotation about x or y axes is 2.
(4) No slip $X = -R\theta$ is 1

From Eq. (2.99), we can find define DOE of an object in Fig. 2.9. That is, $DOF = 6 - 5 = 1$. If slip is allowed, $DOF = 6 - 4 = 2$.

Example 2.11 As seen in Fig. 2.24, there is a sliding ladder that is fixed on the wall. In this case, find DOE, Free Body Diagram (FBD), and motion of equation (MOE).

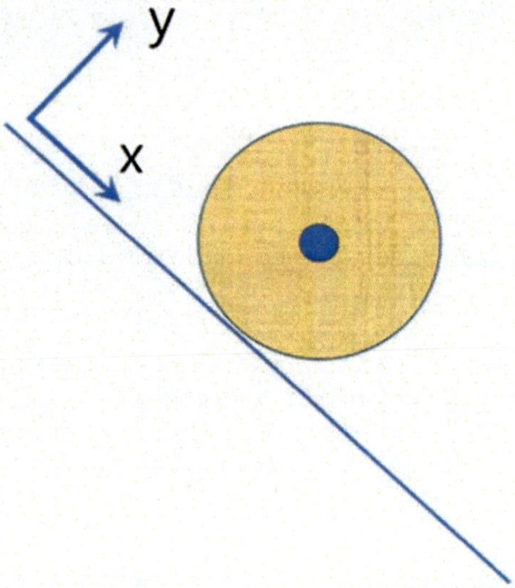

Fig. 2.23 An object in a slope plane

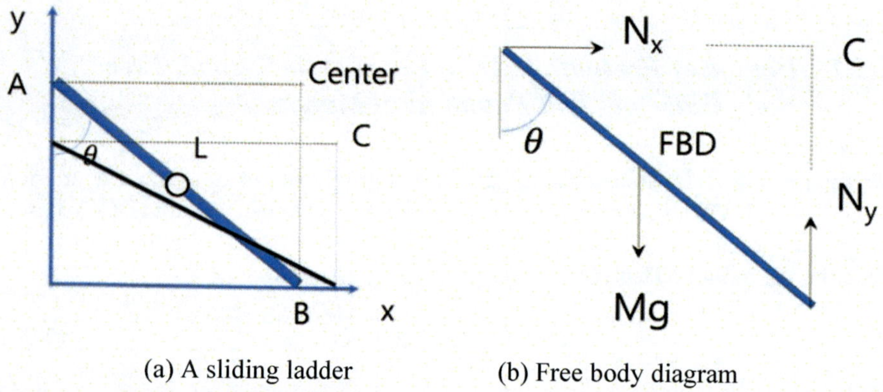

Fig. 2.24 A sliding ladder on the wall

C is 5 as follows:

(1) Constrained in translation x & y is 2
(2) Assume $z = \dot{z} = \ddot{z} = 0$ is 1
(3) No rotation about x & y is 2

$$DOF = 6 - 5 = 1 \quad (2.100)$$

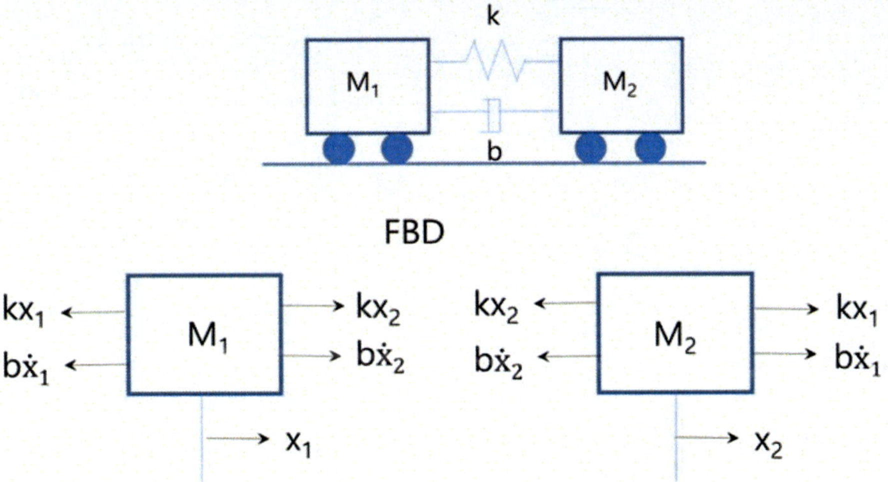

Fig. 2.25 Two bodies on the plane

$$MOE \sum_{ext} \tau_C = \frac{dh_{/C}}{dt} = +Mg\frac{L}{2}\sin\theta \tag{2.101}$$

Example 2.12 As seen in Fig. 2.25, there are two bodies connecting together. Find DOE, Free Body Diagram (FBD), and motion of equation (MOE).

In this problem, only two independent coordinates are needed to describe the motion. Describe the ten constraints that one must assume to be acting on this device, so as to reduce the number of degrees of freedom to two. That is,

C is 10 as follows:

$$DOF = 6 \times 2 - 10 = 2 \tag{2.102}$$

Draw free body diagram, assign two appropriate coordinates, and find the two equations of motion which characterize the system. To get the signs correct on the spring and dashpot forces, one at a time assume positive displacements and velocities of each coordinate and deduce the resulting directions of the spring and damper forces. That is,

Assume x_1, x_2 are position, MOE is.

$$\sum_{ext} F_{1,x_1} = M_1\ddot{x}_1 = k(x_2 - x_1) + b(\dot{x}_2 - \dot{x}_1) \tag{2.103}$$

$$\sum_{ext} F_{2,x_2} = M_2\ddot{x}_2 = k(x_1 - x_2) + b(\dot{x}_1 - \dot{x}_2) \tag{2.104}$$

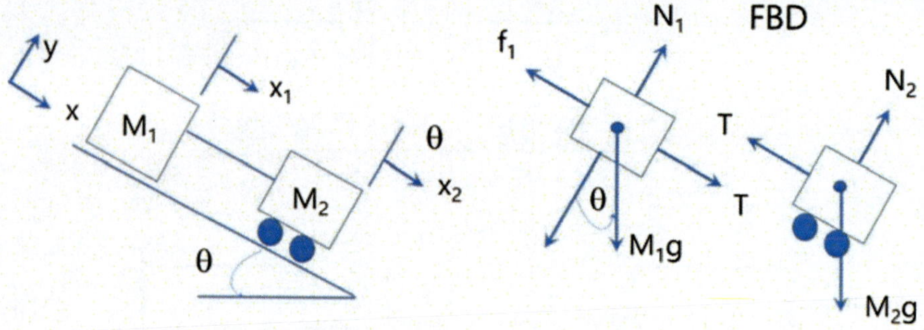

Fig. 2.26 Two bodies connecting together on the sliding plane

Example 2.13 As seen in Fig. 2.26, there are two bodies connecting together on the sliding plane. Find DOE, Free Body Diagram (FBD), and motion of equation (MOE).

C is 9 as follows:

(1) Moving together M_1 & M_2 is 1
(2) No acceleration is 2
(3) No rotation is

$$DOF = 6 \times 2 - C = 12 - 9 = 3 \tag{2.105}$$

Two trivial equations. That is,

$$\sum_{ext} F_{Z_1} = M_1 \ddot{z}_1 = 0 \text{ or } \sum_{ext} F_{Z_2} = M_2 \ddot{z}_2 = 0 \tag{2.106}$$

So, it has 1 significant MOE

$$\sum_{ext} F_x = M \ddot{x}_1 \text{ where } x_1 = x_2 = x \text{ or } \ddot{x}_1 = \ddot{x}_2 = \ddot{x} \tag{2.107}$$

There 5 unknowns such as $N_1, N_2, f_1, T, \ddot{x}$

$$\sum_{ext} F_{y_1} = M_1 \ddot{y}_1 \Rightarrow N_1 \Rightarrow f_1 = \mu N_1 \text{ or } \sum_{ext} F_{y_2} = M_2 \ddot{y}_2 \Rightarrow N_2 \tag{2.108}$$

Leaves T & $\ddot{x}$

$$\sum_{ext} F_x = M_1 \ddot{x} \text{ or } \sum_{ext} F_x = M_2 \ddot{x} \tag{2.109}$$

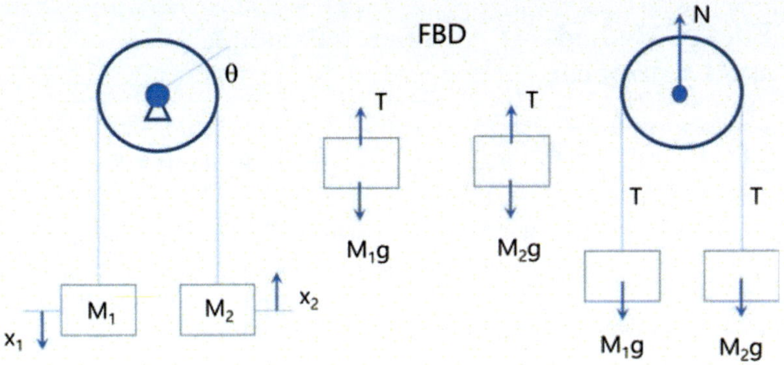

Fig. 2.27 Two bodies in pully hanging

As eliminating T, $\ddot{x}$ can be solved as follows:

$$(M_1 + M_2)\ddot{x} = (M_1 + M_2)g\sin\theta - \mu M_1 g\cos\theta \tag{2.110}$$

Example 2.14 As seen in Fig. 2.27, there are two bodies in pully hanging. Find DOE, Free Body Diagram (FBD), and motion of equation (MOE).

C is 1 as follows:

(1) $x_1 = x_2$ is 1
(2) 4 trivial Equations
(3) 1 significant EOM

$$DOF = 6m + 3n - C = 6 \cdot 0 + 3 \cdot 2 - C = 6 - 1 = 5 \tag{2.111}$$

$x_1 = x_2 = x$ or θ direction: motion of equation (MOE) is

$$\sum_{ext} \tau_A = \frac{dh_{/A}}{dt} + v_{A/O} \times P_{/O} = (M_1 g R - M_2 g R)\hat{k} \tag{2.112}$$

Because $h_{/A} = (M_1 + M_2)R\dot{x}\hat{k}$, $\frac{dh_{/A}}{dt} = (M_1 + M_2)R\ddot{x}\hat{k}$, Eq. (2.112) is $(M_1 g R - M_2 g R)\hat{k} = (M_1 + M_2)R\ddot{x}\hat{k}$.

So, acceleration is

$$\ddot{x} = \left(\frac{M_1 - M_2}{M_1 + M_2}\right)g \tag{2.113}$$

2.4 Energy Method—D'Alembert's Principle and Lagrangian

2.4.1 D'Alembert's Principle for Mechanical Systems

D'Alembert's principle is a clear expression of the basic theories' laws of movement. The scientific theorem of virtual work explains that the total of the incremental virtual works performed by all outer forces F_i acting together with virtual displacements δs_i of the point on which the related force is acting is zero:

$$\delta W = \sum_i F_i \cdot \delta r_i = 0 \tag{2.114}$$

Utilizing D'Alembert's principle and free body diagram, an engineer can model a mechanical system. For example, if there is a car in transit, we can model a straight-forward dynamic system with a mass which is split from a wall by a spring and a dashpot (Fig. 2.28).

The free body diagram is a graphical way of displaying all outer forces applied on an object. There is only one location in this system expressed by the variable "x", which is greater than zero to the right. It is presumed that $x = 0$ when the spring is in its loosened state. There are four forces to evolve a model from the free body diagram: (1) an outer force (F_e), such as air-resistance force and friction force; (2) a spring force, which shall be a force from the spring, $k \cdot x$, to the left; (3) a dashpot force, which shall be a force from the dashpot, $b \cdot v$, to the left; and (4) eventually, there is an inertial force, which is expressed to be opposed to the specified orientation of movement. This is expressed by $m \cdot a$ to the left, which is called as a fictitious force.

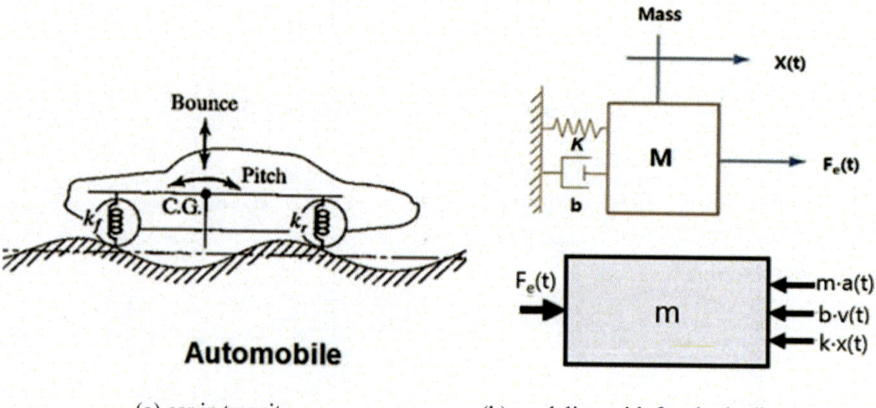

(a) car in transit (b) modeling with free body diagram

Fig. 2.28 Free body diagram for car modeling

It is when one multiplies true acceleration times m and put it on the $\sum_{\text{all lexternal}} F$ side of the equation.

Newton's second law explains that a body accelerates in the orientation of an acted force and that this acceleration is oppositely proportional to the force or taking away the right-hand side outcomes in D'Alembert's principle:

$$\sum_{\text{all lexternal}} F = m \cdot a \text{ or } \sum_{\text{all lexternal}} F - m \cdot a = 0 \tag{2.115}$$

If we think about the $m \cdot a$ expression to be a (fictitious) inertia force (or D'Alembert's force), D'Alembert's principle shall be left

$$\sum_{\text{all}} F \cdot \delta r = 0 \tag{2.116}$$

To imagine this, think about pushing against a mass (in the absence of friction) with your hand in the positive orientation. Your hand undergoes a force in the orientation opposite to that of the direction of the force (this is the $-ma$ term). The inertial force is always in the direction opposite to the prescribed positive direction. We sum all of these forces to zero and obtain the motion of equation as follows:

$$F_e(t) - ma(t) - bv(t) - k \cdot x(t) = 0 \text{ or } m\frac{d^2x}{dt^2} + b\frac{dx}{dt} + kx(t) = F_e(t) \tag{2.117}$$

Furthermore, we can discuss with objects with two-degree freedom (Fig. 2.29).

We can find the motion of equation (MOE) as follows:

For M_1,

$$\sum_{ext} F = M_1\ddot{x}_1 = -(k_1 + k_2)x_1 - (c_1 + c_2)\dot{x}_1 + k_2x_2 + c_2\dot{x}_2 + F_1 \tag{2.118}$$

For M_2,

$$\sum_{ext} F = M_2\ddot{x}_2 = -(k_2 + k_3)x_2 - c_2\dot{x}_2 + k_2x_1 + c_2\dot{x}_1 + F_2 \tag{2.119}$$

So, we can summarize MOE as matrix form,

$$\begin{bmatrix} M_1 & 0 \\ 0 & M_2 \end{bmatrix}\begin{Bmatrix} \ddot{x}_1 \\ \ddot{x}_2 \end{Bmatrix} + \begin{bmatrix} c_1 + c_2 & -c_2 \\ -c_2 & c_2 \end{bmatrix}\begin{Bmatrix} \dot{x}_1 \\ \dot{x}_2 \end{Bmatrix} + \begin{bmatrix} k_1 + k_2 & -k_2 \\ -k_2 & k_2 + k_3 \end{bmatrix}\begin{Bmatrix} x_1 \\ x_2 \end{Bmatrix} \tag{2.120}$$

Example 2.15 As seen in Fig. 2.30, there is a person in elevator, lifting with some acceleration g/4. Find DOE, Free Body Diagram (FBD), and motion of equation (MOE).

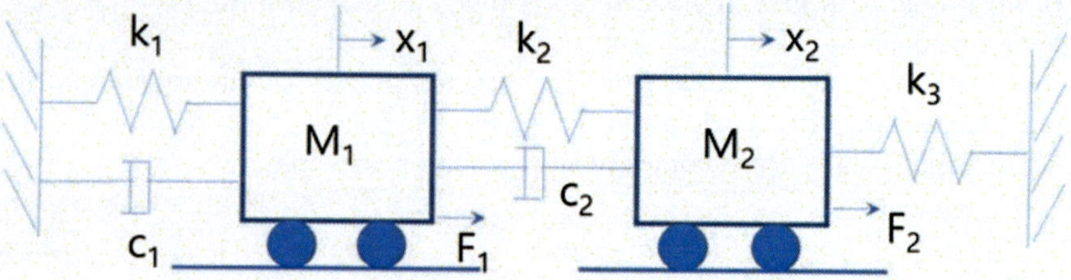

(a) Objects with two-degree freedom

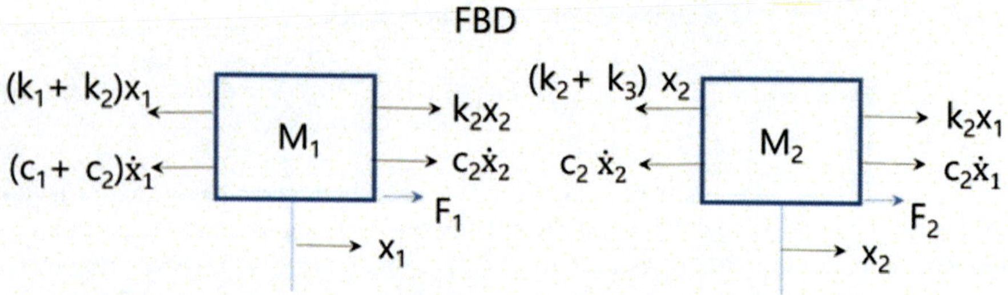

(b) Free body diagram

Fig. 2.29 Objects with two-degree freedom

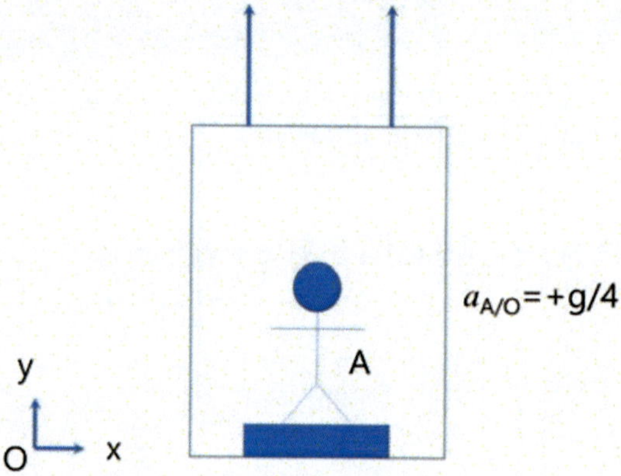

(a) An elevator

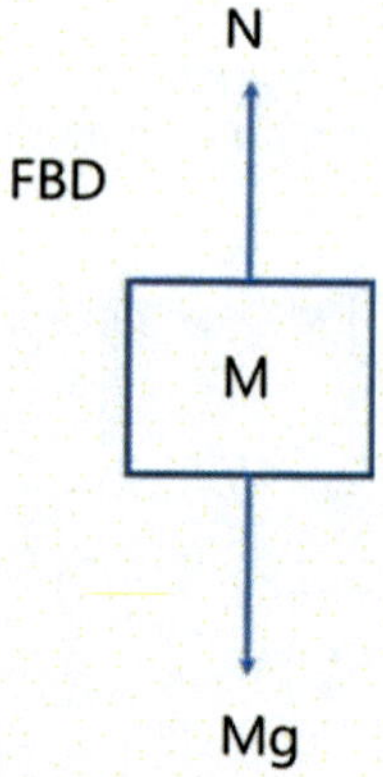

(b) Free body diagram

Fig. 2.30 A lifting motion in the elevator

$$DOF = 1 \tag{2.121}$$

Direction y motion of Equation (MOE):

$$\sum_{ext} F_y = Ma_{A/O} = N - Mg \tag{2.122}$$

If fictitious force is applied and $a_{A/O} = \frac{g}{4}$, it can be

$$\sum_{ext} F_y - Ma_{A/O}(\text{fictitious force}) = 0 = N - Mg - M\frac{g}{4} \text{ or } N$$
$$= Mg + M\frac{g}{4} = \frac{5}{4}Mg \tag{2.123}$$

Example 2.16 As seen in Fig. 2.31, the m kg spool S fits loosely on the rotating inclined rod. Find Free Body Diagram (FBD), and motion of equation (MOE).

$$\sum_{ext} F_y = Ma_{B/O} = M\left(a_{A/O} + (\ddot{r} - r\dot{\theta}^2)\hat{r} + \ddot{z}\hat{k} + (r\ddot{\theta} + 2\dot{r}\dot{\theta})\hat{\theta}\right)$$
$$= M\left(-r\dot{\theta}^2\hat{r} + r\ddot{\theta}\hat{\theta}\right) \tag{2.124}$$

r direction: $\sum_{ext} F_{\hat{r}} = Ma_{B/O} = -Mr\dot{\theta}^2 = N_r \Leftrightarrow \sum_{ext} F_{\hat{r}} = N_r + Mr\dot{\theta}^2(\text{fictitious force}) = 0$

θ direction: $\sum_{ext} F_{\hat{\theta}} = Mr\ddot{\theta}\hat{\theta}$.

Angular momentum is

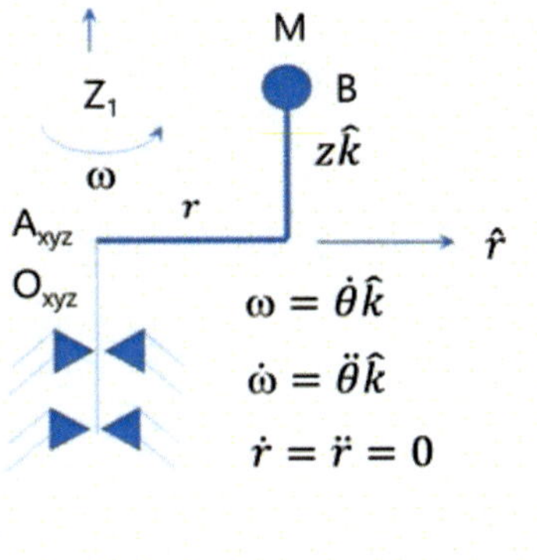

(a) A rotating spool

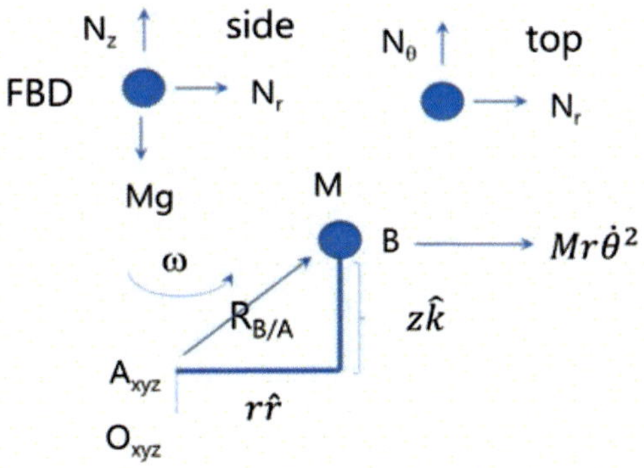

(b) Free body diagram

Fig. 2.31 A rotating spool

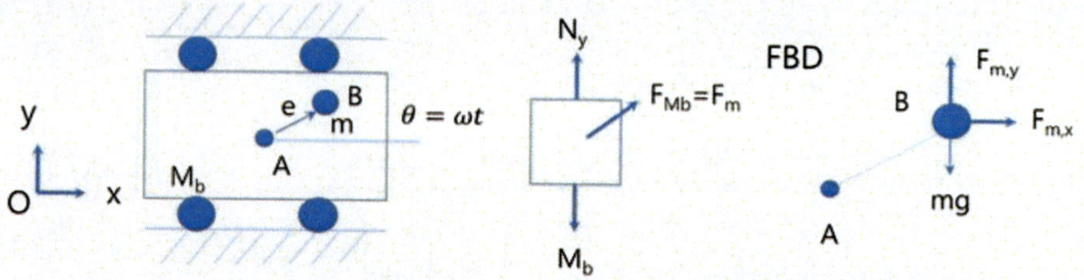

(a) An object constrained (b) Free body diagram

Fig. 2.32 An object constrained in plane, which is rotating with an imbalance mass

$$h_{A/O} = r_{B/A} \times P_{A/O} = \left(r\hat{r} + z\hat{k}\right) \times P_{A/O} = \left(r\hat{r} + z\hat{k}\right) \times Mr\dot{\theta}\hat{\theta}$$
$$= Mr^2\dot{\theta}\hat{k} - Mrz\dot{\theta}\hat{r} \tag{2.125}$$

If Eq. (2.125) is differentiated, it is

$$\sum_{ext} \tau_{A/O} = \frac{dh_{A/O}}{dt} = Mr^2\ddot{\theta}\hat{k} - Mrz\ddot{\theta}\hat{r} - Mrz\dot{\theta}^2\hat{\theta} \tag{2.126}$$

Example 2.17 As seen in Fig. 2.32, this is an object constrained in plane, which is rotating with an imbalance mass. Find Free Body Diagram (FBD), and motion of equation (MOE).

Motion of equation is

$$\sum_{ext} F_{M_b} = M_b\ddot{x} \tag{2.127}$$

$$\sum_{ext} F_m = ma_{B/O} = ma_{A/O} + ma_{B/A} = m\ddot{x} + m\left[\left(\ddot{r} - r\dot{\theta}^2\right)\hat{r} + \left(r\ddot{\theta} + 2\dot{r}\dot{\theta}\right)\hat{\theta}\right] \tag{2.128}$$

$$\sum_{ext} F_m = m\ddot{x}\hat{i} - me\omega^2\left(\cos\omega t\hat{i} + \sin\omega t\hat{j}\right) = F_{m,x}\hat{i} + F_{m,y}\hat{j} - mg\hat{j} \tag{2.129}$$

$$x \text{ direction}: \sum_{ext} F_{m,x} = m\ddot{x} - me\omega^2\cos\omega t = F_{m,x} = -F_{M_b}, x \tag{2.130}$$

$$y \text{ direction}: F_{m,y} = mg - me\omega^2\sin\omega t \tag{2.131}$$

2.4.2 *Derivation of Lagrange's Equations from D'Alembert's Principle*

The movement of particles in space is often confined by constraints. An engineer would work only with independent DOFs (coordinates) without constraints (Fig. 2.33).

Movements of N particles, $n = 3N$ DOFs, are subject to k equations relating coordinates. In classical mechanics, if all constraints of the system are holonomic, a system may be called holonomic. A holonomic constraint should be expressed as follows:

$$f_j(x_1, \ldots, x_n, t) = c_j \quad j = 1, 2, \ldots, k \tag{2.132}$$

That is, a holonomic constraint relies only on the coordinates x_j and time t. The generalized coordinates describe the motion of the system relative to Cartesian coordinates subject to k constraints:

$$\begin{aligned} x_1 &= x_1(q_1, q_2, \ldots, q_{n-k}, t) \\ &\ldots \\ x_n &= x_n(q_1, q_2, \ldots, q_{n-k}, t) \end{aligned} \tag{2.133}$$

where any set of $n - k = 3N - \mathrm{k}$ independent coordinates that totally state the system.

The change of a Cartesian coordinate can be induced from changes in generalized coordinates in dt:

$$dx_1 = \frac{\partial x_1}{\partial q_1} dq_1 + \cdots + \frac{\partial x_1}{\partial q_{n-k}} dq_{n-k} + \frac{\partial x_1}{\partial t} dt \tag{2.134}$$

In a compact way, we can describe as follows:

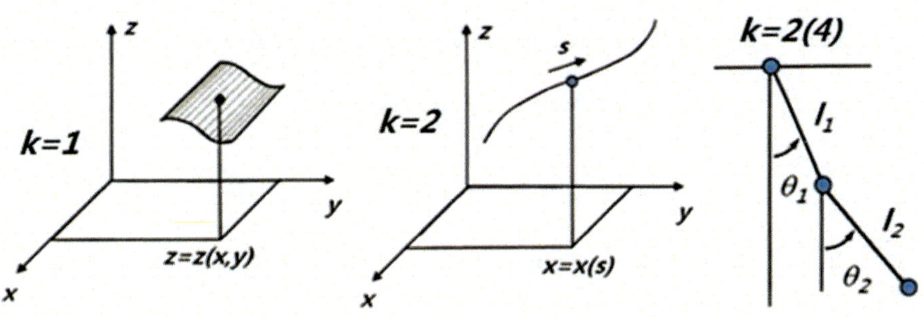

Fig. 2.33 The motion of particles restricted by constraints

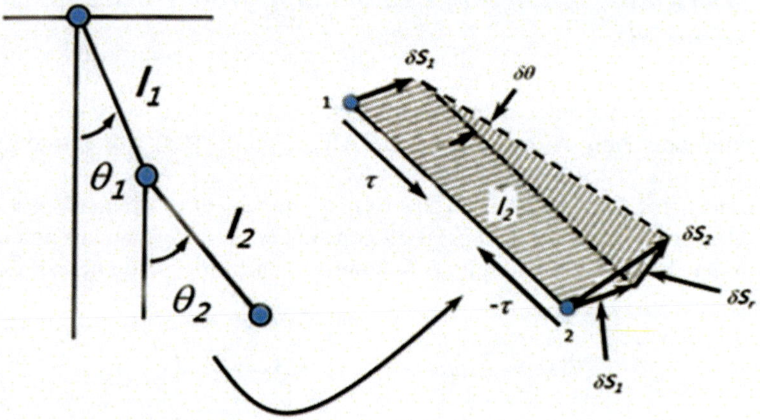

Fig. 2.34 Forces of constraint perpendicular to the direction of motion

$$dx_i = \sum_{\sigma=1}^{n-k} \frac{\partial x_i}{\partial q_\sigma} dq_\sigma + \frac{\partial x_1}{\partial t} dt \quad i = 1, \ldots, n \tag{2.135}$$

Therefore, the virtual displacements are called infinitesimal, instantaneous displacements of the coordinates consistent with the constraints. That is,

$$\delta x_i = \sum_{j=1}^{n-k} \frac{\partial x_i}{\partial q_j} \delta q_\sigma \quad i = 1, \ldots, n \tag{2.136}$$

Because the forces of the constraint are perpendicular to the orientation of movement, the forces of the constraint do not work under a virtual displacement (Fig. 2.34).

$$\delta W = \tau \cdot \delta s_1 - \tau \cdot \delta s_2 = -\tau \cdot \delta s_r = 0 \tag{2.137}$$

We can rewrite Newton's second law as follows:

$$\dot{p}_i = F_i^{(a)} + R_i \quad i = 1, \ldots, n \text{ or } \sum_i \left(F_i^{(a)} + R_i - \dot{p}_i \right) \delta x_i = 0 \tag{2.138}$$

Because $\sum_i R_i \delta x_i = 0$, the virtual work by D'Alembert's law shall be refined as follows:

$$\sum_i \left(F_i^{(a)} - \dot{p}_i \right) \delta x_i = 0 \tag{2.139}$$

We can also rewrite D'Alembert's principle with respect to the generalized coordinates. The virtual work done by applied forces under virtual displacement is defined as follows:

$$\delta W\left(=\sum_i \left(F_i^{(a)}-\dot{p}_i\right)\delta x_i\right)=\sum_i F_i\delta x_i=\sum_{\sigma=1}^{n-k}\left(\sum_{i=1}^{n} F_i\frac{\partial x_i}{\partial q_j}\right)dq_j$$
$$=\sum_{j=1}^{n-k} Q_j\delta q_j \tag{2.140}$$

where $\delta x_i=\sum_{j=1}^{n-k}\left(\frac{\partial x_i}{\partial q_j}\right)dq_j$ and generalized forces $Q_j=\sum_{i=1}^{n} F_i\frac{\partial x_i}{\partial q_j}$

Let Cartesian coordinates be subjected to k constraints $x_i=x_i(q_1,\ldots,q_{n-k},t)$. The change of a Cartesian coordinate induced from changes in generalized coordinates in dt is defined as follows:

$$dx_i=\sum_{j=1}^{n-k}\left(\frac{\partial x_i}{\partial q_j}\right)dq_j+\frac{\partial x_i}{\partial t}dt \quad i=1,\ldots,n \tag{2.141}$$

Taking the time derivatives from Eq. (2.141):

$$\frac{dx_i}{dt}\equiv\dot{x}_i=\sum_{j=1}^{n-k}\left(\frac{\partial x_i}{\partial q_j}\right)\dot{q}_j+\left(\frac{\partial x_i}{\partial t}\right) \quad i=1,\ldots,n \tag{2.142}$$

where $\dot{x}_i=x_i(q_1,\ldots,q_{n-k};\ \dot{q}_1,\ldots,\dot{q}_{n-k};\ t)$.

If other variables are kept constant in taking partial derivatives,

$$\frac{\partial \dot{x}_i}{\partial \dot{q}_j}=\frac{\partial x_i}{\partial q_j}\frac{d}{dt}\left(\frac{\partial x_i}{\partial q_j}\right)=\frac{\partial}{\partial q_j}\left(\frac{dx_i}{dt}\right) \tag{2.143a}$$

$$\text{LHS}=\sum_{\lambda}\left(\frac{\partial}{\partial q_\lambda}\frac{\partial x_i}{\partial q_j}\right)\dot{q}_\lambda+\frac{\partial}{\partial t}\frac{\partial x_i}{\partial q_j} \tag{2.143b}$$

$$\text{RHS}=\sum_{\lambda}\left(\frac{\partial}{\partial q_j}\frac{\partial x_i}{\partial q_\lambda}\right)\dot{q}_\lambda+\frac{\partial}{\partial q_j}\frac{\partial x_i}{\partial t} \tag{2.143c}$$

Here, the order of partial derivatives can be interchanged. The momentum equations might be described as follows:

$$\sum_i \dot{p}_i\delta x_i=\sum_i m_i\ddot{x}_i\delta x_i=\sum_{\sigma}\left(\sum_i m_i\ddot{x}_i\frac{\partial x_i}{\partial q_j}\right)\delta q_j \tag{2.144}$$

Because $\dot{p}_i = m_i\ddot{x}_i$ and $\delta x_i = \sum_{\sigma=1}^{n-k} \frac{\partial x_i}{\partial q_j}\delta q_j$, Eq. (2.144) might be described as follows:

$$\sum_i m_i\ddot{x}_i\frac{\partial x_i}{\partial q_j} \cong \sum_i m_i\frac{d\dot{x}_i}{dt}\frac{\partial x_i}{\partial q_j} = \sum_i m_i\left[\frac{d}{dt}\left(\dot{x}_i\frac{\partial x_i}{\partial q_{\sigma j}}\right) - \dot{x}_i\frac{d}{dt}\frac{\partial x_i}{\partial q_j}\right] \tag{2.145}$$

Because $\frac{\partial \dot{x}_i}{\partial \dot{q}_j} = \frac{\partial x_i}{\partial q_j}$ and $\frac{d}{dt}\left(\frac{\partial x_i}{\partial q_j}\right) = \frac{\partial}{\partial q_j}\left(\frac{dx_i}{dt}\right)$, Eq. (2.145) can be expressed as follows:

$$\sum_i m_i\frac{d}{dt}\left(\dot{x}_i\frac{\partial \dot{x}_i}{\partial \dot{q}_j}\right) = \frac{d}{dt}\left[\frac{\partial}{\partial \dot{q}_j}\left(\frac{1}{2}\sum_i m_i\dot{x}_i^2\right)\right] \tag{2.146a}$$

$$\sum_i m_i\dot{x}_i\frac{\partial}{\partial q_j}\frac{dx_i}{dt} = \frac{\partial}{\partial q_j}\left(\frac{1}{2}\sum_i m_i\dot{x}_i^2\right) \tag{2.146b}$$

From Eq. (2.144), the momentum equations can be expressed as follows:

$$\sum_i \dot{p}_i\delta x_i = \sum_j\left(\frac{d}{dt}\frac{\partial T}{\partial \dot{q}_j} - \frac{\partial T}{\partial q_{\sigma j}}\right)\delta q_j \tag{2.147}$$

where kinetic energy $T = \frac{1}{2}\sum_i m_i\dot{x}_i^2 = T(q_1, \ldots, q_{n-k}; \dot{q}_1, \ldots, \dot{q}_{n-k}; t)$.

Therefore, we can rewrite D'Alembert's principle in terms of the generalized coordinates as follows:

$$\sum_j\left(\frac{d}{dt}\frac{\partial T}{\partial \dot{q}_j} - \frac{\partial T}{\partial q_j} - Q_j\right)\delta q_j = 0 \tag{2.148}$$

If all q are independent and arbitrary, Lagrange's equations are expressed as follows:

$$\frac{d}{dt}\frac{\partial T}{\partial \dot{q}_j} - \frac{\partial T}{\partial q_j} = Q_\sigma \quad \sigma = 1, \ldots, n-k \tag{2.149}$$

The $n-k$ equations for $n-k$ independent generalized coordinates are equivalent to Newton's second law. For conservative forces, the potential energy relies only on the location:

$$V(x_1, \ldots, x_n) = V(q_1, \ldots, q_{n-k}, t) \qquad \frac{\partial V}{\partial \dot{q}_\alpha} = 0 \tag{2.150}$$

Generalized forces, Q_j, are given by negative gradients with respect to the corresponding generalized coordinate:

$$Q_j \equiv \sum_i F_i \frac{\partial x_i}{\partial q_j} = -\sum_i \left[\frac{\partial}{\partial x_i} V(x_i, \ldots, x_n)\right] \frac{\partial x_i}{\partial q_j} = -\frac{\partial}{\partial q_j} V(q_1, \ldots, q_{n-k}, t) \tag{2.151}$$

Lagrange's equations can be described as follows:

$$\frac{d}{dt}\frac{\partial L}{\partial \dot{q}_j} - \frac{\partial L}{\partial q_j} = Q_j \quad j = 1, \ldots, n-k \tag{2.152}$$

If $L = T - V$ is plugging in Eq. (2.152), we can find it as follows:

$$\underbrace{\frac{d}{dt}\frac{\partial T}{\partial \dot{q}_j}}_{①} - \underbrace{\frac{d}{dt}\frac{\partial V}{\partial \dot{q}_j}}_{0} - \underbrace{\frac{\partial T}{\partial q_j}}_{②} + \underbrace{\frac{\partial V}{\partial q_j}}_{③} = \underbrace{Q_j}_{④} \quad j = 1, \ldots, n-k \tag{2.153}$$

That is, we know that ① + ② + ③ = ④.

The generalized coordinates, q_σ, has characteristics as follows: (1) not necessary Cartesian, (2) not even inertial, (3) must be independent and complete, (4) system must be holonomic. Here, independent is that when you fix all but one coordinate, it still has a continuous range of movement in the free coordinate. Complete is capable of locating all parts at all times.

To describe the motion by utilizing Lagrange's Equations, consider an object with damping and spring or double pendulum in Fig. 2.35.

That is, systematic approach by Lagrangian equation can be summarized from Eq. (2.153) as follows:

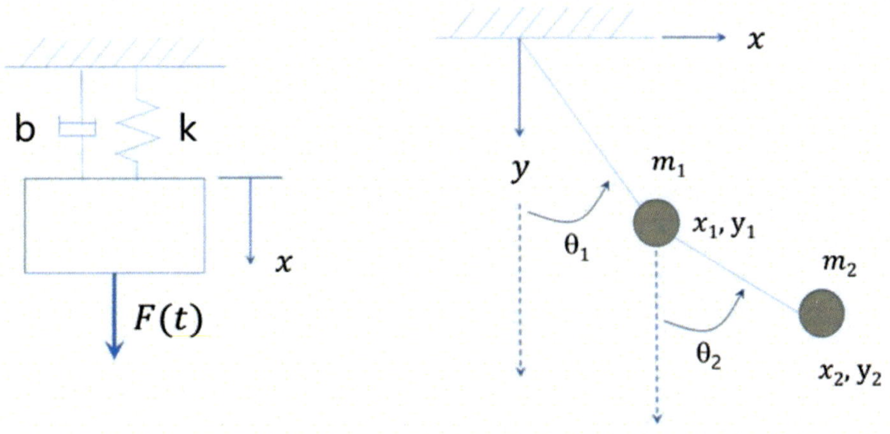

(a) An object with damping and spring (b) Double pendulum

Fig. 2.35 An object with damping and spring or double pendulum

For left-hand side (LHS): (1) determine the number of degree of freedom (DOF) and choose generalized coordinates, q_σ, (2) verify the complete, independent, and holonomic, (3) compute the Lagrangian $L = T + V$, and (4) compute ①, ②, ③ for each generalized coordinates, q_σ.

For Right-hand side (RHS): (1) for generalized coordinates, q_σ, find the Generalized forces, Q_σ, that goes with it, (2) compute the virtual work, δW_j, associated with the virtual displacement δq_σ. That is,

$$\delta W_j = Q_j \delta q_j \tag{2.154}$$

So, we can find the motion of equation (MOE) from Eq. (2.153) as follows:

Kinetic energy:

$$T = \frac{1}{2} M \dot{X}^2 \tag{2.155}$$

Potential energy:

$$V = \frac{1}{2} k X^2 - M g X \tag{2.156}$$

Generalized coordinate:

$$q_j = q_1 = X \tag{2.157}$$

$$\text{①} \quad \frac{d}{dt}\frac{\partial T}{\partial \dot{q}_\sigma} = \frac{d}{dt}\frac{\partial}{\partial \dot{X}}\left(\frac{1}{2} M \dot{X}^2\right) = \frac{d}{dt}(M\dot{X}) = M\ddot{X} \tag{2.158}$$

$$\text{②} \quad -\frac{\partial T}{\partial q_\sigma} = -\frac{\partial T}{\partial X} = 0 \tag{2.159}$$

$$\text{③} \quad \frac{\partial V}{\partial q_\sigma} = \frac{\partial V}{\partial X} = kX - Mg \tag{2.160}$$

$$\text{①} + \text{②} + \text{③} = M\ddot{X} + kX - Mg = \text{④} = Q_j = F(t) - b\dot{X} \tag{2.161}$$

$$\text{④}\, \delta W_j = Q_j \delta q_j = \left(F(t)\hat{i} - b\dot{X}\hat{i}\right) \cdot \delta X = Q_j \delta X \tag{2.162}$$

Example 2.18 There is a baseball bat, seen in Fig. 2.36. Find the motion of equation (MOE), based on the Free Body Diagram (FBD).

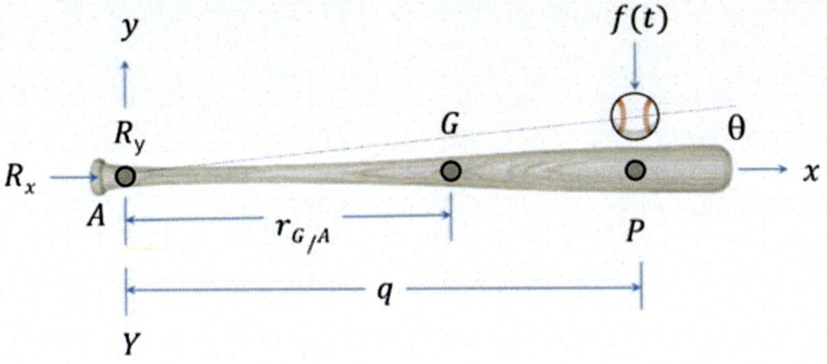

Fig. 2.36 A cart and pendulum

$$a_{G/O} = \underbrace{a_{A/O}}_{0} + \underbrace{\left(a_{G/A}\right)\big|_{\omega=0}}_{0} + \underbrace{\dot{\omega}_{/o} \times r_{G/A}}_{\ddot{\theta}\hat{k}\times r_{G/A}\hat{i}} + \underbrace{\omega_{/o} \times \left(\omega_{/o} \times r_{G/A}\right)}_{-\omega_o^2 r_{G/A}\hat{i}} + \underbrace{2\omega_{/o} \times \left(v_{G/A}\right)\big|_{\omega=0}}_{0} \quad (2.163)$$

$$① \sum_{ext} F_y = R_y\hat{j} - f\hat{j} = Ma_{G/Oy}\hat{j} = Mr_{G/A}\ddot{\theta}\hat{j} \quad (2.164)$$

If $R_y = 0$, $f = R_y - Mr_{G/A}\ddot{\theta} = -Mr_{G/A}\ddot{\theta}$ into ③

$$② \sum_{ext} F_x = R_x\hat{i} = Ma_{G/Ox}\hat{i} \quad (2.165)$$

$$③ \sum_{ext} \tau_{/A} = \frac{dH_{/A}}{dt} = I_{zz/A}\ddot{\theta}\hat{k} = -fq\hat{k} = Mr_{G/A}\ddot{\theta}q \quad (2.166a)$$

$$\ddot{\theta}MK_A^2 = Mr_{G/A}\ddot{\theta}q \text{ or } q = K_A^2/r_{G/A} \text{ where } I_{zz/A} = MK_A^2 \quad (2.166b)$$

Example 2.19 As seen in Fig. 2.37, there is a cart and pendulum. Find the motion of equation (MOE) by utilizing Newton or Lagrange method, based on the Free Body Diagram (FBD).

Newtonian approach:
C is 10 as follows:

(1) No y, z translation on M_1 is 2
(2) No x, y, z rotation on M_1 is 3
(3) No x, y rotation on M_2 is 2
(4) No x, y, z translation on M_2 is 3

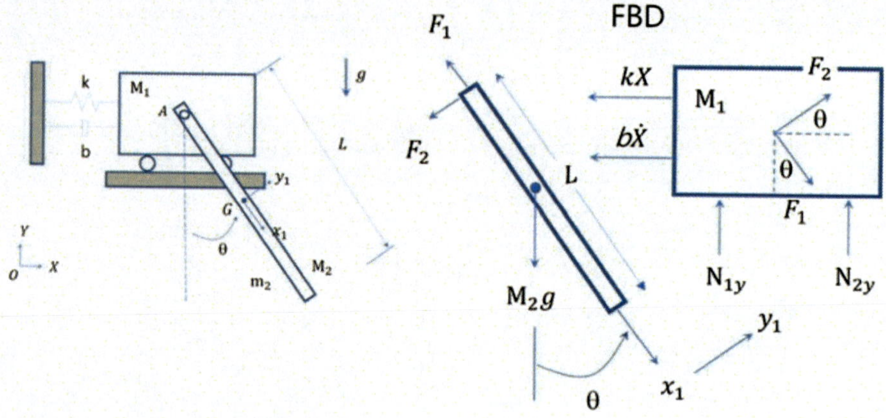

(a) A cart and pendulum

(b) Free body diagram

Fig. 2.37 A cart and pendulum

$$DOF = 6m + 3n - C = 6 \cdot 2 - 10 = 2 \tag{2.167}$$

$$① \sum_{ext} F_{X,M_1} = M_1 \ddot{X}\hat{I} = \left(-kX - b\dot{X} + F_2\cos\theta + F_1\sin\theta\right)\hat{I} \Leftarrow F_1 + F_2 \tag{2.168}$$

$$② \sum_{ext} F_{x_1,M_2} = M_2 \underbrace{a_{G/O}\hat{i}}_{A} = -F_1\hat{i} + M_2 g\cos\theta\hat{i} \Rightarrow F_1 = \tag{2.169}$$

$$③ \sum_{ext} F_{y_1,M_2} = M_2 \underbrace{a_{G/O}\hat{j}}_{B} = -F_2\hat{j} - M_2 g\sin\theta\,\hat{j} \Rightarrow F_2 = \tag{2.170}$$

$$④ \sum_{ext} \tau_{/G} = \frac{dH_{/G}}{dt} = F_2\frac{L}{2}\hat{k} = \frac{d}{dt}\left([I_{/G}]\begin{Bmatrix} 0 \\ 0 \\ \omega_z \end{Bmatrix}\right) = I_{zz/G}\ddot{\theta}\hat{k}$$
$$= \frac{M_2L^2}{12}\ddot{\theta}\hat{k} \Leftarrow F_2 \tag{2.171}$$

We need to calculate velocity $v_{G/O}$ and acceleration $a_{G/O}$

$$v_{G/O} = v_{A/O} + v_{G/A} = \dot{X}\hat{I} + \left(v_{G/A}\right)\Big|_{\omega_{/o}=0} + \omega_{/o} \times r_{G/A}$$
$$= \dot{X}\hat{I} + \omega_z\hat{k} \times \frac{L}{2}\hat{i} = \dot{X}\hat{I} + \dot{\theta}\frac{L}{2}\hat{j} \tag{2.172}$$

$$a_{G/O} = a_{A/O} + \left(a_{G/A}\right)\big|_{\omega_{/o}=0} + \dot{\omega}_{/o} \times r_{G/A} + \omega_{/o} \times \left(\omega_{/o} \times r_{G/A}\right)$$
$$+ 2\omega_{/o} \times \left(v_{G/A}\right)\big|_{\omega_{/o}=0}$$

$$\begin{aligned} a_{G/O} &= \frac{d}{dt}\left(v_{G/O}\right) = \ddot{X}\hat{I} + 0 + \ddot{\theta}\frac{L}{2}\hat{j} - \dot{\theta}^2\frac{L}{2}\hat{i} + 0 \\ &= \underbrace{\left(\ddot{X}\sin\theta - \frac{L}{2}\dot{\theta}^2\right)}_{A}\hat{i} + \underbrace{\left(\ddot{X}\cos\theta + \frac{L}{2}\ddot{\theta}\right)}_{B}\hat{j} \end{aligned} \tag{2.173}$$

$$④' \underbrace{\left(\frac{M_2L^2}{12} + \frac{M_2L^2}{4}\right)}_{\frac{M_2L^2}{3}}\ddot{\theta} + \tfrac{M_2L}{2}\ddot{X}\underbrace{\cos\theta}_{1} + M_2g\tfrac{L}{2}\underbrace{\sin\theta}_{\theta} = 0 \text{ for } \theta \text{ small, } \sin\theta \approx \theta$$

and $\cos\theta \approx 1$

$$①' \ (M_1 + M_2)\ddot{X} + b\dot{X} + kX + \frac{M_2L}{2}\ddot{\theta}\underbrace{\cos\theta}_{1} + M_2\frac{L}{2}\dot{\theta}^2\underbrace{\sin\theta}_{\theta} = 0$$

Lagrange approach from Eq. (146):

$$\frac{d}{dt}\frac{\partial L}{\partial \dot{q}_\sigma} - \frac{\partial L}{\partial q_\sigma} = Q_\sigma \quad \sigma = 1, \ldots, n-k$$

$$\underbrace{\frac{d}{dt}\frac{\partial T}{\partial \dot{q}_\sigma}}_{①} - \underbrace{\frac{d}{dt}\frac{\partial V}{\partial \dot{q}_\sigma}}_{0} - \underbrace{\frac{\partial T}{\partial q_\sigma}}_{②} + \underbrace{\frac{\partial V}{\partial q_\sigma}}_{③} = \underbrace{Q_\sigma}_{④} \quad \sigma = 1, \ldots, n-k$$

$$T = \frac{1}{2}M_1\dot{x}^2 + \frac{1}{2}M_2 v_{G/O} \cdot v_{G/O} + \frac{1}{2}\omega^T H_G\left(= \frac{1}{2}I_{ZZ/G_{M_2}}\dot{\theta}^2 = \frac{1}{2}\frac{M_2L^2}{12}\dot{\theta}^2\right) \tag{2.174}$$

where $v_{G/O} = \left(\dot{x} + \frac{L}{2}\dot{\theta}\cos\theta\right)\hat{i} + \frac{L}{2}\dot{\theta}\sin\theta\,\hat{j}$, $v_{G/O} \cdot v_{G/O} = \left(\dot{x} + \frac{L}{2}\dot{\theta}\cos\theta\right)^2 + \left(\frac{L}{2}\dot{\theta}\sin\theta\right)^2$,

$$\frac{\partial v_{G/O} \cdot v_{G/O}}{\partial \dot{\theta}} = 2\left(\dot{x} + \frac{L}{2}\dot{\theta}\cos\theta\right)\left(\frac{L}{2}\cos\theta\right) + 2\left(\frac{L}{2}\dot{\theta}\sin\theta\right)\left(\frac{L}{2}\sin\theta\right)$$
$$\left(= \frac{L^2}{2}\dot{\theta}\sin^2\theta\right) \tag{2.175}$$

$$V = \frac{1}{2}kx^2 + M_2 g \frac{1}{2} l(1 - \cos\theta) \tag{2.176}$$

θ direction of M_2:

$$\frac{\partial T}{\partial \dot{\theta}} = \frac{1}{2} I_{ZZ/G_{M_2}} 2\dot{\theta} + \frac{1}{2} M_2 \left[2\left(\dot{x} + \frac{L}{2}\dot{\theta}\cos\theta \right)\left(\frac{L}{2}\cos\theta \right) + \frac{L^2}{2}\dot{\theta}\sin^2\theta \right] \tag{2.177}$$

$$\frac{d}{dt}\frac{\partial T}{\partial \dot{\theta}} = I_{ZZ/G_{M_2}}\ddot{\theta} + \frac{M_2 L^2}{12}\ddot{\theta} + M_2\frac{L}{2}\cos\theta\ddot{x} \tag{2.178}$$

$$\frac{\partial V}{\partial \theta} = \frac{M_2 g}{2} L\sin\theta \tag{2.179}$$

$$①+②+③ = I_{ZZ/G_{M_2}}\ddot{\theta} + \frac{M_2 L^2}{12}\ddot{\theta} + M_2\frac{L}{2}\cos\theta\ddot{x} + \frac{M_2 g}{2}L\sin\theta = ④ = Q_\theta \tag{2.180}$$

x direction of M_1:

$$\frac{d}{dt}\frac{\partial T}{\partial \dot{x}} = M_1\ddot{x} + M_2\left(\ddot{x} + \frac{L}{2}\ddot{\theta}\cos\theta - \frac{L}{2}\dot{\theta}^2\sin\theta \right) \tag{2.181}$$

$$\frac{\partial V}{\partial x} = kx \tag{2.182}$$

$$①+②+③ = M_1\ddot{x} + M_2\left(\ddot{x} + \frac{L}{2}\ddot{\theta}\cos\theta - \frac{L}{2}\dot{\theta}^2\sin\theta \right) + kx = ④ = Q_x$$

$$\delta W = Q_x\delta x + Q_\theta\delta\theta = \underbrace{(F_1(t) - b\dot{x})}_{Q_x}\delta x + \underbrace{F_2 L\cos\theta}_{Q_\theta}\delta\theta \tag{2.183}$$

Example 2.20 As seen in Fig. 2.38, there is a sleeve sliding with rod. Find the motion of equation (MOE) by utilizing Newton or Lagrange method, based on the Free Body Diagram (FBD).

Rod: M_1, ${}^A I_{zz_1}$, L_1, G_1.
Sleeve: M_2, ${}^G I_{zz_2}$, L_2, G_2.
where L_o = Unstretched spring length.
If statically chec, it is

$$\underbrace{K\left(x_1 - L_o - \frac{L_2}{2} \right)}_{x_{static}} = M_2 g \tag{2.184}$$

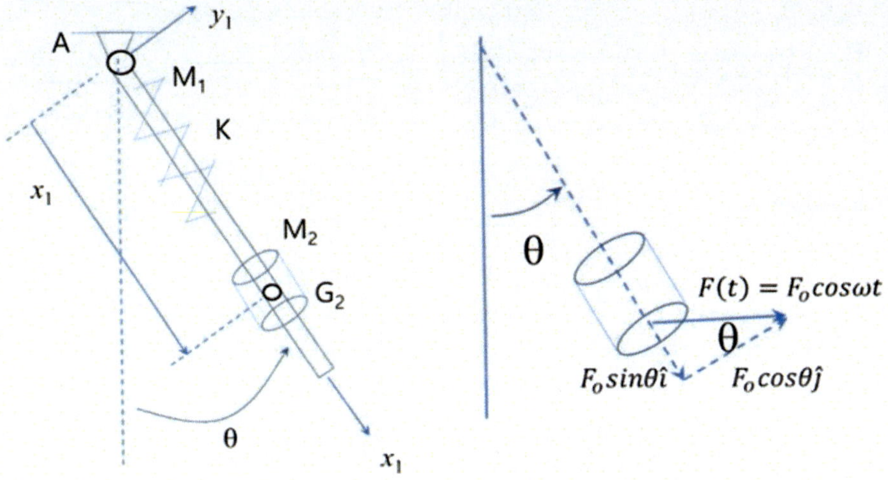

(a) A sleeve sliding with rod (b) Free body diagram

Fig. 2.38 A sleeve sliding with rod

$$T = \left(\frac{1}{2}{}^{A}I_{zz_1} + \frac{1}{2}{}^{G}I_{zz_1}\right)\dot{\theta}^2 + \underbrace{\frac{1}{2}M_2{}^{G}v_{G_2}\cdot{}^{G}v_{G_2}}_{\frac{1}{2}M_2\left(\dot{x}_1^2+x_1^2\dot{\theta}^2\right)} \tag{2.185}$$

$$V = \frac{1}{2}K\left(x_1 - L_o - \frac{L_2}{2}\right)^2 + M_1 g\underbrace{\frac{L_1}{2}(1-\cos\theta)}_{\Delta h} + M_2 g\left(L_o + \frac{L_2}{2}\right) - M_2 g x_1 \cos\theta \tag{2.186}$$

If Lagrange is applied from Eq. (146), it is.

x_1 direction:

$$① \frac{d}{dt}\frac{\partial T}{\partial \dot{x}_1} = \frac{d}{dt}(M_2\dot{x}_1) = M_2\ddot{x}_1 \tag{2.187}$$

$$② -\frac{\partial T}{\partial x_1} = -M_2 x_1\dot{\theta}^2 \tag{2.188}$$

$$③ \frac{\partial V}{\partial x_1} = K\left(x_1 - L_o - \frac{L_2}{2}\right) - M_2 g\cos\theta \tag{2.189}$$

$$① + ② + ③ = M_2\ddot{x}_1 - M_2 x_1\dot{\theta}^2 + K\left(x_1 - L_o - \frac{L_2}{2}\right) - M_2 g\cos\theta = ④ = Q_{x_1} \tag{2.190}$$

$$\delta W_{x_1} = Q_x \cdot \delta x_1 = F(t) \cdot dr = F_o \cos\omega t \left[\sin\theta \hat{i} + \cos\theta \hat{j} \right] \cdot dx\hat{i} = \underbrace{F_o \cos\omega t \sin\theta}_{Q_{x_1}} dx \tag{2.191}$$

θ direction:

$$① \ \frac{d}{dt}\frac{\partial T}{\partial \dot{\theta}} = \left({}^A I_{zz_1} + {}^G I_{zz_1}\right)\ddot{\theta} + M_2 x_1^2 \ddot{\theta} \tag{2.192}$$

$$② \ -\frac{\partial T}{\partial \theta} = 0 \tag{2.193}$$

$$③ \ \frac{\partial V}{\partial \theta} = M_2 g x_1 \sin\theta + M_1 g \frac{L_1}{2} \sin\theta \tag{2.194}$$

$$\begin{aligned} ① + ② + ③ &= \left({}^A I_{zz_1} + {}^G I_{zz_1}\right)\ddot{\theta} + M_2 x_1^2 \ddot{\theta} + M_2 g x_1 \sin\theta + M_1 g \frac{L_1}{2} \sin\theta \\ &= ④ = Q_\theta \end{aligned} \tag{2.195}$$

$$\delta W = F(t) \cdot dr = F_o \cos\omega t \left[\sin\theta \hat{i} + \cos\theta \hat{j} \right] \underbrace{\left(x_1 + \frac{L_2}{2} \right) \delta\theta \, \hat{j}}_{dr}$$

$$= \underbrace{F_o \cos\omega t \cos\theta \left(x_1 + \frac{L_2}{2} \right)}_{Q_\theta} \delta\theta$$

2.5 Bond-Graph Modeling

2.5.1 Introduction

A mechanical system such as an automobile or airplane generates power from an engine and transfers it to transmission via a drive system through a mechanism. When different physical domains—electric, magnetic, mechanical, hydraulic, etc.—are involved in a mechanical system, one of the powerful tools that can model engineering systems is bond graphs. It is a direct graphical tool for modeling dynamic systems involving several academic disciplines, that is, different engineering areas, such as mechanical, electrical, thermal, and hydraulic systems. Designing a newly designed dynamic system is a capable way to use in the form of a graph expression for communicating other engineers on its dynamic modeling.

In 1959, Professor Henry Payner and his former students at MIT suggested the bond graph method, which gave the thoroughgoing concepts for portraying multiport systems in terms of power bonds, attaching the components of the physical system to the presumed junction structures that were demonstrations of the constraints. This power exchange represent of a system is defined as a bond graph. It is suitable to model a mechanical product because a multiport system is composed and power generated from the engine is transmitted to other systems—transmission, drive system, etc.

In 1961, Professor Henry Paynter published a book entitled *Analysis and Simulation of Simulation of Multiport Systems*. Three authors in 2006 issued the fourth edition entitled *System Dynamics—Modeling and Simulation of Mechatronic Systems*. Currently, numerous disciplines of bond graphs have been widely received worldwide as modeling methodologies.

2.5.2 Basic Elements, Energy Relations, and Causality of a Bond Graph

A bond graph is a graphical display of a multiport system. Behavior with respect to energy is domain independent, regardless of the electric or mechanical system. That is, we can conclude that it is the same in all engineering disciplines when contrasting the RLC circuit for an electrical system with the damped mass spring in a mechanical system. Based on mutual energy exchange between different systems, the symbols in the bond graph indicate a bidirectional interchange of physical energy. Therefore, bond graphs might be applied in multiple energy fields such as mechanical, electrical, hydraulic systems, etc.

The analysis of dynamic systems might be comparatively straightforward when it has few DOFs or steady-state behavior. However, it is complex to model a multiport system and derive its governing equation. In most of the scenarios, the principal matter of engineers is to demonstrate a mathematical prototype which expresses the dynamic behavior of the system and how the dissimilar parameters affect the system characteristics because the system dynamic equations are normally partial differential equations, whose solutions necessitate extensive mathematical understanding.

As essential knowledge of bond graph theory, energy flow is a fundamental component in a system. It runs in from one or more sources, is momentarily reserved in system components or partly dissipated in resistances as heat, and last, reaches at "loads" where it creates some wanted effects. Power, therefore, is the rate of energy flow with no orientation.

The bond graph represents the power flow between two systems. For example, consider an electric-hydraulic system. The generated electric power works motor that operates the pump. This energy flow is represented by an arrow (bond) in Fig. 2.39. Each bond depicts the on-the-spot energy flow or power. The flow in each bond is designated by a pair of variables called 'power-conjugated variables' whose product is the instant power of the bond. Because (generated) power in a mechanical system is difficult to immediately compute, engineers can use two transient variables, flow and effort. Every system domain has a pair of effort and flow variables. For instance, in a mechanical system, flow depicts the "velocity" and effort the "force", while in an electrical system, flow depicts the "current" and effort the "voltage". The product of both transient variables, that is, power, is expressed as follows (Table 2.2):

$$P = e(t) \cdot f(t) \tag{2.196}$$

The method enables the simulation of multiple physical domains, such as mechanical, thermal, electrical, and hydraulic domains. Flows and efforts might be recognized with a specific variable for each particular physical domain that is working.

The bond graph is composed of the "bonds" that join in junction with the "1-port", "2-port", and "3-port" elements. Whether or not the power in the bond graph is continuous, every element is depicted by a multiport. Ports are attached by bonds. The fundamental blocs of common bond graph theory are itemized in Table 2.3.

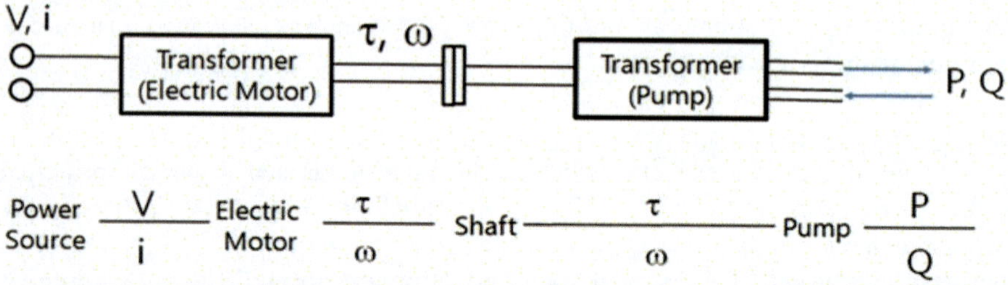

Fig. 2.39 Power flow in the bond graph for the electric-hydraulic system

Table 2.2 Energy flow in the multiport physical system

Modules	Effort, $e(t)$	Flow, $f(t)$
Mechanical translation	Force, $F(t)$	Velocity, $V(t)$
Mechanical rotation	Torque, $\tau(t)$	Angular velocity, $\omega(t)$
Compressor, pump	Pressure difference, $\Delta P(t)$	Volume flow rate, $Q(t)$
Electric	Voltage, $V(t)$	Current, $i(t)$

Table 2.3 Fundamental components of the bond graph

Elements		Symbol	Relation equations
1-Port elements (sources)	Effort	S_e	$S_e = e(t)$
	Flow	S_f	$S_f = f(t)$
1-Port elements	C-type elements	C	$\frac{de(t)}{dt} = \frac{1}{C} f(t)$
	I-type elements	I	$\frac{df(t)}{dt} = \frac{1}{I} e(t)$
	R-type elements	R	$e(t) = R \cdot f(t)$
2-Port elements	Transformer	TF	$e_2(t) = TF \cdot e_1(t)$ $f_2(t) = \frac{1}{TF} \cdot f_1(t)$
	Gyrator	GY	$e_2(t) = GY \cdot f_1(t)$ $f_2(t) = \frac{1}{GY} \cdot e_1(t)$
3-Port junction elements	0-junction	0	$e_2(t) = e_1(t)$
	1-junction	1	$f_2(t) = f_1(t)$

For 1 port, there are effort sources, flow sources, *I*-type elements, *C*-type elements, and *R*-type elements which shall discontinuously attach power. For 2-ports, there are gyrators and transformers which can continuously join power. For 3-ports, there are 1-junction and 0-junction which shall make up the network.

Power bonds can connect at one of the two categories of junctions: a "0" junction and a "1" junction. In these joints, no energy may be produced or dissipated. This is defined as power continuity. In a "0" junction, the flow and the efforts fulfill Eqs. (2.197) and (2.198):

$$\sum \text{flow}_{\text{input}} = \sum \text{flow}_{\text{output}} \tag{2.197}$$

$$\text{effort}_1 = \text{effort}_2 = \ldots = \text{effort}_n \tag{2.198}$$

This matches a node in an electrical circuit (where Kirchhoff's current law exerts). On the other hand, in a "1" junction, the flow and the efforts fulfil Eqs. (2.199) and (2.200):

$$\sum \text{effort}_{\text{input}} = \sum \text{effort}_{\text{output}} \tag{2.199}$$

$$\text{flow}_1 = \text{flow}_2 = \ldots = \text{flow}_n \tag{2.200}$$

This matches the force balance at a mass in a system. An instance of a "1" junction is a resistor in series. In the junction, the assumption of energy preservation is presumed, but no loss is permitted. There are two extra variables that are significant in the account of dynamic systems.

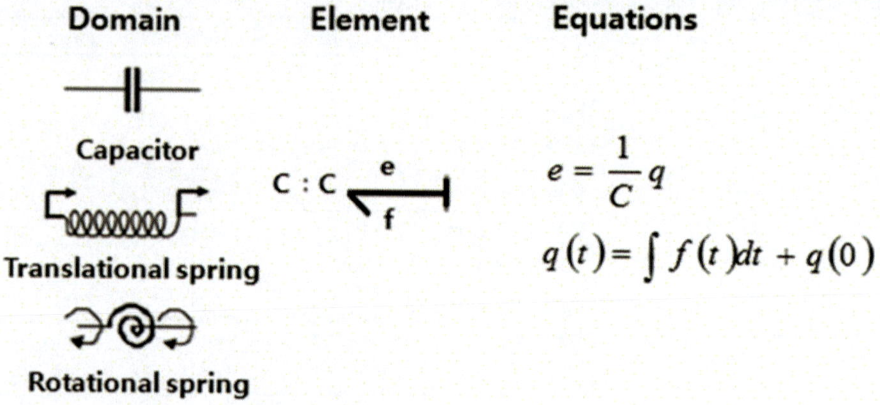

Fig. 2.40 Examples of C elements

For any element with a bond with power variables such as *effort* and *flow*, the energy variation from t_0 to t might be expressed as follows:

$$H(t) - H(t_0) = \int_{t_0}^{t} e(\tau)\, f(\tau)d\tau \tag{2.201}$$

For C-type storage elements, e (effort) is a function of q (displacement) as capacitor or spring (Fig. 2.40).

As displacement is transformed into its derivative, flow is expressed as

$$q(t) = \int f(t)dt \Rightarrow \dot{q} = \frac{dq}{dt} = f(t) \tag{2.202}$$

which is defined as a balance equation and forms a part of the constitutive equations of the storage element. If Eq. (2.202) is changing variables from t to q, the linear case shall be defined as follows:

$$H(q) - H(q_0) = \frac{1}{2C}\left(q^2 - q_0^2\right) \tag{2.203}$$

For I-type storage elements, f (flow) is a function of p (momentum) as an inductor or mass (Fig. 2.41).

As momentum is transformed into its derivative, effort is expressed as

$$p(t) = \int e(t)dt \Rightarrow \dot{p} = \frac{dp}{dt} = e(t) \tag{2.204}$$

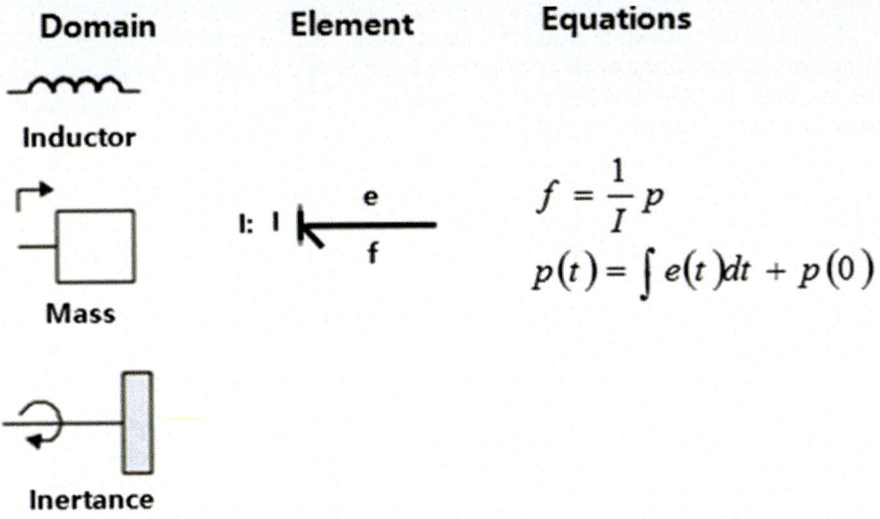

$$f = \frac{1}{I} p$$
$$p(t) = \int e(t)dt + p(0)$$

Fig. 2.41 Examples of I elements

which is called the balance equation. If Eq. (2.204) is exchanging variables from t to p, the linear case shall be defined as follows:

$$H(p) - H(p_0) = \frac{1}{2I}(p^2 - p_0^2) \tag{2.205}$$

Resistor elements depict circumstances where energy dissipates, such as mechanical dampers, electrical resistors, and Coulomb friction. In these kinds of elements, there is a relation between effort and flow. The value of "R" might be stable or a function of any system parameter, including time:

$$e(t) = R \cdot f(t) \tag{2.206}$$

Compliance elements represent the circumstances where energy stores – mechanical springs, electrical capacitors, etc. In these kinds of elements, there is a relation between effort and displacement variables. The value of "K" might be stable or a function of any system parameter, including time:

$$e(t) = K \cdot q(t) \tag{2.207}$$

Inertia elements represent the relation between the "flow" and momentum (mass, moment of inertia, electrical coil, etc.) as Eq. (2.208) shows. The value of "I" tends to be stable

$$p(t) = I \cdot f(t) \tag{2.208}$$

A transformer adds no power but changes it, such as a lever or an electrical transformer. Transformers show those physical phenomena which are differences in the gains of output flow and effort on the gains of input flow and effort. As the transformation proportion is denoted by the "TF" value, the relation between input and output can be expressed as in Eqs. (2.209) and (2.210):

$$e_{\text{output}}(t) = TF \cdot e_{\text{input}}(t) \tag{2.209}$$

$$f_{\text{output}}(t) = \frac{1}{TF} \cdot f_{\text{input}}(t) \tag{2.210}$$

The process of deciding the computational orientation of the bond variables is defined as causal analysis. One is the "half-arrow" sign agreement. It expresses the presumed orientation of positive energy flow. As with electrical circuit diagrams and free-body diagrams, the selection of positive orientation is random, with the warning that the analyst should be compatible throughout with the chosen definition. The other attribute is the "causal stroke". This is a vertical bar put on only one end of the bond. It is not random (Fig. 2.42).

On each bond, one of the two variables should be the cause and the other one the effect. This shall be concluded by the relation designated by the arrow orientation. Effort and flow causalities always act in opposite orientations in a bond. The causality allocation process selects what sets for each bond. Causality allocation is a requisite to change the bond graph into a computable code.

Any port (single, double, or multiple) added to the bond will state either "effort" or "flow" by its causal stroke, but not both. The port added to the ending of the bond with the "causal stroke" states the "flow" of the bond. The bond exposes "effort" upon that port. Similarly, the port on the end with no "causal stroke" exposes "effort" on the bond, while the bond exposes "flow" on that port.

Once the system is depicted in the formation of a bond graph, the state equations which govern its behavior shall be attained straightly as first-order differential

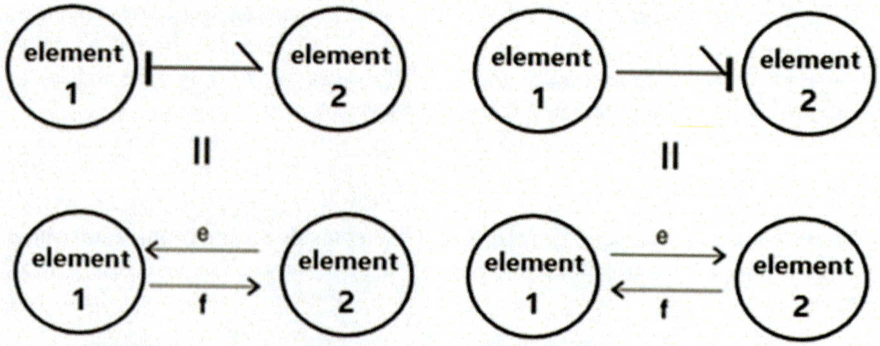

Fig. 2.42 "Half-arrow" sign convention and meaning of the causal stroke

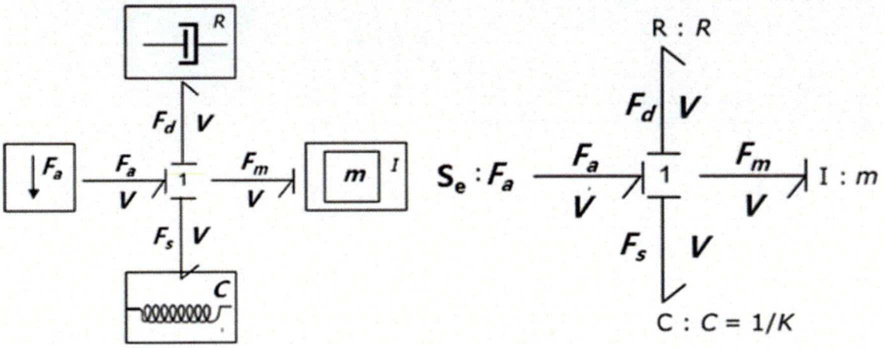

Fig. 2.43 Damped mass–spring system for a mechanical product (example)

formulations in terms of the generalized variables expressed above, utilizing straightforward and standardized processes, regardless of the physical domain to which it associates, even when connected to one another across domains.

Example 2.21 The damped mass–spring system for a mechanical product is represented in Fig. 2.43. In mechanical diagrams, the port variables of the bond graph elements are the *force* on the element port and *velocity* of the element port. These two variables are associated with each other. The *power* being changed by a port with the rest of the system is the multiplication of force and velocity. That is, $P = Fv$. The equations of a damper, spring, and mass are as follows (we utilize damping coefficient a, spring coefficient K_s, mass m, and exerted force F_a):

$$F_d = \alpha\, v,\ \ F_s = K_s \int v dt = \frac{1}{C_s} \int v dt,\ \ F_m = m\frac{dv}{dt} \tag{2.211}$$

Because the loose ends of the instance all have the same velocity, the common v is altered to a '1', a so-called 1-junction. This junction element also involves that the forces sum up to zero, considering the sign (related to the power direction). The force is depicted as an *effort,* and the velocity is depicted as a *flow.*

Example 2.22 To understand a transformer, a mass-less ideal lever is considered. As seen in Fig. 2.44, the lever is rigid to establish a linear relation between power variables at both ends.

From the geometry, we have

$$V_2 = (b/a)V_1 \tag{2.212}$$

The power transmission indicates

$$F_2 = (a/b)F_1 \text{ so that } V_2F_2 = V_1F_1 \tag{2.213}$$

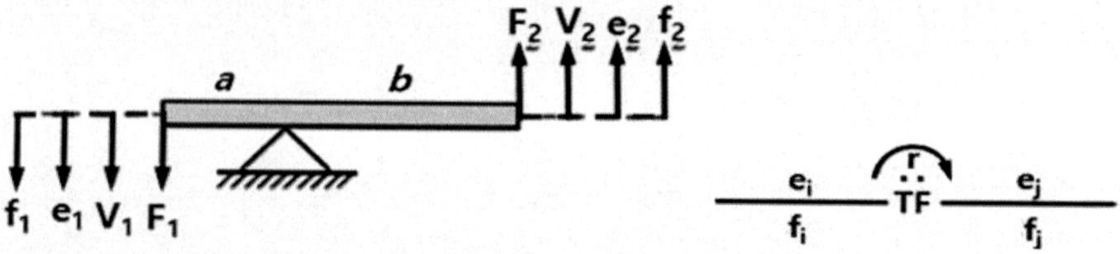

Fig. 2.44 Ideal lever system (example)

In bond graphs, such a circumstance can be expressed as TF (transformer). The 'r' above the transformer denotes the modulus of the transformer, which may be a constant ('b/a'):

$$f_j = r f_i \quad \text{and} \quad e_j = (1/r) e_i \tag{2.214}$$

Thus, the following representation establishes the conservation of power:

$$e_j f_j = e_i f_i \tag{2.215}$$

2.5.3 *Case Study: Failure Analysis and Redesign of a Helix Upper Dispenser*

The icemaker in a domestic refrigerator comprises many mechanical parts. In the ice-making process, these components are subjected to some mechanical loads. Ice making requires some mechanical procedures: (1) the filtered water through a tap line is pumped to provide the tray; (2) the air chilled in the heat exchanger supplies the water tray; (3) ice cubes are produced in the bucket until it is full; and (4) when the customer pushes the lever by force, cubed or crushed ice will be dispensed. In the United States, customers usually require an SBS refrigerator to harvest 10–200 cubes a day. Ice harvest might be affected by uncontrollable consumer use circumstances such as ice consumption, water pressure, refrigerator notch settings, and the number of cycles the door is open. As the refrigerator is plugged in, the cubed ice mode is chosen by itself without direct human control. A crusher shatters the cubed ice in the crushed mode. Usually, the load of the icemaker is small because it is functioned with no fused or webbed ice.

However, for Asian customers, fused or webbed ice shall regularly crystallize in the tray because in cubed mode, they rarely dispense ice. As ice is dispensed under the situations, a significant mechanical overload happens in the ice crusher. However, in Europe or the United States, the icemaker system functions without stop as it is repeatedly utilized in both crushed and cubed ice modes. This might generate a mechanical overload.

Figure 2.45 gives a general summary of the ice maker. Figures 2.20 and 2.21 depict simplified diagram of the mechanical load transfer in the ice bucket assembly and its bond graphs. An AC auger motor produces power. To obtain sufficient torque to crush the ice, motor power is moved through the gear system to the ice bucket fabrication—in other words, to the helix upper dispenser, the blade dispenser, and the ice crusher (Fig. 2.46).

The bond graph in Fig. 2.47 might be expressed as follows:

$$dfE_2/dt = 1/L_a \times eE_2 \tag{2.216}$$

$$dfM_2/dt = 1/J \times eM_2 \tag{2.217}$$

The junction from Eq. (2.216)

$$eE_2 = e_a - eE_3 \tag{2.218}$$

$$eE_3 = R_a \times fE_3 \tag{2.219}$$

The junction from Eq. (2.217)

$$eM_2 = eM_1 - eM_3 \tag{2.220}$$

$$eM_1 = (K_a \times i) - T_{\text{Pulse}} \tag{2.221}$$

$$eM_3 = B \times fM_3 \tag{2.222}$$

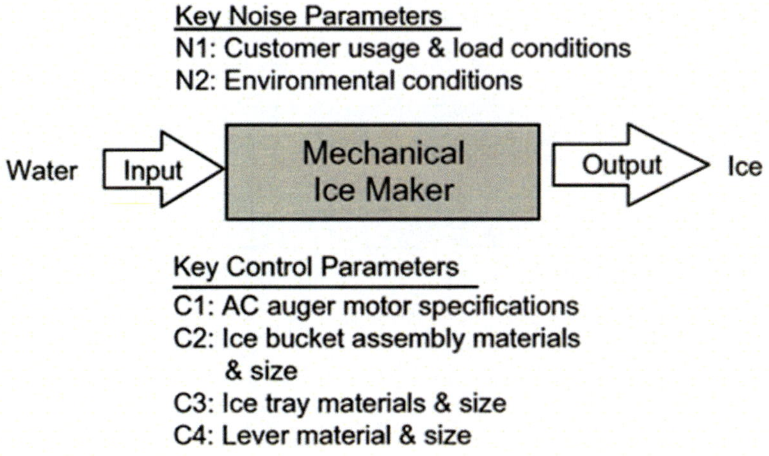

Fig. 2.45 Robust design simplified illustration of ice-maker

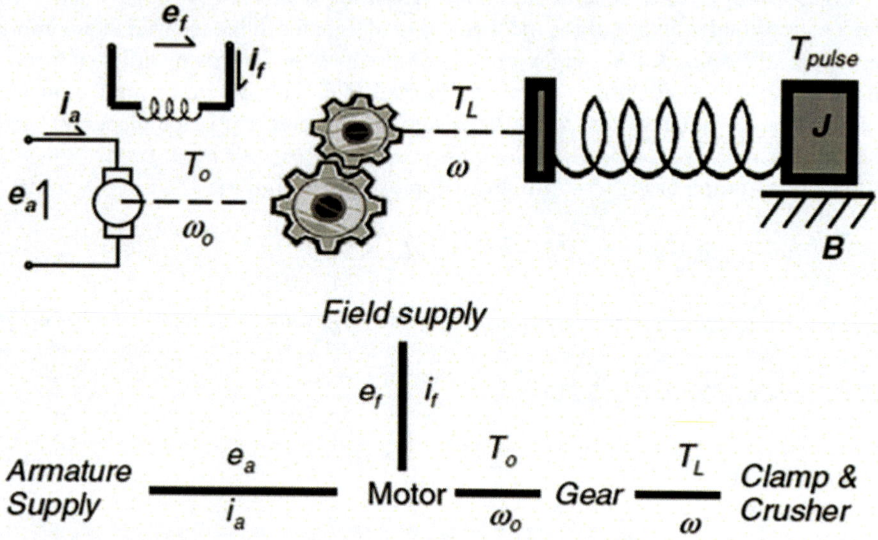

Fig. 2.46 Simplified illustration for mechanical ice bucket assembly

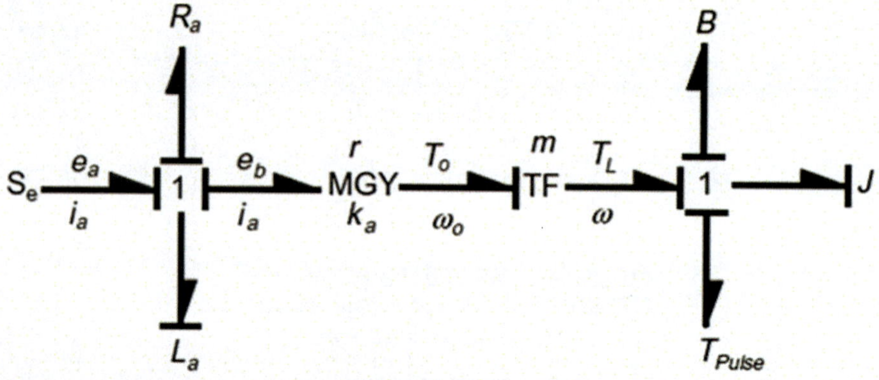

Fig. 2.47 Bonding graph of ice bucket assembly

Because $fM_1 = fM_2 = fM_3 = \omega$ and $i = fE_1 = fE_2 = fE_3 = i_a$. From Eqs. (2.218) and (2.219),

$$eE_2 = e_a - R_a \times fE_3 \tag{2.223}$$

$$fE_2 = fE_3 = i_a \tag{2.224}$$

If substituting Eqs. (2.223) and (2.224), then

$$di_a/dt = 1/L_a \times (e_a - R_a \times i_a) \tag{2.225}$$

We can obtain

$$eM_2 = [(K_a \times i) - T_{\text{Pulse}}] - B \times fM_3 \tag{2.226}$$

$$i = i_a \tag{2.227}$$

$$fM_3 = fM_2 = \omega \tag{2.228}$$

If substituting Eq. (2.226–2.228), then

$$d\omega/dt = 1/J \times [(K_a \times i) - T_{\text{Pulse}}] - B \times \omega \tag{2.229}$$

Therefore, the governing equation (or state equation) might be as follows:

$$\begin{bmatrix} di_a/dt \\ d\omega/dt \end{bmatrix} = \begin{bmatrix} -R_a/L_a & 0 \\ mk_a & -B/J \end{bmatrix} \begin{bmatrix} i_a \\ \omega \end{bmatrix} + \begin{bmatrix} 1/L_a \\ 0 \end{bmatrix} e_a + \begin{bmatrix} 1 \\ -1/J \end{bmatrix} T_{Pulse} \tag{2.230}$$

When Eq. (2.230) is integrated, the angular velocity of the ice bucket mechanical assembly is attained as

$$y_p = \begin{bmatrix} 0 & 1 \end{bmatrix} \begin{bmatrix} i_a \\ \omega \end{bmatrix} \tag{2.231}$$

Chapter 3
Probability and Its Distribution in Statistics

3.1 Fundamentals of Probability

As seen in Fig. 3.1, the happening of an occurrence is somewhere between impossible and certain. It may be defined as the chance that something will happen. In addition to words, we may utilize numbers as follows: (1) Impossible has zero, (2) Certain has one. That is, because the chance of an event shall not be less than 0, 0 has impossible (certain that thing unspecified shall not occur). As the probability of an event shall not be more than 1, this is because 1 is certain that thing unspecified shall occur. For example, we manifest the chance that (a) The sun shall arise tomorrow, (b) I shall not have to memorize mathematics at school, (c) If a coin flips, it shall get heads up, and (d) Selecting a red ball from a bag with 1 red ball and 3 green balls.

Choosing a particular event $A_i \in S$ provides an observation. Definite set theory concepts have special terminology and meaning when utilized in the conditions of probability.

1. A subset A of S is specified as an event.
2. For any event $A, B \in S$:

- The complement event A^c is the event that is not in A or not in A'.
- Union $A \cup B$: an outcome if it is either in A or in B.
- Intersection $A \cap B$: an outcome if it is in both A and B.
- $A \subseteq B$: the occurrence of A means the event of B.
- $A \cap B = \phi$: A and B are mutually exclusive or disjoint events.

If the number of possible outcomes is to be counted, the sample space is $S = \{A_1, A_2, A_3, \ldots\}$.

In probability, the sample space composes a finite number of equally likely (or mutually exclusive) outcomes, $S = \{A_1, A_2, A_3, \ldots\}$. The probability can classically be defined as follows:

S. Woo, *Design of Mechanical Systems*, Springer Series in Reliability Engineering,
https://doi.org/10.1007/978-3-031-28938-5_3

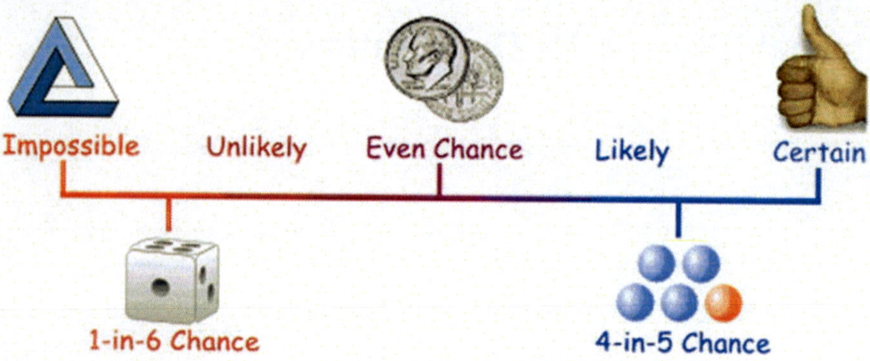

Fig. 3.1 Probability line

$$P(A) = \frac{\text{number of outcomes in } A}{\text{number of outcomes in } S} = \frac{|A|}{N} \tag{3.1}$$

This classical definition for probability in Eq. (3.1) is instinctively logical, but it makes several practical presumptions about the existence of the limit and the fact that the ratio merges to a single value for all possible orders of experimental outcomes. That is, if the trial number n increases infinitely, the relative frequency $\frac{r}{n}$, where r is the frequency for event A, will approach $P(A)$. For instance, the chance of a tail in a single flip of a fair coin is $P(A) = lim_{n\to\infty} \frac{r}{n} = \frac{1}{2}$ as $n \to \infty$.

Rather than making these kinds of presumptions, modern probability theorists prefer to explain a more fundamental set of axioms about the essence of the probability measure P. Using the following axioms, we can obtain more complicated results, including the intuitive meaning of probability as a fraction of occurrence.

The chance of event A is a number $P(A)$ assigned to A that satisfies the following axiom conditions:

1. $0 \leq P(A) \leq 1$ for sample space S and event E.
2. $P(S) = 1,\ P(\phi) = 0.$ (3.2)
3. For any sequence of events $A_1, A_2, \ldots, A_n$ that are mutually exclusive, satisfy the following:

$$P(A_1 \cup A_2 \cup \cdots \cup A_n) = P(A_1) + P(A_2) + \cdots + P(A_n) \tag{3.3}$$

The first axiom explains that the chance of an event is a number between 0 and 1. The second axiom explains that the event expressed by the whole sample space has a probability of 1. The third axiom explains how to integrate the probabilities of mutually exclusive events.

Permutation: From a set of n elements, the number of ways to obtain an ordered subset of k elements is given by

$$ {}_nP_k = n(n-1)(n-2)\cdots(n-r+1) = \frac{n!}{(n-r)!} \tag{3.4}$$

The numeral of permutations of n discrete objects is n factorial, expressed as $n!$, which means that the product of all positive integers is less than or equal to n. That is,

$$ {}_nP_n = n(n-1)(n-2)\cdots 1 = n! \tag{3.5}$$

Combination: In permutation, the order is important. However, combination is a choice of entries from a collection, so that the sequence of choice does not have any importance. For instance, given three fruits, say an apple, an orange, and a pear, there are three combinations of two which might be taken out from this set: an apple and a pear; an apple and an orange; or a pear and an orange. In other words, it is the number of methods of picking k unordered consequences from n possibilities. It is also familiar as the binomial coefficient or choice number that can read 'n choose k',

$$ {}_nC_k = \binom{n}{k} = \frac{{}_nP_r}{k!} = \frac{n!}{(n-k)!} \tag{3.6}$$

Addition theorem of probability: The chance of event A or B shall be attained by attaching the chances of both events A and B and subtracting any intersection of the two events. That is,

$$P(A \cup B) = P(A) + P(B) - P(A \cap B) \tag{3.7}$$

If $A \cap B = \phi$, $P(A \cup B) = P(A) + P(B)$. That is, events A and B are mutually exclusive events.

Conditional probability: The conditional probability of an event B is the chance that the event shall happen given the knowledge that an event A has already happened. This chance is expressed as $P(B|A)$, notation for the chance of B given A. In the case where events A and B are independent (where event A has no impact on the chance of event B), the conditional chance of event B given event A is straightforwardly the chance of event B, that is, $P(B)$. If events A and B are dependent on each other, the chance of the intersection of A and B (the chance that both events happen) is expressed by

$$P(A \cup B) = P(A)P(B|A) \tag{3.8}$$

From this meaning, the conditional chance $P(B|A)$ is simply attained by spliting by $P(A)$:

$$P(B|A) = \frac{P(A \cap B)}{P(A)} \tag{3.9}$$

Multiplication theorem of chance: If $P(A) > 0$ and $P(B) > 0$, thus

$$P(A \text{ and } B) = P(A \cap B) = P(A)P(B|A) = P(B)P(A|B) \tag{3.10}$$

Complement rule: If A is an event and A^c is its complement, the probability of A is equal to one minus the chance of A^c:

$$P(A^c) = 1 - P(A) \tag{3.11}$$

This will apply to all events and their complements.

3.2 A Short History of Statistics and Probability

The motive of statistics is to evolve a systematic method for removing practical understanding from both test and data. Additionally, its basic role in data examination, statistical reasoning is also very practical in data collection (design of experiments and surveys) and in guiding proper scientific inference. Statistics is neither truly a science nor a division of mathematics. It is regarded as a meta-science for handling data collection, analysis, and interpretation. As such, its range is huge, and it supplies much leading intuition in many areas of science and business.

The word statistics was earliest utilized by a German professor Gottfried Achenwall in the middle of the eighteenth century as the science of statecraft covering the collection and usage of data by the state. The word statistics from the Latin word "Status" or Italian word "Statista" or German word "Statistik" or the French word "Statistique", signifying a political state, initially meant information useful to the state, such as information about the sizes of the population (human, animal, products, etc.) and armed forces.

In accordance with pioneer statistician Yule, the word 'statistics' happened the earliest in the book "the element of universal erudition" by Baron (1770). In 1787, a wider meaning utilized by Zimmermann in "A Political survey of the present state of Europe". It seemed to appear in the encyclopedia of Britannica in 1797 and was utilized by Sir John Sinclair in Britain in a series of volumes issued between 1791 and 1799, giving a statistical account of Scotland. In the nineteenth century, the word statistics obtained a wider definition covering numerical data of almost any subject and interpretation of data through suitable analysis. That's all about the brief history of Statistics. That is,

(1) Seventeenth century—probability theory
(2) Eighteenth century—inferential statistics
(3) Nineteenth century—statistics applied to education and sociology
(4) Twentieth century—regression and correlation; computer analyses

As the earliest statistics and probability theory were developed in the nineteenth century, statistical reasoning and probability prototypes utilized by social scientists also proceeded with the modern sciences of experimental sociology and psychology. It was applied to thermodynamics and statistical mechanics by physical scientists in the last century. Early on, Laplace suggested the central limit theorem. He also read Gauss's work and made the revolutionary connection between the least square estimation and central limit theorem. The Gauss–Laplace combination is considered one of the crucial milestones in the history of science. Karl Pearson (1857–1936) established the *P* value, Pearson's correlation coefficient, Pearson's chi-square test, and principal component analysis, among many other things.

At the beginning of the twentieth century, Ronald A. Fisher (1890–1962) implemented statistical procedures in the design of scientific experiments—for example, ANOVA. Fisher also summarized his statistical effort in Statistical Methods and Scientific Inference (1956). Based on small sets of data, William S. Gosset established methods for decision-making. In the manufacture of mechanical products, there is a possibility of recalls from the marketplace. A new reliability methodology might be developed in the future based on the probability and its statistics (Fig. 3.2).

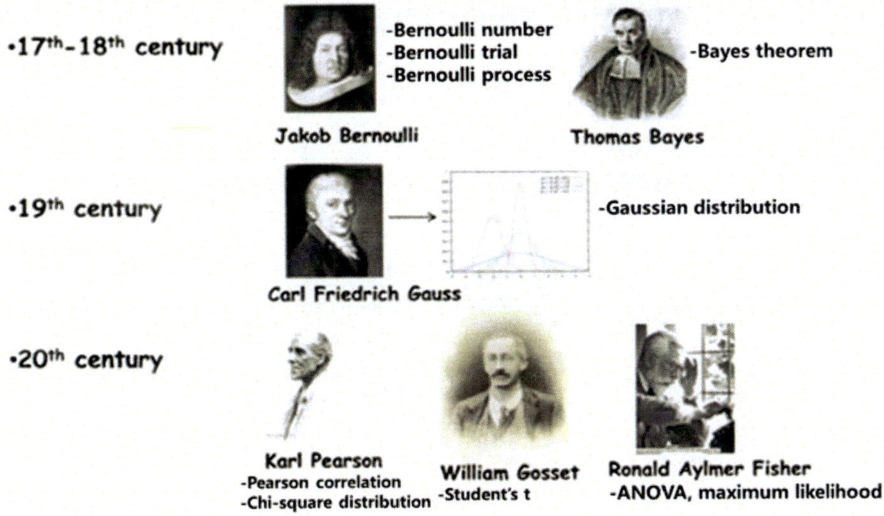

Fig. 3.2 A brief history of statistics

3.3 Statistics Terminologies

We can define statistical terminologies such as population, sample, parameter, and statistics as follows (Fig. 3.3).

- A *population* is a group of all things or items under analysis in a statistical investigation. Population shall be attributed as the set of persons or things where an engineer is interested and from which the engineer wants to draw conclusions. For example, if an engineer is about to design an automobile engine, the whole population will be related to the engine.
- A *sample* is the subset of the population that is truly being noticed or learned. The reasons to choose a sample from the population are as follows: (1) population investigation is impossible because of time and economics, (2) investigation results should be provided in time, and (3) there is the possibility of inaccuracy in total sampling because of investigation error.
- *Random sampling* is a way to choose a well-representative sample from a population. For a sample to be precise, it might be random with a similar possibility of every representative of the population possessing a probability of being chosen. This can lessen bias.
- A *parameter* is a quantitative attribute of the populace in guessing or testing such as a population mean or proportion. For instance, if you want to determine which

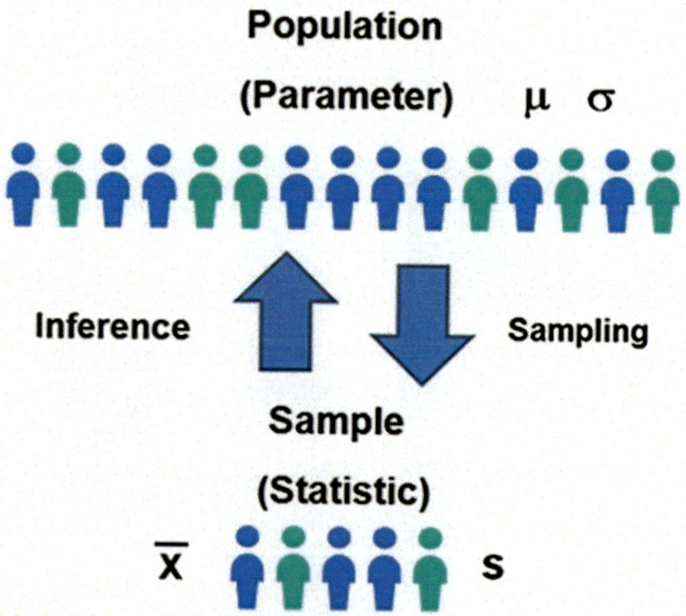

Fig. 3.3 Population and sampling

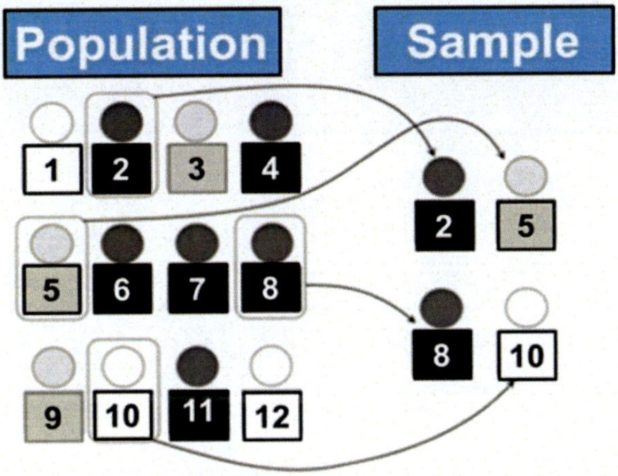

Fig. 3.4 Concept of confidence interval

classes freshmen choose at a college, you can query everyone (via email) and obtain a parameter such as the mean weight (or height) of freshmen.

- A *statistic* is a quantitative attribute of a sample that helps to guess or test the population parameter such as a sample mean or proportion.
- A confidence interval (CI) explains this problem because it supplies a span of values that are likely to hold the population parameter of interest. That is, it stands for the frequency of feasible CIs that hold the true value of the unspecified population parameter (Fig. 3.4).
- A CI supplies a span of values for undefined parameters of the population by computing a statistical sample. This is defined as an interval and the level of confidence that the parameter is in the interval.
- Hypothesis testing is where scientists assert the population by inspecting a statistical sample. By design, there is unspecified unpredictability in this process. This shall be defined as a level of significance.
- A CI is distinguished as the chance that a random value lies in a certain range. CI is expressed as a percentage. For instance, a 90% CI suggests that in 90 out of 100 occasions, the noticed value drops in this certain interval. As a certain sample is chosen, the population parameter is either or not in the interval achieved. The required level of confidence is determined by the researcher. A 90% CI means a significance level of $\alpha = 10\%$. The confidence level also relies on the product field.
  - Sixty percent CI is widely used as an international standard organization.
  - For IEC and GM, 50% CI is used.
  - The MIL standard adapts 60–90% CI.
  - A random variable, X, is a variable whose feasible values are numerical results of a random occurrence from a statistical test. A random variable is a function from a sample space S into the real numbers (Fig. 3.5).

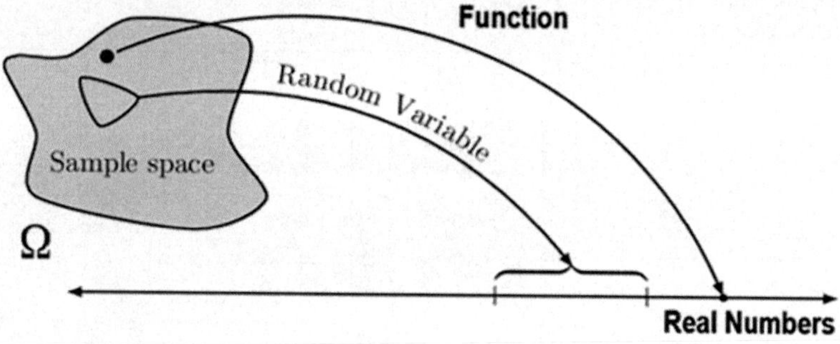

Fig. 3.5 Random variable

There are two kinds of random variables: continuous and discrete. A discrete random variable is one that shall put on only a countable number of definite values such as 0, 1, 2, 3, 4, …. Instances of discrete random variables are the number of typos on a page, the number of children in a family, the number of faulty light bulbs in a box of ten, and the number of problematic products per lot in manufacturing. On the other hand, a continuous random variable is one that puts on an infinite number of possible values in a real interval. It includes the departing time of airplanes, the time necessary to run a mile, height, and weight.

A random variable X is continuous if there is a function $f(x)$, such that for any $a \le b$, we have

$$P(a \le X \le b) = \int_a^b f(x)\mathrm{d}x \tag{3.12}$$

The function $f(x)$ is specified as the probability density function (*pdf*).

The *pdf* always fulfills the following properties:

1. $f(x) \ge 0$ (f is nonnegative) (3.13)
2. $\int_{-\infty}^{\infty} f(x)dx = 1$ (This is equivalent to: $P(-\infty < X < \infty) = 1$) (3.14)

It is also specified as the probability distribution function or probability function. It is denoted by $f(x)$.

Assume X is a continuous random variable with range $[a, b]$ and probability density function $f(x)$. The expected value of X is expressed as

$$E(X) = \int_a^b xf(x)dx \tag{3.15}$$

However, for a discrete random variable, the expected value of X is

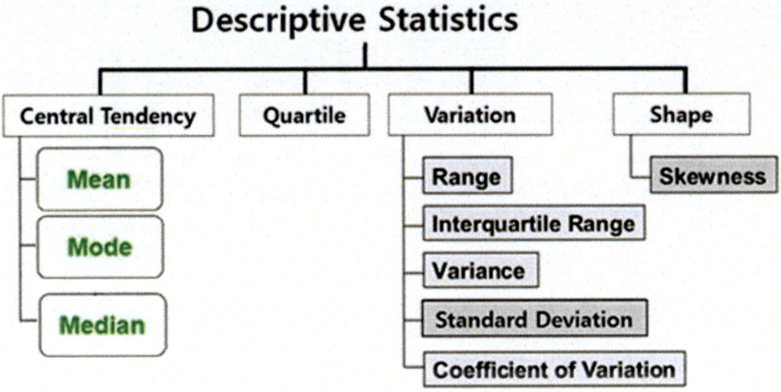

Fig. 3.6 Measures summary of center and variation

$$E(X) = \sum_{i=1}^{n} x_i f(x_i) \tag{3.16}$$

The discrete formula takes a weighted summation of the values x_i of X, where the weights are the chances $f(x_i)$. $f(x)dx$ stands for the chance that X is in an infinitesimal range of width dx around x. Therefore, we may explain the formula for $E(X)$ as a weighted integral of the values x of X, where the weights are the chances $f(x)dx$. As mentioned before, the expected value is also defined as the average or mean.

All collected data might be expressed by a measure of central tendency or dispersion in descriptive statistics. Different patterns of data are abridged by distinct measures of central tendency and dispersion (Fig. 3.6).

The central tendency means where the observed data are located densely. There are three statistics which are commonly utilized to represent the middle of a collected set of measurements (or data):

- Mean (also called the average or arithmetic mean): a quantity which has an intermediate value between those of the utmost members of some set. Some kinds of means pre-exist, and the way of computing a mean relies on the relation known or presumed to rule the other members. For lifetime, the central tendency of product lifetime can be described as Mean Time To Failure. If n observed data are x_1, x_2, x_3, …, x_n, the arithmetic mean of the variable x is expressed as follows:

$$\overline{x} = \frac{1}{n}(x_1 + x_2 + \cdots + x_n) = \frac{\sum_{i=1}^{n} x_i}{n} \tag{3.17}$$

- Median (the middle of a data set): Median, in statistics, is the center value of the given list of data, when organized in an order. The median is not influenced by utmost values because it only responds to the number of observations, not the size of the observations. Ordinal data are best expressed by the median. Continuous data with extreme values can also be expressed by the median. The positioning

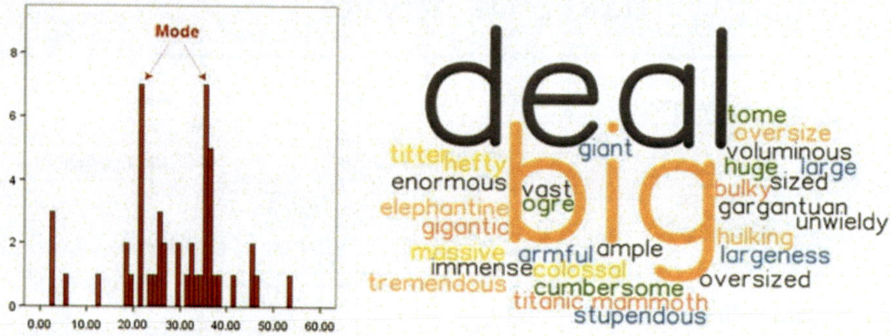

Fig. 3.7 Mode in this particular data set

of data or observations shall be done either in ascending sequence or descending sequence.

- Mode: The number that arises most frequently in a set of numbers. It is often utilized to express nominal data. It is not influenced by extreme values.

The mode is the variable which happens with the largest frequency in the collected data. It can be differently defined that a mode for quantitative data is the observed value, whereas a mode for qualitative data is the name itself (Fig. 3.7).

To completely describe all collected data, we have to report not only a measure of the central tendency of the data but also the measure of dispersion of that data. Measures of dispersion, therefore, tell how much data are spread. Look at the following frequency graph to see why the central tendency of data (mean) is the same, but the dispersion of data (standard deviation) is obviously somehow different (Fig. 3.8).

We shall explain three of the most often utilized standards of variation: range, interquartile range (IQR), and standard deviation.

- Range: The range of the variable is the difference between the lowest and highest values in a data set. That is, Range = Max − Min. The sample range of the variable is quite uncomplicated to calculate and affected by outliers. However, in utilizing the range, much information is disregarded because the other observed values are disregarded (Fig. 3.9).
- Interquartile range (IQR): a measure of statistical dispersion, which is the scatter of the data. The most often utilized percentiles are quartiles. The quartiles of the variable split the observed values into quarters. The three quartiles, Q_1, Q_2, and Q_3, can be defined as follows: the first quartile Q_1 is at location $(n + 1)/4$, the second quartile Q_2 (the median) is at location $(n + 1)/2$, and the third quartile Q_3 is at location $3(n + 1)/4$. Therefore, the interquartile range, denoted as IQR, is the difference between the first and third quartiles of the variable, that is, IQR $= Q_3 – Q_1$. The IQR gives the span of the middle 50% of the observed values (Fig. 3.10).

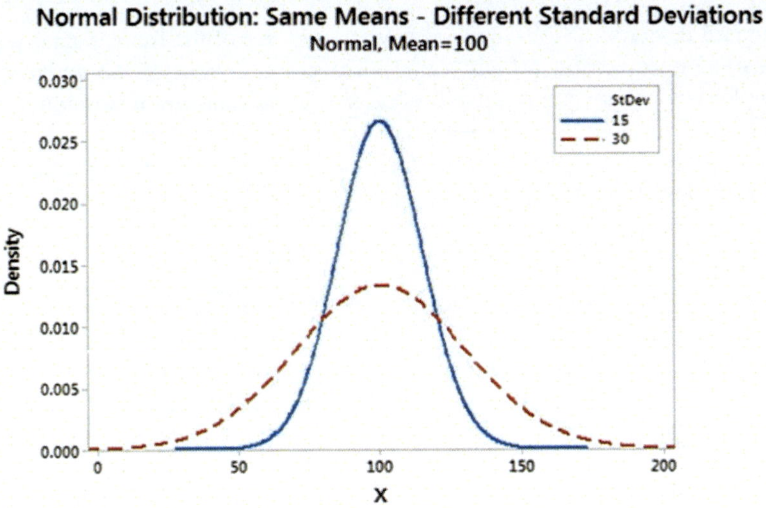

Fig. 3.8 Normal distribution with same central tendency but different variation

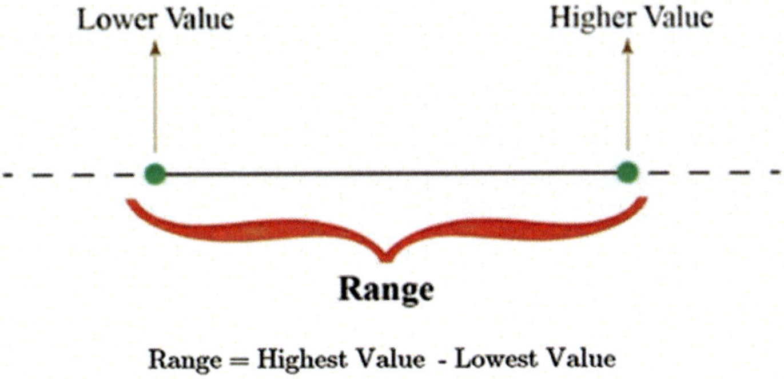

Fig. 3.9 Range concept

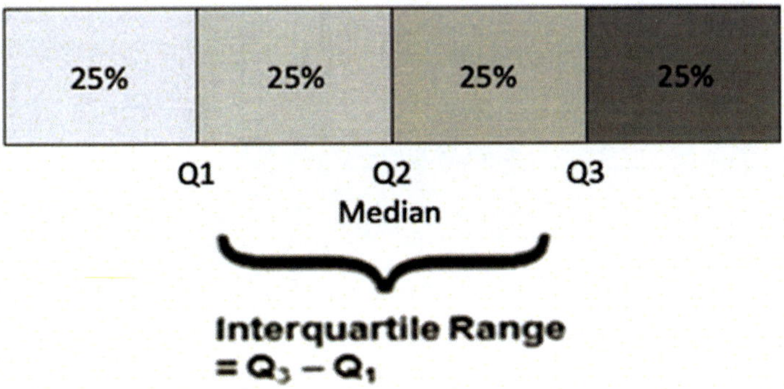

Fig. 3.10 Interquartile range (IQR)

- Standard deviation (SD): Standard deviation is one of the fundamental ways of statistical examination, which is the positive square root of the variance. The sample standard deviation is the most often utilized standard of variability. A unit of standard has to do with variance around an average with continuous data. SD can only be utilized with normally distributed data. For a variable x, the sample standard deviation, designated by s_x, is

$$s_x = \sqrt{\frac{\sum_{i=1}^{n} (x_i - \overline{x})^2}{n-1}} = \sqrt{\frac{\sum_{i=1}^{n} x_i^2 - n\overline{x}^2}{n-1}}. \tag{3.18}$$

Since the standard deviation is expressed as the sample mean $\overline{x}$ of the variable x, it is a favored standard of variation when the mean is utilized as the standard of center. Bear in mind that the standard deviation is always a positive number, i.e., $s_x \geq 0$.

In a formula of the standard deviation, the total of the squared deviations from the mean is called the total of squared deviations and supplies a standard of whole deviation from the mean for all the observed values of the variable. Once the total of squared deviations is split by $n - 1$, we can obtain

$$s_x^2 = \frac{\sum_{i=1}^{n} (x_i - \overline{x})^2}{n-1} \tag{3.19}$$

which is called the sample variance.

Sixty-eight percent of all data fall in 1 SD of a mean, and 95% of all data fall in 2 SD of a mean (Fig. 3.11).

Therefore, if we obtain a mean of 80 and a standard deviation of 10 for heart rate in a population, 68% of all people should have a heart rate between 70 and 90, and 95% of all population should have a heart rate between 60 and 100.

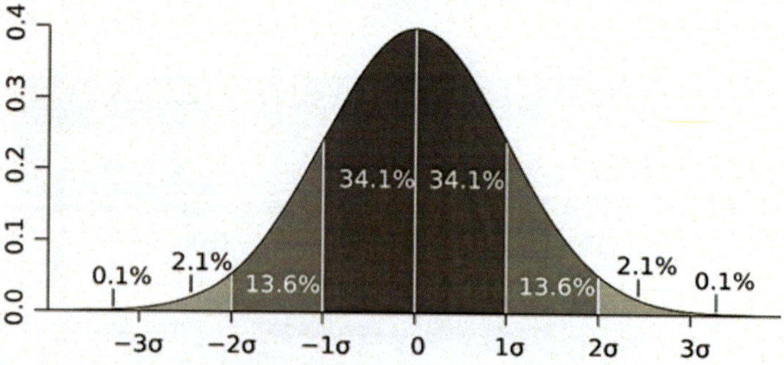

Fig. 3.11 Standard deviation in normal distribution

3.4 Probability Distributions

In real life, we can find that certain probability distributions occur regularly. A probability distribution is a statistical expression which computes the chances of events with dissimilar possible results in a statistical test. Probability distributions are a table or an equation that connects random variables (or outcomes of a statistical experiment) with their chances of events. There are two kinds of probability distributions: continuous and discrete.

A discrete probability distribution shall be applied to the plots where the set of possible results is discrete, for instance, Poisson distribution and binomial distribution. On the other hand, a continuous probability distribution shall be applied to the scenarios where the set of possible results might take values in a continuous span, such as the temperature on a specified day. The normal or Weibull distribution is a usually experienced continuous probability distribution. There are general lifetime distributions that model failure times arising from a wide span of products, such as the Weibull distribution and exponential distribution.

3.4.1 Binomial Distribution

Binomial distribution occurs in daily life. The following are a few general examples: (1) vote counts for two dissimilar applicants in an election, (2) the number of tails/heads in a sequence of coin flips, (3) the number of successful sales calls, (4) the number of female/male employees in a company, and (5) the number of problematic products in a manufacturing line.

There are some presumptions that explain a binomial distribution: (1) n predetermined statistical tests are performed, (2) each trial has one of the two results—a success or a failure (Bernoulli trial), (3) the chance of 'success' p is the same for each result, (4) the results of individual trials are independent, and (5) we are interested in the entire number of successes in these n trials.

Under the above presumptions, assuming that the random variable X is the whole number of successes, the probability distribution of X is defined as the binomial distribution. Probability is defined as follows:

$$\begin{aligned} P(X = x) &= \binom{n}{x} p^x (1-p)^{n-x} \\ &= \frac{n!}{x!(n-x)!} p^x (1-p)^{n-x} \quad \text{for } x = 0, 1, 2, \ldots, n \end{aligned} \tag{3.20}$$

The binomial mean and variance are expressed,

$$\mu = E(X) = np, \quad Var(X) = np(1-p) \tag{3.21}$$

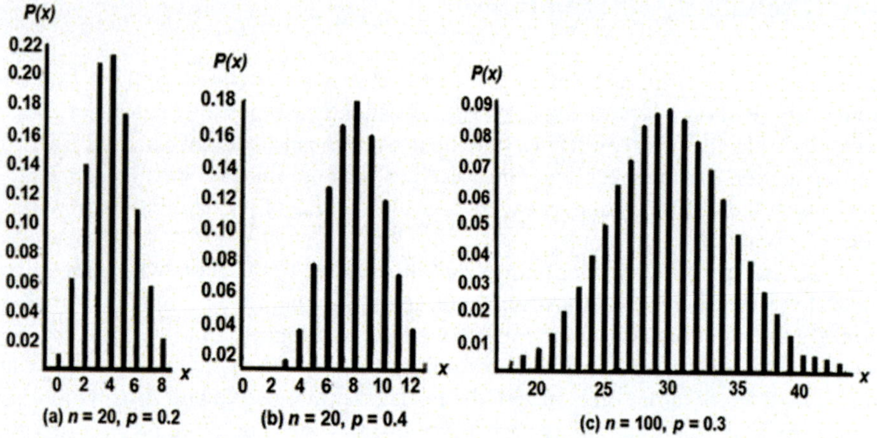

Fig. 3.12 Form of the binomial distribution according to n and p

where the values of n and p are the parameters of the binomial distribution.

As $p = 0.5$, the binomial distribution is uniform—the median and the mean are identical. Even when $p < 0.5$ (or $p > 0.5$), the larger the value of N *is*, the more uniform the form of the distribution. Because the binomial distribution shall be cumbersome, there are estimations of the binomial distribution, which shall be much easier to utilize when N is large (Fig. 3.12).

3.4.2 *Poisson Distribution*

Poisson distribution arises based on how many times one should expect an event to occur in a fixed time period. The Poisson distribution is coined after Simeon Poisson (1781–1840), a French mathematician, and utilized in circumstances where large declines happen in a period with a particular mean rate, regardless of the elapsed time. More precisely, this distribution is utilized when the number of possible events is large, but the event probability over a particular period is small.

The probability distribution comes from a Poisson experiment: (1) the test results shall be categorized as successes or failures, (2) the mean number of successes (μ) which happen in a particular area is recognized, (3) the chance that a success shall happen is proportionate to the magnitude of the area, and (4) the chance that a success shall happen in a very small region is almost zero.

The Poisson distribution occurs in (1) the number of hurricanes hitting Hawaii each year, (2) the hourly number of consumers coming at a bank, (3) the number of battery failures and replacements, (4) the number of hummingbirds seen while going to school each morning, (5) the everyday number of accidents on a specific stretch of highway, (6) the everyday number of emergency calls in a city, (7) the monthly number of employees who had a nonattendance in a big company, (8) the number of typos in a book, and (9) the hourly number of accesses to a specified web server.

Poisson distribution also has the following assumptions: (1) independence: events must be independent, (2) homogeneity: the average number of targets scored is presumed to be the same for all teams, and (3) time period (or space) must be set. If probability p is tiny and trial is far enough, Poisson probability from Eq. (3.11) means that binomial distribution shall be estimated. That is,

$$\begin{aligned} {}_nC_x p^x (1-p)^{n-x} &= \frac{n(n-1)\cdots(n-x+1)}{x!}\left(\frac{m}{n}\right)^x\left(1-\frac{m}{n}\right)^n/\left(1-\frac{m}{n}\right)^x \\ &= \frac{m^x}{x!}\underbrace{1\left(1-\frac{1}{n}\right)\left(1-\frac{2}{n}\right)\cdots\left(1-\frac{x-1}{n}\right)}_{A}\underbrace{\left(1-\frac{m}{n}\right)^n}_{B} \\ &\quad /\underbrace{\left(1-\frac{m}{n}\right)^x}_{C} \end{aligned} \tag{3.22}$$

As n increases, A–C shall be altered.

$$A = 1\left(1-\frac{1}{n}\right)\left(1-\frac{2}{n}\right)\cdots\left(1-\frac{x-1}{n}\right) \underset{n\to\infty}{\to} 1 \tag{3.23a}$$

$$B = \left(1-\frac{m}{n}\right)^n \underset{n\to\infty}{\to} \mathrm{e}^{-m} \tag{3.23b}$$

$$C = \left(1-\frac{m}{n}\right)^x \underset{n\to\infty}{\to} 1 \tag{3.23c}$$

Therefore, we can summarize the results as follows (Fig. 3.13):

$$P(X=x) = \frac{(m)^x \mathrm{e}^{-m}}{x!} \quad \text{where parameter m} = \lambda\text{t} \tag{3.24}$$

3.4.3 Poisson Process

In probability theory, the Poisson process is one of the main random processes. It is popularly utilized to model random 'points' in space and time. Some critical probability distributions normally result from the Poisson process. In other words, the exponential distribution with a stable failure rate in the bathtub curve might be expressed by a Poisson process (Fig. 3.14).

- Time homogeneity: $\mathrm{P}(k, \tau)$ = Probability of k arrivals in interval of duration τ
- The numbers of arrivals in disjoint time intervals are independent

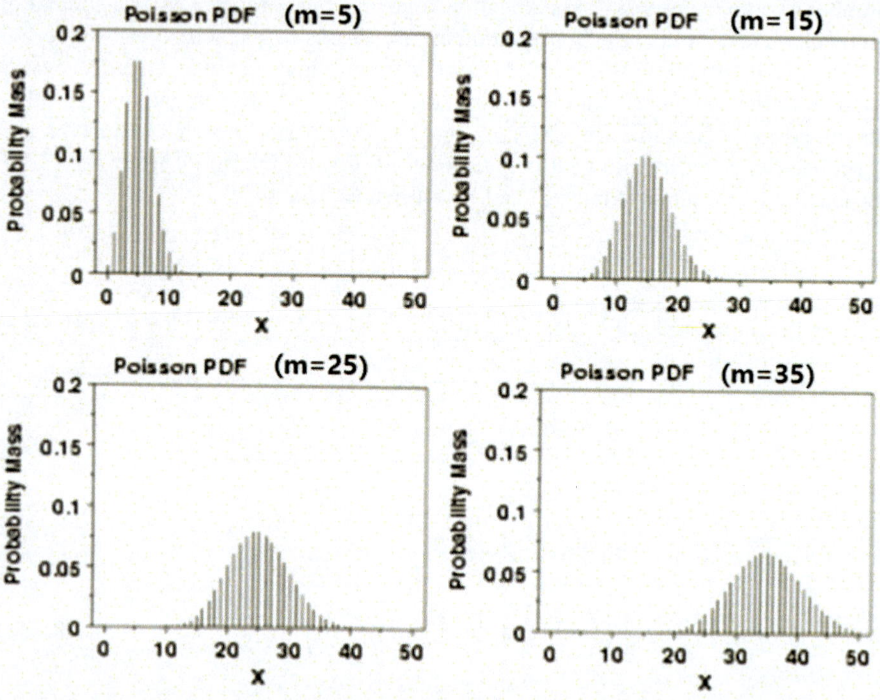

Fig. 3.13 Shape of the Poisson distribution according to m

Fig. 3.14 Definition of the Poisson process

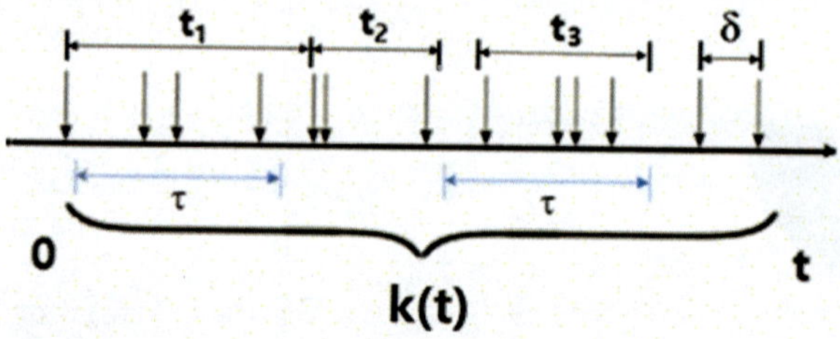

– Small interval probabilities: $P(k, \tau) = \begin{cases} 1 - \lambda\delta & if\ k = 0 \\ \lambda\delta & if\ k = 1 \\ \lambda\delta & if\ k > 1 \end{cases}$ where λ is the arrival rate

In other words, the process is expressed by the so-called counter process $k(t)$. The counter tells the number of failures which have happened in the interval $(0, t)$ or, more usually, in the interval (t_1, t_2).

$N(t)$ = number of failures in the interval $(0, t)$ (the stochastic process).

$N(t_1, t_2)$ = number of arrivals in the interval (t_1, t_2) (the increment process $N(t_2) - N(t_1)$).

The counting process $\{N(t), t \geq 0\}$ is a Poisson process with rate λ if all the following terms hold:

$$\text{i. } N(0) = 0, \tag{3.25a}$$

$$\text{ii. } N(t) \text{ has independent increments,} \tag{3.25b}$$

$$\text{iii. } N(t) - N(s) \sim \text{Poisson}(\lambda(t - s)) \text{ for } s < t. \tag{3.25c}$$

Example 3.1 You email in accordance with a Poisson process at a rate of $\lambda = 5$ messages per hour. You examine your email every thirty minutes. Find the probability of no new message and one new message.

$$\lambda \text{t} = 5 \cdot \frac{1}{3} = 2.5$$

$$\boldsymbol{P(0, 1/2) = e^{-2.5} \frac{(2.5)^0}{0!} = 0.082}$$

$$\boldsymbol{P(1, 1/2) = e^{-2.5} \frac{(2.5)^1}{1!} = 0.205}$$

Example 3.2 The number of typographical mistakes in a 'large' textbook is Poisson distributed with an average of 1.5 per 100 pages. Assume that 100 pages of the book are randomly chosen. What is the probability that there are no typos?

$$P(X = 0) = \text{e}^{-m} \frac{m^x}{x!} = \text{e}^{-1.5} \frac{1.5^0}{0!} = 0.2231$$

Example 3.3 In a certain area, the average number of traffic accidents occurs in one per two days. Discover the chance of $x = 0, 1, 2$ accidents occurring in a specified day.

Solution: One per two days means the mean of transport mishaps, $m = \lambda t = 0.5$,

$$X = 0, \ f(0) = \frac{(0.5)^0 \text{e}^{-0.5}}{0!} = 0.606$$

Accident days = 365 days × 0.606 = 221 days

$$X = 1, \ f(1) = \frac{(0.5)^1 \text{e}^{-0.5}}{1!} = 0.303$$

Accident days = 365 days × 0.303 = 110 days

$$X = 2,\ f(2) = \frac{(0.5)^2 e^{-0.5}}{2!} = 0.076$$

Accident days = 365 days × 0.076 = 27 days.

3.4.4 Exponential Distribution

In reliability engineering, the exponential distribution with only one unspecified parameter is the easiest of all life distribution models. Many engineering modules show a stable failure rate in the product life if they fit the exponential distribution. Additionally, it is comparatively incomplicated to manage in carrying out reliability analysis. The critical equations for the exponential distribution are expressed as follows:

$X(t)$ = the time it takes for one additional arrival, assuming that someone arrived at time t.

By definition, the following conditions are identical:

$$(X(t) > x) \equiv (N(t) = N(t + x)) \tag{3.26}$$

The occurrence on the left-hand side of Eq. (3.26) catches the possibility that no one has arrived in the time interval $[t,\ t + x]$, which indicates that the number of arrivals at time $t + x$ is similar to the number of arrivals at time t, which is the possibility on the right side.

By the complement rule, we also have

$$P(X(t) \leq x) = 1 - P(X(t) > x) \tag{3.27}$$

Utilizing the equivalence of the two events that we expressed above, we shall rewrite the above as follows:

$$P(X(t) \leq x) = 1 - P(N(t + x) - N(t) = 0) \tag{3.28}$$

where $P(N(t + x) - N(t) = 0) = P(N(x) = 0)$.

Utilizing the Poisson distribution Eq. (3.24), where λ is the mean number of arrivals per time unit and x is the quantity of time units, the right side in Eq. (3.28) shall be defined as follows:

$$P(N(t + x) - N(t) = 0) = \frac{(\lambda x)^0}{0!} e^{-\lambda x} \tag{3.29}$$

Exchanging Eq. (3.29) into Eq. (3.27), we have

$$P(X(t) \leq x) = 1 - e^{-\lambda x} \tag{3.30}$$

In other words, let X_1 be the time of the first failure. We can find the reliability function $R(t)$ as follows:

$$R(t) = P(X_1 > t) = P(\text{no failure in } (0, t]) = \frac{(m)^0 e^{-m}}{0!} = e^{-m} = e^{-\lambda t} \quad (3.31)$$

Therefore, the accumulative distribution function as a complement, $F(t)$, is also defined as follows:

$$F(t) = 1 - \mathrm{e}^{-\lambda t} \quad (3.32)$$

If the accumulative distribution function is distinguished, the probability density function is attained as follows:

$$f(t) = \lambda \mathrm{e}^{-\lambda t} \quad t \geq 0,\ \lambda > 0 \quad (3.33)$$

Failure rate $\lambda(t)$ is expressed by

$$\lambda(t) = \frac{f(t)}{R(t)} = \frac{\lambda \mathrm{e}^{-\lambda t}}{\mathrm{e}^{-\lambda t} = \lambda} \quad (3.34)$$

Consider that the failure rate lessens the constant λ for any period. The exponential distribution is the only distribution to have a stable failure rate. Additionally, the other name for the exponential mean is the mean time to fail or MTTF, and we have MTTF $= 1/\lambda$. Generally, if a product suits the exponential distribution, its MTTF is 0.63 at $1/\lambda$ (Fig. 3.15).

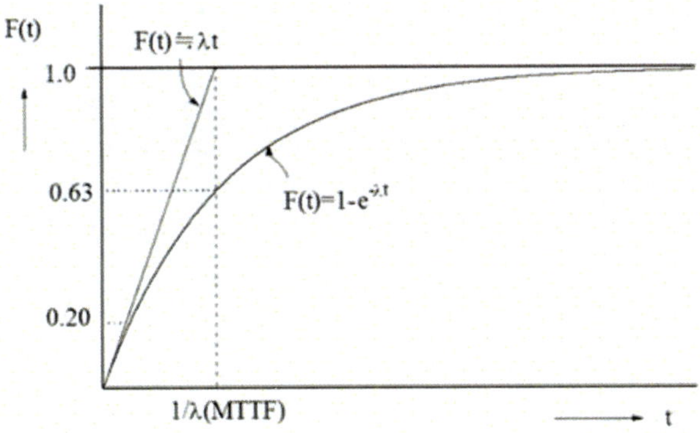

Fig. 3.15 Accumulative distribution function F(t) of the exponential distribution

3.4.5 Normal Distribution

The normal distribution function is expressed as follows:

$$f(x) = \frac{1}{\sqrt{2\pi}\sigma} \exp\left\{-\frac{1}{2}\left(\frac{x-\mu}{\sigma}\right)^2\right\} \quad \text{for } -\infty < x < \infty \tag{3.35}$$

Therefore, the Gaussian distribution curve has two parameters, namely, mean μ and standard deviation σ, which are symmetric around the mean μ and have a bell shape. As a random variable X is normally distributed with mean μ and variance σ^2, it shall be defined as $X \sim N(\mu, \sigma^2)$. Its characteristics may be abridged as follows:

- Continuous for all values of X between $-\infty$ and ∞ and bell shaped.
- Symmetric around the average μ. The chance for the left and the right of the mean is 0.5.
- Relies on parameters μ and σ, and there are boundless normal distributions.
- The probability for interval $[\mu - \sigma \leq X \leq \mu + \sigma]$ is 0.6826. The probability for interval $[\mu - 2\sigma \leq X \leq \mu + 2\sigma]$ is 0. 9544. The probability for interval $[\mu - 3\sigma \leq X \leq \mu + 3\sigma]$ is 0.997. In other words, most of the data in the normal distribution are located around the mean, and there are very few data at more than three times the standard deviation.

In statistics, the normal distribution is the most critical distribution since it results normally in many areas. The major reason is that large sums of (small) random variables usually prove to be normally distributed. As random variable X suits $N(\mu, \sigma^2)$, the chance for interval $[a, b]$ shall be the region of $f(x)$ which is surrounded by a and b on the x-axis. The arithmetic region is expressed as follows:

$$P(a \leq X \leq b) = \int_a^b \frac{1}{\sqrt{2\pi}\sigma} \exp\left\{-\frac{1}{2}\left(\frac{x-\mu}{\sigma}\right)^2\right\} dx \tag{3.36}$$

However, Eq. (3.36) is very hard to determine. Luckily, in the case of a normal random variable X with arbitrary parameters μ and σ, we shall alter it into a standardized normal random variable $Z = \frac{X-\mu}{\sigma}$ with parameters 0 and 1 (Fig. 3.16).

3.4.6 Sample Distributions

The sample mean $\overline{x}$, which is calculated from a large sample, tends to be closer to μ than does $\overline{x}$ based on a small n. Assume that $X_1, \ldots, X_n$ are random variables with the same distribution as the mean μ and population standard deviation σ. Now consider the random variable $\overline{X}$.

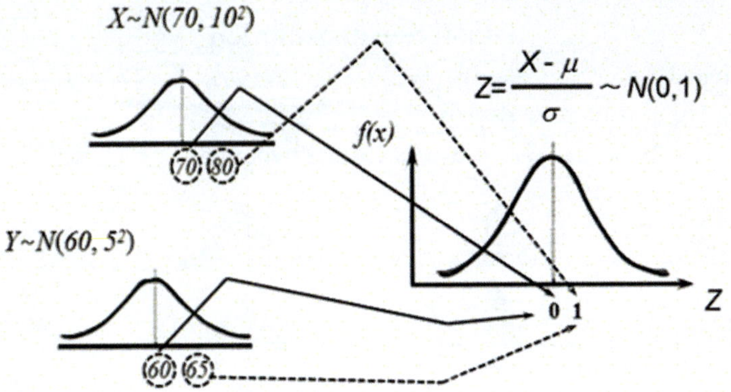

Fig. 3.16 Normal distribution standardization

$$\overline{X} = \frac{1}{n}(X_1 + X_2 + \cdots + X_n) \tag{3.37}$$

$$E(\overline{X}) = \frac{1}{n}(E(X_1) + E(X_2) + \cdots E(X_n)) = \frac{1}{n} \cdot n\mu = \mu \tag{3.38}$$

$$V(\overline{X}) = \frac{1}{n^2}(V(X_1) + V(X_2) + \cdots + V(X_n)) = \frac{1}{n^2} \cdot n\sigma^2 = \frac{\sigma^2}{n} \tag{3.39}$$

The expectation value is $E(\overline{X}) = \mu$, and the variation is $V(\bar{X}) = \sigma^2/n$. That is,

1. The population mean of $\overline{X}$, indicated as $\mu_{\overline{X}}$, is identical to μ.
2. The population standard deviation of $\overline{X}$, indicated as $\sigma_{\overline{X}}$, is equal to $\sigma_{\overline{X}} = \frac{\sigma}{\sqrt{n}}$.

This signifies that the sampling distribution of $\overline{x}$ is pivoted at μ. As n increases, the second statement means that the spread of the sampling distribution (sampling variability) will decrease. The standard deviation of a statistic is specified as the standard error of the statistic. The standard error takes the exactness of the statistic for approximating a population parameter. The smaller the standard error is, the higher the precision. The standard error of the mean $\overline{X}$ is $SE(\overline{X}) = \sigma / \sqrt{n}$.

We discovered about the average and the standard deviation of the sampling distribution for the sample mean; we might ask, if there is anything, we shall tell about the shape of the density curve of this distribution.

If the population is boundless and sample size n is large enough, we are aware that the distribution of the sample mean is roughly normally distributed regardless of the population characteristics. Under rather common circumstances, the central limit theorem states that the means of random samples drawn from one population tend to have a roughly normal distribution. We discover that it is not important which sort of distribution we discover in the population. It can even be discrete or very skewed (Fig. 3.17).

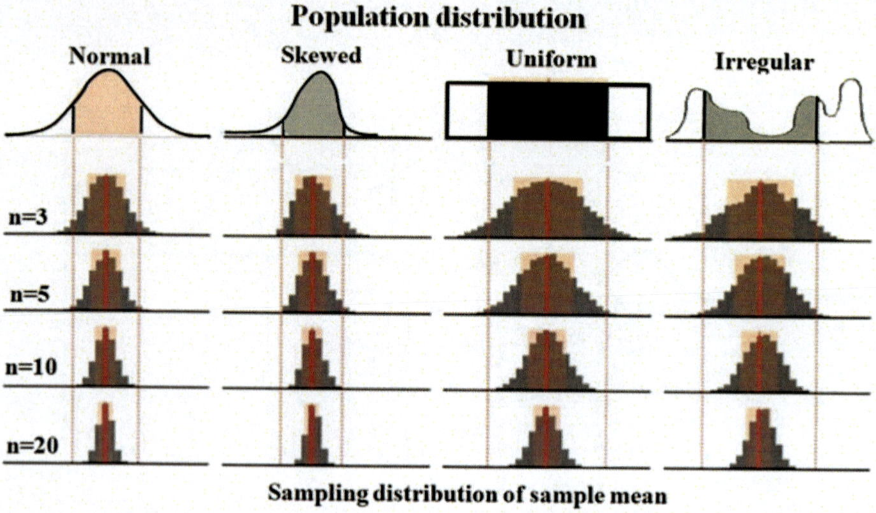

Fig. 3.17 Central limit theorem

3.4.7 Central Limit Theorem

From any population with finite mean μ and standard deviation σ, as n is big, if random samples of n observations are selected, the sampling distribution of the mean $\overline{X}$ is roughly normally distributed, with mean μ and standard deviation $\sigma/\sqrt{n}$. That is,

$$\overline{X} \sim N\left(\mu, \frac{\sigma^2}{n}\right) \tag{3.40}$$

The central limit theorem has become a considerable one in modern statistics. Glance at a binomially distributed random variable X. With probability p (Bernoulli trial), we shall perform the experiment of n trials. According to this theorem, as n increases immensely, the random variable X shall be distributed normally.

We can apply a binomial distribution. That is, fix p where $0 < p < 1$. The random variable $S_n = X_1 + X_2 + \cdots + X_n$ is binomially distributed $B(n, p)$ with mean pn and standard deviation $\sqrt{p(1-p)}$. As n increases immensely, we know that the random variable S_n is normally distributed. That is,

$$\frac{S_n - np}{\sqrt{np(1-p)}} \sim N(0, 1) \tag{3.41}$$

Additionally, Poisson arrivals during unit intervals are equivalent to the sum of n (independent) Poisson arrivals during n intervals during n intervals of length $1/n$.

For Binomial (n, p), p fixed, $n \to \infty$: normal. On the other hand, np fixed, $n \to \infty$, $p \to 0$: Poisson. For example, $p = 1/100$, $n = 100$: Poisson. $p = 1/10$, $n = 500$: normal.

Example 3.4

When n = 36, p = 0.5, find $\mathrm{P}(S_n \le 21)$.

Solution: $\mathrm{E}[S_n] = np = 18$, $\sigma_{S_n}^2 = np(1-p) = 9$. $\sigma_{S_n} = 3$.

Therefore, $S_n \le 21 \quad Z = \frac{S_n - 18}{3} \le \frac{21-18}{3} = 1$

$$\mathrm{P}(S_n \le 21) = P(Z \le 1) = 0.843$$

Because S_n is an integer, $\mathrm{P}(S_n \le 21) = \mathrm{P}(S_n < 22)$.

Consider $\mathrm{P}(S_n \le 21.5) = P\left(Z = \frac{S_n - 18}{3} \le \frac{21.5-18}{3} = 1.17\right) = 0.879$.

Exact answer is

$$\sum_{k=0}^{21} \binom{36}{k} \left(\frac{1}{2}\right)^{36} = 0.8785$$

3.5 Weibull Distributions and Reliability Testing

If a random variable X is characterized as the individuals of a population, the distribution function of X, designated $F(x)$, can be expressed as the number of all individuals possessing an $X \le x$, divided by the entire number of individuals. This function also takes the probability P of selecting at random an individual possessing a value of X is identical to or less than x, and therefore, we have

$$P(X \le x) = F(x) \tag{3.42}$$

Any distribution function shall be defined as follows:

$$F(x) = 1 - e^{-\phi(x)} \tag{3.43}$$

Although this seems to be a complication, this formal change relies on the relation

$$(1 - P)^n = e^{-n\phi(x)} \tag{3.44}$$

The advantages of Eq. (3.44) can be explained by a product failure. Presume that we have a product composing some parts. If we have found, by testing, the chance of failure P at any load x applied to a 'single' part (or link), and if we want to find the chance of failure P_n of a product (or chain) composing n parts (or links), we have to base our subtraction upon the proposition that the whole chain fails if any one of its parts fails. Therefore, the probability of failure of the chain $(1 - P_n)$ is identical to the probability of the simultaneous failure of all the links. Thus, we have $(1 - P_n) = (1 - P)^n$.

If the distribution function of a single link has, we obtain

$$P_n = 1 - e^{-n\phi(x)} \tag{3.45}$$

Equation (3.45) takes the proper arithmetic formulation for the principle of the weakest chain, or, more usually, for the size effect on failures in solids. The same way of reasoning may be applicable to a large group of problems, where the occurrence of an event in any part of an object shall be said to be happened in the object as a whole, e.g., static or dynamic strengths, electrical insulation breakdowns, the phenomena of yield limits, life of electric bulbs, or even death of a man, as the chance of not having died from many different causes.

Now we have to state the function $\phi(x)$. The only requisite common condition this function has to assure is to be a positive, nondecreasing function. The simplest function fulfilling this condition is

$$F(x) = \int_0^\infty f(x)dx = 1 - e^{-\left(\frac{x}{\eta}\right)^\beta} \tag{3.46}$$

where η and β are characteristic life and shape parameters, separately.

This is mostly due to the product's weakest chain premises, but the other reasons for its growing failure rate are, for example, the age of its parts and variety of distribution shapes. The growing failure rate explains to some extent the fatigue failures. The density function depends on the shape parameter β. For low β values ($\beta < 1$), the failure behavior shall be alike to the exponential distribution. For $\beta > 1$, the density function starts at $f(t) = 0$, increases to a maximum as the lifetime passes, and decreases slowly again. A two-parameter fit is more common in reliability testing and more well-organized with the same sample size. If the shape parameter is presumed to be well known, it shall lessen the fit to one parameter (Fig. 3.18).

By using the maximum likelihood, we shall approximate the characteristic life as follows:

$$\hat{\eta} = \left[\sum_{i=1}^{N}\left(\frac{t_i^\beta}{r}\right)^{\frac{1}{\beta}}\right] \tag{3.47}$$

where t is the testing time, r is the number of failed samples, and N is the total number of failures.

There are two generally utilized general ways which might approximate life distribution parameters from a particular data set: (1) graphical estimation in the Weibull plotting and (2) median rank regression (MRR) and maximum likelihood estimation (MLE). The Weibull plotting is a graphical way for unofficially examining the

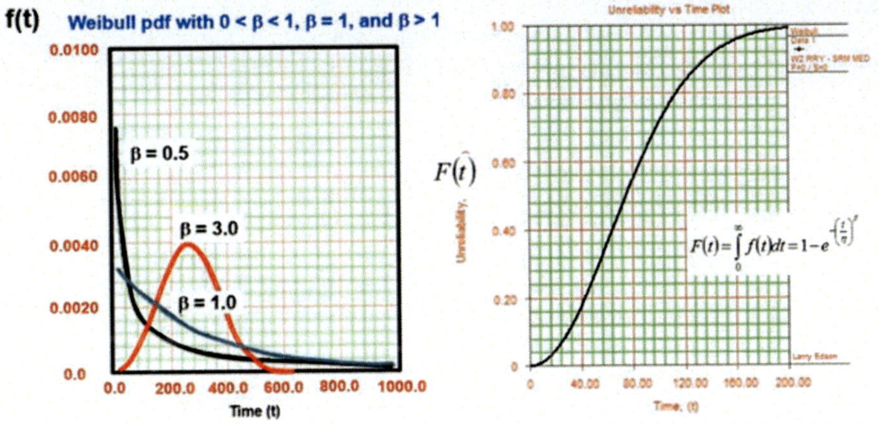

(a) Probability density function (b) Cumulative distribution function

Fig. 3.18 Probability density and cumulative distribution function on the Weibull distributions

presumption of the Weibull distribution prototype and for approximating the two Weibull parameters—characteristic life and shape parameter.

For a Weibull probability chart, sketch a parallel line from the y-axis to the fitted line at the 62.3 percentile point. That estimate line bisects the line through the points at a time which is the approximate of the characteristic life parameter η. To approximate the gradient of the fitted line (or the shape parameter β), select any two points on the fitted line and split the change in the y variable by the change in the x variable.

The median ranks way is utilized to obtain an approximation of the unreliability for each failure.

First, we analyze the fitting of the two-parameter Weibull model utilizing the MRR method. MRR decides the best-fit straight line by least squares regression curve fitting. This way begins as follows:

Classify the times-to-failure in accordance with increasing order $t_1 < t_2 \ldots < t_n$

i	1	2	3	…	$r-1$	r
t_i	t_1	t_2	t_3		t_{r-1}	t_r

The cumulative distribution function $F(t)$ has an S-like curve (Fig. 3.10a). If the function $F(t)$ is marked in a Weibull probability paper, it is practical to assess the life of a mechanical system in reliability testing.

As putting an inverse number and logarithmic transformation from reliability Eq. (3.36), it might be defined as follows:

$$\ln(1 - F(t))^{-1} = \left(\frac{t}{\eta}\right)^{\beta} \tag{3.48}$$

After one more time taking logarithmic transformation, it might be expressed as follows:

$$\ln\left(\ln\frac{1}{1-F(t)}\right) = \beta \cdot \ln\ t - \beta\ \ln\ \eta \tag{3.49}$$

If F is vert small, Eq. (3.49) may be restated as follows:

$$\ln\left(\ln\frac{1}{1-F(t)}\right) \cong \ln\ F(t) = \beta \cdot \ln\ t - \beta\ \ln\ \eta \tag{3.50}$$

Equation (3.50) is equivalent to a linear equation. That is,

$$y = ax + b \tag{3.51}$$

with the variables

$$a = \beta \quad \text{(slope)} \tag{3.52a}$$

$$b = -\beta\ \ln\ \eta \quad \text{(axis intersection)} \tag{3.52b}$$

$$x = \ln\ t \quad \text{(abscissa scaling)} \tag{3.52c}$$

$$y = \ln(-\ln(1 - F(t))) \quad \text{(ordinate scaling)} \tag{3.52d}$$

Two-parametric Weibull distribution might be stated as a straight line on the Weibull probability paper. The gradient of its straight line is the shape parameter β (Fig. 3.19).

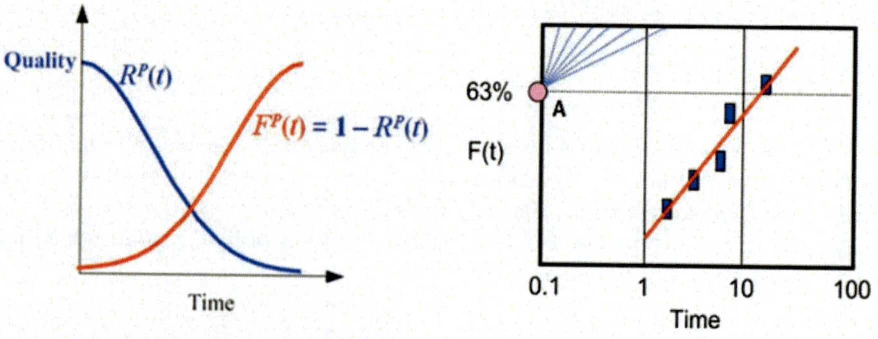

Fig. 3.19 A plot of Weibull probability paper

Compute median ranks: Classify failure times in an increasing order. Mean ranks are less correct for the skewed Weibull distribution; thus, median ranks are better. Median ranks might be computed as follows:

$$\sum_{k=i}^{N} \binom{N}{k} (MR)^k (1 - MR)^{N-k} = 0.5 = 50\% \tag{3.53}$$

Bernard utilized an estimation of it as follows:

$$F(t_i) \approx \frac{i - 0.3}{n + 0.4} \times 100 \tag{3.54}$$

where i is the failure-order number and N is the total sample size.

Engineers determine that reliability test criteria may be stated in units of time (Type I censoring) or failure number (Type II censoring). In Type I censoring, the items shall be tested for a certain amount of time, whereas in Type II censoring, the items shall be tried out until a certain number of failures. Because it is common in time-terminated tests to have many surviving units, engineers prefer to utilize the failure number (Type II censoring) (Fig. 3.20).

To forecast the product life, MLE is a favored way based on the observed data of reliability testing. MLE is a statistical way for approximating the parameters of a statistical prototype—some unspecified mean and variance that are specified to a data set. It chooses a set of values of the model parameters which make the likelihood function as large as possible.

For a fixed set of data and a fundamental statistical prototype, the maximum likelihood method approximates a set of values of the model parameters which maximize the likelihood function. Assume that there is a sample t_1, t_2, …, t_n of n independent and identically distributed failure times originating from a distribution with an unknown probability density function $f_0(\cdot)$. On the other hand, suppose that the function f_0 belongs to a certain family of distributions $\{f(\cdot| \theta), \theta \in \Theta\}$ (where θ

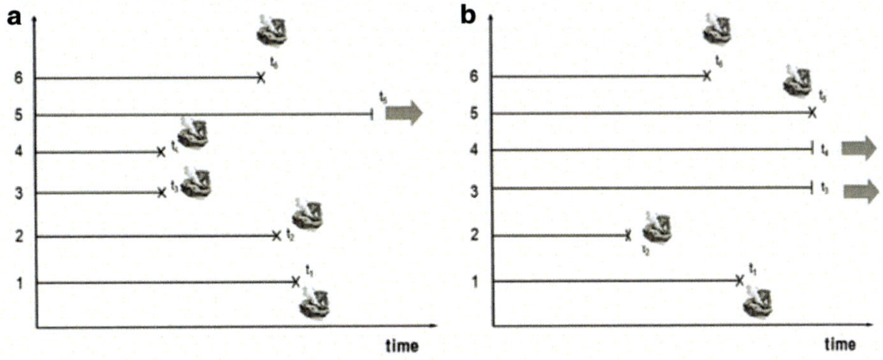

Fig. 3.20 Simplified illustration of type I (**a**) and type II censoring (**b**)

is a vector of parameters), specified as the parametric model, so that $f_0 = f(\cdot | \theta_0)$. The unknown value θ_0 is anticipated to be a true value of the parameter vector. An estimator $\hat{\theta}$ might be very close to the true value θ_0. The observed variables t_i and the parameter θ are vector parts:

$$L(\Theta; t_1, t_2, \ldots, t_n) = f(t_1, t_2, \ldots, t_n | \Theta) = \prod_{i=1}^{n} f(t_i | \Theta) \tag{3.55}$$

This function is defined as the likelihood. The plan of this procedure is to discover a function f for which product L is maximized. Here, the function should have high values of the density function f in the corresponding region with some failure times t_i. Simultaneously, only low values of f in regions with few failures are found. Therefore, the real failure behavior is correctly depicted. If decided in this manner, function f offers the best chance to explain the test samples.

It is frequently more successful to utilize the log-likelihood function. Therefore, the product equation becomes an addition equation, which greatly clarifies the differentiation. Since the natural log is a monotonic function, this footstep is mathematically logical:

$$\ln L(\Theta; t_1, t_2, \ldots, t_n) = \sum_{i=1}^{n} \ln f(t_i | \Theta) \tag{3.56}$$

Differentiating Eq. (3.56), the maximum of the log-likelihood function and therefore the statistically optimal parameters Θ_l may be found as follows:

$$\frac{\partial \ln(L)}{\partial \theta_l} = \sum_{i=1}^{n} \frac{1}{f(t_i; \Theta)} \cdot \frac{\partial f(t_i; \Theta)}{\partial \theta_l} = 0 \tag{3.57}$$

Because these equations can be nonlinear, it is usually useful to apply the parameter estimates in the Weibull distribution by numerical procedures such as the Newtion-Raphson method. By the likelihood function value, the chance is given to approximate the quality of the adaptation of a distribution to the failure data. The greater the likelihood function value is, the better the conclusive distribution function, which represents the actual failure behavior. However, based on MLE, the characteristic life η_{MLE} from the reliability test may be approximated on the Weibull plot.

3.6 Reliability and Bathtub Curve

3.6.1 *Product Reliability*

Let the survival time, T, for individuals in a population have a density function $f(t)$. The equivalent distribution function is the fraction of the population being unsuccessful by time t. That is,

$$F(t) = \int_{-\infty}^{t} f(s)ds \tag{3.58}$$

The complementary function $1 - F(t)$, often defined as the reliability function, is the fraction still living at time t. The failure rate function $\lambda(t)$ calculates the instantaneous risk, in which $\lambda(t)\delta t$ is the chance of failing in the following small interval δt specified survival to time t. From the conclusion,

$$\text{pr(survival to } t + dt) = \text{pr(survival to } t) \text{ pr(survival to } \delta t|\text{survival to } t) \tag{3.59}$$

where pr is the probability.

We have the following equations:

$$1 - F(t + \delta t) = \{1 - F(t)\}\{1 - \lambda(t)\delta(t)\} \tag{3.60}$$

$$\delta t F'(t) = \{1 - F(t)\}\lambda(t)\delta(t) \tag{3.61}$$

so that the failure rate function is defined as:

$$\lambda(t) = \frac{\frac{\delta(t)F'(t)}{\{1-F(t)\}}\delta(t)(= f(t))}{\{1 - F(t)\}} = \frac{f(t)}{\{1 - F(t)\}} \tag{3.62}$$

A distribution for survival times should have a failure rate function with proper properties. Therefore, for a large t, the failure rate function will not lessen because at a definite point in time, the probability of breakdown does not normally lessen. For a small t, various forms may be justified, including the one that initially declines with t, for such distribution might express the behavior of a mechanical part with a settling-in period, where reliability grows once the initial period is over.

The straightforward failure rate function involves an exponential distribution of survival times with a Poisson process. For if T has the density

$$f(t) = \lambda e^{-\lambda t}; \quad t \geq 0, \tag{3.63}$$

then

$$F(t) = 1 - e^{-\lambda t}, \tag{3.64}$$

and from Eq. (3.52a–d), we can find the hazard rate function:

$$\lambda(t) = h(t) = \frac{f(t)}{\{1 - F(t)\}} = \frac{\lambda e^{-\lambda t}}{e^{-\lambda t}} = \lambda \tag{3.65}$$

3.6.2 *Bathtub Curve*

To reach defensible conclusions from an experiment's outcomes, mechanical engineers would analyze data after a new product is tested under customer usage (or environmental conditions) for its lifetime. To design a product, an engineer must determine the meaning and interpretation of numerical and diagrammatic statements requiring data. At that time, we rely on the fundamental tabular and diagrammatic techniques for displaying the central trends and spread of data. We can find another trend, not a typical histogram, that is expressed as data and its number. Many product failures occur initially. As time passes, the number of failures will decrease (Table 3.1). We know that the curve is a right-skewed shape. The basic question is how to effectively describe its trends. The final answer is the bathtub curve. That is, if the failure rate is represented in time, we can find another trend of the product life—the bathtub curve.

On the bathtub curve, we can graphically represent the product life. That is, it comprises three regions: (1) an early mortality time with a lessening failure rate, (2) followed by a normal life time (also known as 'useful life') with a low and stable failure rate, and (3) concluding with a wear-out period which shows a growing failure rate. Bathtub therefore supplies an outline of how early mortality, normal life failures, and wear-out modes integrate to produce the general product failure distributions (Fig. 3.21).

Table 3.1 Quality defects and failures

Time	Number of failure	$\lambda(t)$
1	5	5/100
2	10	10/95
3	35	35/85
4	30	30/50
5	15	15/20
6	2	2/5
7	2	2/3
8	1	1/1
Total	100	

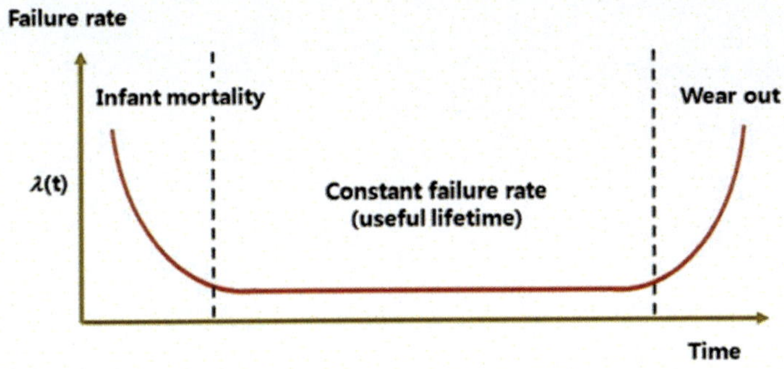

Fig. 3.21 Bathtub curve

When a product is initially designed and manufactured, the life features of the problematic product, such as a vacuum tube, can be described as the bathtub curve because it has a high failure rate and short lifetime. It was produced by functioning the rate of initial "infant mortality" failures ($\beta < 1$), the rate of random failures with a stable failure rate during its "useful life" ($\beta = 1$), and eventually the rate of "wear-out" failures ($\beta > 1$), which pursue the Weibull distribution.

Early potential origins of initial failure, such as handling, storage, and installation error, are dominant in bathtub curves. The aging test may immediately eliminate them. In the middle of bathtubs, the failure rate of a product is stable because it follows an exponential distribution. In this period, the product can endure random failures due to, for example, overuse by customers and exposure to overstress. The product will experience wear failure due to long usage in the latter part of bathtub. However, if any design problems exist in the system, the failure rate of the product might increase suddenly and catastrophically in useful life. Therefore, this area will be analyzed by the Weibull distribution.

If a product pursues the bathtub curve, it shall face difficulties in being successful in the field. Because of the higher failure rates and short lifetime due to inherited design faults, companies may experience financial losses throughout the total product life cycle. Therefore, they should emphatically enhance the design of a product by planning reliability targets for new products to (1) reduce initial failures, (2) lessen random failures for the product operating period, and (3) grow product life.

3.6.3 *Cumulative Distribution Function F(t)*

Reliability is the chance that a mechanical system shall satisfactorily work for a design lifetime under the operating/environmental circumstances. As T is the random variable indicating the time to failure, the reliability at time t might be defined as

$$R(t) = P(T > t) \tag{3.66}$$

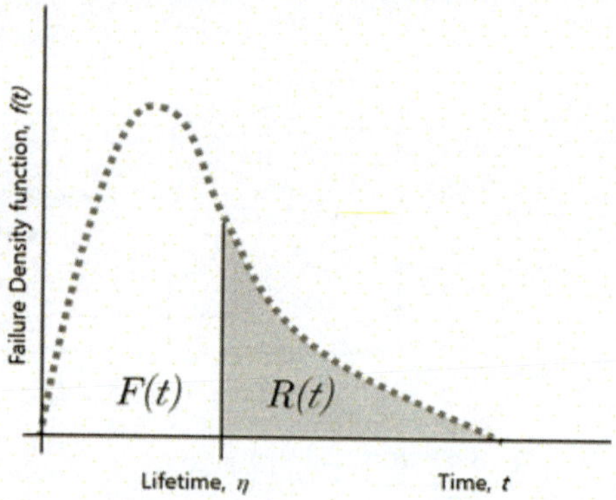

Fig. 3.22 Reliability function $R(t)$ and cumulative distribution function $F(t)$

The failure rate in a time interval $[t_1, t_2]$ might be expressed as the chance that a failure rate unit time happens in the interval specified that no failure has happened earlier to t_1, the starting of the interval (Fig. 3.22).

Therefore, the failure rate is defined as follows:

$$\frac{R(t_1) - R(t_2)}{(t_2 - t_1)R(t_1)} \tag{3.67}$$

If we take the place of t_1 by t and t_2 by $t + \Delta t$, we might redefine Eq. (3.67) as follows:

$$\frac{R(t) - R(t + \Delta t)}{\Delta t R(t)} \tag{3.68}$$

As Δt approaches zero, the instant failure rate may be defined as follows:

$$\lambda(t) = \lim_{\Delta t \to 0} \frac{R(t) - R(t + \Delta t)}{\Delta t R(t)} = \frac{1}{R(t)}\left[-\frac{d}{dt}R(t)\right] = \frac{f(t)}{R(t)} \tag{3.69}$$

A hazard function is used to examine the anticipated duration of time till one or more events occur, such as failure in mechanical systems. The accumulative hazard rate function $\Lambda(t)$ is expressed as follows:

$$\Lambda(t) = \int_0^t \lambda(x)dx \tag{3.70}$$

Assume that the failure rate $\lambda(t)$ is recognized. Thus, $f(t)$, $F(t)$, and $R(t)$ can be obtained:

$$f(t) = \frac{dF(t)}{dt} = -\frac{dR(t)}{dt} \Rightarrow \lambda(t) = -\frac{\frac{dR}{dt}}{R} \tag{3.71}$$

If Eq. (3.71) is integrated, the reliability function can be defined as follows:

$$R(t) = \exp\left[-\int_0^t \lambda(\tau)d\tau\right] \tag{3.72}$$

Therefore, the cumulative distribution function and density function are redefined as follows:

$$f(t) = \lambda(t)\exp\left[-\int_0^t \lambda(\tau)d\tau\right] \tag{3.73}$$

$$F(t) = 1 - \exp\left[-\int_0^t \lambda(\tau)d\tau\right] \tag{3.74}$$

The relation between the accumulative distribution function $F(t)$ and reliability function $R(t)$ in reliability engineering can be abridged, as shown in Fig. 3.23.

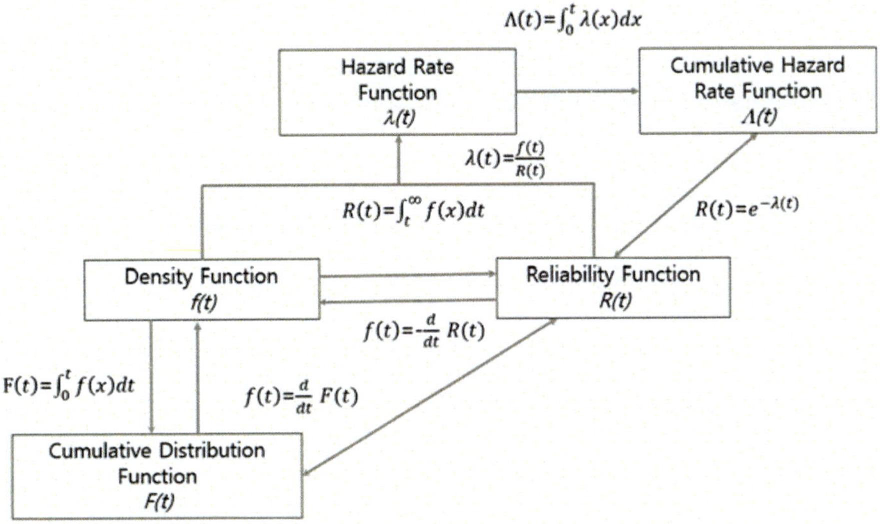

Fig. 3.23 Relation between the reliability function R(t) and accumulative distribution function F(t) in reliability engineering

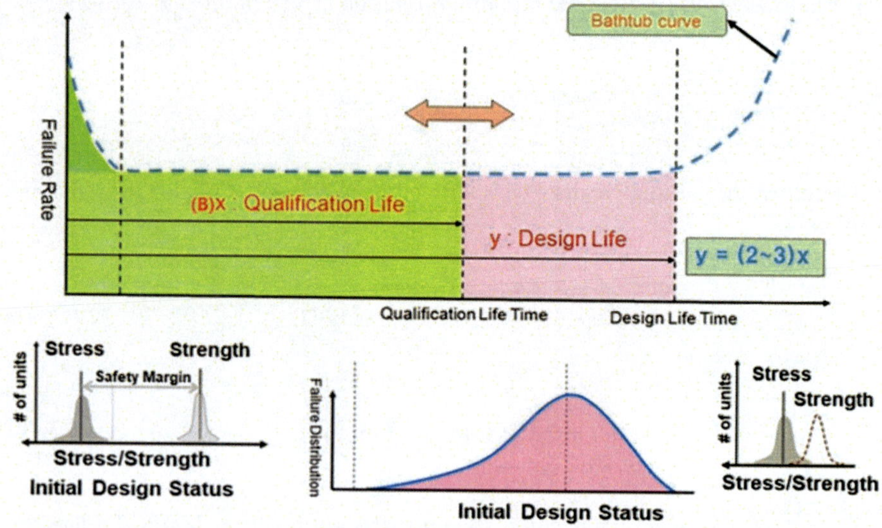

Fig. 3.24 BX life (qualification life) and design life

3.7 Lifetime Metrics for Design

When a customer utilizes a product, failure is revealed in the weakest part of the product due to a design fault. As time passes, the trend of product reliability shall be defined as the failure rate in the bathtub curve. For useful lifetimes, the product pursues an exponential distribution with a constant failure rate. Engineers usually explain a design's lifetime to represent an engineer's statements of product usage under which the reliability shall be proven. Generally, the design's lifetime (or useful life) is 2–3 times longer than the BX life (qualification life) (Fig. 3.24).

3.7.1 *Mean Time to Failure (MTTF)*

MTTF is the span of time that a product is anticipated to end in operation. MTTF is a fundamental lifetime standard of reliability to state the life of nonrepairable products such as "one-shot" tools such as light bulbs. A nonrepairable system is one for which single items unsuccessful are eliminated forever from the population. It is the average time till a piece of equipment is unsuccessful at the beginning statistically. MTTF is the average over a long period of time with a big unit (Fig. 3.25):

$$\mathrm{MTTF} = \frac{t_1 + t_2 + \cdots + t_n}{n} \tag{3.75}$$

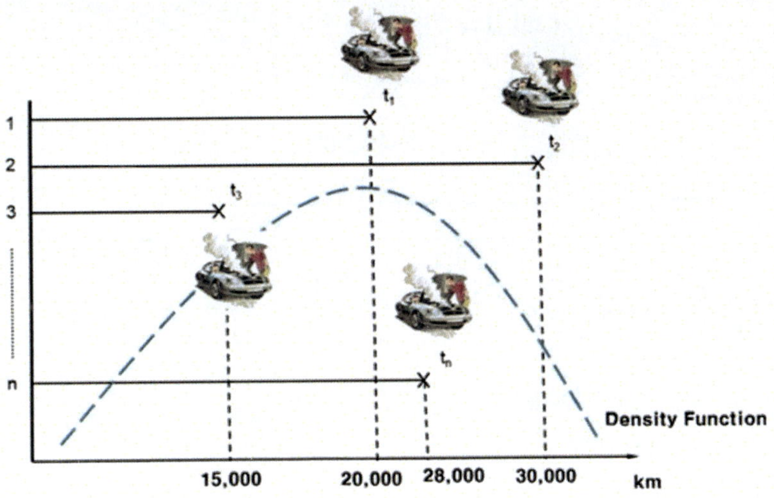

Fig. 3.25 Idea of mean time to failure

3.7.2 *Mean Time Between Failures (MTBF)*

MTBF is the forecasted passed period between the failures of a mechanical system for normal system operation. MTBF may be calculated as the mean time between failures of a system. MTBF is a reliability standard utilized to report the average life of repairable parts such as airplanes, automobiles, construction machines, and refrigerators. A repairable system is the one that may be replaced with adequate operation by any steps, including component replacements or changes to adaptable settings.

MTBF is a fundamental standard of system reliability for most systems, although it is altered. MTBF is more critical for integrators and industries than for customers (Fig. 3.26).

$$\mathrm{MTBF} = \frac{T}{n} \tag{3.76}$$

The MTBF value corresponds to the anticipated number of working hours (service life) before a product is unsuccessful. There are some variables which may affect failures. Aside from part failures, consumer utilization/installation may also result in a failure. MTBF is often computed based on a calculation process in which factors in all of a product's parts are summed to find the product's life cycle in hours. MTBF is regarded as a system failure. It is still considered a useful tool when thinking about the purchase and installation of a product. For repairable complex systems, failures are regarded as those out of design conditions that put the system out of

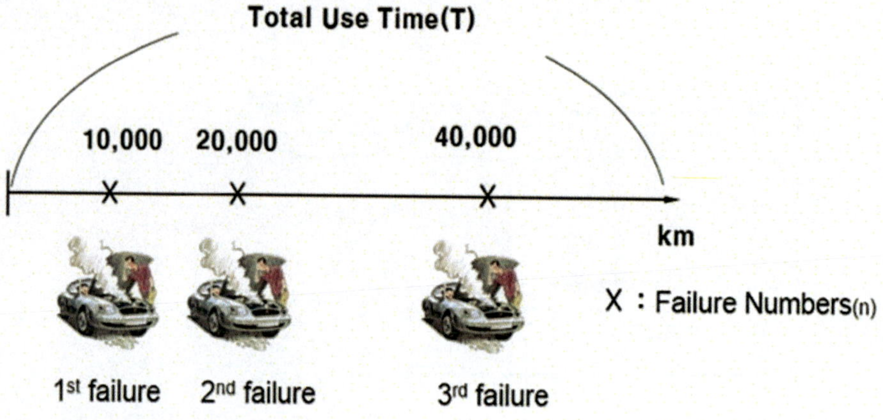

Fig. 3.26 Concept of MTBF

service and into a state for repair. Strictly, MTBF is utilized only in reference to a repairable component, while MTTF is utilized for nonrepairable parts such as electric components.

3.7.3 BX *Life*

A proper measure of product life, such as *BX* life, must be selected for performing parametric ALT. BX life is defined as the period at which *X% of the samples in a population will be unsuccessful.* For example, if the lifetime of a system has a B20 life of 10 years, then 20% of the population will be unsuccessful for a 10-year operation period. 'BX life Y years' can satisfactorily determine the accumulative failure rate of a system and its life with respect to field requirements. The MTTF as the inverse of the failure rate cannot be used for the lifetime because it equates to a B60 life that is too large. BX life indicates a more adequate estimate of lifetime compared to MTTF.

The design life of products, BX, shall disagree with the life metrics—MTTF or MTBF. The use of the BX life metric started from the ball and roller bearing manufacturing, but it has grown today as a product lifetime standard utilized across some industries. It is especially functional in defining warranty time for a system. The BX % life or "Bearing Life" is the life standard that fails *X%* of the samples in a population. For instance, as a sample has a B10 life of 1000 km, 10% of the population shall fail by 1000 km of operation.

On the other hand, the B10% lifetime has the 90% reliability of a population at a particular end in product life. The B10 life metric became favorable among product industries due to the manufacture's severe need. Now, the B1, B10, and B50 lifetime

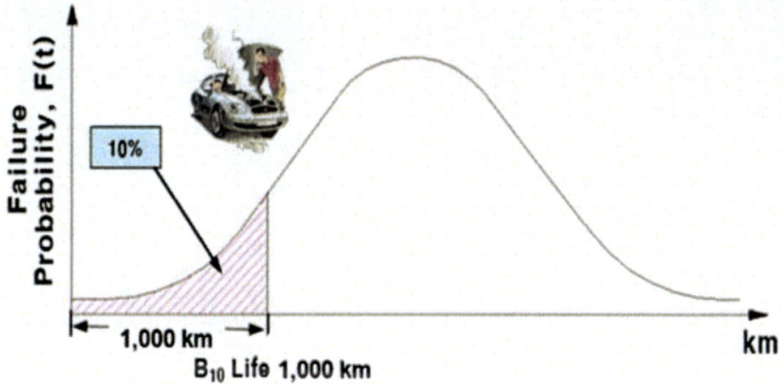

Fig. 3.27 Concept of BX % life

values serve as measurements for the reliability of a product. For instance, if the MTTF of a product is 100,000 km, then its B10 life shall be 10,000 km (Fig. 3.27).

The particular design life supplies a usage or time frame for reliability tests, such as parametric ALT. Some organizations might straightforwardly select to design a product to be reliable over a stated warranty (qualification) time. Because it is equal with how long the product is anticipated to be utilized in the marketplace, an improved organization should select a design lifetime. Relying on the designer's view, the lifetime statement should be based on any of the design lifetimes—BX life.

Chapter 4
Design of Mechanical Structure Including Mechanisms

4.1 Introduction

The mechanical systems convert (generated) power to fulfil its purpose that requires forces (effort) and velocity (flow), which provides mechanical advantages by adapting product mechanisms. A product mechanism is a kind of kinematic chain that makes motions such as rotary motion, oscillation, and linear motion. There are several methods of power transmission as follows: (1) gear trains (Ex: automobiles, engines, etc.), (2) chain drive (Ex: bicycle, motor cycle, etc.), (3) rope drive (Ex: lift, crane, etc.), (4) belt drive (Ex: Rice mills, sewing machine, etc.). For example, a car is also a wheeled motor vehicle utilized for transit. It comprises engine, transmission, drive, electric, and body systems [1]. Under the Carnot cycle, an engine goes through four stages: (1) induction (fuel enters), (2) compression, (3) ignition, and (4) emission (exhaust out). The engine is designed to convert fuel into mechanical energy through the slider-crank mechanism, transmission including gear system, rear drive shaft, rear differential, and vehicle axle. The automobile obtains the power to move forward through transmission and drive systems (Fig. 4.1).

In power transmission, a product including several modules will be subjected to a variety of stresses due to repeated loads. If there is a design defect in a product that brings an inadequacy in the product's stiffness (or strength) when subjected to loads, it may cause a structural failure of parts in a mechanical system, i.e., the (low/high) cycle fatigue that is characterized by repeated plastic deformation. While designing a mechanical system, an engineer should recognize the mechanism and the structural loading in the mechanical system.

S. Woo, *Design of Mechanical Systems*, Springer Series in Reliability Engineering,
https://doi.org/10.1007/978-3-031-28938-5_4

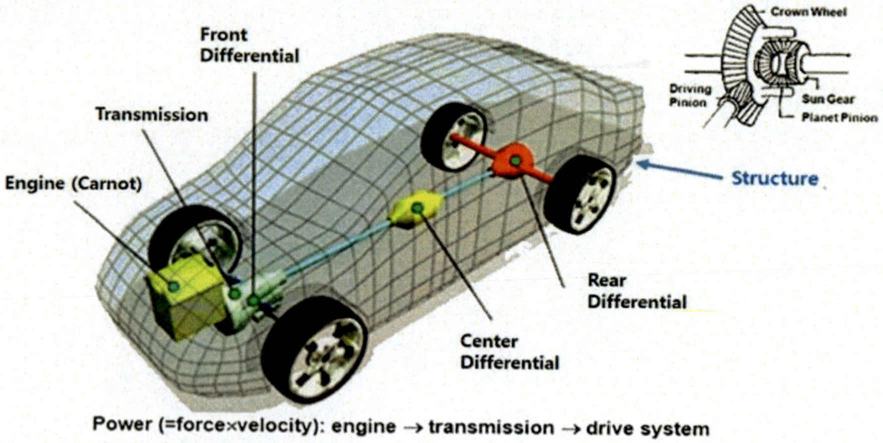

Fig. 4.1 Automobile: cycle, structure, and mechanisms (example)

4.2 Mechanical Structure Including Mechanisms

4.2.1 Introduction

Structures such as building, frame, and bridge is a core part of live modern life, which is the assembly of rigid bodies connected by joints with no mobility and involving no mechanism. They do not convert the available energy into work and carry only loads. Early architects and engineers designed the structure for its own purposes as a facility (or product) with the capacity to withstand loads. For example, many historical structures, supervised by the Pharaohs of ancient Egypt and the Caesars of Rome, are still standing in historical places. In Europe, many buildings and bridges built in the Renaissance Period are still functioning. These structures are built in thousands of shapes and different sizes in nature, which provides them with inherent robustness (Fig. 4.2).

Various mechanisms, mounted on the mechanical structure of a product, convert available power to fulfil its useful work that requires forces and movement. They are also mounted on mechanical systems from internal combustion engines to helicopters and machine tools. Many tools, such as wrenches, screwdrivers, hammers, and jacks, have their own mechanisms. In addition, the hands and feet, legs, arms, and jaws play a critical part (mechanisms) in human bodies, as do the paws and legs, wings, tails, and flippers in animal bodies.

A mechanism is an assembly of transporting components which execute some task, which generates some motions, such as a slider-crank. The mechanical mechanism and structure are joined together in a product as an arrangement of parts and provide enough strength and stiffness to bear loadings. For example, the bicycle is operated by a big driver gear wheel with pedals added. When the back-wheel revolves, the

Fig. 4.2 Golden gate bridge (Wikipedia)

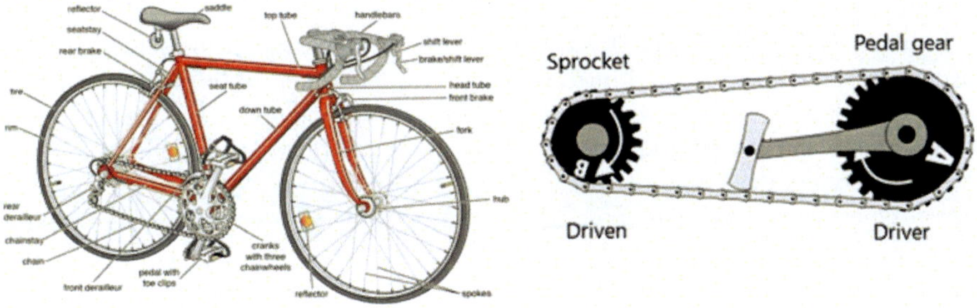

(a) Bicycle (structure + mechanism) (b) Chain & sprocket mechanism

Fig. 4.3 Bicycle and its mechanism

bicycle moves forward. We know that a mechanism such as the chain and sprocket in a bicycle is distinctly constructed around a frame structure (Fig. 4.3).

4.2.2 *Mechanical Mechanisms*

A variety of mechanisms in airplanes, automobiles, refrigerators, bicycles, etc., are mechanical portions that can transfer movement and forces from one power origin to the other. Most mechanisms are designed to provide a mechanical advantage. They also form a kinematic chain where no less than one link is attached to a framework of reference (ground) and makes four different motions, such as rotary motion, oscillation, linear motion, and reciprocating motion. Rotary motion is the starting point for many mechanisms, mostly provided by an engine or dc motors. Oscillation is a back and forth movement about a pivot location. Linear motion is a one-dimensional motion along a straight line. Reciprocating movement is again a back and forth motion.

4.2.2.1 (Lever) Mechanisms and History

From the old ages, the lever is the most basic mechanism, with a rigid beam rotating on a fulcrum used for thousands of years. "If you had a lever long enough and a fit fulcrum, you might carry the earth". A small effort from one ending of the beam shall lift a heavy load at the other ending. That is, by moving the fulcrum of a long beam closer to the load, the lever can move a large load with minimal effort.

As seen in Fig. 4.4, there are three types of levers. The Class 1 lever is positioned along the distance of the lever in which the effort is put. For example, the crowbar is a lever, but the pliers/scissors utilize two levers connected together at the fulcrum. Bottle opener and wheel borrow are examples of Class 2. Forearm and tongs are examples of Class 3 levers.

Leonard Euler (1707–1783) was one of the earliest experts in mathematics to investigate the mathematics of linkage design (synthesis). Most linkages are planar; their movement is enclosed to a plane. The general investigation of linkage movements, planar and spatial, is defined as screw theory. Sir Robert Stawell Ball (1840–1913) is deemed the father of screw theory. During the near industrial revolution, many of the weaving of cloth suggested the necessity for more complicated apparatus

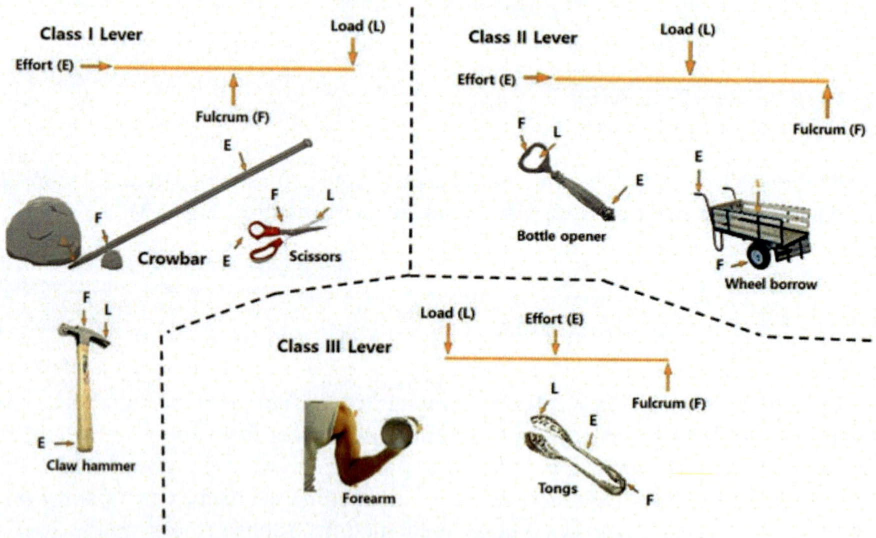

Fig. 4.4 Three classes of levers

Fig. 4.5 James Watt's industrial steam engine

to change waterwheels' rotary movement into complicated movements. The design of the steam engine generated a huge necessity for newly designed mechanisms and aparatus. Lengthy linear motion travel was necessitated to robustly use steam power. In particular, although he did not understand it, James Watt (1736–1819) applied thermodynamics and rotary joints and long links to generate structured straight-line movement (Fig. 4.5).

4.2.2.2 Mechanical Advantage

When a worker uses a stake to raise a rock by a lever mechanism, the worker applies a little quantity of effort at the ending of the lever to move the rock (class 1). Levers are utilized to change the orientation, distance, or velocity of motion or to lessen the effort necessary to raise a load. This lessening of effort is known as mechanical advantage. With its fulcrum at X, the lever is a bar of length AB, splitting the length of the bar into components L_1 and L_2. To raise a load W through a height of h, a force F should be employed downward through a distance s. The triangles AXC and BXD are alike and proportionate; thus, ignoring friction:

In the instance, the effort of 10 kg_f is necessary to raise a load of 50 kg_f. Thus, we can calculate the mechanical advantage. That is, mechanical advantage (MA) = load/effort = $s/h = L_1/L_2 = 50\ kg_f/10\ kg_f$. Thus, this system has an MA of 5. There are no units because it is a ratio of the same units (Fig. 4.6).

Fig. 4.6 Mechanical advantage

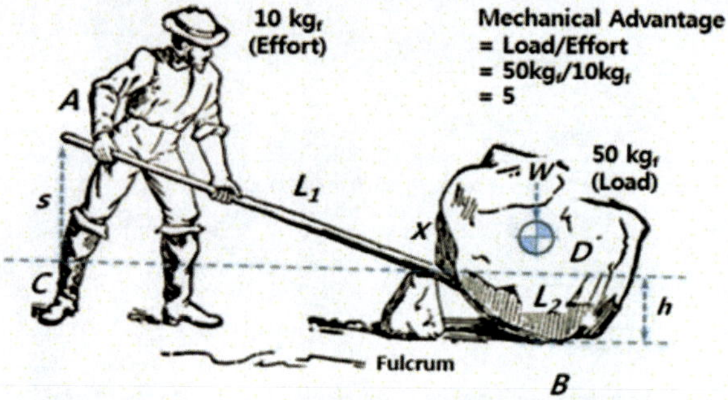

4.3 Design of Mechanisms

To fulfil a set of performance needs for the machine, a mechanism design is required to prescribe the sizes, forms, material configurations, and positioning of parts by a proper kinematic analysis and synthesis. The resulting machine shall carry out the recommended tasks.

4.3.1 Classification of Mechanisms

There are three applications of mechanisms: (1) path generation mechanisms, (2) function generation mechanisms, and (3) motion generation mechanisms (Fig. 4.7).

Analysis is a technique that permits an engineer to critically inspect an already existing, or suggested, design to decide its suitability for the assignment. The examination of mechanisms and machines therefore aims to understand the relations between the movements of the machine components (kinematics) and the forces which produce the movements (kinetics). In a mechanism design, kinematic analysis is necessitated to determine the displacement, position, rotation, velocity, speed, and acceleration of the mechanism. On the other hand, synthesis is a procedure of managing a scheme or a way of fulfilling a given purpose. In other words, it is a process of evolving a mechanism to fulfil a set of performance needs for the machine. Especially in mechanisms, synthesis is the design of a linkage to generate a wanted output movement for a specified input movement (Fig. 4.8).

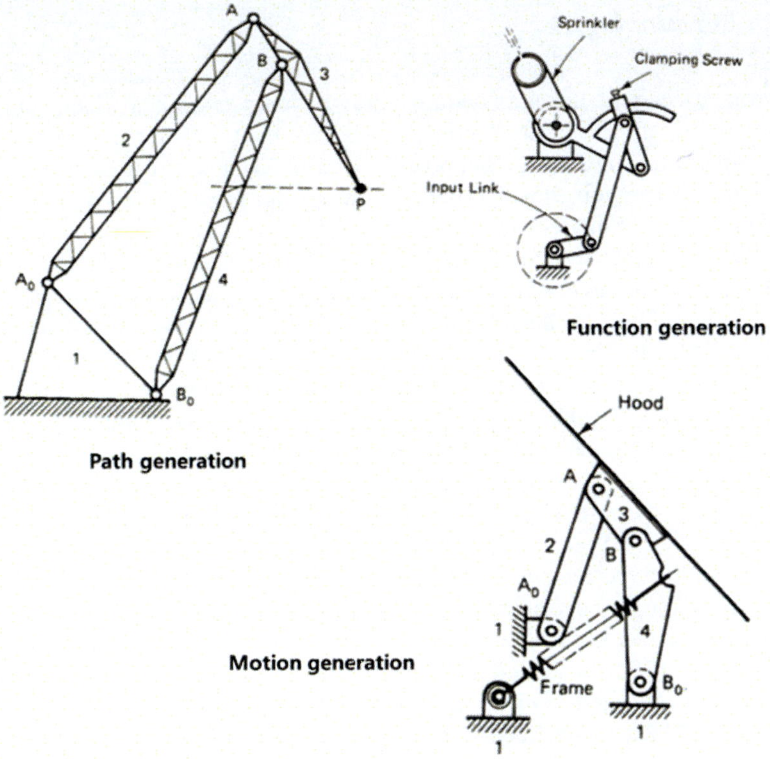

Fig. 4.7 Classifications of mechanisms in application

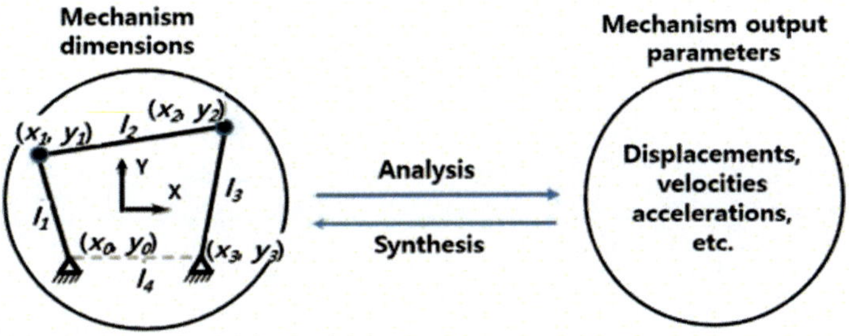

Fig. 4.8 Kinematic synthesis versus analysis

4.3.2 Terminologies

A kinematic chain or linkage is an assembly of links and joints which supplies a required output movement in response to a stated input movement. Links are individual parts of the mechanism that possess at least two nodes. They are regarded as rigid bodies and are joined with other links to transfer forces and movement.

A link is a machine part that has relative motion to some other parts. It is an assembly of resistant structures attached in such way that they make a single part of it and have no relative motion between them. The types of link are classified as follows: (1) rigid link, (2) flexible link, (3) fixed link, based on function. On the other hand, it is (1) binary link, (2) ternary link, (3) quaternary link, based on end vertices.

A joint (or pair) is a portable attachment between links and permits relative motions between the links. Especially, in a slider crank mechanism, crank (link no.1) has motion relative to the connecting rod (link no. 2) and the frame (link no. 4) (Fig. 4.9).

Kinematic links are classified as follows: (1) lower joint and higher joint (contacts), (2) closed joint and unclosed joint (constraint), (3) sliding and turning joint. If a joint in motion has a surface contact between its elements, it is called a lower joint. Example is shaft rotating in a bearing. If a joint in motion has a line or point contact between its element, it is called a higher joint. Example is ball and shaft bearing. Revolute and sliding joints are primary joints. If the elements of the joint are held together mechanically, they form a closed joint. Example is all lower joints. When two elements of the joint are not held together mechanically,

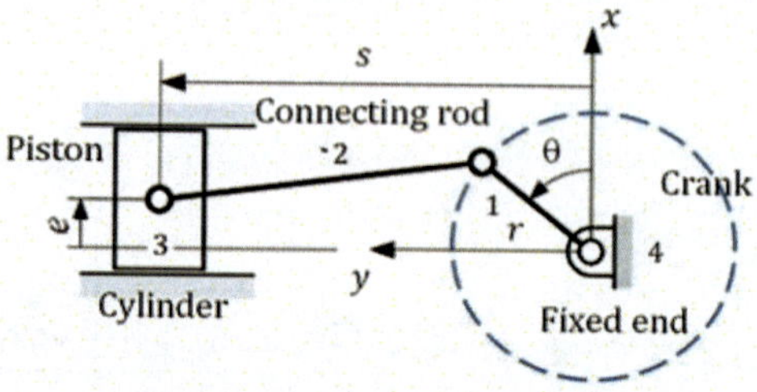

(a) Slider crank mechanism (example)

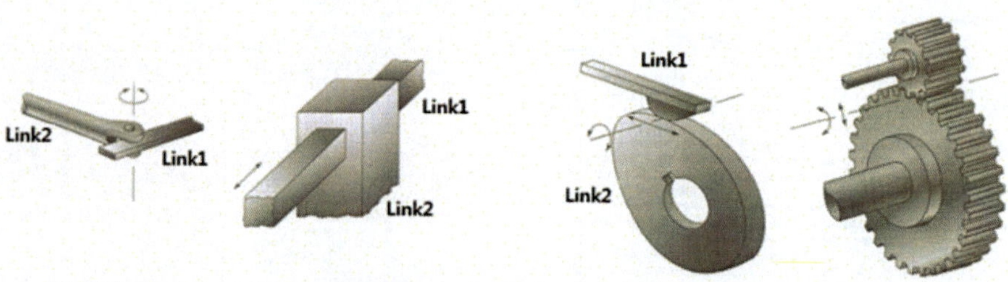

(b) Primary joints: pin and sliding (c) Higher-order joints: cam joint and gear joint

Fig. 4.9 Slider crank mechanism and joints

they constitute an unclosed joint. Example is cam and follower. Two elements in a joint have sliding motion relative to each other. Example is piston and cylinder joint. When two elements are connected such that one element revolves around the other, it constitutes a turning point. Turning joint is also known as a revolute pair. Example is slider crank mechanism.

4.3.3 *Kinematic Chain and Mobility*

Kinematic chain is a combination of kinematic joints joined in such a way that the relative motion between them is completely constrained. The kinematic chain is classified as follows: (1) four bar chain, (2) slider crank chain, and (3) double slider crank chain.

The degree of freedom (DOF) is the number of self-determining inputs needed to exactly position all links of a mechanism with reference to the ground or the number of actuators required to work the mechanism. It is given by the mobility M:

$$M = \text{degree of freedom} = 3(n - 1) - 2jp - jh \tag{4.1}$$

where n is the total number of links, jp is the number of primary joints, and jh is the number of higher-order joints.

4.3.4 *Kinematic Model/Diagram*

A kinematic prototype is a representation of a mechanism which only manifests the dimensions which affect the movement. To draw the kinematic model from a mechanism, remove any superfluous detail, simplify the drawing of the mechanism for further examination, and carry out the displacement, velocity, and acceleration analysis of a mechanism in machines (Fig. 4.10).

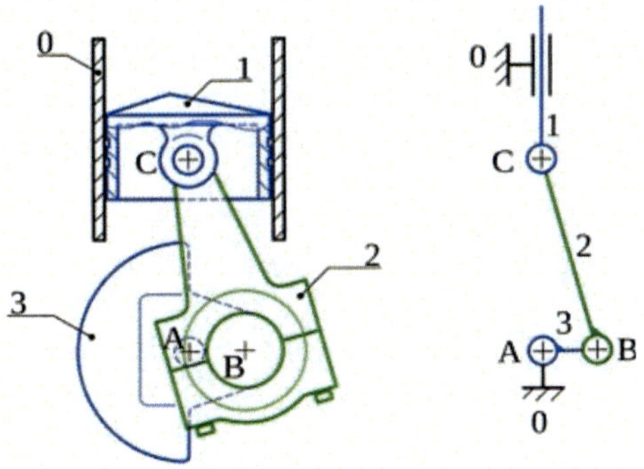

Fig. 4.10 Engine and its kinematic mechanism

Example 4.1 A kinematic diagram is drawn, and a shear press is shown, which is utilized to cut and trim electronic circuit board laminates (Fig. 4.11).

Solution: The first stage in building a kinematic diagram is to determine the component which shall be appointed as the frame. The movement of all other links shall be decided relative to the frame. In some instances, its choice is clear as the frame is securely fastened to the ground. In this problem, the big base which is bolted to the table is appointed as the frame. The movement of all other links is decided relative to the base.

The base is numbered as link 1. Cautious watching shows three other moving parts: (2) handle, (3) cutting blade, and (4) bar which attaches the cutter with the handle. Pin joints are utilized to attach these three different parts. Pin joints are labeled *A* through *C*. Additionally, the cutter glides up and down along the base. This gliding joint is labeled *D*. Eventually, the movement of the ending of the handle is desired. This is appointed as the point of interest *X*. The kinematic diagram is specified in Fig. 4.12.

Example 4.2 Figure 4.13 represents a pair of vice grips. Sketch a kinematic diagram.

Solution: The first stage is to determine the component which shall be appointed as the frame. In this problem, no components are added to the ground. Thus, the choice of the frame is rather random. The top handle is appointed as the frame. The movement of all other links is decided relative to the top handle.

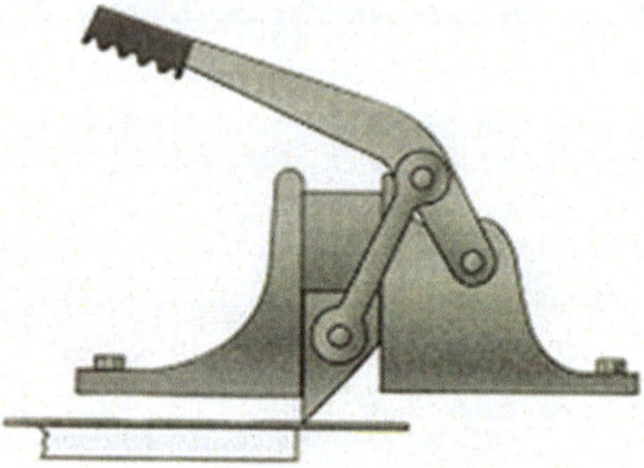

Fig. 4.11 Shear press (example)

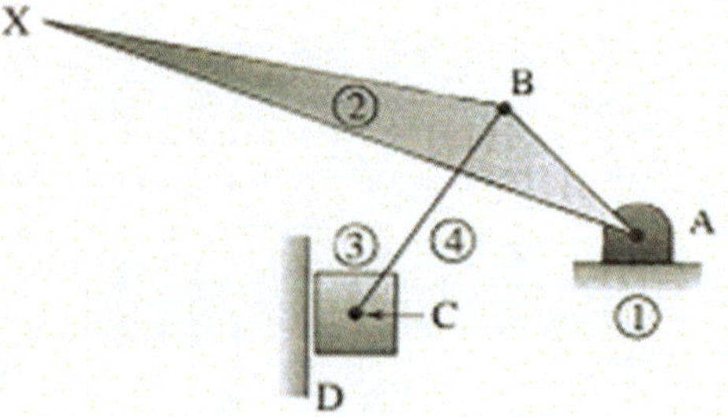

Fig. 4.12 A kinematic diagram of the shear press

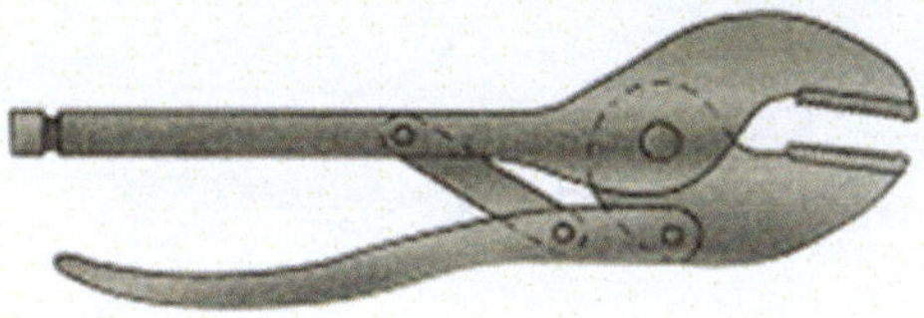

Fig. 4.13 A pair of vice grips (example)

The top handle is numbered as link (1) Cautious observation shows three other moving parts: (2) the bottom handle, (3) the bottom jaw, and (3) the bar which attaches the top and bottom handles. Four-pin joints are utilized to attach these different components. These joints are labeled A through D. In addition, the movement of the ending of the bottom jaw is wanted. This is appointed as the point of interest X. Finally, the movement of the ending of the lower handle is also needed. This is appointed as the point of interest Y. The kinematic diagram is specified in Fig. 4.14

Example 4.3 Figure 4.15 displays a toggle clamp. Sketch a kinematic diagram utilizing the clamping surface and the handle as points of interest. Additionally, calculate the DOFs for the clamp.

Four-pin joints are utilized to attach these dissimilar parts. These joints are labeled A through D. In addition, the movement of the clamping surface is required. This is appointed as the point of interest X. Eventually, the movement of the ending of the handle is also required. This is signified as the point of interest Y. The kinematic diagram is shown in Fig. 4.16.

The figure below displays a toggle clamp. Sketch a kinematic diagram utilizing the clamping surface and the handle as points of interest. Additionally, calculate the DOFs for the clamp. Calculating the mobility of the mechanism, it is seen there are four links. There are also four pin joints. Therefore,

$$n = 4,\ jp = 4\text{pins},\ jh = 0 \text{ and:}$$

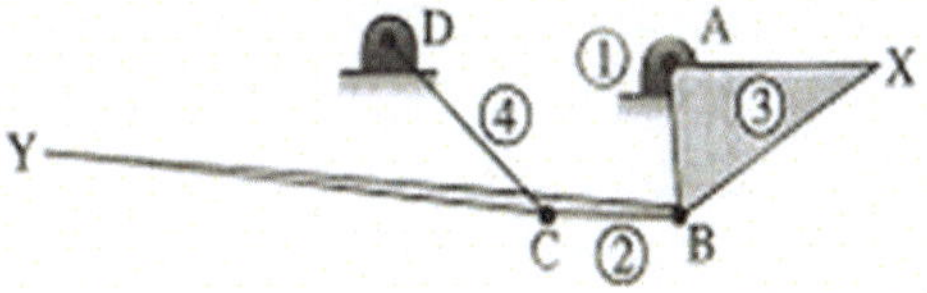

Fig. 4.14 A kinematic diagram of vice grips

Fig. 4.15 A toggle clamp (instance)

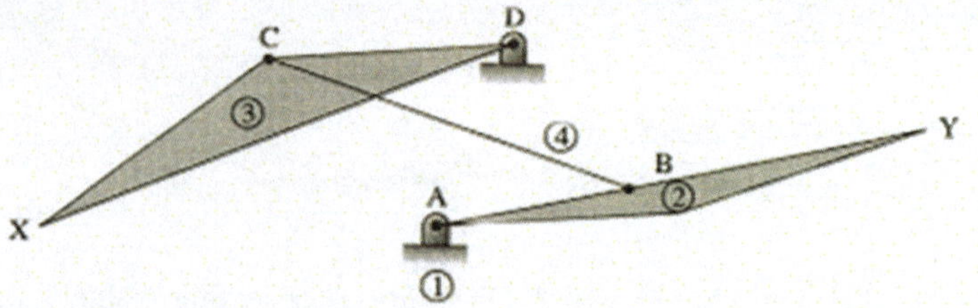

Fig. 4.16 A kinematic diagram of toggle clamp

$$M = 3(n - 1) - 2jp - jh = 3(4 - 1) - 2(4) - 0 = 1$$

With one DOF, the clamp mechanism is restricted. Moving only one link, the handle exactly positions all other links in the clamp.

4.3.5 Grashof's Law and Some Inversion Mechanisms

Grashof's Law states that for a planar four-bar linkage system, the sum of the shortest and longest link lengths cannot be greater than the sum of the remaining two link lengths if there is to be a continuous relative rotation between two members.

Mathematically, Suppose s = length of the shortest link, l = length of the longest link, p & q = lengths of other links (Fig. 4.17).

$$s + l \leq p + q \tag{4.2}$$

Fig. 4.17 A planar four-bar linkage

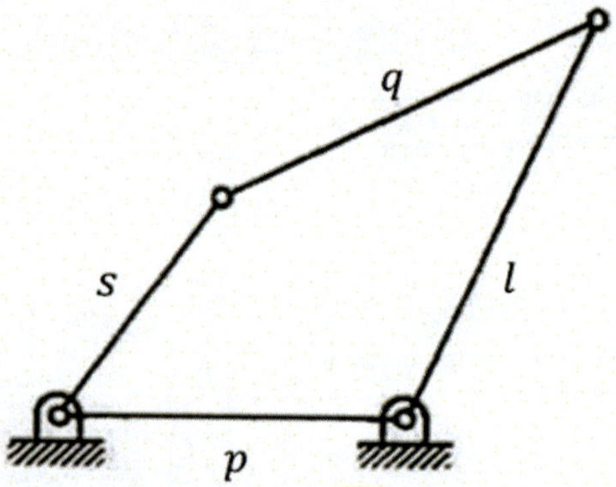

4.3.5.1 Inversions of Four bar Mechanisms

Some important inversions of four bar mechanisms are as follows: (1) Coupled wheel locomotive, (2) Beam engine, (3) Pantograph, (4) Watt mechanism. Coupled wheel locomotive comprises of four links.

A coupling rod connects the driving wheels of a locomotives. The opposite links are equal in lengths. Because link 1 and 3 are cranks, the mechanism is also known as double crank mechanism (Fig. 4.18).

A beam engine, known as crank and lever mechanism, a type of steam engine where a pivoted overhead beam is used to apply the force from a vertical piston to a vertical connecting rod. The mechanism is used to convert rotary motion into reciprocating motion. Link 3 oscillates about point c. Link 4 is fixed (Fig. 4.19).

Pantograph is a mechanical linkage connected in a manner based on parallelogram so that it will reproduce a displacement in a reduced or enlarged scale. It is used for duplicating the drawings, maps, plans, etc. All the four links are turning in nature and the mechanism is basically a quadratic cycle in the form of a parallelogram according to geometry ($AC/AC' = AE/AE'$) (Fig. 4.20).

Watt mechanism is a type of mechanical linkage invented by James Watt in which his steam engine will guide the piston rod. Lines OA and BC are parallel in the mean position of the mechanism. They are the levers of the mechanism and connected with

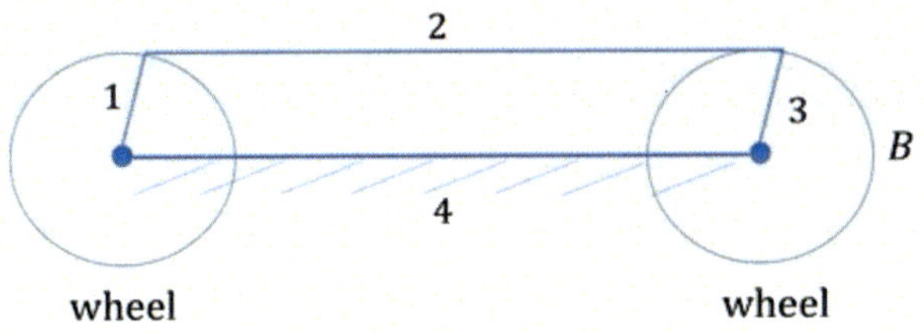

Fig. 4.18 Coupled wheel locomotive

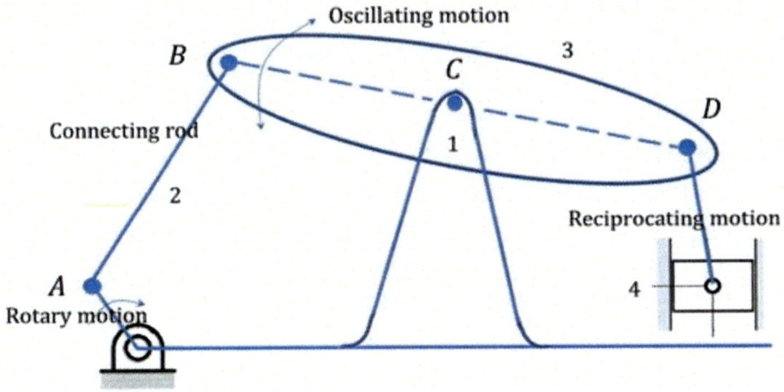

Fig. 4.19 Beam engine

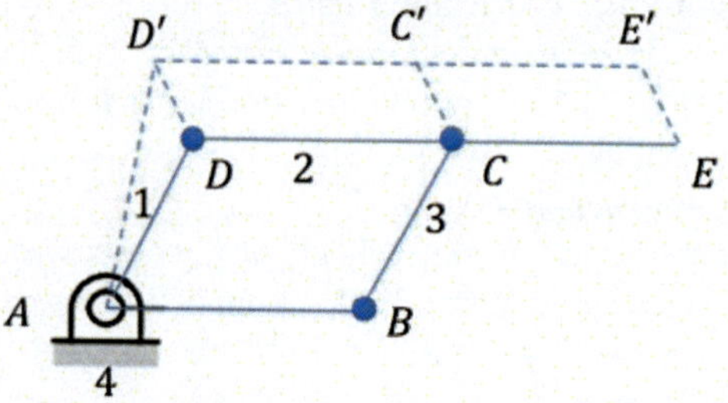

Fig. 4.20 Pantograph

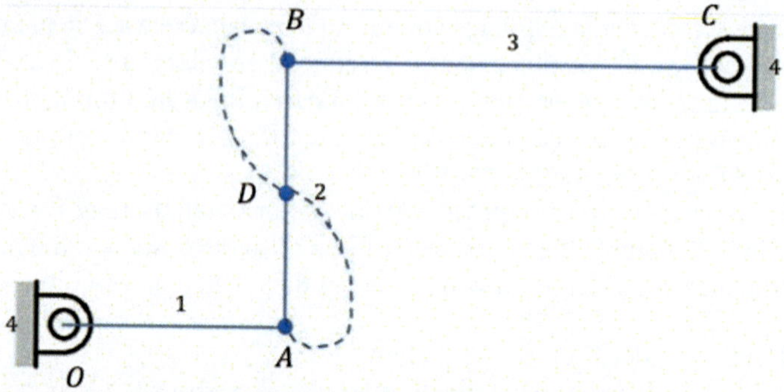

Fig. 4.21 Watt mechanism

a link AB. For small displacement of levers OA and BC, D will trace an approximately straight line on AB, such that ($AD/DB' = OA/BC'$) (Fig. 4.21).

4.3.5.2 Inversions of Slider Crank Chain

Some important inversions of slider crank chain are as follows: (1) Slider crank mechanism, (2) Hand pump mechanism, (3) Oscillating cylinder engine, (4) Crank and slotted lever quick return motion mechanism, (5) Whitworth quick return motion, and (6) rotary engine. Slider crank mechanism is an arrangement of mechanical parts that is designed to convert rotary motion to straight-line motion. Link 2 in slider crank mechanism is the crank. Link 3 is the connecting rod between link 2 and link 4, which is slider. Link 1 is the frame fixed. Fixing link 1 and rotating link 2 will make link 4 slide within link 1 thus transmitting rotary motion into oscillatory motion (Fig. 4.22).

Hand pumps are a type of human operated positive displacement pump. Hand pumps have a simple and robust design, require little maintenance, and are easy to operate, this has led to their widespread application throughout the world (Fig. 4.23).

An oscillating cylinder engine is a simple steam-engine design that requires no valve gear. When link 2 of a slider crank mechanism is fixed, the inversion obtained

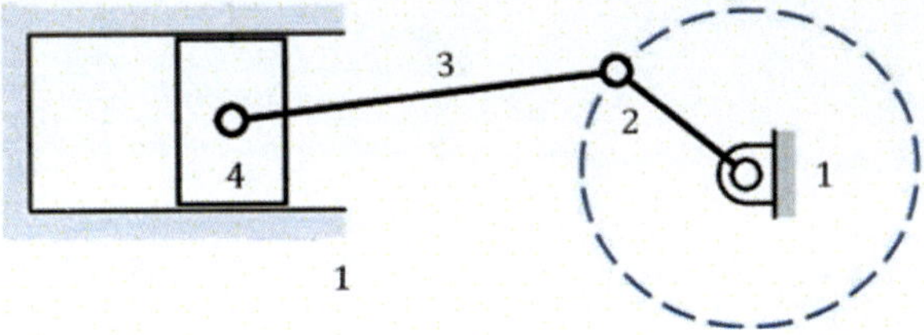

Fig. 4.22 Slider crank mechanism

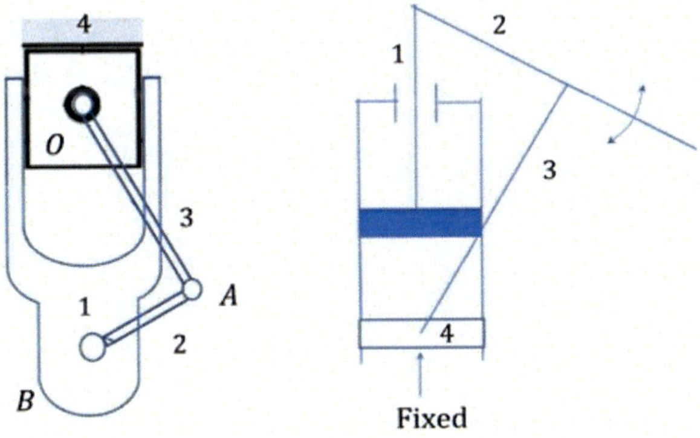

Fig. 4.23 Hand pumps

is oscillating cylinder mechanism. Link 1 acts as crank. Link 4 reciprocates inside the link 3. Link 3 is pivoted to fixed link 2 and oscillates when link 1 rotates (Fig. 4.24a). The crank and slotted lever quick return mechanism converts the rotary motion into reciprocating motion. Link 2 in crank and slotted lever quick return motion mechanism is fixed. Link 4, slider reciprocates in oscillating lever, link 3. Link3 oscillates about O, pivoted to fixed link 2 when link 1 rotates about point A. This mechanism is used in shaping and slotting machines (Fig. 4.24b).

The Whitworth Quick Return Mechanism is a mechanism that can transform circular movement into reciprocating movement. Link 3 is fixed. Link 1 rotates about point B. Link 4 slider within the slotted link2. Link 2 oscillates about pivot point A when link 1 rotates (Fig. 4.24c). The rotary engine is an early type of internal combustion engine, usually designed with an odd number of cylinders per row in a radial configuration In rotary engine, the slider is replaced by six pistons, pivoted at point A. Here crank (link 3) is fixed. All cylinders, i.e., link 2 rotate about the same fixed-point A. The pistons (link 4) oscillate in these rotating cylinders symmetric arrangement of seven or nine (Fig. 4.24d).

4.3.5.3 Inversions of Slider Crank Chain

Some important inversions of double slider crank chain are as follows: (1) Elliptical trammel, (2) Scotch yoke mechanism, and (3) Oldham's coupling. A trammel of Archimedes is a mechanism that generates the shape of an ellipse. It consists of two

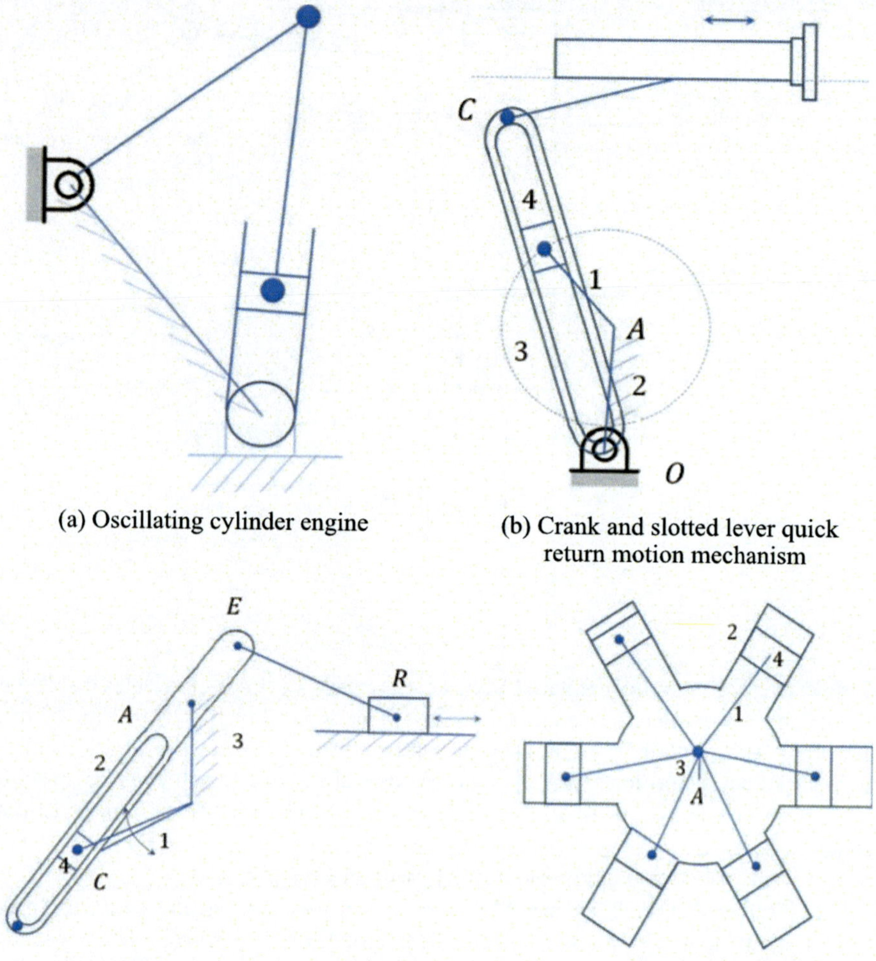

(a) Oscillating cylinder engine

(b) Crank and slotted lever quick return motion mechanism

(c) Whitworth quick return motion

(d) Rotary engine

Fig. 4.24 Inversions of slider crank chain (example)

shuttles which are confined to perpendicular channels or rails and a rod which is attached to the shuttles by pivots at fixed positions along the rod. Link 4 is the frame that is slotted and fixed. Link 1 and 3 are sliders that move vertically and horizontally. Link 2 is the connecting link 1 and 3 and is slightly extended to a point C (Fig. 4.25a). The Scotch yoke mechanism is a reciprocating motion mechanism, converting the linear motion of a slider into rotational motion of a crank or vice versa. It is obtained when link 1 (slider) is fixed and pivoted to frame at point A. Link2 is a crank that rotates completely. The rotation of link 2 makes link 3 to slide in the frame. Link 4, the frame thus reciprocates. Rotary motion thus is converted to reciprocating motion (Fig. 4.25b). The third inversion of double slider crank mechanism is Oldham's

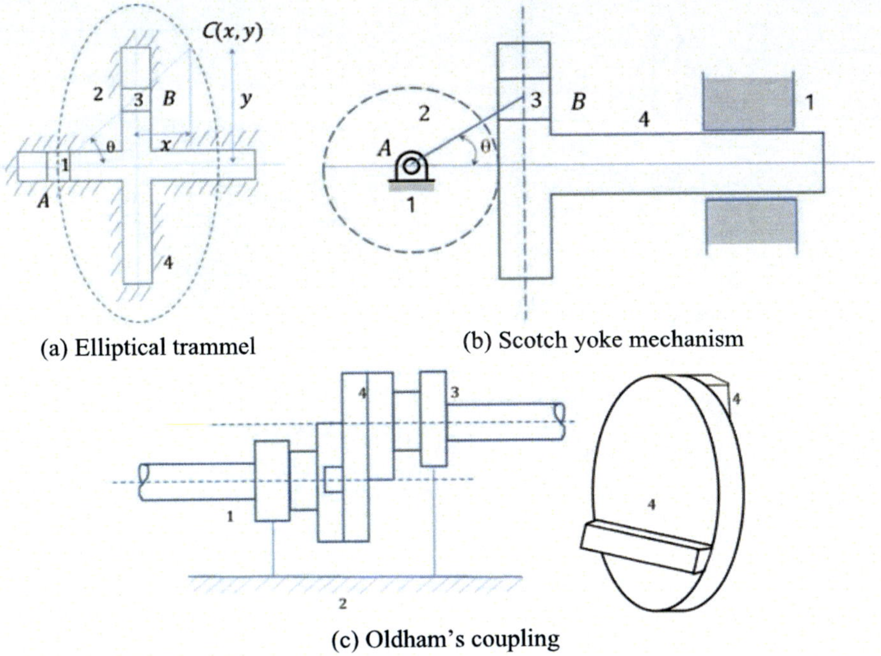

Fig. 4.25 Inversions of double slider crank chain (example)

coupling. It is a form of flexible coupling designed for applications that must be free from backlash. This is used to connect two parallel shafts whose axes are at small distance apart. Link 2, the frame is fixed. Link 3, comprises of a shaft fitted (Fig. 4.25c).

There are other intermittent motion mechanisms as follows: (1) Geneva wheel, (2) Ratchet and Pawl, and (3) Ackermann steering. Geneva wheel mechanism translates a continuous rotation movement into intermittent rotary motion. The rotating wheel (crank) is equipped with a pin that reaches into a slot located in the other wheel (driven wheel) that advances it by one step at a time (Fig. 4.25a). Ratchet and Pawl mechanism is to allow the shaft to rotate in one direction only. A ratchet is a wheel with a shape similar to a circular saw blade or a horizontal milling cutter (Fig. 4.26a).

4.3.5.4 Inversions of Straight-Line Motion Mechanisms

There are straight line motion mechanisms as follows: (1) Peaucellier mechanism and (2) Robert's mechanism. The conditions for exact straight-line motion are as follows (Fig. 4.27):

$$OB \cdot OC = constant \quad (4.3)$$

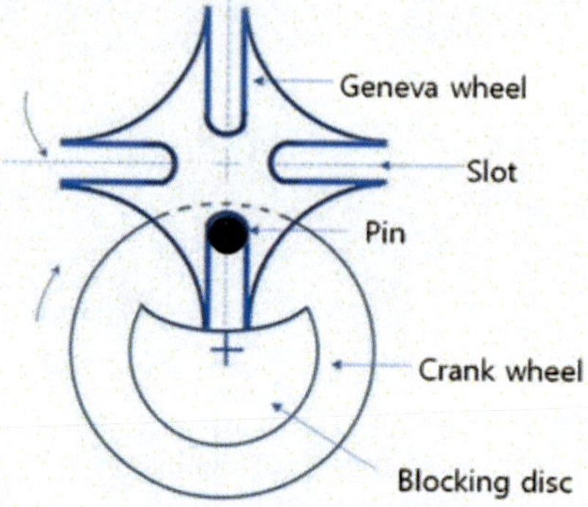

(a) Geneva wheel

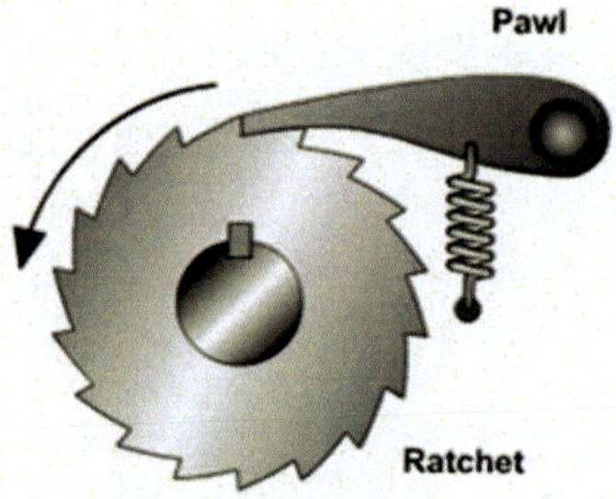

(b) Ratchet and Pawl

Fig. 4.26 Intermittent motion mechanisms (example)

Fig. 4.27 Condition for exact straight-line motion

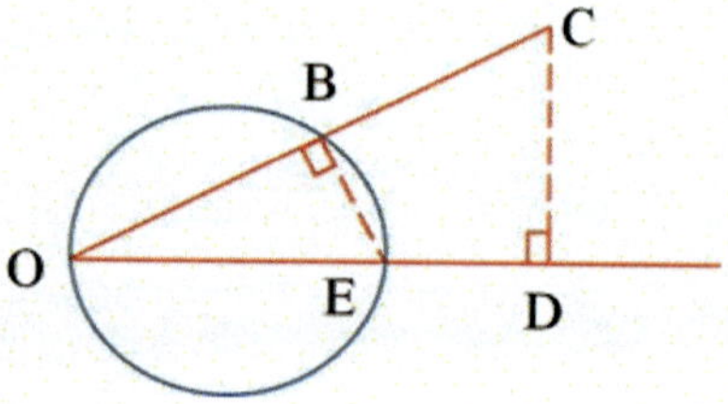

From $\triangle OBE$ and $\triangle OCD$

$$\frac{OB}{OE} = \frac{OD}{OC} \text{ or } OD = \frac{OB}{OE} \cdot OC \tag{4.4}$$

Because $OE \rightarrow Diameter$, we know that $OD = constant$.

Peaucellier mechanism is an eight-link mechanism as shown in Fig. 4.28a. Link 2 is the driving link. Link 4, 5, 6 and 7 are all of equal lengths, forming a closed four bar linkage. The joints P & Q are at the opposite corner of the four-bar linkage. Since OB & BP are of constant length $OP \cdot OQ = constant$, P traces a straight path.

Robert mechanism is a six-link mechanism as shown in Fig. 4.28b. Link 1, the fixed link, link 2, 3 and 4 comprises a four-bar chain. Link 5 and 6 are joined to point P from joints A and B respectively. As link2 oscillates, point P traces approximately a straight line.

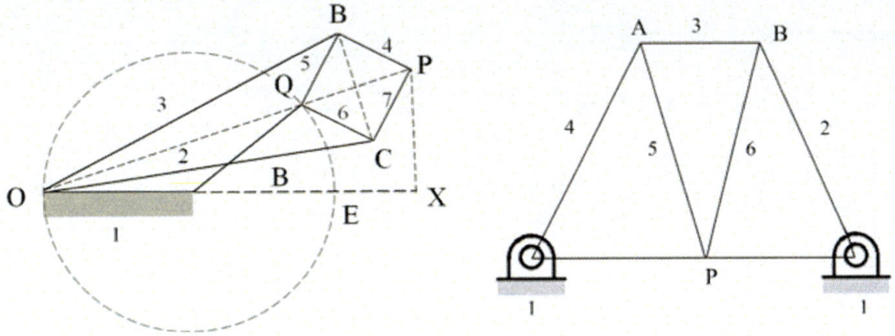

(a) Peaucellier mechanism (b) Robert mechanism

Fig. 4.28 straight-line motion mechanisms

4.3.6 *Kinematic Analysis of Mechanisms*

4.3.6.1 Position Analysis Using the Absolute Cartesian Method

There are two typical coordinate systems: (1) Cartesian (R_x, R_y) and Polar (R_A, θ). We can convert between Cartesian and Polar (Fig. 4.29). That is,

$$R_A = \sqrt{R_x^2 + R_y^2} \text{ or } \theta = \arctan\left(R_y/R_x\right) \tag{4.5}$$

where $R_x = R_A\cos\theta$ or $R_y = R_A\sin\theta$

Relative position can be defined as follows:

$$R_{B/A} = R_B - R_A \tag{4.6}$$

The position analysis of a kinematic chain necessitates the decision of the joint positions, the position of the centers of gravity, and the angles of the links with the horizontal axis. A rigid body in two-dimensional space shall be expressed by two

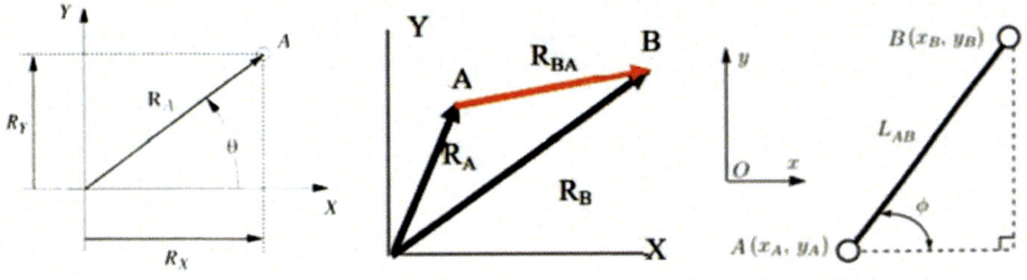

Fig. 4.29 Coordinate systems

points, A and B, on the rigid body.

$$(x_A - x_B)^2 + (y_A - y_B)^2 = AB^2 = L_{AB}^2 \tag{4.7}$$

$$m = \tan\phi = \frac{y_B - y_A}{x_B - x_A} \tag{4.8}$$

where $y = mx + n$

4.3.6.2 Graphical Method for Analyzing Velocity and Acceleration

First of all, we can graphically analyze the velocity of each points B and C in four-link bar (Fig. 4.30).

As link AB rotates constantly, we can define the velocity at point B as follows:

$$V_B = ab = \omega \cdot AB \text{ where } AB \text{ is the length of link} \tag{4.9}$$

From velocity diagram, we find velocity V_{CB} at the intersection. So, angular velocity is found as follows:

$$\omega_C = \frac{V_{CB}}{CB} \quad \text{where CD is the length of link} \tag{4.10}$$

We know that at point B, link BC rotates counter clockwise. The acceleration of four-link bar is graphically calculated as follows (Fig. 4.31).

The tangential component of acceleration at link AB, shown in Fig. 4.32, is,

$$f_{BA}^t = 0 \tag{4.11}$$

On the other hand, the radial component of acceleration at link AB is,

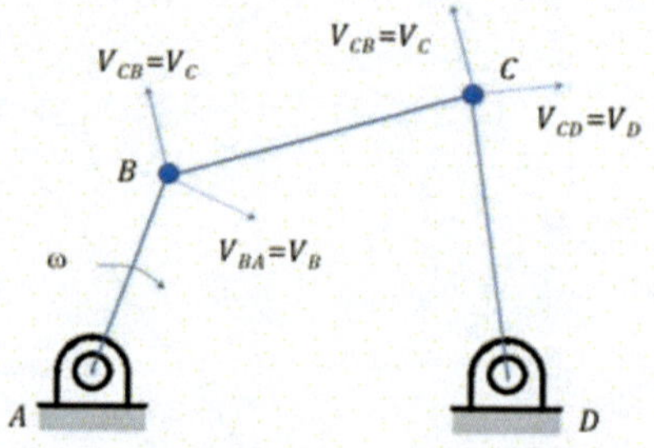

(a) Velocity in four-link bar

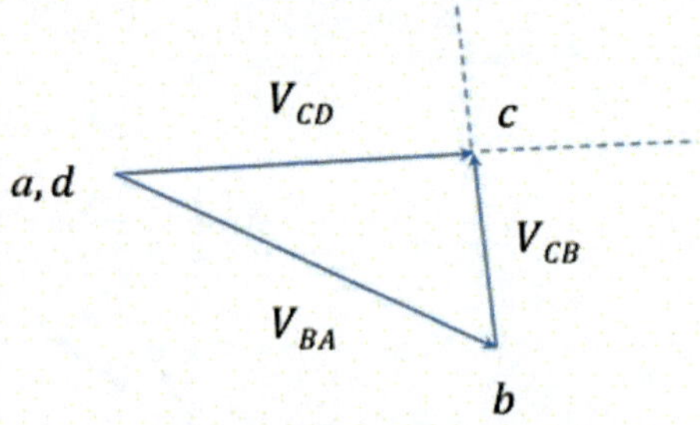

(b) Velocity diagram

Fig. 4.30 Velocity at four-link bar

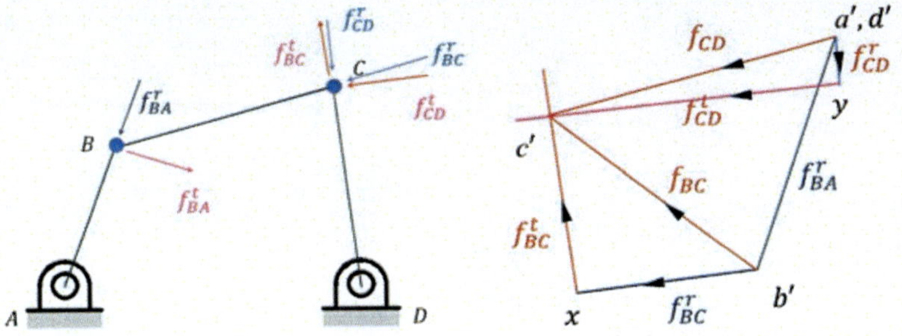

(a) Acceleration in four-link bar (b) Acceleration diagram

Fig. 4.31 Acceleration in four-link bar

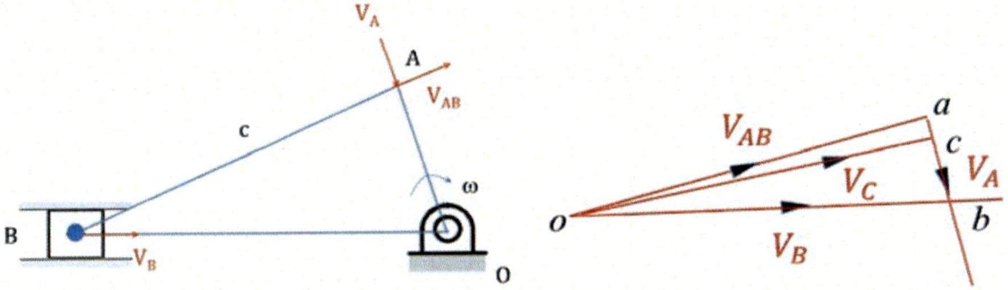

(a) Velocity in slider crank (b) Velocity diagram

Fig. 4.32 Velocity at slider crank mechanism

$$a'b' = f_{BA}^r = \omega^2 \cdot AB \tag{4.12}$$

The radial component of acceleration at link BC is,

$$b'x = f_{BC}^r = \frac{V_{BC}^2}{BC} \tag{4.13}$$

On the other hand, the radial component of acceleration at link C is,

$$a'y = f_{CD}^r = \frac{V_{CD}^2}{CD} \tag{4.14}$$

The angular acceleration at link BC is,

$$\alpha_{BC} = \frac{f_{BC}^t}{BC} = \frac{x_c}{BC} \tag{4.15}$$

On the other hand, the angular acceleration at link CD is,

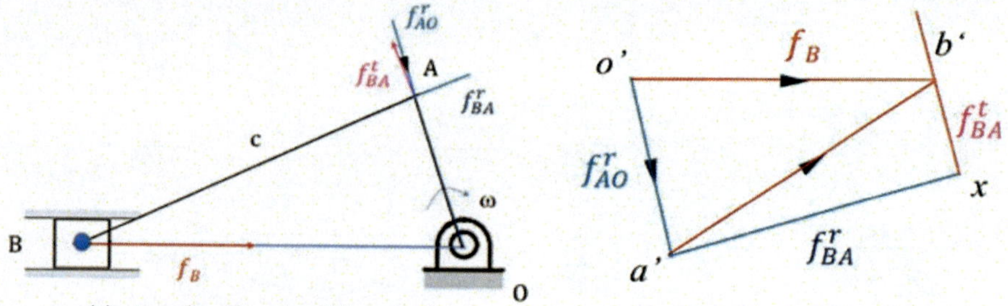

Fig. 4.33 Acceleration in slider crank

$$\alpha_{CD} = \frac{f^t_{CD}}{CD} = \frac{y_c}{CD} \tag{4.16}$$

As link OA rotates constantly, we can define the velocity at point A as follows (Fig. 4.33):

$$oa = V_A = \omega \cdot OA \text{ where } OA \text{ is the length of link} \tag{4.17}$$

From velocity diagram, $\frac{ac}{ab} = \frac{AC}{AB}$ *or* $ac = ab \cdot \frac{AC}{AB}$. The angular velocity at link AB is

$$\omega_{AB} = \frac{V_{AB}}{AB} \text{ or } \omega_C = \frac{V_C}{AC} \quad \text{where AB and AC are the length of link} \tag{4.18}$$

The tangential component of acceleration at link AO is,

$$f^t_{AO} = 0 \tag{4.19}$$

On the other hand, the radial component of acceleration at link AO is,

$$o'a' = f^r_{AO} = \omega^2 \cdot OA \tag{4.20}$$

The radial component of acceleration at link AB is,

$$a'x = f^r_{AB} = \omega^2_{AB} \cdot AB \tag{4.21}$$

Example 4.4 A four-link bar mechanism is as shown in the Fig. 4.34. It has AD 120 mm ling fixed. The crank AB is 30 mm long and rotates at 100 RPM, clockwise. The link CD, 50 mm long oscillates about D, BC = AD. Find the angular velocity of link CD, and angular acceleration of link CD, graphically when angle BAD = 60°.

Because N_{BA} = 100 RPM, the velocity of link AB is

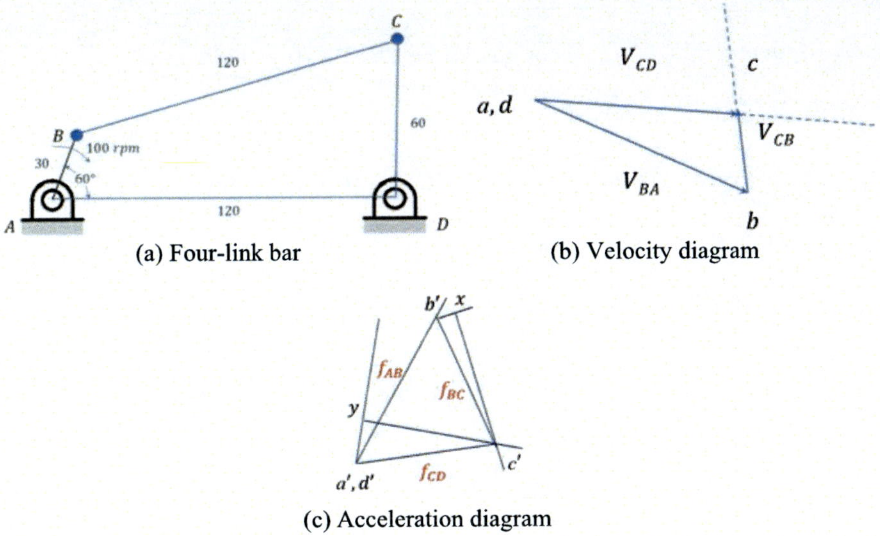

Fig. 4.34 A four-link bar mechanism

$$\omega_{AB} = \frac{2\pi N_{BA}}{60} = \frac{2\pi \cdot 100}{60} = 10.47 \text{ rad/s}$$

$$V_{AB} = \omega_{AB} \cdot AB = 10.47 \cdot 30 = 314.1 \text{ mm/s}$$

From velocity diagram, because 100 mm/s = 1 mm, V_{AB} = 31.41 mm long (vector length).

Velocity V_{BC} is

$$V_{BC} = 12 \text{ mm} \cdot 10 = 120.0 \text{ mm/s}$$

$$\omega_{BC} = \frac{V_{BC}}{BC} = \frac{120}{120} = 1 \text{ rad/s}$$

On the other hand, velocity V_{CD} is

$$V_{CD} = 26 \text{ mm} \cdot 10 = 260.0 \text{ mm/s}$$

$$\omega_{CD} = \frac{V_{CD}}{CD} = \frac{260}{60} = 4.33 \text{ rad/s}$$

The acceleration at link AB is,

$$f_{BA}^{t} = 0$$

$$vector\ a'b' = f_{BA}^{r} = \omega_{AB}^{2} \cdot AB = 10.47^2 \cdot 30 = 3288.63\ \text{mm/s}^2$$

The acceleration at link BC is,

$$vector\ b'x = f_{BC}^{r} = \omega_{BC}^{2} \cdot BC = 1.0^2 \cdot 120 = 120\ \text{mm/s}^2$$

On the other hand, acceleration at link C is,

$$vector\ a'y = f_{CD}^{r} = \omega_{CD}^{2} \cdot CD = 4.33^2 \cdot 60 = 1124.93\ \text{mm/s}^2$$

The angular accelerations at links BC and CD are,

$$\alpha_{BC} = \frac{f_{BC}^{t}}{BC} = \frac{x_c}{BC} = \frac{2600}{120} = 21.66\ \text{rad/s}^2$$

$$\alpha_{CD} = \frac{f_{CD}^{t}}{CD} = \frac{y_c}{CD} = \frac{2500}{60} = 41.66\ \text{rad/s}^2$$

Example 4.5 For the configuration of the slider crank mechanism is as shown in the Fig. 4.35. The crank OA rotates at 200 RPM clockwise. Determine acceleration of the slider, acceleration of point E, angular acceleration of link AB.

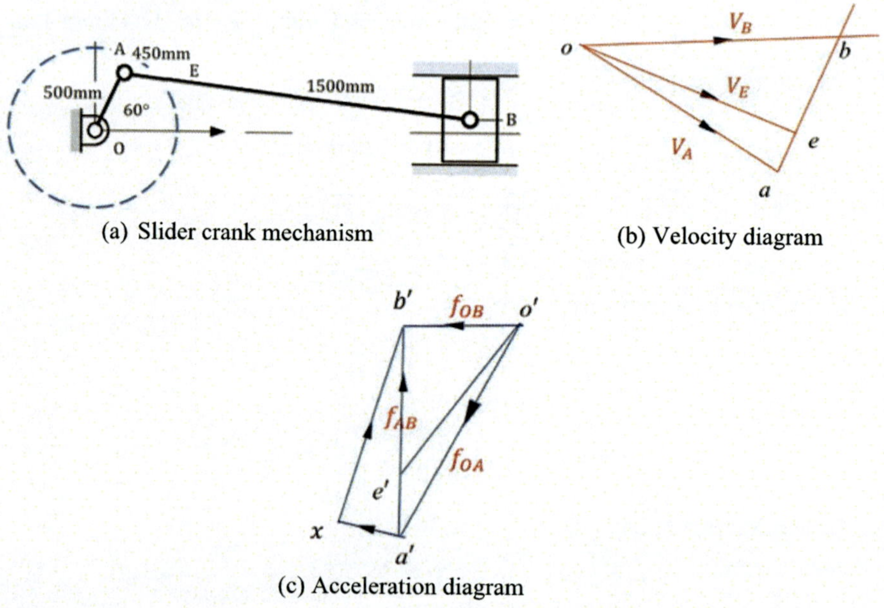

(a) Slider crank mechanism

(b) Velocity diagram

(c) Acceleration diagram

Fig. 4.35 A slider crank mechanism

Because $N_{OA} = 200$ RPM, the velocity of link OA is

$$\omega_{OA} = \frac{2\pi N_{OA}}{60} = \frac{2\pi \cdot 200}{60} = 20.94 \text{ rad/s}$$

$$V_{OA} = \omega_{OA} \cdot OA = 20.94 \cdot 0.5 = 10.47 \text{ m/s}$$

From velocity diagram, velocities at E and AB are

$$vector\ oe = V_E = 10.25 \text{ m/s}$$

$$vector\ ab = V_{AB} = 5 \text{ m/s}$$

The acceleration at link AB is,

$$f_{OA}^t = 0$$

$$vector\ o'a' = f_{OA}^r = \frac{V_{OA}^2}{OA} = \frac{10.47^2}{0.5} = 219.3 \text{ m/s}^2$$

The acceleration at link AB is,

$$vector\ a'b' = f_{AB}^r = \frac{V_{AB}^2}{AB} = \frac{5^2}{1.5} = 16.67 \text{ m/s}^2$$

Because $\frac{a'e'}{a'b'} = \frac{AE}{AB}$, acceleration at link E is,

$$vector\ o'e' = f_E = 158 \text{ m/s}^2$$

Angular accelerations at links AB is,

$$xb' = f_{AB}^t = 206.25 \text{ m/s}^2$$

$$\alpha_{AB} = \frac{f_{AB}^t}{AB} = \frac{206.25}{1.5} = 137.5 \text{ rad/s}^2$$

Example 4.6 For the configuration of the toggle mechanism is as shown in the Fig. 4.36. D is constrained to move on a horizontal path. The crank OA is rotating in CCW direction at 180 RPM and increasing at the rate of 50 rad/s^2. The dimensions of various links are as follows: OA = 180 mm; CB = 240 mm; AB = 360 mm and BD = 540 mm. Determine the velocity of the slider, angular velocity of BD, acceleration of slider, and angular acceleration of AB.

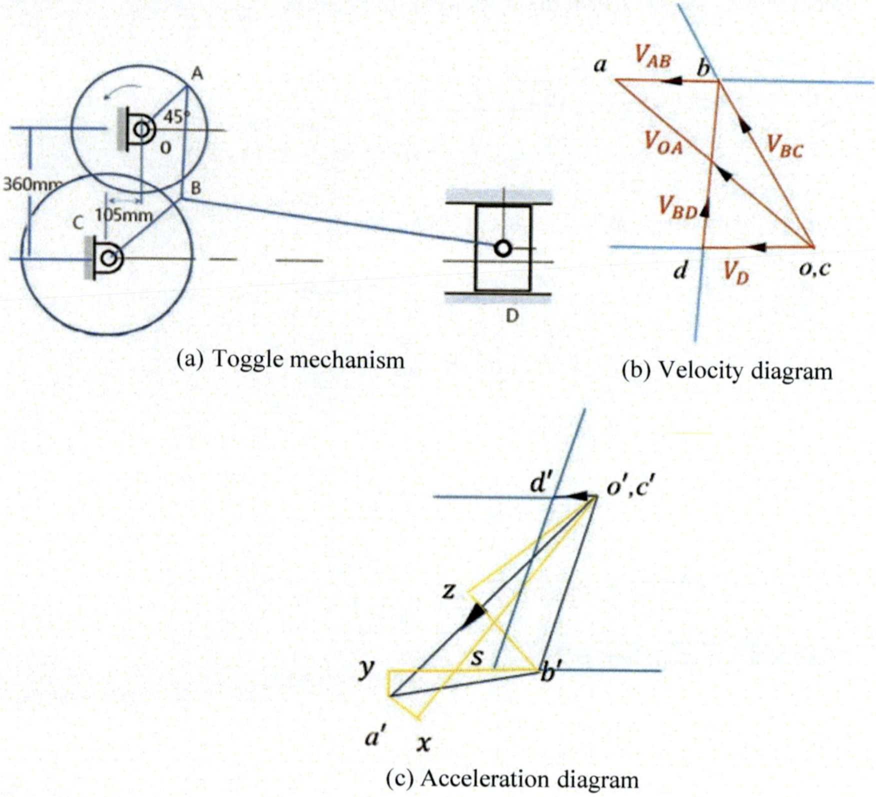

(a) Toggle mechanism

(b) Velocity diagram

(c) Acceleration diagram

Fig. 4.36 Toggle mechanism

Solution

Because $N_{OA} = 200$ RPM, the velocity of link OA is

$$\omega_{OA} = \frac{2\pi N_{OA}}{60} = \frac{2\pi \cdot 180}{60} = 18.85 \text{ rad/s}$$

$$V_{OA} = \omega_{OA} \cdot OA = 18.85 \cdot 0.18 = 3.4 \text{ m/s}$$

From velocity diagram, velocities at AB, BC and BD are

$$vector\ ab = V_{AB} = 0.9 \text{ m/s}$$

$$vector\ bc = V_{BC} = 2.8 \text{ m/s}$$

$$vector\ bd = V_{BD} = 2.4 \text{ m/s}$$

$$vector\ cd = V_D = 2.05 \text{ m/s}$$

$$\omega_{BD} = \frac{V_{BD}}{BD} = \frac{2.4}{0.54} = 4.5 \text{ rad/s}$$

The acceleration at link AO is,

$$a^t_{AO} = \alpha_{AO} \cdot OA = 50 \cdot 0.18 = 9.0 \text{ m/s}^2$$

$$a^r_{AO} = \frac{V^2_{AO}}{OA} = \frac{3.4^2}{0.18} = 63.9 \text{ m/s}^2$$

The (angular) acceleration at link DB is,

$$a^r_{DB} = \frac{V^2_{DB}}{DB} = \frac{2.4^2}{0.54} = 10.8 \text{ m/s}^2$$

$$a^t_{DB} = vector\ sd' = 38.5 \text{ m/s}^2$$

$$\alpha_{DB} = \frac{a^t_{DB}}{DB} = \frac{38.5}{0.54} = 71.3 \text{ rad/s}^2$$

The accelerations at links D, AB and BC are,

$$vector\ c'd' = a_D = 13.3 \text{ m/s}^2$$

$$a^r_{BA} = \frac{V^2_{BA}}{AB} = \frac{0.9^2}{0.36} = 2.25 \text{ m/s}^2$$

$$a^r_{BC} = \frac{V^2_{BC}}{BC} = \frac{2.8^2}{0.24} = 32.5 \text{ m/s}^2$$

4.3.6.3 Coriolis Component or Acceleration

First of all, we think the sliding object on the link (Fig. 4.37).

As a sliding object moves $d\theta$ for dt, the arc EF that is additionally moved due to (angular) velocity is

$$\begin{aligned} Arc\ EF &= Arc\ CF - Arc\ CE = Arc\ CF - Arc\ BD \\ &= OC\ d\theta - OB\ d\theta = (OC - OB)d\theta = BC\ d\theta = DE\ d\theta \end{aligned} \tag{4.22}$$

Because DE is the distance of slider for dt, Eq. (4.22) is

Fig. 4.37 A sliding object on the link

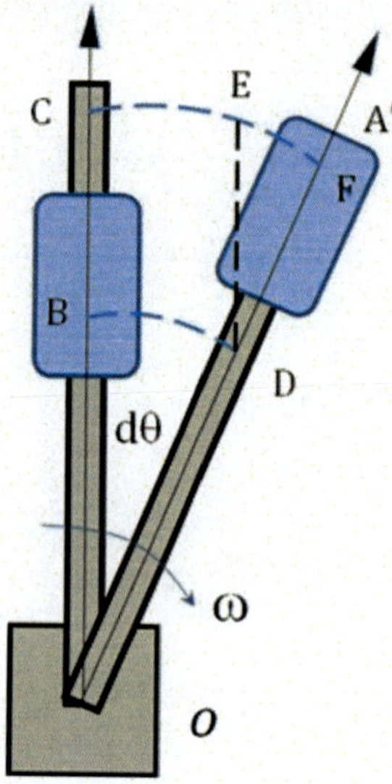

$$Arc\ EF = v\omega dt^2 \tag{4.23}$$

where $DE = motion\ of\ slider \times time = vdt,\ d\theta = \omega dt$

Because $h = ut + \frac{1}{2}ft^2\ and\ u = 0$, acceleration, f^{CC}, due to Coriolis Component is

$$v\omega dt^2 = \frac{1}{2}f^{CC}(dt)^2 \text{ or } f^{CC} = 2\omega v \tag{4.24}$$

If vector form is represented, it is

$$f^{CC} = 2\omega \times v \tag{4.25}$$

(compared with Eq. (2.18) in Chap. 2 as reference).

4.3.6.4 Instantaneous Center of Velocity

For anybody undergoing planar motion, there always exists a point in the plane of motion at which the velocity is instantaneously zero (if it were rigidly connected to the body). This point is called the instantaneous center of zero velocity, or IC. If the location of this point can be determined, the velocity analysis can be simplified because the body appears to rotate about this point at that instant.

Example 4.7 A four-link bar joined by pin is as shown in the Fig. 4.38. As link AB rotates at 100 RPM, find the angular velocity of the link BC using instantaneous center of rotation when angle BAD is 60°.

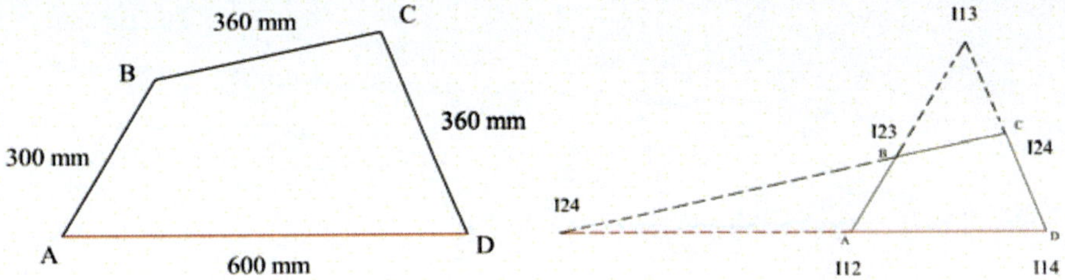

(a) Four-link bar (b) Instantaneous center

Fig. 4.38 A four-link bar joined by pin

$$\omega_{AB} = \frac{2\pi N_{AB}}{60} = \frac{2\pi \cdot 100}{60} = 10.47 \text{ rad/s}$$

Because AB = 300 mm = 0.30 m, velocity at point B is

$$V_{BA} = \omega_{AB} \cdot AB = 10.47 \cdot 0.30 = 3.141 \text{ m/s}$$

$$V_B = \omega_{BC} \cdot \text{I13B}$$

$$\omega_{AB} = \frac{V_B}{\text{I13B}} = \frac{3.141}{0.495} = 6.345 \text{ rad/s}$$

Example 4.8 A slider crank mechanism is as shown in the Fig. 4.39. The crank BC rotates at 100 at 100 RPM. Find the angular velocity of the link BC using instantaneous center of rotation when angle BCA is 45°.

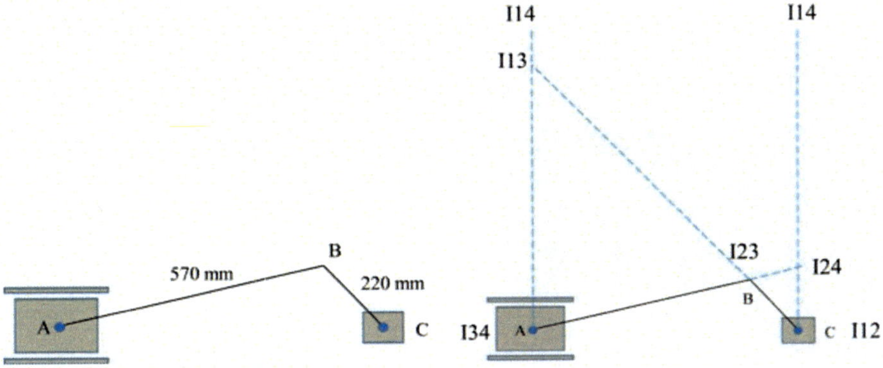

(a) Slider crank (b) Instantaneous center

Fig. 4.39 A slider crank mechanism

$$\omega_{BC} = \frac{2\pi N_{BC}}{60} = \frac{2\pi \cdot 100}{60} = 10.47 \text{ rad/s}$$

Because BC = 220 mm = 0.22 m, velocity at point B is

$$V_B = \omega_{BC} \cdot BC = 10.47 \cdot 0.22 = 2.303 \text{ m/s}$$

$$V_A = V_B \cdot \frac{\text{I13A}}{\text{I13B}} = 2.303 \cdot \frac{0.68}{0.76} = 2.061 \text{ m/s}$$

$$\omega_{AB} = \frac{V_B}{\text{I13B}} = \frac{2.303}{0.76} = 3.030 \text{ rad/s}$$

Example 4.9 A mechanism of wrapping machine is as shown in the Fig. 4.40. O_1A = 100 mm, AC = 700 mm, BC = 200 mm, O_3C = 200 mm, O_2E = 400 mm, O_2D = 200 mm and BD = 150 mm. Crank O_1A rotates at uniform speed of 100 rad/s. Find the velocity of the point E of the bell crank lever by IC method.

By measurement, we know that

$I_{13}A = 910$ mm = 0.91 m; $I_{13}B = 820$ mm = 0.82 m; $I_{15}B = 130$ mm = 0.13 m;

$I_{15}D = 50$ mm = 0.05 m; $I_{16}D = 200$ mm = 0.2 m; $I_{16}E = 400$ mm = 0.4 m;

Let v_E = velocity of point E on the belt crank lever, v_B = velocity of point B, v_D = velocity of point D.

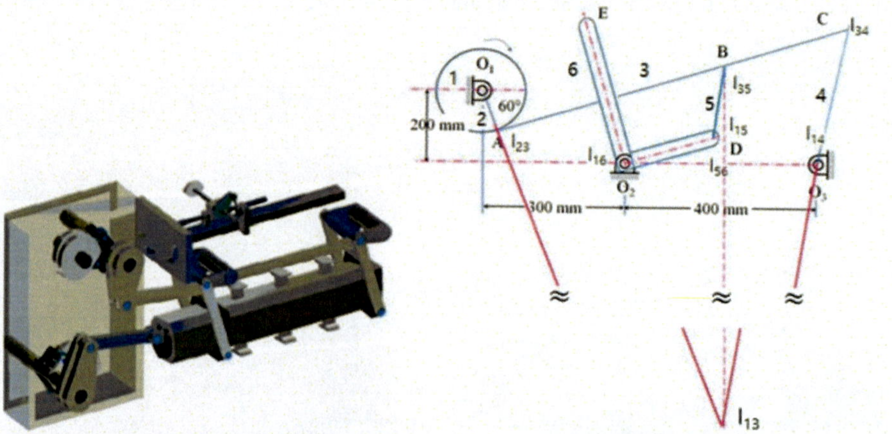

(a) Wrapping machine (b) Instantaneous center

Fig. 4.40 A wrapping machine mechanism

We know that $\frac{v_A}{I_{13}A} = \frac{v_B}{I_{13}B}$, $v_B = \frac{v_A}{I_{13}A} \times I_{13}B = \frac{10}{0.91} \times 0.82 = 9.01$ m/s.
$\frac{v_B}{I_{15}B} = \frac{v_D}{I_{15}D}$, $v_B = \frac{v_B}{I_{15}B} \times I_{15}D = \frac{9.01}{0.13} \times 0.05 = 3.46$ m/s.
$\frac{v_D}{I_{16}D} = \frac{v_E}{I_{16}E}$, $v_E = \frac{v_D}{I_{16}D} \times I_{16}E = \frac{3.46}{0.2} \times 0.4 = 6.92$ m/s.

4.3.6.5 Klein's Construction

Klein's construction is used to draw the velocity and acceleration diagrams for a single slider crank mechanism. The velocity and acceleration of piston of a reciprocating engine mechanism can be determined by Fig. 4.41.

The acceleration of the piston C with respect to crank pin B (i.e. f_{BC}) may be obtained as follows:

$$\omega^2 = \frac{f_{BC}}{BQ} \tag{4.26}$$

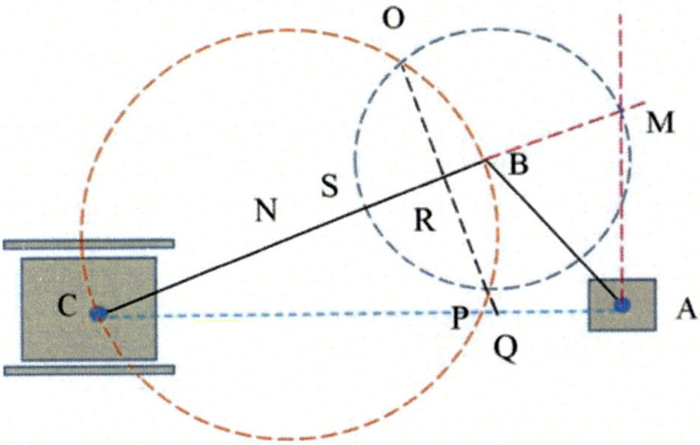

(a) Reciprocating engine mechanism

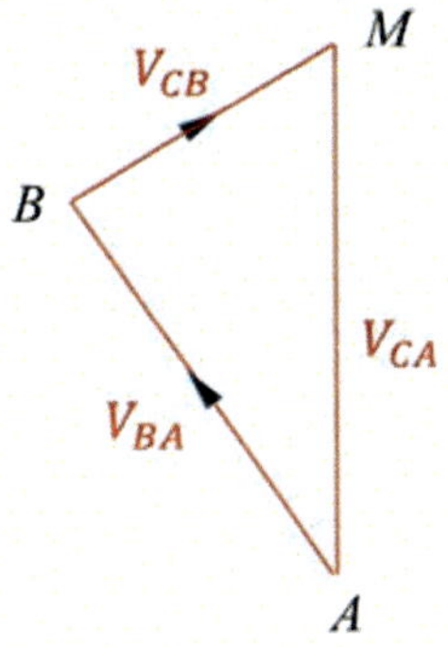

(b) Velocity diagram

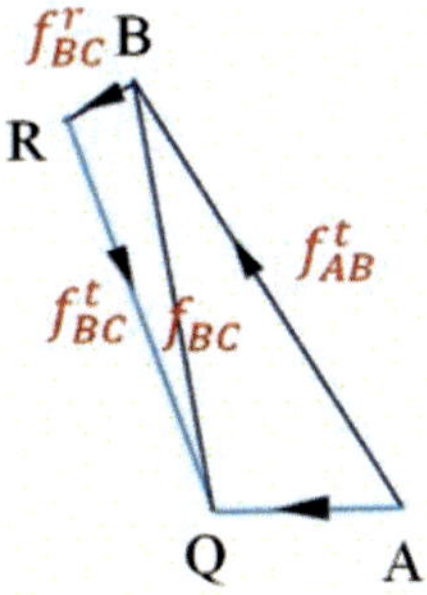

(c) Velocity diagram

Fig. 4.41 Klein's construction

To find the velocity of any point D on the connecting rod BC. Divide BM at f_{BC} in the same ratio as D divides BC (Fig. 4.42a).

$$\frac{BD_1}{BM} = \frac{BD}{BC} \tag{4.27}$$

Velocity at point D is

$$V_D = \omega \times AD \tag{4.28}$$

To find the acceleration of any point D on the connecting rod BC. Draw a line from point D || to AC which intersects BQ_2 of D_2 (Fig. 4.42b).

$$f_D = \omega^2 \times AD_2 \tag{4.29}$$

Example 4.10 Determine the velocity and acceleration of the piston in a slider crank mechanism by Klein's construction to the following specification: (1) stroke = 300 mm, (2) ratio of length of connecting rod to crank = 4, (5) speed of the engine = 300 RPM, and (6) piston of crank = 45° with inner dead center (Fig. 4.43).

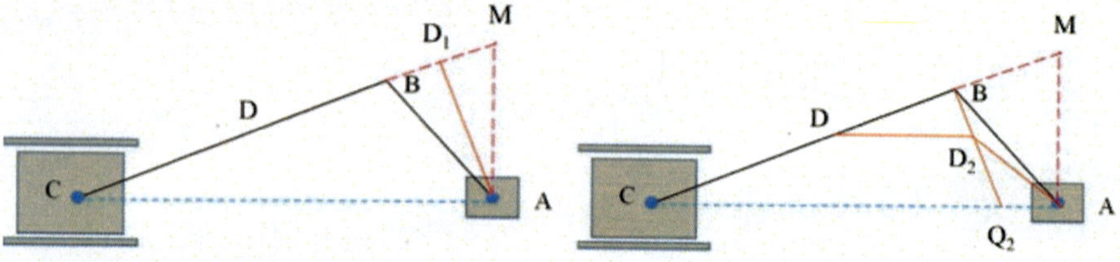

(a) Velocity (b) Acceleration

Fig. 4.42 Klein's construction in reciprocating engine

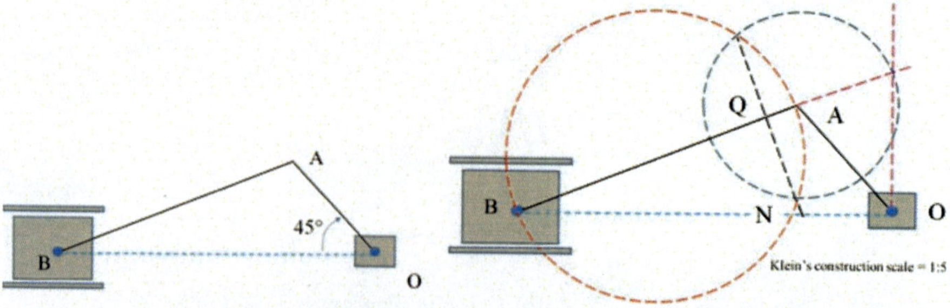

(a) Slider crank mechanism (b) Klein's velocity polygon

Fig. 4.43 A slider crank mechanism by Klein's construction

Solution

$$\omega_{OA} = \frac{2\pi N}{60} = \frac{2\pi 300}{60} = 31.41 \text{ rad/s}$$

Because engine speed is 300 RPM, ΔOAM is Klein's velocity polygon

$$V_B = OM \times \omega \times SF = 24.5 \times 31.41 \times 5 = 3847.725 \text{ rad/s}$$

To find acceleration of piston, it is

$$f_B = \omega^2 \times ON \times SF = 31.41^2 \times 21 \times 5 = 1.035 \times 10^5 \text{ mm/s}^2$$

Example 4.11 The lengths of the crank and connecting rod of a reciprocating engine are 200 mm and 800 mm respectively. The crank is rotating at a uniform speed of 480 RPM. Using Klein's construction, find acceleration of the piston (Fig. 4.44).

Solution

The acceleration of the middle point of the connecting rod and angular acceleration of the connecting rod when the crank had turned through 45° from the inner dead center.

$$\omega_{crank} = \frac{2\pi N}{60} = \frac{2\pi \times 480}{60} = 50.26 \text{ rad/s}$$

Acceleration of piston is

$$f_c = \left[AN - \omega^2\right] = \left[14 \times 10 \times \omega^2\right] = 3537 \times 10^5 \text{ mm/s}^2$$

Angular acceleration of connecting rod is

Fig. 4.44 Klein's velocity polygon

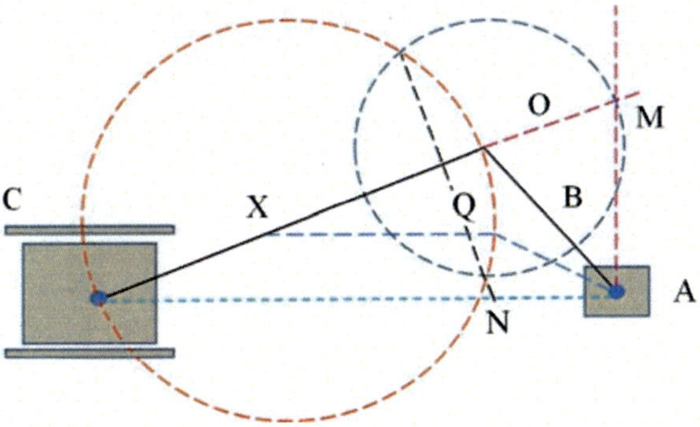

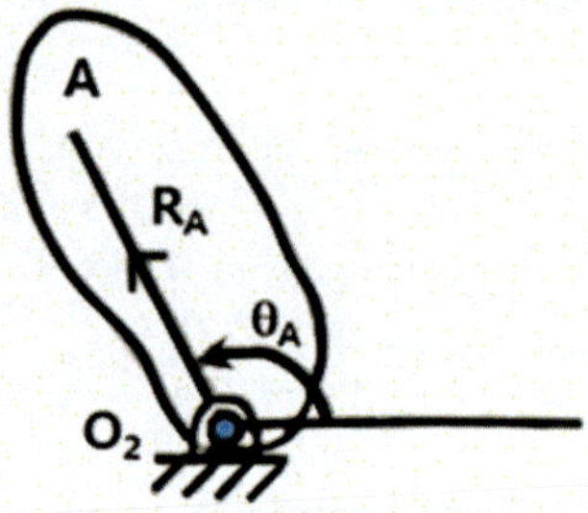

Fig. 4.45 Analytical solution (example)

$$a_{ac} = \frac{f'BC}{BC} = \frac{QN \times \omega^2}{BC} = 14 \times 10 \times 50.26^2/800 = 442.06 \text{ rad/s}^2$$

4.3.6.6 Velocity and Acceleration Analysis of Mechanisms (Example)

To complete the kinematics analysis of mechanisms, we will discuss the most common and currently practiced methods that describe the motion through velocity and acceleration. First, position and velocity analysis must be completed before acceleration analysis. All acceleration components should be expressed as one coordinate system—the inertial frame of reference of the fixed link of the mechanism. For example, we simply consider that a link at center O_2 rotates with radius R_A. Find velocity and acceleration at point *An* in Fig. 4.45.

If this link is represented as a polar coordinate, we can describe the displacement of point A:

$$R_A = r_A e^{i\theta_A} \tag{4.30}$$

If Eq. (4.30) is differentiated, we shall discover the velocity of the link:

$$V_A = \frac{dR_A}{dt} = i\omega_2 r_A e^{i\theta_A} \tag{4.31}$$

If Eq. (4.31) is differentiated, we can find the acceleration of the link:

$$\begin{aligned} a_A = \frac{dV_A}{dt} &= i\frac{d\omega_2}{dt} r_A e^{i\theta_A} + i\omega_2 r_A \frac{d\left(e^{i\theta_A}\right)}{dt} \\ &= i\alpha_2 r_A e^{i\theta_A} + i\omega_2 r_A i\omega_2 e^{i\theta_A} \\ &= \underbrace{i\alpha_2 R_A}_{a_A^t} - \underbrace{\omega_2^2 R_A}_{a_A^n} = \left(-\omega_2^2 + i\alpha_2\right) R_A \end{aligned} \tag{4.32}$$

4.4 Design of the Belt Drive

A belt is a flexible mechanical component which transfers power from one shaft to another. The quantity of power transferred by the belt relies on (1) the belt velocity, (2) the belt tension, (3) the contact arc between the belt and smaller pulley, and (4) the circumstance under which the belt is utilized. That is, we can choose the belt drive as follows: (1) speed of the driving and driven shafts, (2) power to be transferred, (3) positive drive needs, (4) space available, (5) speed reduction proportion, (6) center interval between the shafts, (7) shaft arrangement, and (8) service circumstances.

According to the matter utilized in belts, they are grouped as (1) leather belts, (2) cotton or fabric belts, and (4) rubber belts. The belt type is as follows: (1) flat belt, (2) V-belt, and (3) circular belt (Fig. 4.46).

There are two belt drive types as follows: (1) Open belt drive: The open belt drive is utilized with shafts arranged parallel and rotating in the same orientation. (2) Crossed belt drive: The crossed belt drive is employed with shafts arranged parallel and rotating in the opposite direction (Fig. 4.47).

Other belt drives are compound belt drives that are utilized when power is transferred from one shaft to another through a number of pulleys and stepped or cone pulley drives that are utilized for altering the speed of the driven shaft while the main shaft moves at constant speed.

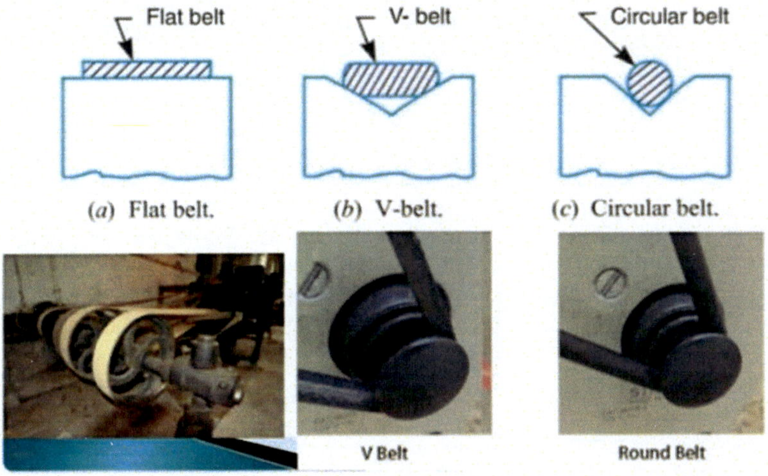

Fig. 4.46 Belt type

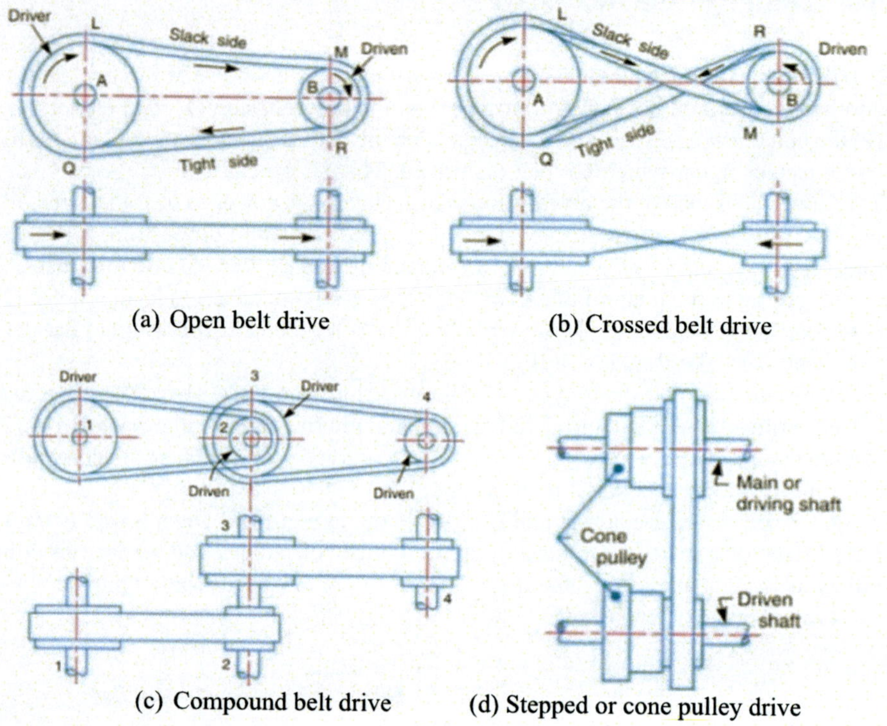

Fig. 4.47 Belt drive types

4.4.1 Length of the Belt and Contact Angle

Because the contact angle can be approximated as $\alpha \approx \frac{(r_1 - r_2)}{x}$, we can calculate the distance of the open belt drive as follows (Fig. 4.48):

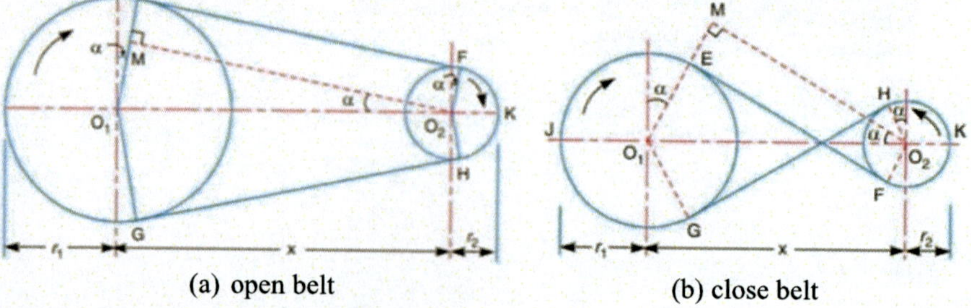

Fig. 4.48 Schematic diagram of open (or closed) belt

$$L = \pi(r_1 + r_2) + 2x + (r_1 - r_2)\sin\alpha = \pi(r_1 + r_2) + 2x + \frac{(r_1 - r_2)^2}{x} \quad (4.33)$$

where $\sin\alpha \approx \frac{(r1-r2)}{x}$

4.4.2 *Proportion of Driving Tensions for Flat Belt*

Think about a driven pulley revolving in the clockwise orientation, as depicted in Fig. 4.49.

Think about a little part of the belt PQ, subtending an angle $\delta\theta$ at the center of the pulley. The belt PQ is in equilibrium under a driven pulley under the succeeding forces: (1) Tension T at P, (2) Tension $(T + \delta T)$ at Q, (3) Normal reaction R_N, (4) Friction force, $F = \mu\, R_N$.

When solving all the forces horizontally and equating the same, we can find the reaction force. That is,

$$R_N = (T + \delta T)\sin\frac{\delta\theta}{2} + T\sin\frac{\delta\theta}{2} \quad (4.34)$$

Because the angle $\delta\theta$ is very little and $\sin\frac{\delta\theta}{2} = \frac{\delta\theta}{2}$, Eq. (4.23) can be simplified as follows:

$$R_N = (T + \delta T)\sin\frac{\delta\theta}{2} + T\sin\frac{\delta\theta}{2} = \frac{T\delta\theta}{2} + \frac{\delta T\delta\theta}{2}(\approx 0) + \frac{T\delta\theta}{2} = T\delta\theta \quad (4.35)$$

Solving the forces in vertical, we have

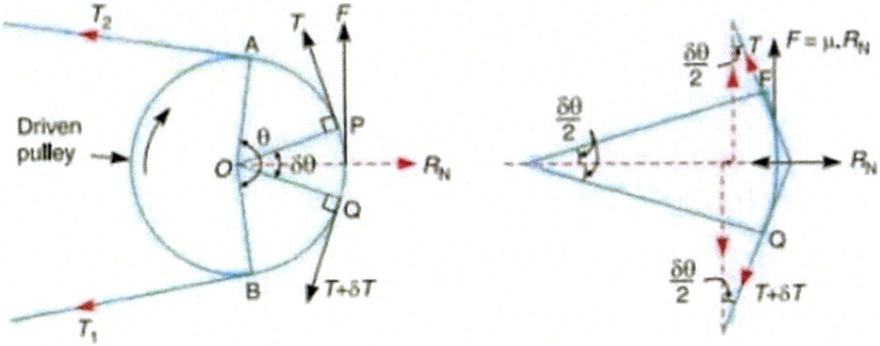

Fig. 4.49 Free body diagram near the belt contact point

$$\mu R_N = (T + \delta T)\cos\frac{\delta\theta}{2} - T\cos\frac{\delta\theta}{2} \tag{4.36}$$

Because the angle $\delta\theta$ is very little, $\cos\frac{\delta\theta}{2} = 1$.

$$\mu R_N = T + \delta T - T = \delta T \tag{4.37}$$

When equating the values of R_N, we can obtain the following:

$$T\delta\theta = \delta T/\mu \quad \text{or} \quad \delta T/T = \mu\delta\theta \tag{4.38}$$

Integrating both sides between the limits T_2 and T_1 and from 0 to θ, separately, we can obtain the following:

$$\int_{T_2}^{T_1} \frac{\delta T}{T} = \mu \int_0^{\theta} \delta\theta \quad \text{or} \quad \log\left(\frac{T_1}{T_2}\right) = \mu\theta, \quad \frac{T_1}{T_2} = e^{\mu\theta} \tag{4.39}$$

Therefore, belt V over the Flat Belt has the following advantages: (1) the V-belt drive takes compactness due to the little distance between the centers of pulleys, (2) the drive is positive because the slip between the belt and the pulley groove is trivial, (3) because the V-belts are made endless and there is no joint trouble, the drive is thus smooth, (4) it supplies a longer life, 3 to 5 years, (5) it can be simply mounted and eliminated, (6) the functioning of the belt and pulley is silent, (7) the belts have the capacity to protect the shock when machines are begun, and (8) the high velocity proportion (maxi. 10) can be attained.

Example 4.12 An open belt operating over two pulleys 240 and 600 mm attaches two parallel shafts 3 m apart. The power to be transferred is 4 kW from the smaller pulley, which revolves at 300 RPM. The coefficient of friction between the belt and pulley is 0.3, and the safe working tension is 10 N per mm width. Decide (1) least width of belt, (2) beginning belt tension, (3) length of belt necessitated (Fig. 4.50).

Fig. 4.50 An open belt revolving over two pulleys

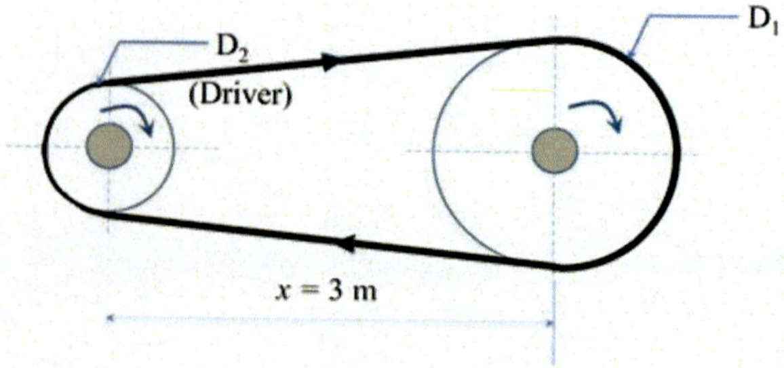

Solution

Given data: $D_1 = 600$ mm, $D_2 = 240$ mm, P = 4 kW, $\mu = 0.3$, safe tension = 10 N/mm width

The velocity of the belt going over a smaller pulley is specified by the following:

$$V_2 = \frac{\pi D_2 N_2}{60} = \frac{\pi \times 0.24 \times 300}{60} = 3.77 \text{ m/s}$$

Now, the power transferred by the belt driver

$$P = (T_1 - T_2) \cdot V_2 \text{ That is, } 4 \times 10^3 = (T_1 - T_2) \cdot 3.77 \text{ or } (T_1 - T_2) = 1061$$

The angle of the lap for the open belt drive is given by the following:

$$\theta = 180 - 2\alpha = 180 - 2 \times 3.44 = 173.12^\circ \times \frac{\pi}{180} = 3.022 \text{ rad}$$

where $\alpha = \sin^{-1}\left(\frac{D_1 - D_2}{2x}\right) = \sin^{-1}\left(\frac{0.6 - 0.24}{2 \times 3}\right) = 3.44^\circ$

The belt tension ratio is given by the following:

$$\frac{T_1}{T_2} = e^{\mu\theta} = e^{0.3 \times 3.022} = 2.5$$

$(2.5T_2 - T_2) = 1061$. That is, $T_2 = 718N$, $T_1 = 1779N$

Because 1 mm $width$ = 10 N, b = Width of belt required = 1779/10 = 177.9 mm

Initial tension in the belt (T_o)

$$T_0 = (T_1 + T_2)/2 = (1779 + 718)/2 = 1248.5\text{N}$$

Length of open belt drive

$$\text{L} = \pi(0.3 + 0.12) + (2 \times 3) + (0.3 - 0.12)2/3 = 7.33 \text{ m}$$

4.5 Design of the Gear Drive

4.5.1 Introduction

A gear is a part of a transmission implement which transfers rotational force to another gear or device. The advantages of a gear drive are as follows: (1) it transfers a precise velocity proportion, (2) it can be utilized to transfer large powers, (3) it has high efficiency, (4) it has good service, and (5) it has a close-packed arrangement. On the other hand, the disadvantages of Gear Drive are as follows: (1) the manufacture of gears necessitates particular tools and apparatus, (2) the error in cutting teeth can create vibration and noise in functioning.

The different types of gears are present in many sectors, such as:

- The agricultural sector, in which they play a key role in carrying out mechanized tasks, such as sowing, ploughing or irrigation, as well as in the tractors themselves.
- In the automotive field, their function is usually to act as transmitters of forces and to regulate speed.
- As for naval vehicles, gears operate on fishing boats, submarines, workboats or yachts.
- In the generation of wind power, gears increase the speed of generators, a function that is also used by cement manufacturing industries. Roller mills are used for the transport of slabs and for wire rolling mills.
- Hydraulic pump: It converts rotary mechanical energy into hydraulic energy. It consists of a pair of coupled gears and has the driven shaft and the driver, which is driven by the motor shaft. This one, due to the displacement caused by the contact between the teeth of the gears, rotates the driven shaft.
- Speed reducer: They used circular and toothed gear pairs to power the motor speed efficiently and safely. In addition, they use gears with very different diameters to reduce the speed of rotation.
- Differential: Widely used in the automotive sector, it makes it easier for the two driving wheels of a vehicle to turn at different speeds than the others. A differential is made up of two planetary gears attached to the ends of the semi-axles of the wheels, and two other satellites or conical pinions located at the ends of their satellite-carrying axle.
- It couples the motor with the transmission system through gear ratios. It also reduces the engine speed. Some ball bearings support the gear shafts. It is coupled to the motor flywheel via the clutch or torque converter.

4.5.2 *Types of Gears*

In accordance with the position of axes of the shafts, we can be classified as follows (Fig. 4.51).

There are several types of gears as follows:

(1) Spur gear: in a pair of mating spur gears the axes of these two gears are parallel. The gear teeth are straight along the length and are parallel to the axis of the gear. Hence they are not subjected to axial thrust due to tooth load. Spur gears can have both external and internal teeth. It has high power transmission efficiency (98–99%). It transmits a large quantity of power (≈50,000 kW), is compact, is uncomplicated to mount, and has no slip. The teeth are parallel to the axis of rotation. It transmits power from one shaft to another parallel shaft. It can be utilized in metal cutting, power plants, and fuel pumps. oscillating sprinkler, windup alarm clock, washing machine and clothes dryer.

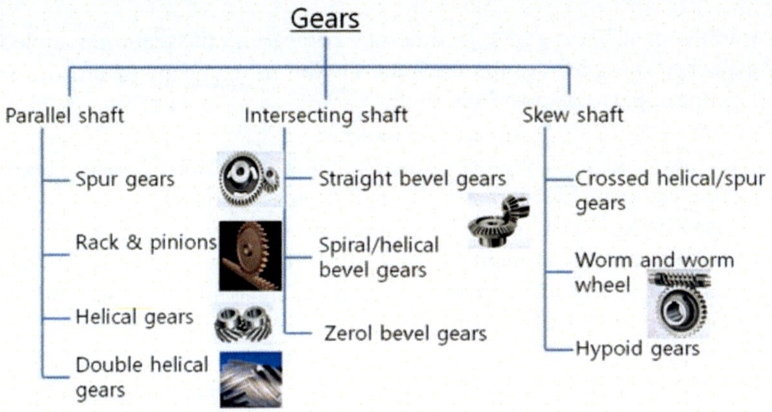

Fig. 4.51 Gears in accordance with the position of axes of the shafts

(2) For external and internal gears, an external gear is one with the teeth shaped on the outside surface of a cylinder or cone. On the other hand, an internal gear is one with the teeth shaped on the inner surface of a cylinder or cone. Worm gears are utilized when large gear reductions are necessitated (20:1). Even up to 300:1 or greater (less space/good meshing), worm gear materials are expensive.

(3) Spur rack and pinion: a spur rack is a straight sided gear and are special cases of a spur gear that is made of infinite diameter so that the pitch surface is a plane. The common application of this arrangement is in translating rotary motion to linear motion. It is utilized in a lathe, cranes and actuators, and rack railways. They do not have a transmission ratio, but a length ratio. In this case, it is referred to as the distance between axis, since the rack falls in the category of gears of infinite diameter.

(4) Helical gears: In helical gears the teeth are at a helix angle α with respect to the axis of the gear. It shows a pair of opposite hand helical gear in mesh. Their axes are parallel and have same helical angle. Helical gears are less noisy and more load, but costly. This gradual engagement makes helical gears work much more evenly and silently than spur gears. It can be applied as machine tools, turbine drives, feed drives, sand mullers, rolling mills and marine applications, and actuators.

(5) Double helical gears: these are formed by joining two helical gears of identical pitch and diameter but of opposite hand on the same shafts. These two sets of teeth are often cut on the same gear blank. Axial thrust that occurs in single helical gears is eliminated in double helical gears hence.

(6) Straight bevel gears: in the type of gearing, the axes are intersecting at 90° to each other, known as straight bevel gears. This 90° of axis intersection is known as shaft angle.

(7) Spiral/helical bevel gears: in this type of gearing, the teeth are angled with respect to the axis, i.e. the teeth are curved in the shape of a spiral so that the contact between their intersecting/intermeshing teeth begins gradually and continues smoothly from one end to other.
(8) Zerol bevel gears: They are special form of spiral/helical bevel gears with curved teeth, and having a zero degree mean spiral angle.
(9) Crossed helical/spiral gears: In case of non-parallel, non-intersecting shafts known as skew shafts, pure rolling contact is not possible unlike parallel and intersecting shafts. The action of crossed helical/spiral gear consists primarily of screwing or wedging as a resultant high degree of sliding on the tooth flanks.
(10) Worm gears: in worm gears, the axis are non-intersecting and the planes containing the axes containing the axes are normally at right angle to each other. It can effortlessly revolve the gear, but the gear cannot revolve the worm. It may be applied as gates and conveyor belts, elevators/lifts, material handling and transportation machinery, machine tools, and cars.
(11) Hypoid gears: It is similar to spiral bevel gearing, but have non-intersecting axes, i.e. the axis of the pinion is offset relative to the gear axis. The planes containing the axis are to each other. The teeth in mesh have line contact.
(12) For herringbone gears, to stop from axial thrust, two helical gears of opposite hands shall be installed side by side to eliminate the resulting thrust forces. Herringbone gears are mainly utilized on heavy machinery. A noncircular gear (NCG) is a special gear design with particular attributes and purpose. Examples are textile machines, potentiometers, CVTs (continuously variable transmissions), window shade, panel drives, mechanical presses, and high torque hydraulic engines.
(13) Main implementations of epicyclic gears are in transmission, calculating devices, wrists watches, hoists, back gear of lathe, etc. Bevel gears are functional when the orientation of a shaft's rotation needs to be altered. The teeth on bevel gears shall be straight, spiral or hypoid. Examples are differential drives in automobiles, mining machine equipment, spiral bevel gear in rotorcraft, marine implementations, printing processes, cooling towers, power plants, steel plants, railway tracks, etc.

4.5.3 Nomenclature of (Spur) Gears

There are some nomenclature used in (spur) gears as follows (Fig. 4.52):

(1) Pitch surface: the surface of the imaginary rolling cylinder (cone, etc.)
(2) Pitch circle: a tight section of the pitch surface
(3) Addendum circle: a circle bounding the ends of the teeth
(4) Root circle: the circle bounding the spaces between the teeth
(5) Addendum: radial distance between pitch circle and addendum circle
(6) Dedendum: radial distance between the pitch circle and the root circle

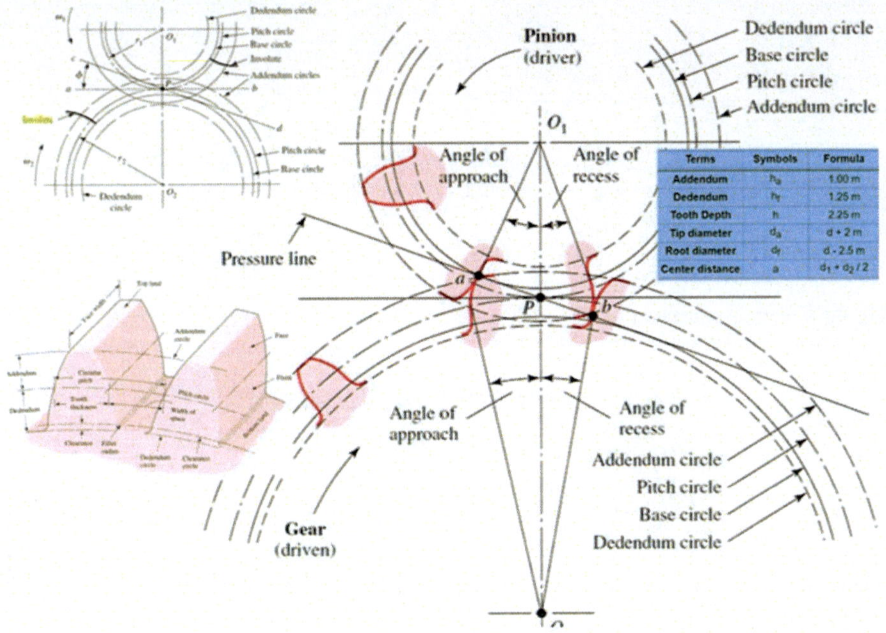

Fig. 4.52 Nomenclature of (spur) gears

(7) Clearance: difference between the dedendum of one gear and the addendum of the mating gear
(8) Face of a tooth: Part of tooth surface lying outside the pitch surface
(9) Flank of a tooth: Part of tooth surface lying inside the pitch surface
(10) Circular thickness: Thickness of tooth measured on the pitch circle
(11) Tooth space: Distance between adjacent teeth measured on pitch circle
(12) Backlash: Distance between the circle thickness of one gear and the tooth space of the mating gear
(13) Circular pitch (P_c): Width of a tooth and a space, measured on the pitch circle. $p = \pi d/T$ where T = number of teeth, d = pitch diameter.
(14) Diametral pitch (P_d): Number of teeth of a gear unit pitch diameter. It is defined as the number of teeth divided by the pitch diameter. That is, $P = T/d$
(15) Module: (m): Pitch diameter divided by the number of teeth. That is, m = d / T.

4.5.3.1 Length of Path of Contact

Path of contact (CD) is defined as path of approach plus path of recess (Fig. 4.53). That is,

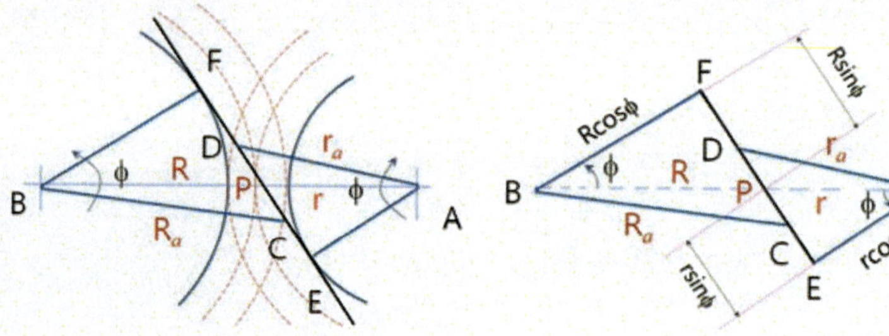

Fig. 4.53 Path of contact of (spur) gears

$$\begin{aligned} CD &= CP + PD = (FC - FP) + (ED - EP) \\ &= \sqrt{R_a^2 - R^2\cos^2\phi} - R\sin\phi + \sqrt{r_a^2 - r^2\cos^2\phi} - r\sin\phi \\ &= \sqrt{R_a^2 - R^2\cos^2\phi} - (R + r)\sin\phi + \sqrt{r_a^2 - r^2\cos^2\phi} \end{aligned} \tag{4.40}$$

where $FC = \sqrt{R_a^2 - R^2\cos^2\phi}$, $FC = R\sin\phi$, $ED = \sqrt{r_a^2 - r^2\cos^2\phi}$, $EP = r\sin\phi$

4.5.3.2 Derivation of Arc of Contact

If t_a is time to traverse the arc of approach, arc of approach PP' is

$$\begin{aligned} PP' &= [\text{tangential velocity of P}'] \times t_a = \omega_a r \times t_a = \omega_a(r\cos\phi)\frac{1}{\cos\phi} \times t_a \\ &= [\text{tangential velocity of H}] \times \frac{t_a}{\cos\phi} \\ &= \frac{HK}{\cos\phi} = \frac{FK - FH}{\cos\phi} = \frac{FP - FC}{\cos\phi} = \frac{CP}{\cos\phi} \end{aligned} \tag{4.41}$$

On the other hand, arc of recess PP" is

$$\begin{aligned} PP'' &= [\text{tangential velocity of P}] \times t_a \\ &= \omega_a r \times t_a = \omega_a(r\cos\phi)\frac{1}{\cos\phi} \times t_a \\ &= [\text{tangential velocity of K}] \times \frac{t_a}{\cos\phi} \\ &= \frac{KL}{\cos\phi} = \frac{FL - FK}{\cos\phi} = \frac{FD - FP}{\cos\phi} = \frac{PD}{\cos\phi} \end{aligned} \tag{4.42}$$

Arc of contact is defined as arc of approach plus arc of recess (Fig. 4.54). That is,

Fig. 4.54 Arc of contact

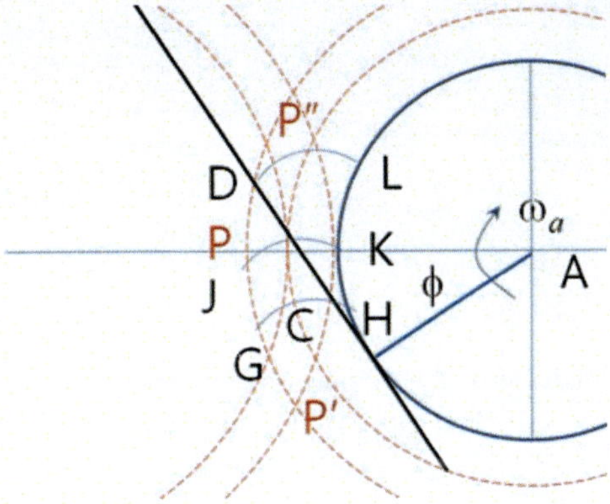

$$\text{Length of arc of contact} = PP' + PP'' = \frac{CP}{\cos\phi} + \frac{PD}{\cos\phi}$$

$$= \frac{CD}{\cos\phi}\left(= \frac{\text{Length of Path of contact}}{\cos\phi}\right) \quad (4.43)$$

$$\text{Contact ratio(CR)} = \frac{\text{Length of arc of contact}}{circular\ pitch}$$

$$= \frac{\frac{CD}{\cos\phi}}{\pi m} = \frac{CD}{\pi \cdot m \cdot \cos\phi} \quad (4.44)$$

Example 4.13 The number of teeth on each of the two equal spur gears in mesh are 40. The teeth have 20° involute profile and the module is 6 mm. If the arc of contact is 1.75 times the circular pitch, find the addendum.

Solution

Given: T = t = 40, ϕ = 20°, m = 6 mm.

WKT the circular pitch,

$$p = \pi \text{m} = \pi \times 6 = 18.85 \text{ mm}$$

$$\text{Lengh of arc of contact} = 1.75 \times 18.85 = 33 \text{ mm}$$

Hence length of path of contact = Lengh of arc of contact $\times \cos\phi$ = 33 cos 20° = 31 mm.

Let $R_a = r_a = radius\ of$ addendum circle $of\ each\ wheel$

$$R = r = \text{m} \cdot T/2 = 6 \times 40/2 = 120 \text{ mm}$$

Because $R = r$ and $R_a = r_a$, we know that

$$CD = \sqrt{R_a^2 - R^2\cos^2\phi} - (R + r)\sin\phi + \sqrt{r_a^2 - r^2\cos^2\phi}$$

$$= 2\left[\sqrt{R_a^2 - R^2\cos^2\phi} - R\sin\phi\right]$$

$31 = 2\left[\sqrt{R_a^2 - 120^2\cos^2 20} - 120\sin 20\right]$ or $R_a = 126.12$ mm.
So, the addendum of the wheel $= R_a - R = 126.12 - 120 = 6.12$.

Example 4.14 A pair of gears that have 40 teeth and 20 teeth are respectively rotating in mesh. The speed of the smaller gear is 2000 RPM. Determine the velocity of sliding between the gear teeth faces at the point of engagement, at the pitch point and at the point of disengagement, if the smaller gear is the driver. Assume that the gear teeth are 20° involute form, module is 5 mm and addendum length is 1 module. Also find the angle through the pinion turns while any pairs of teeth are in contact and the contact ratio.

Solution

Given: T = 40, t = 20, N_1 = 2000 RPM, ϕ = 20°, addendum = 5 mm, m = 5 mm.
Velocity of smaller gear (pinion),

$$\omega_1 = \frac{2\pi N_1}{60} = \frac{2\pi \times 2000}{60} = 209.5 \text{ rad/s}$$

Velocity of larger gear,

$$\omega_2 = \omega_1 \frac{t}{T} = 209.5 \times \frac{20}{40} = 104.75 \text{ rad/s}$$

Pitch circle radius of pinion

$$\mathrm{r} = \mathrm{m} \cdot \frac{t}{2} = 5 \times \frac{20}{2} = 50 \text{ mm}$$

Pitch circle radius of gear

$$\mathrm{R} = \mathrm{m} \cdot \frac{T}{2} = 5 \times \frac{40}{2} = 100 \text{ mm}$$

Addendum circle radius of pinion

$$r_a = r + \text{Addendum} = 50 + 5 = 55 \text{ mm}$$

Addendum circle radius of gear

$$R_a = R + \text{Addendum} = 100 + 5 = 105 \text{ mm}$$

Length of path of contact $CD = CP + PD$

$$CP = \sqrt{R_a^2 - R^2\cos^2\phi} - R\sin\phi = \sqrt{105^2 - 100^2\cos^2 20} - 100\sin 20$$
$$= 12.65 \text{ mm}$$

$$PD = \sqrt{r_a^2 - r^2\cos^2\phi} - r\sin\phi = \sqrt{55^2 - 50^2\cos^2 20} - 50\sin 20$$
$$= 11.5 \text{ mm}$$

Velocity of sliding at point of engagement C

$$V_{sc} = (\omega_1 + \omega_2)CP = (209.5 + 104.75)12.65 = 3975 \text{ mm/s}$$

Velocity of sliding at pitch point P

$$V_{sp} = 0$$

Velocity of sliding at point of disengagement D

$$V_{sc} = (\omega_1 + \omega_2)PD = (209.5 + 104.75)11.5 = 3614 \text{ mm/s}$$

$$\text{Length of arc of contact} = \frac{\text{Length of Path of contact}}{\cos\phi} = \frac{CP + PD}{\cos\phi}$$
$$= \frac{12.65 + 11.5}{\cos 20} = 25.69 \text{ mm}$$

$$\text{Contact ratio} = \frac{\text{Length of arc of contact}}{\cos\phi} = \frac{25.69}{\cos 20} = 27.34 \text{ i.e.27(approximately)}$$

Angle through which pinion turns

$$\text{Angle through which pinion turns}$$
$$= \text{Length of arc of contact} \times \frac{360°}{2\pi r_a}$$
$$= \frac{25.69 \times 360^{\circ}}{2\pi 55} = 29.45°$$

4.5.3.3 Minimum Number of Teeth on Pinion and Gear to Avoid Interference

From triangle O_1PN (Fig. 4.55)

$$\begin{aligned}(O_1N)^2 &= (O_1P)^2 + (PN)^2 - 2 \times O_1P \times PN \cos O_1PN \\ &= r^2 + R^2\sin^2\phi - 2rR\sin\phi\cos(90^\circ + \phi) = r^2 + R^2\sin^2\phi + 2rR\sin^2\phi \\ &= r^2\left[1 + \frac{R^2\sin^2\phi}{r^2} + \frac{2R\sin^2\phi}{r}\right] = r^2\left[1 + \frac{R}{r}\left(\frac{R}{r} + 2\right)\sin^2\phi\right]\end{aligned} \tag{4.45}$$

where t is number of teeth on the pinion, T is number of teeth on the wheel, m is module of the teeth, r is pitch circle radius of pinion, G is gear ratio, ϕ is pressure angle

Limiting radius of the pinion addendum circle is

$$O_1N = r\sqrt{1 + \frac{R}{r}\left[\frac{R}{r} + 2\right]\sin^2\phi} = \frac{mt}{2}\sqrt{1 + \frac{T}{t}\left[\frac{T}{t} + 2\right]\sin^2\phi} \tag{4.46}$$

where $O_1P = r = \frac{mt}{2}$

$$\begin{aligned}A_pm &= \frac{mt}{2}\sqrt{1 + \frac{T}{t}\left[\frac{T}{t} + 2\right]\sin^2\phi} - \frac{mt}{2} \\ &= \frac{mt}{2}\left[\sqrt{1 + \frac{T}{t}\left[\frac{T}{t} + 2\right]\sin^2\phi} - 1\right]\end{aligned} \tag{4.47}$$

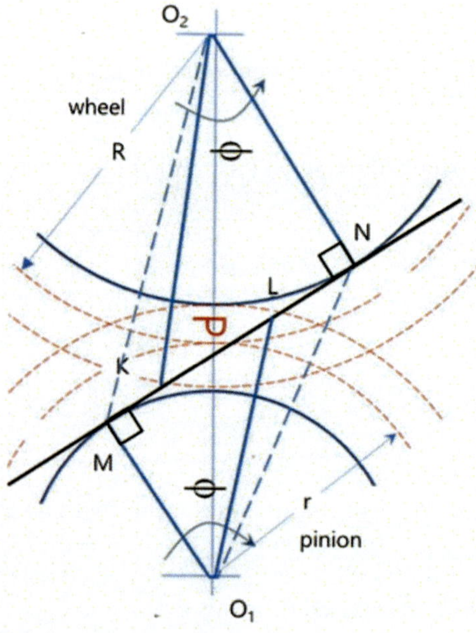

Fig. 4.55 Minimum number of teeth on pinion and gear to avoid interference

Table 4.1 Minimum number of teeth on the pinion

No	System of gear teeth	Minimum number of teeth on the pinion
1	14 1/2″ composite	12
2	14 1/2″ full depth involute	32
3	20″ full depth involute	18
4	20″ sub involute	14

$$A_p = \frac{t}{2}\left[\sqrt{1+\frac{T}{t}\left[\frac{T}{t}+2\right]\sin^2\phi} - 1\right] \tag{4.48}$$

$$t = \frac{2A_p}{\left[\sqrt{1+\frac{T}{t}\left[\frac{T}{t}+2\right]\sin^2\phi} - 1\right]} = \frac{2A_p}{\left[\sqrt{1+G[G+2]\sin^2\phi} - 1\right]} \tag{4.49}$$

See Table 4.1.
Using the same notation, we know that $PM = O_2P\sin\phi = R\sin\phi$.
From triangle O_2PM

$$\begin{aligned}(O_2M)^2 &= (O_2P)^2 + (PM)^2 - 2\times O_2P\times PM\cos O_1PM\\ &= R^2 + r^2\sin^2\phi - 2Rr\sin\phi\cos(90^\circ+\phi)\\ &= r^2 + R^2\sin^2\phi + 2Rr\sin^2\phi\\ &= R^2\left[1+\frac{r^2\sin^2\phi}{R^2}+\frac{2r\sin^2\phi}{R}\right] = R^2\left[1+\frac{r}{R}\left(\frac{r}{R}+2\right)\sin^2\phi\right]\end{aligned} \tag{4.50}$$

Limiting radius of the wheel addendum circle is

$$O_1M = R\sqrt{1+\frac{r}{R}\left[\frac{r}{R}+2\right]\sin^2\phi} = \frac{MT}{2}\sqrt{1+\frac{t}{T}\left[\frac{t}{T}+2\right]\sin^2\phi} \tag{4.51}$$

where $O_1P = R = \frac{mT}{2}$

$$A_wm = \frac{mT}{2}\sqrt{1+\frac{T}{t}\left[\frac{T}{t}+2\right]\sin^2\phi} - \frac{mT}{2} = \frac{mT}{2}\left[\sqrt{1+\frac{T}{t}\left[\frac{T}{t}+2\right]\sin^2\phi} - 1\right] \tag{4.52}$$

$$A_w = \frac{T}{2}\left[\sqrt{1+\frac{t}{T}\left[\frac{t}{T}+2\right]sin^2\phi} - 1\right] \tag{4.53}$$

$$T = \frac{2A_w}{\left[\sqrt{1+\frac{t}{T}\left[\frac{t}{T}+2\right]\sin^2\phi} - 1\right]} = \frac{2A_w}{\left[\sqrt{1+\frac{1}{G}\left[\frac{1}{G}+2\right]\sin^2\phi} - 1\right]} \tag{4.54}$$

4.5.4 *Design of (Spur) Gear Trains*

A gear train is two or more gears operating jointly by meshing their teeth and turning each other in a system to produce power and speed. That is, as electric motors are utilized as power source, the gear systems transmit motion from one shaft and another and lessen speed and grow torque.

Types of gear trains have as follows: (1) simple gear train: the most usual gear train is the gear pair attaching parallel shafts. The teeth of this type shall be spur, helical or herringbone, (2) compound gear train: for large velocities, compound arrangement is preferred, and (3) planetary gear train: all automatic transmissions rely on it (Fig. 4.56).

Based on kinematic relations such as system inputs (RPM1 → RPM2, power, space constrains, and material), we can obtain the gear design—number of teeth, T, pitch circle diameter, D_p, addendum. A_w and dedendum, d_w, face width, b, tooth thickness, t (Fig. 4.57).

Because the circular pitch of gear 1 is equal to the circular pitch of gear 2, we can determine the number of teeth, N and pitch diameter, $d_{p,}$ as follows:

$$P_1 = P_2 \tag{4.55a}$$

$$\frac{N_1}{d_1} = \frac{N_2}{d_2} \tag{4.55b}$$

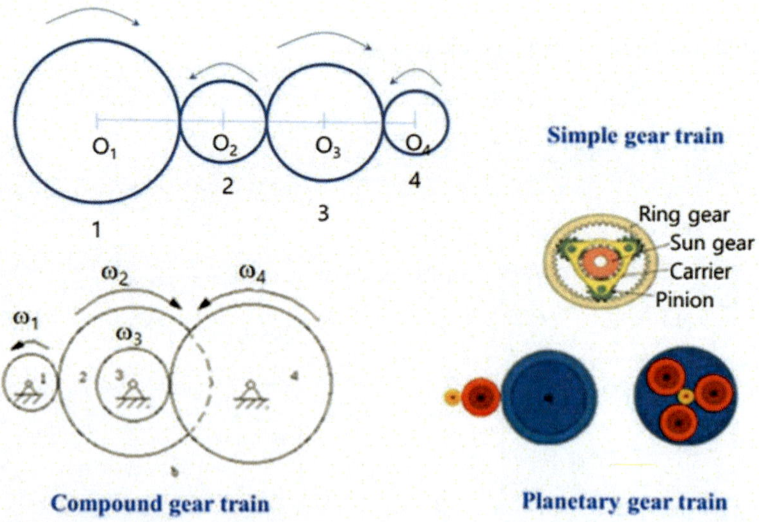

Fig. 4.56 Types of gear trains

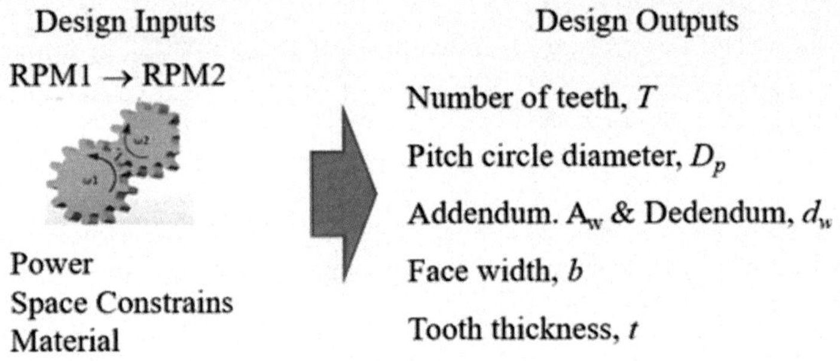

Fig. 4.57 Design concept of gear trains

$$\frac{N_1}{N_2} = \frac{d_1}{d_2} \tag{4.55c}$$

Because the elapsed path at a given time is the same $\widehat{AB} = \widehat{AC}$ or $\frac{\widehat{AB}}{t} = \frac{\widehat{AC}}{t}$, we can determine the rotational velocity. That is, Kennedy's theorem states that if three bodies have plane motions relative to one another, their instantaneous centers lie in a straight line.

$$V_1 = V_2 \tag{4.56a}$$

$$\omega_1 \cdot r_1 = \omega_2 \cdot r_2 \tag{4.56b}$$

$$\frac{2}{2} \cdot \frac{r_1}{r_2} = \frac{\omega_2/2\pi}{\omega_1/2\pi} \tag{4.56c}$$

$$\frac{d_1}{d_2} = \frac{n_2}{n_1} \tag{4.56d}$$

Because the power of gear 1 is equal to the power of gear 2 $H_1 = H_2$, we can determine the torque and revolution per minute as follows:

$$\omega_1 \cdot T_1 = \omega_2 \cdot T_2 \tag{4.57a}$$

$$\left[\omega_1 \cdot \left(\frac{60}{2\pi}\right)\right] \cdot T_1 = \left[\omega_2 \cdot \left(\frac{60}{2\pi}\right)\right] \cdot T_2 \tag{4.57b}$$

$$n_1 \cdot T_1 = n_2 \cdot T_2 \tag{4.57c}$$

$$\frac{n_1}{n_2} = \frac{T_2}{T_1} \tag{4.57d}$$

From Eqs. (4.55) to (4.57), we can determine the gear relation as follows:

$$\frac{N_1}{N_2} = \frac{d_1}{d_2} = \frac{n_2}{n_1} = \frac{T_1}{T_2} \tag{4.58}$$

Example 4.15 When mechanical power including gear train and belt transfers as the following Fig. 4.58 and specification (input speed: 600 rpm), determine the final output speed n_9.

Solution

$$n_9 = \frac{N_8 \cdot N_6 \cdot N_4 \cdot N_2}{N_9 \cdot N_7 \cdot N_5 \cdot N_3} \cdot n_2$$

$$\frac{N_6}{N_7} = \frac{d_6}{d_7}$$

$$n_9 = \frac{N_8 \cdot d_6 \cdot N_4 \cdot N_2}{N_9 \cdot d_7 \cdot N_5 \cdot N_3} \cdot n_2$$

$$n_9 = \frac{(3)(4)(18)(6)}{(36)(28)(36)(10)} \cdot 1050 \text{ RPM}$$

$$n_9 = 3.75 \text{ RPM}$$

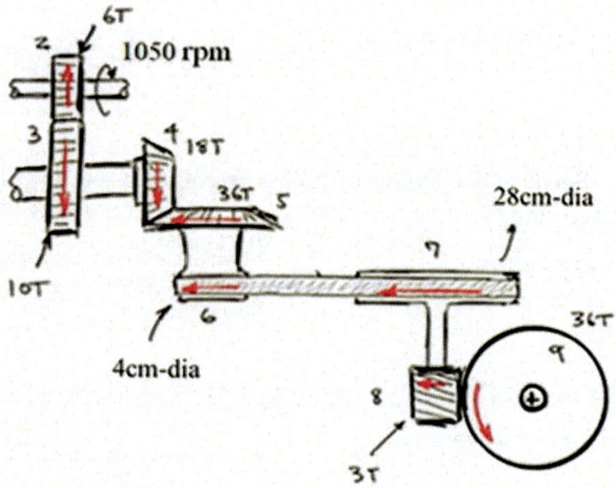

Fig. 4.58 Design concept of gear trains

4.5.5 Force Components and Reaction Forces of (Spur) Gears

If gear force is defined as W, the tangential force W^t, radial force W^r, and axial force W^a are shown as follows (Fig. 4.59).

For two gear forces (W_{32} and W_{23}) in power transmission, the magnitude is the same, but the direction is opposite. That is,

$$W_{32} = W_{23} \tag{4.59a}$$

$$W_{23}^{t} = W_{23} \cos \emptyset \tag{4.59b}$$

$$W_{23}^{r} = W_{23} \sin \emptyset \tag{4.59c}$$

Thus, power can be defined as follows:

$$H = T \cdot \omega = T \cdot \left(\frac{2\pi}{60} \cdot n \right) \tag{4.60}$$

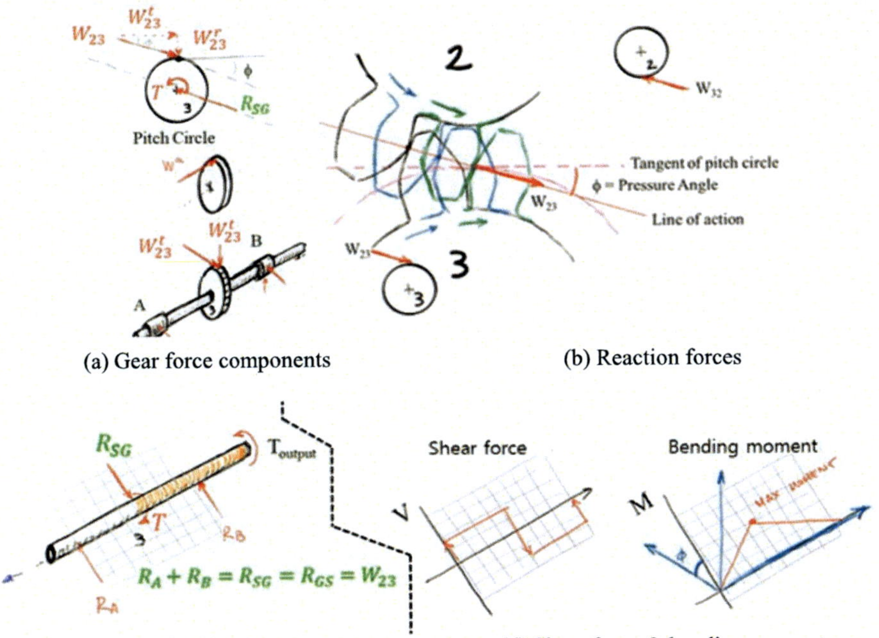

(a) Gear force components (b) Reaction forces

(c) Reaction force (d) Shear force & bending moment

Fig. 4.59 Design concept of gear trains

If we know the power and speed, torque is defined as:

$$T = \frac{H \cdot 60}{2\pi \cdot n} \tag{4.61a}$$

$$T = r_2 \cdot W_{23}^t \tag{4.61b}$$

From Eq. (4.61), we can redefine the tangential force. That is,

$$W_{23}^t = \frac{H \cdot 60}{2\pi \cdot n \cdot r} \tag{4.62}$$

If the tangential force in Eq. (4.62) is known, we can find the reaction forces on the bearing.

$$\mathrm{W}_{23} = \frac{W_{23}^t}{\cos \varnothing} \tag{4.63a}$$

$$R_{SG} = R_{GS} = \mathrm{W}_{23} \tag{4.63b}$$

Example 4.16 When the gear train is given as the following Fig. 4.60 and specification (input power: 50 kW and speed: 200 rpm), the output torque is determined.

Solution

$$\begin{aligned}\mathrm{T_a} &= \frac{H \cdot 60}{2\pi \cdot n_a} = 50{,}000\ \mathrm{W} \cdot \left(\frac{60\ \mathrm{s}}{1\ \mathrm{min}}\right)\left(\frac{1\ \mathrm{rev}}{2\pi\,[\mathrm{rad}]}\right)\left(\frac{1}{200\ \mathrm{rev/\,min}}\right) \\ &= 2.39\ \mathrm{kNm}\ (input)\end{aligned}$$

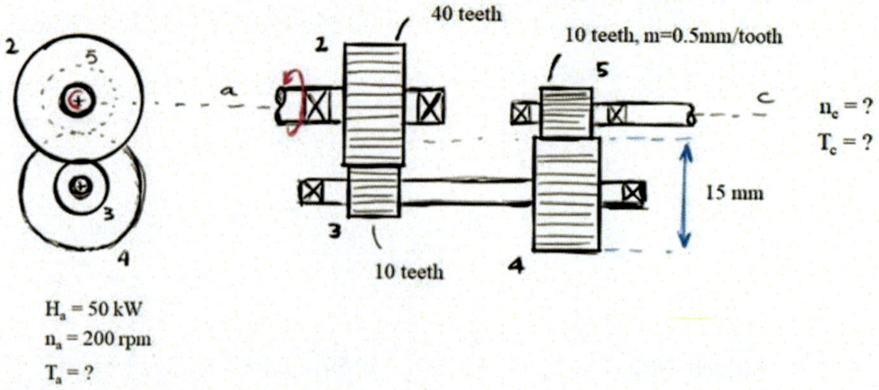

Fig. 4.60 Gear trains

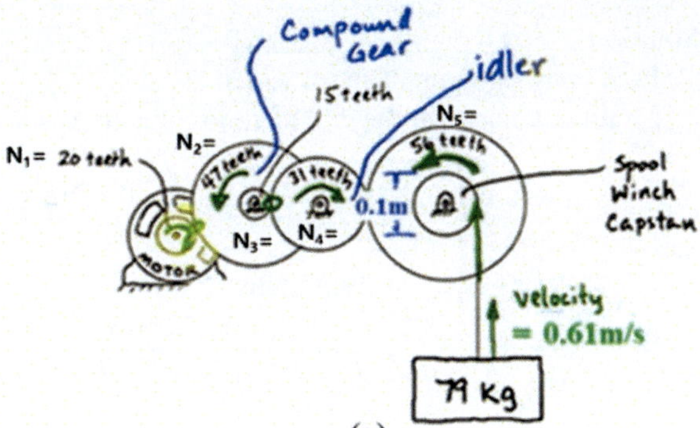

(a)

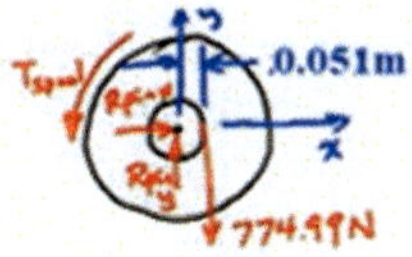

(b)

Fig. 4.61 Compound gear system

$$n_c = n_5$$

$$n_2 = n_a = 200 \text{ RPM}$$

$$\frac{n_2}{n_3} = \frac{N_3}{N_2} \rightarrow n_3 = \frac{N_2}{N_3} \cdot n_2 = \frac{40}{10} \cdot 200 = 800 \text{ RPM}$$

$$n_4 = n_3 = 800 \text{ RPM}$$

$$\text{m} = 0.5 = \frac{d_5}{N_5} \rightarrow d_5 = 0.5 \cdot N_5 = 0.5 \cdot 10 = 5 \text{ mm}$$

$$\frac{d_5}{d_4} = \frac{n_4}{n_5} \rightarrow n_5 = \frac{d_4}{d_5} \cdot n_4 = \frac{15}{5} \cdot 800 = 2400 \text{ RPM}$$

$$T_c = \frac{H \cdot 60}{2\pi \cdot n_c} = 50{,}000 \text{ W} \cdot \left(\frac{60s}{1 \text{ min}}\right)\left(\frac{1\text{rev}}{2\pi\,[\text{rad}]}\right)\left(\frac{1}{200 \text{ rev/min}}\right)$$
$$= 198.9 \text{ Nm (output)}$$

Example 4.17 The engineer would be designed to lift up the following weight by the compound gear system (Fig. 4.61). Assume there are no losses. How much torque does the motor need to apply to lift the weight? What direction should the motor turn to lift the weight?

How fast should the motor turn to lift the weight?

Solution

Free body diagram at spool winch capstan

Moment balance at the end of spool winch capstan

$$\sum M = T_{spool} - \left(79\ \text{kg} \cdot \frac{9.81\ \text{m}}{s^2}\right)(0.051\ \text{m}) = 0.$$

$$T_{spool} = 39.37\ \text{Nm}$$

$$T_{spool} = \frac{N_5 \cdot N_4 \cdot N_2}{N_4 \cdot N_3 \cdot N_1} \cdot T_{motor} = \frac{56 \cdot 31 \cdot 47}{31 \cdot 15 \cdot 20}(= 8.77) \cdot T_{motor}$$

$$T_{motor} = \frac{T_{spool}}{8.77} = 4.487\ \text{Nm}$$

Because Speed · πd = V, we can obtain the speed of the spool winch capstan. That is,

$$Speed_{spool} = \frac{0.61\ \text{m/s}}{\pi \cdot 0.1\ \text{m}} = 1.91\frac{\text{rev}}{\text{s}}\left(\frac{60\ \text{s}}{\text{min}}\right) = \text{RPM}$$

$$\begin{aligned} Speed_{motor} &= Speed_{spool}\frac{N_5 \cdot N_4 \cdot N_2}{N_4 \cdot N_3 \cdot N_1} \\ &= 114.59\frac{56 \cdot 31 \cdot 47}{31 \cdot 15 \cdot 20}(= 8.77) = 1005.34\ \text{RPM} \end{aligned}$$

4.5.6 Lewis Bending Equation for Design of (Spur) Gear

If $\sigma > S_y$ (moderate use) or S_e (fatigue), gear will start the bending failure of the teeth. The simplified model of the cantilever beam can be expressed as follows (Fig. 4.62):

$$|\sigma| = \frac{My}{I} = \frac{M(t/2)}{\frac{1}{12}Ft^2} = \frac{(W^t l)(t/2)}{\frac{1}{12}Ft^3} = \frac{6W^t l}{Ft^2} \tag{4.64}$$

Because $\frac{t/2}{X} = \frac{l}{t/2}$ or $l = \frac{t^2}{4X}$, Eq. (4.64) is redefined. That is,

$$\sigma = \frac{6W^t(t^2/4X)}{Ft^2} = \frac{6W^t}{4FX} = \frac{W^t p}{F\left(\frac{2}{3}X\right)p} = \frac{W^t}{Fpy} \tag{4.65}$$

where $y = \frac{2x}{3p} = Lewis\ form\ factor$ as a function of N (number of teeth) and ϕ (pressure angle).

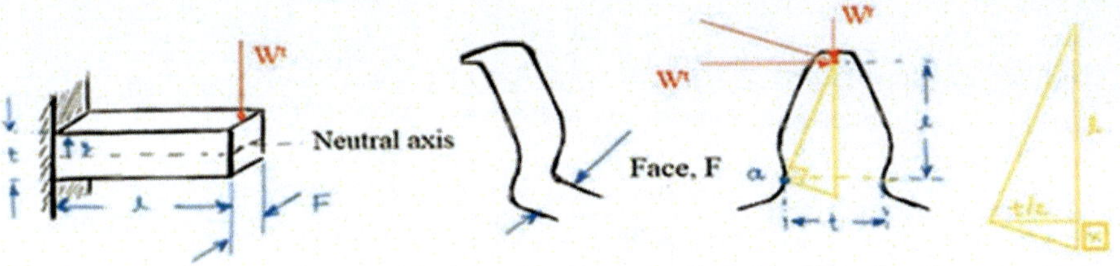

(a) Simplified model for Lewis form factor calculation

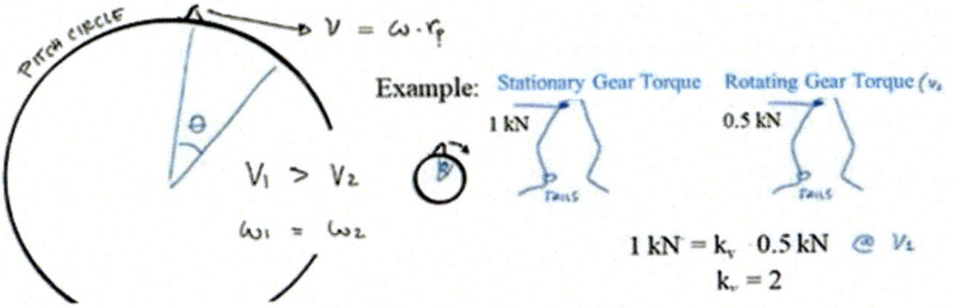

(b) Stress k_v- (dynamic) factor

Fig. 4.62 Simplified model for Lewis form factor calculation

Because p = circular pitch $= \frac{\pi d}{N} = \pi m$ and = diametral pitch $= \frac{N}{d} = \frac{\pi}{p}$, Eq. (4.65) can be stated as follows:

$$\sigma = \frac{W^t}{F(m \cdot \pi)y} = \frac{W^t}{FmY} \text{ (metric)} \tag{4.66a}$$

$$\sigma = \frac{W^t}{F\left(\frac{\pi}{p}\right)y} = \frac{W^t P}{F[\pi y]} = \frac{W^t P}{FY} \text{ (English)} \tag{4.66b}$$

where $Y = y\pi = \frac{2XP}{3}$ (Lewis Form Factor Y that shall be attained from table)

When the bending stress is calculated from the Lewis bending equation for the design of a (spur) gear, we should determine the stress k_v-(dynamic) factor from the experimental data.

For example, the rotating gear force is 0.5 kN at a certain line velocity, which is defined as $v = \omega \cdot r_p$. If failure occurs at a stationary gear force of 1 kN, the stress k_v-(dynamic) factor is 2 (see Fig. 4.32b).

Therefore, k_v can be expressed as follows:

$$\text{Cast } k_v = \frac{3.05 + v}{3.05} \text{ (metric) or } k_v = \frac{600 + v}{600} \text{ (english)} \tag{4.67a}$$

$$\text{Cut or milled } k_v = \frac{6.1 + v}{6.1} \text{ (metric) or } k_v = \frac{1200 + v}{1200} \text{ (english)} \tag{4.67b}$$

$$\text{Hobbed or shaped } k_v = \frac{3.56 + \sqrt{v}}{3.56} \text{ (metric) or } k_v = \frac{50 + \sqrt{v}}{50} \text{(english)} \quad (4.67c)$$

$$\text{Shaped or ground } k_v = \sqrt{\frac{5.56 + \sqrt{v}}{5.56}} \text{ (metric) or } k_v = \sqrt{\frac{78 + \sqrt{v}}{78}} \text{ (english)} \quad (4.67d)$$

Example 4.18 When a tooth in the gear train is given as the following figure and specification, determine the bending stress

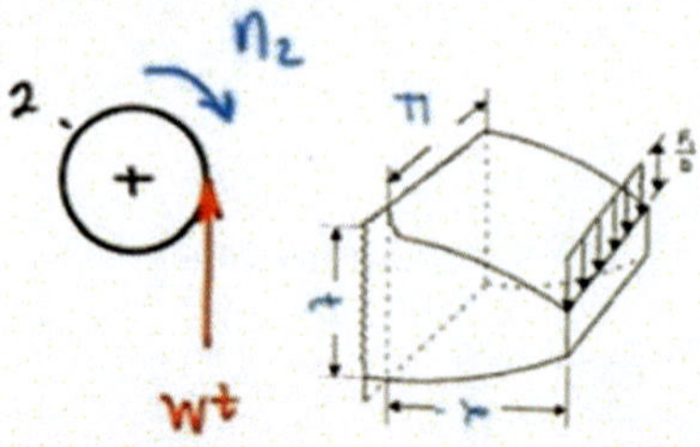

See Table 4.2.
Module: m = 1.25 mm/teeth ($=\frac{d}{N}$).
N_2 = 18 teeth, $\phi = 20°$, F = 12 mm.
n_2 = 1800 rpm, H = 0.5 kW = 500 W.

Solution

Bending stress? → $\sigma = K_v \frac{W^t}{FmY}$

$$\sigma = K_v \frac{W^t}{FmY} = 1.348 \cdot \frac{235.8}{0.012 \cdot 0.00125 \cdot 0.309} = 68.6 \text{ MPa}$$

$$K_v = \frac{6.1 + V}{6.1} = \frac{6.1 + 2.121}{6.1} = 1.348$$

$$V = \omega \cdot \mathrm{r} = n \cdot \frac{2\pi}{60} \cdot r = 1800 \times \frac{2\pi}{60} \cdot 11.25$$

$$V = 2121 \text{ mm/s} \rightarrow V = 2.121 \text{ m/s}$$

$$\mathrm{d} = \mathrm{N} \cdot \mathrm{m} = 18 \cdot 1.25 = 22.5 \text{ mm} \rightarrow \mathrm{r} = 11.25$$

$$\mathrm{H} = \mathrm{T} \cdot \omega = (W^t \cdot r) \cdot \left(n \frac{2\pi}{60}\right)$$

Table 4.2 Values of the Lewis form factor Y (these values are for a normal pressure angle of 20°, full-depth teeth, and a diametral pitch of unity in the plane of rotation)

Number of teeth	Y
12	0.245
13	0.261
14	0.277
15	0.290
16	0.296
17	0.303
18	**0.309**
19	0.314
20	0.322
21	0.328
22	0.331
24	0.337
26	0.346
28	0.353
30	0.359
34	0.371
38	0.384
43	0.397
50	0.409
60	0.422
75	0.435
100	0.447
150	0.460
300	0.472
400	0.480
Rack	0.485

$$\rightarrow W^t = \frac{60 \cdot H}{r \cdot n \cdot 2\pi} = \frac{60 \cdot 500}{0.01125 \cdot 1800 \cdot 2\pi} = 235.8 \text{ N}$$

4.5.7 Design of Epicyclic Gear Train (Planetary Gear)

As seen in Fig. 4.63, a planetary gear includes two gears installed so that the center of one gear rotates around the center of the other. A carrier joins the centers of the two gears and revolves the planet and sun gear mesh so that their pitch circles spin with no slip. A point on the pitch circle of the planet gear tracks down an epicycloid curve. In this clarified case, the sun gear is secured, and the planetary gear(s) roll around the sun gear. Therefore, the speed relation between the input gear and output gear is defined as follows:

$$n_5 = \left[\frac{N_4 N_2}{N_5 N_4}\right] n_2 \tag{4.68a}$$

$$n_5 = e \cdot n_2 \tag{4.68b}$$

$$\frac{n_2}{n_5} = \frac{1}{e} \tag{4.68c}$$

where $e = \left[\frac{N_4 N_2}{N_5 N_4}\right]$

The relative angular velocity of ARM in the planetary gear set shall be defined as follows:

$$\frac{n_{23}}{n_{53}} = \left[\frac{n_2 - n_3}{n_5 - n_3}\right] = \frac{n_2}{n_5} = \frac{1}{e} = \left[\frac{N_4 N_2}{N_5 N_4}\right] \tag{4.69a}$$

$$e = \left[\frac{n_5 - n_3}{n_2 - n_3}\right] \tag{4.69b}$$

$$e = \left[\frac{n_L - n_A}{n_F - n_A}\right] \tag{4.69c}$$

where n_L = last gear, n_A = ARM, and n_F = last gear.

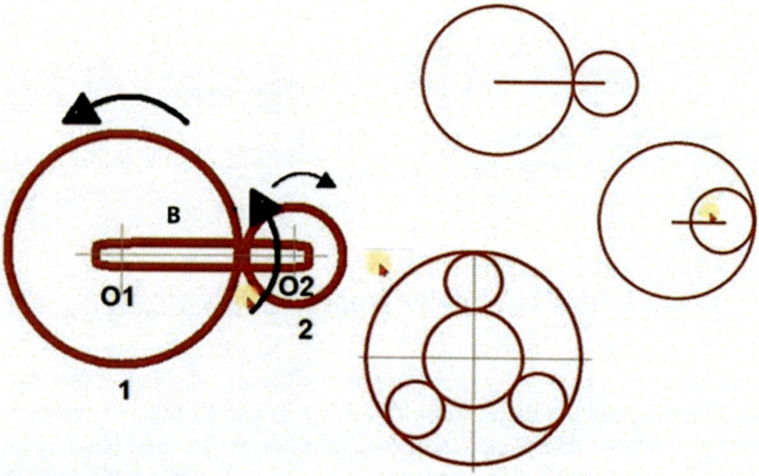

Fig. 4.63 Epicyclic gear train

Tabular method of determining velocity ratio of epicyclic gear train

The conditions are as follows: (1) Arm B is fixed, (2) Gear 1 rotates counterclockwise, (3) +ve counterclockwise, and (4) – ve clockwise (left in Fig. 4.63)

$$N_B = 0 \Leftrightarrow N_1 = +x \Leftrightarrow N_2 = -x(T_1/T_2)$$

$$(N_1/N_2) = T_1/T_2 \Leftrightarrow (x/N_2) = T_2/T_1$$

$$N_2 = x(T_1/T_2) \Leftrightarrow N_2 = -x(T_1/T_2)$$

$$N_B = y \Leftrightarrow N_1 = x + y \Leftrightarrow N_2 = y - x(T_1/T_2)$$

Tabulation

No	Conditions of motion	Motion of elements		
		N_B	N_1	N_2
1	Arm B is fixed, gear 1 rotates through + 1 revolutions in CCW direction	0	+ 1	$-(T_1/T_2)$
2	Arm B is fixed, gear 1 rotates through + x revolutions in CCW direction	0	+ x	$-x(T_1/T_2)$
3	Add y to call elements	y	x + y	$y - x(T_1/T_2)$

Example 4.18 This gear is a planetary gear (see the following figure). The sun gear and planet rotate around the ring gear. The sun gear is – 100 RPM (clockwise). The ring gear is 10 RPM (counterclockwise). Find the speed of arm and planet.

Solution

$n_2 = -\ 100$ RPM, $n_5 = 10$ RPM, $n_A = n_3 = 12$ RPM

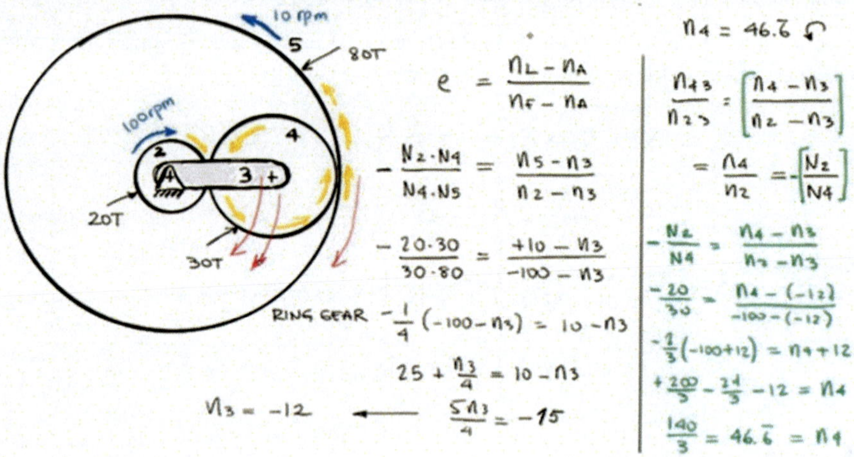

Example 4.19 An epicyclic gear train contains an arm and two gears A and B having 30 and 40 teeth, separately. The arm revolves about the center of gear A at a counterclockwise speed of 80 RPM. Decide the speed of gear B if (i) gear A is fixed and (ii) gear A revolves at 240 RPM clockwise instead of gear A fixed.

Solution

Given data: $N_A = 30$, $N_B = 40$. When arm speed $a = 80$ RPM, $n_B = ?$

By using the tabular method, we can calculate it as follows:

S.N	C.O.M	Arm a	Gear A	Gear B
1	Arm fixed rotation from gear A +1	0	+1	$-\frac{N_A}{N_B}$
2	$+x$	0	$+x$	$-x\frac{N_A}{N_B}$
3	$+y$	y	y	y
4	Net R. Speed	$+y$	$x+y$	$y - x\frac{N_A}{N_B}$

$n_A = 0$

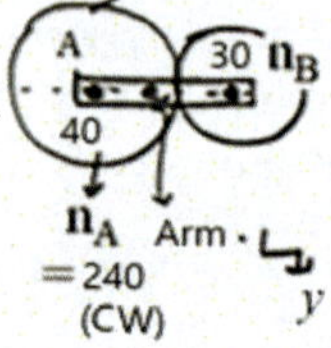

$$\frac{n_B}{n_A} = -\frac{N_A}{N_B}$$

$y = +80$ RPM (CCW)

$x + y = 0$

$x = -y = -80$ RPM $= 80$ (CW)

$n_B = y - x\frac{N_A}{N_B} = 80 + 80 \times \frac{30}{40}$

$n_B = 140$ RPM (CCW)

4	Net R. Speed	y	$x+y$	$y - x\frac{N_A}{N_B}$

$n_A = -240$ (CW)

$y = a = 80$ (CCW)

$y = 80$ RPM (CCW)

$x + y = -240$ (CW)

$x + 80 = -240$

$x = -240 - 80$

$x = -320$ RPM

$n_B = y - x\frac{N_A}{N_B}$

$= 80 + 320 \times \frac{30}{40}$

$n_B = 320$ RPM (CCW)

①

$\frac{n_A}{n_B} = -\frac{N_B}{N_A}$

$\frac{n_A - n_C}{n_B - n_C} = -\frac{N_B}{N_A}$

$n_A = 0$

$n_C = 80$

$\frac{0 - 80}{n_B - 80} = -\frac{40}{30}$

$n_B = \frac{80 \times 30}{40} + 80$

$n_B = 140$ RPM

②

$n_A = 240$

$\frac{-240 - 80}{n_B - 80} = -\frac{40}{30}$

$n_B = 320$ RPM

Example 4.20 An epicyclic gear train contains three gears A, B and C. Gear A has 72 internal teeth, and gear C has 32 external teeth. Gear B meshes with both A and C and is carried on arm ‘ff’, which revolves about the center of ‘A’ at 18 RPM. If gear A is fastened, the speed of gears B and C is determined.

Solution

Given data: $N_A = 72$, $N_C = 32$. $n_{EF} = 18$ RPM.

S·N	C·O·M	Arm EF	Gear C	Gear B	Gear A
1	Arm fixed +1	0	+1	$-\frac{N_C}{N_B}$	$-\frac{N_C}{N_A}$
2	Arm fixed $+x$	0	$+x$	$-x\frac{N_C}{N_B}$	$-x\frac{N_C}{N_A}$
3	Add $+y$	$+y$	$+y$	$+y$	$+y$
4	Total	$+y$	$x+y$	$y-x\frac{N_C}{N_B}$	$y-x\frac{N_C}{N_A}$

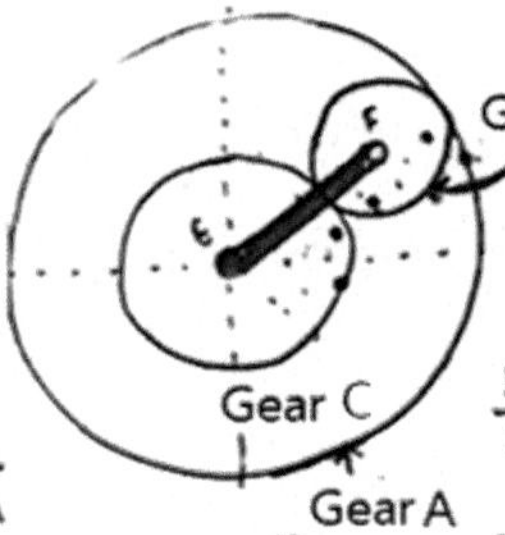

$\frac{n_B}{n_C} = -\frac{N_C}{N_B}$

$\frac{n_A}{n_C} = -\frac{N_C \times N_B}{N_B \times N_A}$

→ Gear A is fixed

$n_A = 0$

$Y - x\frac{T_C}{T_A} = 0$

$18 - x \times \frac{32}{72} = 0$

$x = 40.5$ RPM

→ Speed of gear C

$n_C = x + Y$

$= 40.5 + 18$

$n_C = 58.5$ RPM

Same direction

$Y = 18$ RPM

Speed of Gear B

$d_A = d_c + d_B$.

$d_A = d_c + 2d_B$

$d_A = d_c + 2d_B$

$N_A = N_c + 2N_B$

$72 = 32 + 2N_B$

$N_B = 20$

$n_B = Y - x\cdot\frac{N_c}{N_B} = 18 - 40.5 \times \frac{32}{20}$

$= -46.8$ RPM

46.8 RPM

Opposite

Example 4.21 An epicyclic gear train is shown in the figure. The numbers of teeth on A and B are 80 and 200, respectively. Decide the speed of arm *a* as follows: (i) if A revolves at 100 RPM clockwise and B at 50 RPM and (ii) if A revolves at 100 RPM (clockwise) and B is fixed.

Solution

Given data: $N_A = 80$, $N_B = 200$

→ S·N	C·O·M	a	Gear A	Gear C	Gear B
1	Arm fixed +1	0	+1	$-\frac{N_A}{N_C}$	$-\frac{N_B}{N_A}$
2	$+x$	0	$+x$	$-x\frac{N_A}{N_C}$	$-x\frac{N_B}{N_A}$
3	$+y$	$+y$	$+y$	$+y$	$+y$
4	Net	$+y$	$x+y$	$y - x\frac{N_A}{N_C}$	$y - x\frac{N_B}{N_A}$

$\frac{n_C}{n_A} = \frac{-N_A}{N_C}$

$\frac{n_A}{n_B} = \frac{-N_C \times N_B}{N_A \times N_C}$

$\frac{n_A}{n_B} = -\frac{N_B}{N_A}$

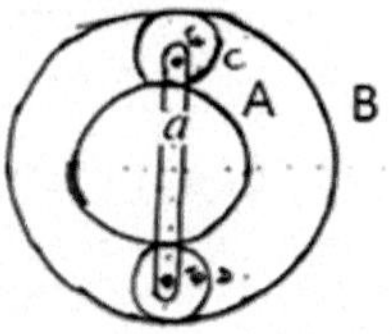

① $n_A = 100$ RPM (CW)
$n_B = 50$ RPM (CCW)

$$x + y = 100$$

$$y - x\frac{N_A}{N_B} = -50$$

$$\rightarrow y - x\frac{80}{300} = -50$$

$$\rightarrow y - 0.4x = -50 \quad \text{—①}$$

$$x + y = 100$$

$$-0.4x + y = -50$$

$$1.4x = 150$$

$$x = 107.14$$

$$x + y = 100$$

$$y + 107.14 = 100$$

$$y = -7.14$$

$$y = 7.14 \text{ RPM}$$

(CCW)

② $n_A = 100$ RPM (CW)
$n_B = 0$

$$x + y = 100$$

$$y - x\frac{N_A}{N_B} = 0$$

$$y - 0.4x = 0$$

$$-0.4x + y = 0$$

$$x + y = 100$$

$$1.4x = 100$$

$$x = 71.42$$

$$x + y = 100$$

$$71.42 + y = 100$$

$$y = 28.6 \text{ RPM}$$

(CW)

4.5.8 Design of Bevel Gears

Bevel gears are gears where the axes of the two shafts intersect and the tooth-bearing faces of the gears themselves are conically formed. Bevel gears are installed on shafts which are 90° apart but shall be designed to function at other angles as well. The pitch surface of bevel gears is a cone. Therefore, the tangential, radial, and axial components of helical gears are expressed as follows (Fig. 4.64):

$$\text{Tangential } W^t = W^t \tag{4.70a}$$

$$\begin{aligned} \text{Radial } W^r &= F\cos\Upsilon = W\sin\phi\cos\Upsilon \\ &= \frac{W^t}{\cos\phi}\cdot\sin\phi\cos\Upsilon = W^t\tan\phi\cos\Upsilon \end{aligned} \tag{4.70b}$$

$$\begin{aligned} \text{Axial(thrust) } W^a &= Fsin\Upsilon = W\sin\phi sin\Upsilon \\ &= \frac{W^t}{\cos\phi\cdot}\cdot\sin\phi\sin\Upsilon = W^t\tan\phi\sin\Upsilon \end{aligned} \tag{4.70c}$$

The pitch angle is determined as follows:

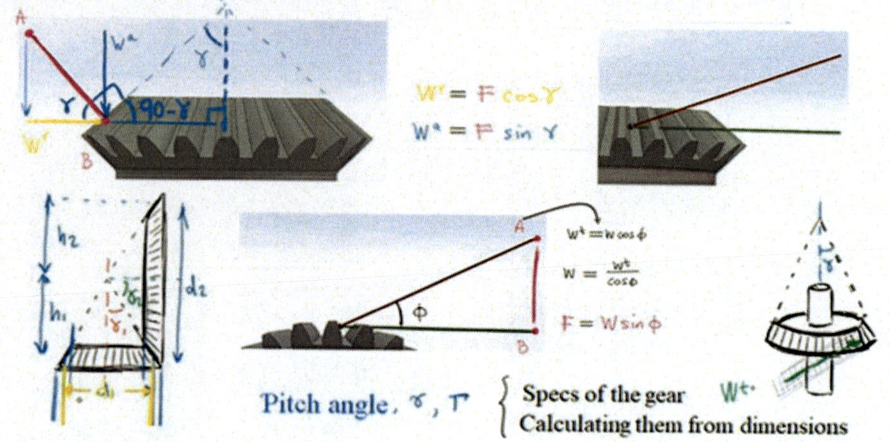

Fig. 4.64 Force components of bevel gears

$$h_1 = h_2 = \frac{1}{2} d_2 \tag{4.71a}$$

$$\Upsilon_1 = \tan^{-1}\left(\frac{d_1}{d_2}\right) \tag{4.71b}$$

$$\Upsilon_2 = \tan^{-1}\left(\frac{d_2}{d_1}\right) \tag{4.71c}$$

Example 4.22 Gear 3 has 32 teeth. The power to be transferred is 2.5 hp at 240 RPM. On the other hand, gear 3 has 16 teeth and pressure angle 20°. If the following figure is given, find the reaction forces at bearings A and B.

Solution

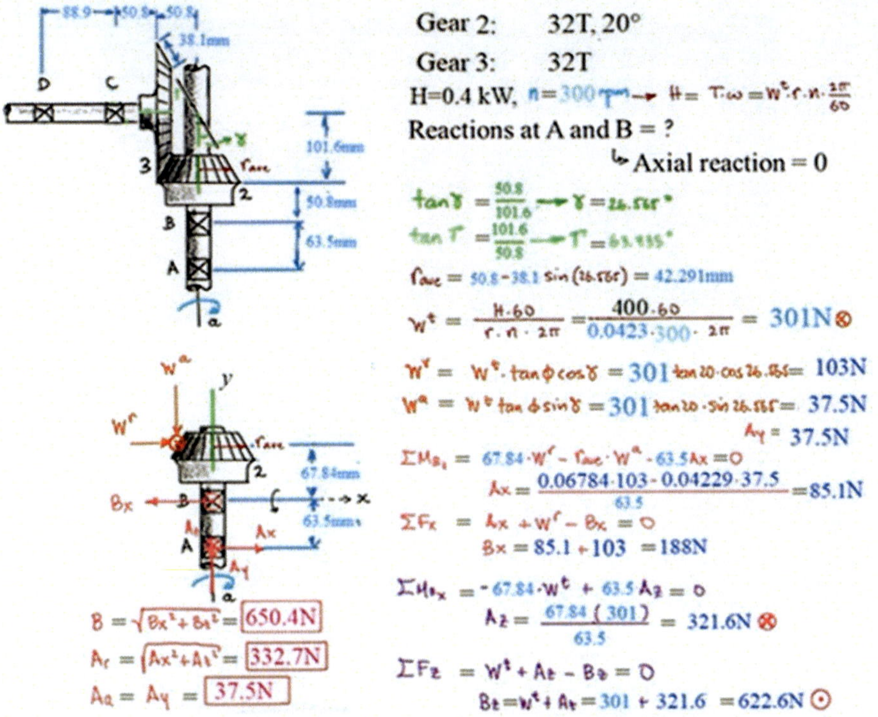

4.6 Design of Bearing

4.6.1 Introduction

A bearing is a machine part that bears another moving machine element (known as a journal). It needs support to ensure stability and frictionless rotation (restrict motion) and permits a relative movement between the contact surfaces of the members. If the rubbing surfaces are in direct contact, there shall be rapid wear. All bearings provide some lubrication arrangement to reduce friction. The lubricant is usually a mineral oil, vegetable oils, silicon oils, grease, etc.

The invention of the rolling bearing, in the shape of wooden roller bearings, causes the invention of the wheel. Leonardo da Vinci invented ball bearing in his design for a helicopter. Today, ball and roller bearings are utilized in rotating components (Fig. 4.65).

Fig. 4.65 Leonardo da Vinci's ball bearing

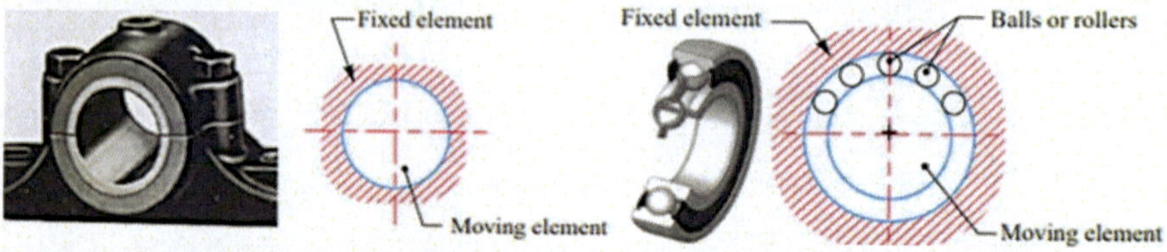

(a) Sliding contact bearing (b) Rolling contact bearing

Fig. 4.66 Sliding contact bearings and rolling contact bearings

4.6.2 *Classification of Bearings*

Relying on the characteristics of contact, there are two types as follows: (a) sliding contact bearings and (b) rolling contact bearings. In sliding contact bearings, sliding occurs along the outside of the contact between the moving component and the fastened component (plain bearings). On the other hand, the steel balls or rollers are interposed between the moving and fastened components. The balls provide rolling friction at two points for each ball or roller (Fig. 4.66).

Relying upon the orientation of load to be maintained, there are two types as follows: (a) radial bearings and (b) thrust bearings. In radial bearings, the load acts perpendicular to the movement of the moving element. On the other hand, in thrust bearings, the load acts along the axis of rotation (Fig. 4.67).

4.6.3 *Bearing Life and Load Ratings*

Bearing failure is spalling or pitting of an area of 0.01 in^2 (ABMA). Life for one bearing shall be expressed as the number of revolutions (or hours @ given speed) necessitated for failure. For a group of bearings, the rating life is one that is necessary

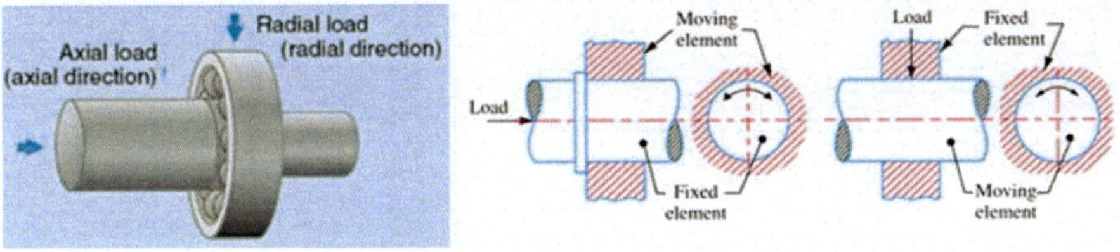

(a) Radial bearing (b) Thrust bearing

Fig. 4.67 Radial bearing and thrust bearing

for 10% of the sample to be unsuccessful and is also called the minimum life or L10 Life. On the other hand, the median life for many groups of bearings is the mean life necessitated for 50% of the sample to be unsuccessful and is also called the mean life or median life. Therefore, median life is typically 4 or 5 times that of L10 Life (Fig. 4.68).

There are four load ratings as follows: (1) Catalog Load Rating, C_{10}, (2) Basic Load Rating, C, (3) Static Load Rating, C_o, and (4) Equivalent Radial Load, F_e.

- Catalog Load Rating, C_{10}, is a constant radial load which brings 10% of a group of bearings to be unsuccessful at the bearing manufacturer's rating life. It relies on the type, geometry, accuracy of fabrication, and material of the bearing. In addition, it is specified as Basic Dynamic Load Rating, and Basic Dynamic Capacity.
- Basic Load Rating, C, is a catalog load rating based on a rating life of 10^6 revolutions of the inner ring. The radial load that shall be necessary to bring failure at such a low life is unrealistically high. The Basic Load Rating is a reference value, not an real load.
- Static Load Rating, C_o, is a static radial load that equates to a permanent deformation of rolling element and race at the most heavily stressed contact of 0.0001d (= diameter of roller). It is utilized to examine for permanent deformation. It is utilized in integrating radial and thrust loads into an equivalent radial load.

Fig. 4.68 Bearing life definitions

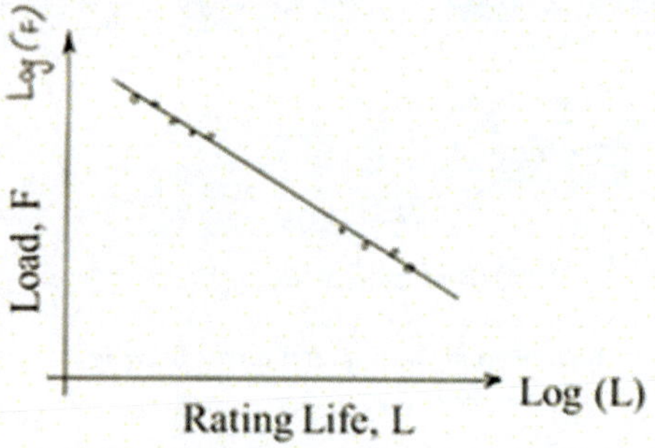

Fig. 4.69 Load-life relationship

- The equivalent radial load, F_e, is a constant stationary load exerted to the bearing with a rotating inner ring, which brings the same life as the real load and rotation circumstances.

4.6.4 *Bearing Design*

Identical groups of bearings are tested to determine the life-failure standard at some loads. We can plot load vs. life on a log–log scale that is roughly linear. Utilizing a regression equation, we can express it as follows (Fig. 4.69):

$$FL^{\frac{1}{a}} = constant \tag{4.72a}$$

$$F_R L_R^{\frac{1}{a}} = F_D L_D^{\frac{1}{a}} \tag{4.72b}$$

where R = Rated, D = Desired, a = 3 for ball bearings, a = 10/3 for roller bearings (cylindrical and tapered roller)

Example 4.23 When mechanical power is driven by the gear system, the reaction forces at A and B are calculated as 0.89 and 1.33 kN, separately. The rated life of each bearing is 106 (rating life: # revolutions until 90% of the samples have not yet failed). The desired bearing life is 8000 h@1600 RPM. In this case, determine the dynamic (or catalog) load rating at (roller) bearing A and B

SKF Bearings : $L_R = 10^6$

Bearing A : $\begin{cases} F_D = 0.89\text{ kN} \\ L_D = 7.68 \times 10^8 \end{cases}$

Bearing B : $\begin{cases} F_D = 1.33\text{ kN} \\ L_D = 7.68 \times 10^8 \end{cases}$

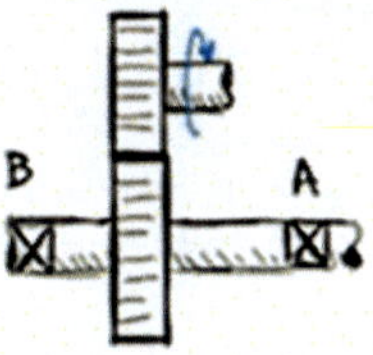

Desired Bearing Life : 8000 hrs @ 1600 rpm

$$L_D = 8000\,hr \cdot \frac{1600\,rev}{min} \cdot 60\frac{min}{hr} = 7.68 \times 10^8 \text{ cycles}$$

$$C_{10} L_R^{1/a} = F_D \cdot L_D^{1/a} \qquad C_{10} = F_D \left(\frac{L_D}{L_R}\right)^{1/a} \qquad a = \frac{10}{3}$$

$$C_{10} = 0.89 \cdot \left(\frac{7.68 \times 10^8}{10^6}\right)^{1/(10/3)} \qquad C_{10} = 1.33 \left(\frac{7.68 \times 10^8}{10^6}\right)^{1/(10/3)}$$

$$\boxed{C_{10} \geq 1468\text{ lbf} = 6.53\text{ kN}} \qquad \boxed{C_{10} \geq 2202\text{ lbf} = 9.8\text{ kN}}$$

If bearings are subjected to combined axial and radial loads, we should find the equivalent radial load, F_e because $F_e \neq F_a + F_r$ or $F_e \neq \sqrt{F_a^2 + F_r^2}$ (Fig. 4.70).

The relationship of the equivalent radial load, F_e, on the graph can be expressed as follows (Fig. 4.68):

$$\frac{F_e}{V \cdot F_r} = 1 \text{ or } F_e = V \cdot F_r \text{ for } \frac{F_a}{V \cdot F_r} \leq e \tag{4.73a}$$

$$\frac{F_e}{V \cdot F_r} = Y\left(\frac{F_a}{V \cdot F_r}\right) + X \text{ for } \frac{F_a}{V \cdot F_r} \geq e \tag{4.73b}$$

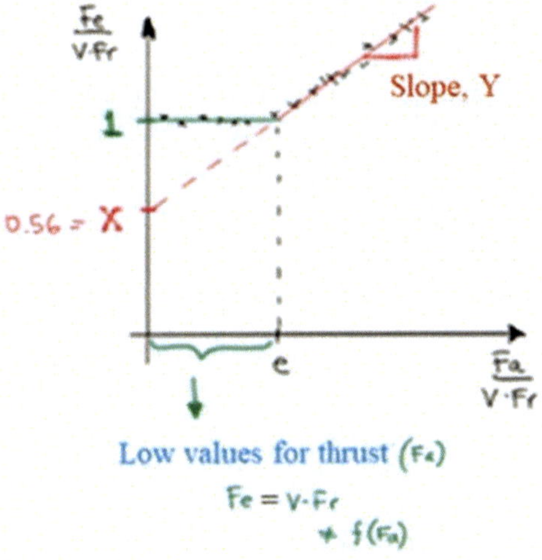

Fig. 4.70 Relationship of the equivalent radial load, F_e

We know that Eq. (4.73b) has the relationship $y = \mathrm{m} \cdot \mathrm{x} + \mathrm{b}$ (straight line). Therefore, if we redefine Eq. (4.73b), we can obtain it as follows:

$$\mathrm{F_e} = X_i \cdot \mathrm{V} \cdot F_r + Y_i \cdot F_a \tag{4.74}$$

where V = 1 (inner ring rotates), V = 1.2 (outer ring rotates) obtained from experiment

For ball bearings, $X_i = 0.56$ and $Y_i = \mathrm{f}\left(\frac{F_a}{C_o}\right)$ can be obtained from the table (equivalent radial load factors). That is, we can summarize the procedure of bearing design subjected to combined axial and radial loads as follows:

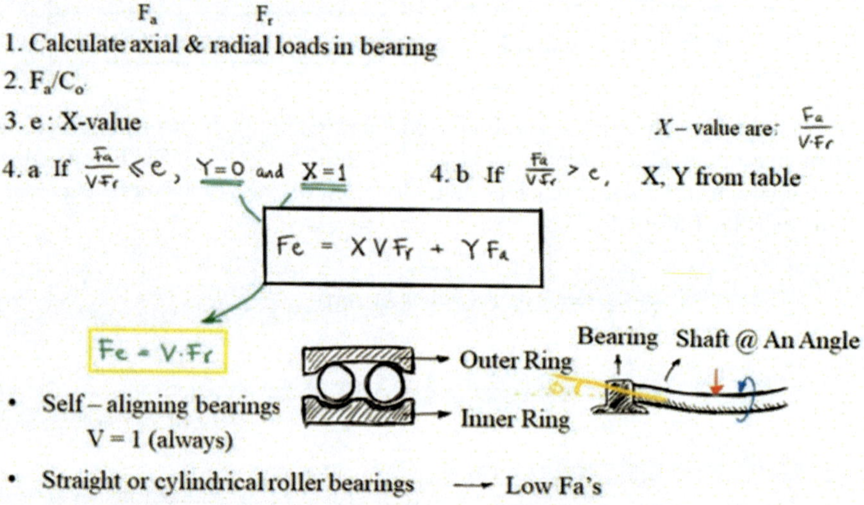

Example 4.24 For the (angular contact) ball bearing, if $\mathrm{C_o} = 20$ kN, $\mathrm{C_{10}} = 35.6$ kN, and $\mathrm{L_{10}} = 10^6$ are given, find the lifetime for $\mathrm{F_a} = 1.7$ kN, $\mathrm{F_r} = 2.1$ kN, and n = 680 RPM.

$$F_R \cdot L_R^{1/a} = C_{10} \cdot L_{10}^{1/a} = F_D \cdot L_D^{1/a} = F_e L^{1/a}$$

$$F_e L^{1/a} = C_{10} L_{10}^{1/a}$$

$$L^{1/a} = \frac{C_{10}}{F_e} \cdot L_{10}^{1/a}$$

Life in # of cycles $= n \cdot \mathcal{L} \cdot 60 = L = \left(\frac{C_{10}}{F_e}\right)^a \cdot L_{10}$

$$\mathcal{L} = \frac{L_{10}}{n \cdot 60}\left(\frac{C_{10}}{F_e}\right)^a = \frac{10^6}{680 \cdot 60}\left(\frac{35.6}{3.81}\right)^3 = \boxed{20{,}077 \text{ hrs}}$$

$$F_e = X \cdot V \cdot F_r + Y \cdot F_a \Rightarrow F_e = (0.56)(1)(2.1) + 1.55(1.7) = 3.81 \text{ kN}$$

$\frac{F_a}{V \cdot F_r}$ vs. e

$$\frac{F_a}{C_o} = \frac{1.7 \text{ kN}}{20 \text{ kN}} = 0.084 \rightarrow e = 0.28$$

$$\frac{F_a}{V \cdot F_r} = \frac{1.7}{(1)\,2.1} = 0.8 > e$$

Table 11-1 Equivalent Radial Load Factors for Ball Bearings

		$F_a/(VF_r) \le e$		$F_a/(VF_r) > e$	
F_a/C_0	e	X_1	Y_1	X_2	Y_2
0.014*	0.19	1.00	0	0.56	2.30
0.021	0.21	1.00	0	0.56	2.15
0.028	0.22	1.00	0	0.56	1.99
0.042	0.24	1.00	0	0.56	1.85
0.056	0.26	1.00	0	0.56	1.71
0.070	0.27	1.00	0	0.56	1.63
0.084	0.28	1.00	0	0.56	1.55
0.110	0.30	1.00	0	0.56	1.45
0.17	0.34	1.00	0	0.56	1.31
0.28	0.38	1.00	0	0.56	1.15
0.42	0.42	1.00	0	0.56	1.04
0.56	0.44	1.00	0	0.56	1.00

*Use 0.014 if $F_a/C_0 < 0.014$

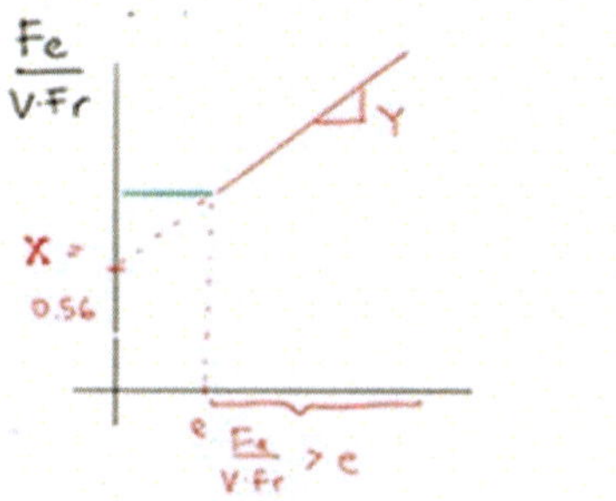

Reference

1. Sonntag RE, Borgnakke C (2007) Introduction to engineering thermodynamics. Wileys, York

Chapter 5
Mechanical System Design (Strength and Stiffness)

5.1 Introduction

The design of a mechanical system subjected to various (random) loads is required to possess a fine quality, such as sufficient stiffness or strength. Strength—the resistance against permanent deformation—is continually necessitated to be high throughout the design of material or product form because a permanent deformation of material due to loads can bring to crack, fracture, and eventually the losing of (intended) functions in a product. On the other hand, the need for stiffness—the resistance against reversible deformation—can change over a wide span based on the product's application.

If there is a design flaw in a structure, it can cause an inadequacy of strength (or stiffness). That is, when exposed to loads, the product structure can undergo permanent deformation, crack, or fracture at that location because it does not have enough strength to bear stress in the elasticity range. The failure mechanics—especially fatigue or fracture—of parts which could no longer work shall be distinguished by two elements: (1) the loads (or stress) on the construction and (2) the pattern of product matters and their form utilized in the construction, including the mechanism. To stop mechanical failures, mechanical engineers reproduce them and modify product designs by selecting a proper material and product shape so that the products can have proper strength and stiffness. Consequently, an engineer would prefer to make a better design of products for (repeated) loads. As a result, the mechanical system shall be redesigned to have a better design that can robustly endure loads in product life (Fig. 5.1).

S. Woo, *Design of Mechanical Systems*, Springer Series in Reliability Engineering,
https://doi.org/10.1007/978-3-031-28938-5_5

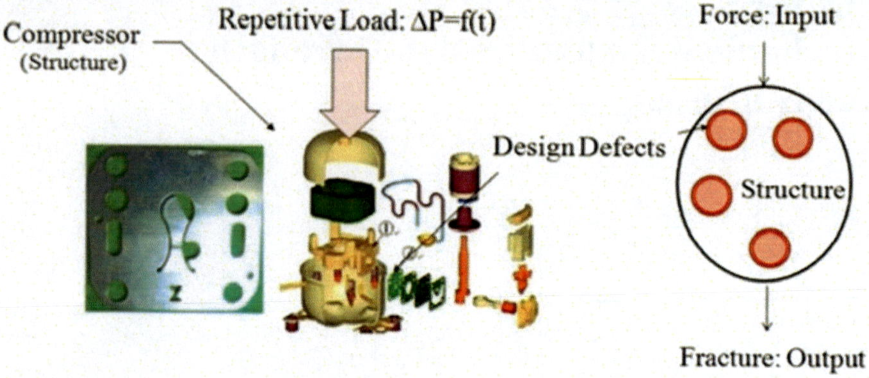

Fig. 5.1 Failure mechanics produced by a load on a component made from a specific material

5.2 Strength of Mechanical Product

5.2.1 *Introduction*

The area of strength of materials, also specified as the mechanics of materials, usually mentions numerous ways of computing the stresses and strains in constructional elements, such as beams, columns, and shafts. The ways utilized to forecast the response of a structure under loading and its vulnerability to numerous failure modes considers the behaviors of the materials, such as its yield strength, ultimate strength, Young's modulus, and Poisson's ratio. Additionally, the mechanical element's macroscopic behaviors (geometric attributes), such as its length, width, thickness, boundary constraints and sudden modifications in geometry, such as holes, are contemplated.

The theory started with the deliberation of the functioning of one- and two-dimensional components of structures, whose conditions of stress can be closed as two dimensional, and was thus generalized to three dimensions to evolve a more entire theory of the elastic and plastic behavior of materials. A main pioneer in the mechanics of materials was Stephen Timoshenko.

The strength of materials is a theme that addresses the attribute of solid bodies subject to stresses and strains. The strength of a matter is its capacity to endure an exerted load without failure or plastic deformation. It handles with forces and deformations which result from their acting on a material. A load exerted on a mechanical component shall cause inner forces in the member called stresses when those forces are defined on a unit basis. The stresses exerting on the matter bring deformation of the matter in a variety of way, including breaking them totally. Deformation of the matter is specified as strain when those deformations are also expressed as a unit basis.

A matter has a rest shape, which leaves from the rest shape on the implementation of an external force. The quantity of departure from the rest shape is called deformation. Force and deformation are made a connection together by the stiffness relation. The ratio of deformation to original size is called strain. Deformation and strain are made a connection together by compatibility equations. The matter tries to resist this deformation and generates internal resistance that is termed stress.

Stress and strain are made a connection together by constitutive equations (Figs. 5.2 and 5.3).

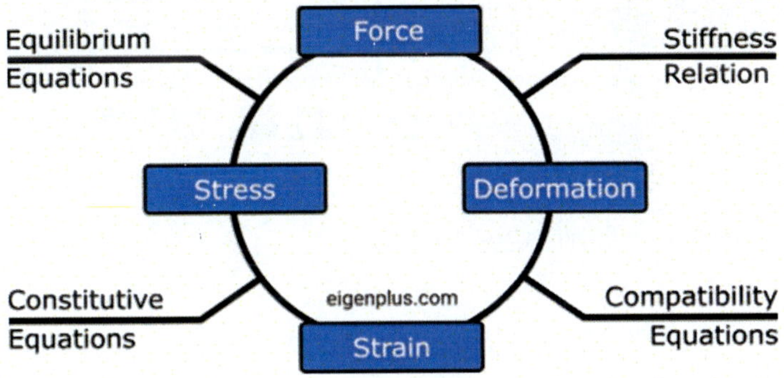

Fig. 5.2 Relationship between force, stress, strain, and deformation

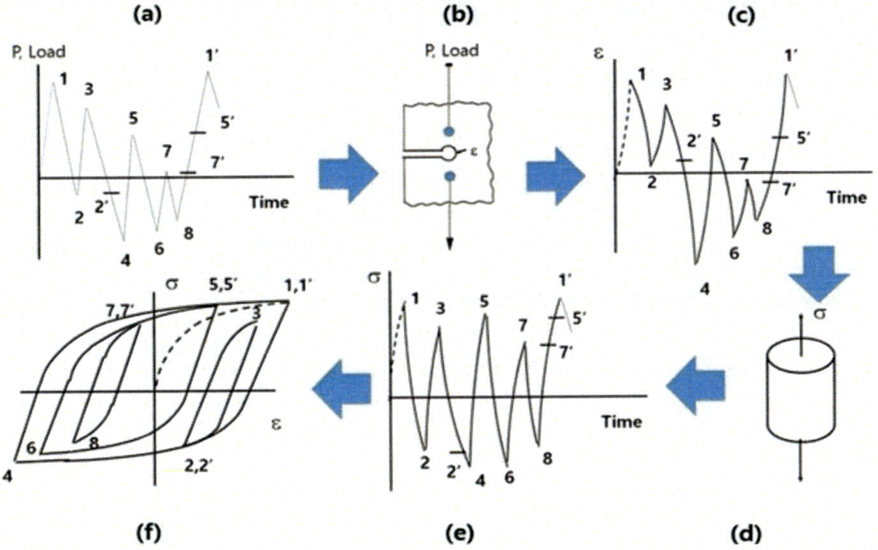

Fig. 5.3 Load history subjected to tensile and compression in the field

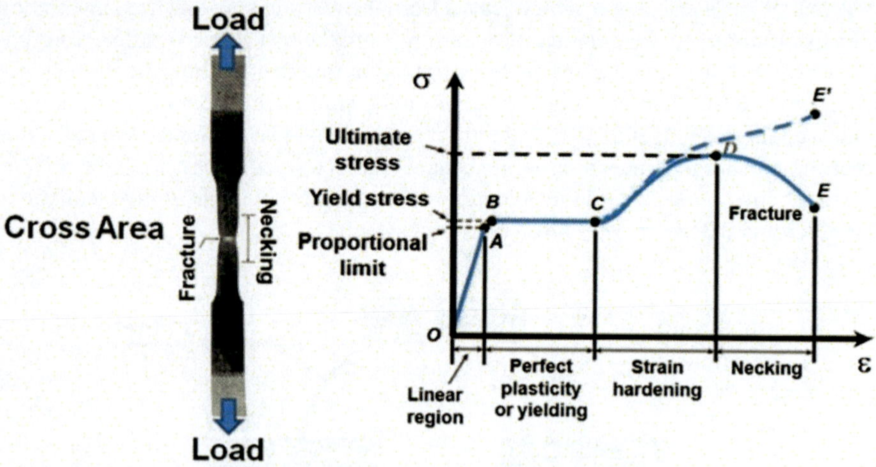

Fig. 5.4 Terminologies in the strength of materials

5.2.1.1 A Brief History

Galileo Galilei (1564–1642) and Leonardo da Vinci (1452–1519) made the earliest endeavors at finding a beam theory. Although Leonardo da Vinci made the major findings, to finish his theory with the help of calculus in mathematics and Hooke's law, he still needed to be developed and compensated. Galileo was also held back by an insufficient presumption he proposed. They carried out tests to decide the strength of wires, beams, and bars, although they did not suggest proper theories to describe their experimental outputs.

The well-known mathematician Leonhard Euler (1707–1783) suggested the concept of columns and computed the major load of a column in 1744. With inappropriate experiments to support his theories, the Euler–Bernoulli beam theory endured unemployment for over a hundred years. Buildings and bridges were continually designed by exemplars until the latter of part of the nineteenth century, while the Ferris wheel and Eiffel Tower explained the effectiveness of the theory on large scales. Here, they are the foundation for the analysis and design of most columns, such as tip-loaded cantilever beams (Fig. 5.4).

5.2.1.2 Terminologies in the Strength of Materials

Stress is the exerted force or system of forces which tends to deform a body.

$$\text{Stress} = \frac{External\ defroming\ force}{Area} = \frac{F}{A} \tag{5.1}$$

where the dimensional formula of the stress $= \frac{[MLT^{-2}]}{L^2} = [ML^{-1}T^{-2}]$, and the SI unit of stress is $Nm^{-2} = Pascal$, which is the same as that of pressure

The stress evolved in a body relies upon how the external forces are exerted over it. On this basis, there are two kinds of stress as follows: (1) normal stress and (2) tangential stress. Normal stress is a stress which happens when the surface of the body is loaded by an axial force.

$$\sigma = \frac{Axial\ force}{Cross\ sectional\ area} = \frac{F}{A} \tag{5.2}$$

The normal stress is of two kinds: (1) tensile stress and (2) compression stress. Tensile stress is the stress state bringing to expansion; that is, the length of a material tends to increase in the tensile orientation. Compressive stress is a force which tries to squeeze or compress a matter.

There are two material behaviors: (1) ductile behavior and (2) brittle behavior. Ductility is a solid matter's capacity to deform under tensile stress. The brittle behavior breaks without insignificant deformation when subjected to stress. Glass is a good instance.

Strain, ε, is the change in the size or shape of a body due to the deforming force. Type equation here.

$$\text{Strain} = \frac{Change\ in\ dimension}{Original\ dimension} \tag{5.3}$$

where strain is dimensionless because there is no unit.

Because the deforming force can generate three deformations (i.e., change in length, volume, or shape) in a body, there are three kinds of strain as follows: (1) longitudinal strain, (2) volumetric strain, and (3) shearing strain. Longitudinal strain, ε, is when the deform force generates a change in length. The volumetric strain, ε, is when the deform force produces a change in the volume. Shearing strain, ε, is when the deforming force generates a change in the form of the body.

Elasticity is the inclination of solid materials to go back to their original form after being forced onto them. As the forces are eliminated, the body shall return to its original form and size if the material is elastic. That is, the deformation totally vanishes after elimination of outer forces. Steel cables, rubber bands, and springs are instances of elastic matter.

The greatest of proportionality is the position on the graph where extension is no longer proportional to the exerted force. As a tensile force exerted on a uniform bar of mild steel is slowly increased and the corresponding extension of the bar is measured, the exerted force supplied is not too big. Since the graph is a straight line, extension is immediately proportional to the exerted force (Hook's Law).

Elastic limit, ε, is the greatest stress from which an elastic body shall retrieve its initial state after the elimination of the deforming force. Just beyond this position, the matter shall act in a nonlinear manner till the elastic limit is reached. It varies extensively for different matters. It is very big for a substance such as steel and low for a substance such as lead.

The yield limit is known as the yielding of matter. As the sample is stressed beyond the elastic limit, the strain grows more quickly than the stress. Unexpected elongation of the specimen occurs without a considerable grow in the stress. The part between the upper yield point and lower yield point is specified as the yield stage. The stress corresponding to the point of the upper yield point is specified as yield stress.

Strain hardening is the occurrence of an increase in stress after yielding. The ultimate stress is the greatest load taken by the specimen. Because of the plastic deformation, the matter strain hardens, and further strain beyond the lower yield point necessitates an increase in stress.

Young's modulus, also known as the tensile modulus or elastic modulus, is a quantity of the stiffness of an elastic material and was called after a British Scientist THOMAS YOUNG.

$$\text{Stress} \propto \text{Strain or } \frac{Stress}{Strain} = constant = \text{Modulus of Elasticity} \tag{5.4}$$

If the modulus of elasticity of a matter is big, it signifies that a bigger stress shall generate only a small strain. Corresponding to kinds of strain, there are three moduli of elasticity: (1) Young's modulus of elasticity, Y $= \frac{Normal\ Stress}{Longitudinal\ Strain}$, (2) Bulk modulus of elasticity, K $= \frac{Normal\ Stress}{Volumetric\ Strain}$, (3) Modulus of rigidity, $\eta = \frac{Shearing\ Stress}{Shearing\ Strain}$.

Poisson's ratio is the quantity of transversal elongation split by the quantity of axial compression. When a body is subjected to axial tensile force, it elongates and contracts laterally. Likewise, it shall contract, and its sides expand laterally when subjected to an axial compressive force. Most materials have Poisson's ratio values varying between 0.0 and 0.5.

5.2.2 *Elasticity*

Elasticity is a property of (solid) matters to become distorted when exposed to an outer force and to recover their initial shape after the force is eliminated. To comprehend the design of the mechanical product, we will discuss the elasticity. After starting one-dimensional elasticity, we can expand the three-dimensional elasticity.

We can define a simple body of volume, V, and surface, S. The action of deforming at local point $x = [x, y, z]^T$ is given by the three components of its displacement $u = [u, v, w]^T$, where $u = u(x, y, z)$, i.e., each displacement component is a function of position. In this case, there are two basic patterns of outer forces that act on a body: (1) body force (force per unit volume), e.g., weight, and (2) surface traction (force per unit surface area), e.g., friction (Fig. 5.5).

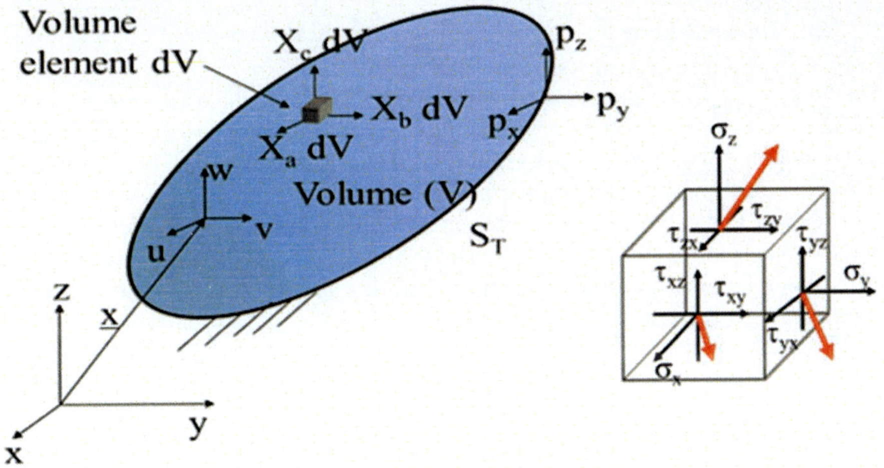

Fig. 5.5 A simple body subjected to body force and surface traction

5.2.2.1 Body Force

For volume element dV, the body force can be defined as $\underline{X} = [X_a, X_b, X_c]^T$. If the body is accelerating, the inertia force might be considered part of $\underline{X}$. That is,

$$\underline{X} = \underline{\tilde{X}} - \rho\underline{\ddot{u}} \tag{5.5}$$

where $\rho\underline{\ddot{u}} = \begin{Bmatrix} \rho\ddot{u} \\ \rho\ddot{v} \\ \rho\ddot{w} \end{Bmatrix}$.

5.2.2.2 Surface Traction

As the distributed force per unit surface area, the surface traction can be defined as follows:

$$\underline{T}_S = \begin{Bmatrix} p_x \\ p_y \\ p_z \end{Bmatrix} \tag{5.6}$$

There is an internal force that is reacted in a body. If it is expressed as the internal reaction forces per unit area, we can say it stress.

5.2.2.3 Internal Forces

If a thick, solid piece of material from the body is removed, there are reaction forces due to the outer forces exerted on it. For the hexahedron, on each surface, the internal reaction force per unit area (black arrows) is known as stress and may be decomposed into three orthogonal components.

Assume a body aligned in the Cartesian coordinate system with a number of forces applied to it, such that the vector total of all the forces is zero. Take a piece orthogonal to the x direction and explain a small area on this slice as ΔA_x. Let the total force applied on this small area be

$$\underline{\Delta F} = \Delta F_x \cdot \hat{i} + \Delta F_y \cdot \hat{j} + \Delta F_z \cdot \hat{k} \tag{5.7}$$

We can explain the following scalar amounts:

$$\sigma_{xx} = \lim_{\Delta A_x \to 0} \frac{\Delta F_x}{\Delta A_x}, \quad \tau_{xy} = \lim_{\Delta A_x \to 0} \frac{\Delta F_y}{\Delta A_x}, \quad \tau_{xz} = \lim_{\Delta A_x \to 0} \frac{\Delta F_z}{\Delta A_x} \tag{5.8}$$

Equivalently, considering slices orthogonal to the y and z directions, we find

$$\tau_{yx} = \lim_{\Delta A_y \to 0} \frac{\Delta F_x}{\Delta A_y}, \quad \sigma_{yy} = \lim_{\Delta A_y \to 0} \frac{\Delta F_y}{\Delta A_y}, \quad \tau_{yz} = \lim_{\Delta A_y \to 0} \frac{\Delta F_z}{\Delta A_y} \tag{5.9}$$

and

$$\tau_{zx} = \lim_{\Delta A_z \to 0} \frac{\Delta F_x}{\Delta A_z}, \quad \tau_{zy} = \lim_{\Delta A_z \to 0} \frac{\Delta F_y}{\Delta A_z}, \quad \sigma_{zz} = \lim_{\Delta A_z \to 0} \frac{\Delta F_z}{\Delta A_z} \tag{5.10}$$

We can say that σ_x, σ_y, and σ_z are normal stresses. On the other hand, the remaining six are the shear stresses. Because $\tau_{xy} = \tau_{yx}$, $\tau_{yz} = \tau_{zy}$ and $\tau_{zx} = \tau_{xz}$, we can say that only six stress components are independent.

The stress vector is described as follows:

$$\underline{\sigma} = \left[\sigma_x\, \sigma_y\, \sigma_z\, \tau_{xy}\, \tau_{yz}\, \tau_{zx}\right]^T \tag{5.11}$$

On the other hand, the quantity of deformation is defined as the *strain* and may be decomposed into three orthogonal components. Strains have six independent strain components. That is,

$$\underline{\varepsilon} = \left[\varepsilon_x\, \varepsilon_y\, \varepsilon_z\, \gamma_{xy}\, \gamma_{yz}\, \gamma_{zx}\right]^T \tag{5.12}$$

Assume the equilibrium of a differential volume element to obtain the three equilibrium equations of elasticity:

$$\begin{aligned}\frac{\partial\sigma_x}{\partial x}+\frac{\partial\tau_{xy}}{\partial y}+\frac{\partial\tau_{xz}}{\partial z}+X_a=0\\ \frac{\partial\tau_{xy}}{\partial x}+\frac{\partial\sigma_y}{\partial y}+\frac{\partial\tau_{yz}}{\partial z}+X_b=0\\ \frac{\partial\tau_{xz}}{\partial x}+\frac{\partial\tau_{yz}}{\partial y}+\frac{\partial\sigma_z}{\partial z}+X_c=0\end{aligned} \tag{5.13}$$

Concisely, we can describe the equilibrium equations as follows:

$$\underline{\partial}^T\underline{\sigma}+\underline{X}=\underline{0} \tag{5.14}$$

where $\underline{\partial}=\begin{bmatrix}\frac{\partial}{\partial x} & 0 & 0\\ 0 & \frac{\partial}{\partial y} & 0\\ 0 & 0 & \frac{\partial}{\partial z}\\ \frac{\partial}{\partial y} & \frac{\partial}{\partial x} & 0\\ 0 & \frac{\partial}{\partial z} & \frac{\partial}{\partial y}\\ \frac{\partial}{\partial z} & 0 & \frac{\partial}{\partial x}\end{bmatrix}$.

5.2.2.4 Strong Formulation for the 3D Elasticity Problem

If the external exerted loads (on S_T and in V) and the stated displacements (on S_u) are given, we can resolve for the resultant displacements, strains, and stresses necessary to maintain equilibrium of the body.

Equilibrium equations in a certain volume are described as follows:

$$\underline{\partial}^T\underline{\sigma}+\underline{X}=\underline{0}\quad\text{in}\quad V \tag{5.15}$$

As displacement boundary states, displacements are stated on the boundary of S_u:

$$\underline{u}=\underline{u}^{\text{specified}}\quad\text{on}\quad S_u \tag{5.16}$$

If we set $\sin\,\theta\; = \frac{dx}{ds} = n_y$ and $\cos\,\theta\; = \frac{dy}{ds} = n_x$ in Fig. 5.6, elasticity traction in 2D can be derived as follows:

$$p_x ds=\sigma_x dy+\tau_{xy}dx \text{ or } p_x=\sigma_x\frac{dy}{ds}+\tau_{xy}\frac{dx}{ds}=\sigma_x n_x+\tau_{xy}n_y \tag{5.17}$$

As traction (force) boundary circumstances, tractions are stated on the boundary of S_T. Traction can be described as a distributed force per unit area. That is,

$$\underline{T}_S=\begin{Bmatrix}p_x\\ p_y\\ p_z\end{Bmatrix} \tag{5.18}$$

If $\underline{\text{n}}$ is the unit external normal to S_T in Fig. 5.7, traction can be defined as follows:

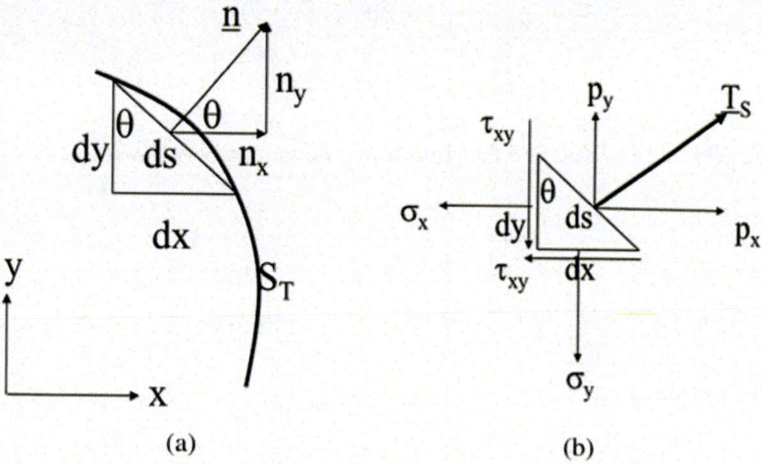

Fig. 5.6 A simple wedge body subjected to 2D surface traction

Fig. 5.7 A simple wedge body subjected to 3D surface traction

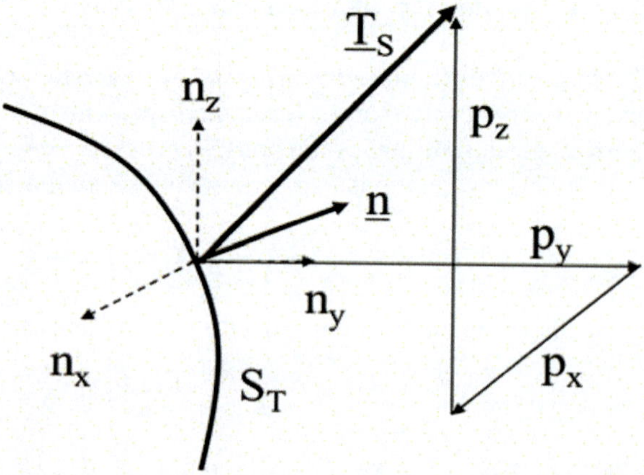

$$
\begin{aligned}
p_x &= \sigma_x n_x + \tau_{xy} n_y + \tau_{xz} n_z \\
p_y &= \tau_{xy} n_x + \sigma_y n_y + \tau_{yz} n_z \\
p_z &= \tau_{xz} n_x + \tau_{zy} n_y + \sigma_z n_z
\end{aligned} \tag{5.19}
$$

5.2.2.5 Strain–Displacement Relationships

As seen in Fig. 5.8, in 2D elasticity, the strain–displacement relationships from small fragments can be obtained as follows:

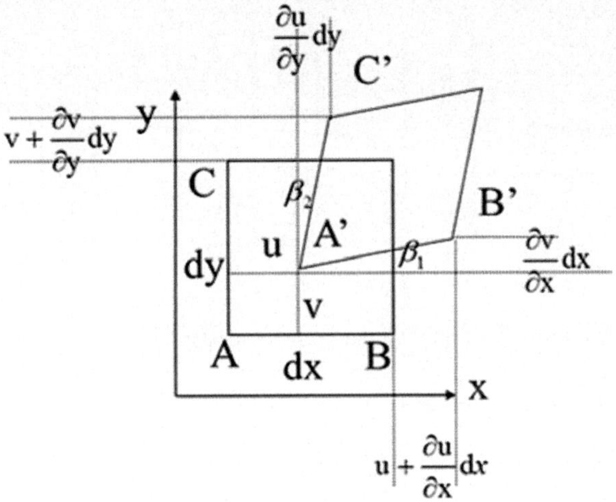

Fig. 5.8 Strain–displacement relationships of small fragments in 2D elasticity

$$\varepsilon_x = \frac{A'B' - AB}{AB} = \frac{\left(dx + \left(u + \frac{\partial u}{\partial x}dx\right) - u\right) - dx}{dx} = \frac{\partial u}{\partial x} \tag{5.20a}$$

$$\varepsilon_y = \frac{A'C' - AC}{AC} = \frac{\left(dy + \left(v + \frac{\partial v}{\partial y}dy\right) - v\right) - dy}{dy} = \frac{\partial v}{\partial y} \tag{5.20b}$$

$$\gamma_{xy} = \frac{\pi}{2} - \text{angle}\left(C'A'B'\right) = \beta_1 + \beta_2 \approx \tan\beta_1 + \tan\beta_2 \approx \frac{\partial v}{\partial x} + \frac{\partial u}{\partial x} \tag{5.20c}$$

For 3D elasticity, strain–displacement relationships can be expanded as follows:

$$\begin{aligned}
\varepsilon_x &= \frac{\partial u}{\partial x} \\
\varepsilon_y &= \frac{\partial v}{\partial y} \\
\varepsilon_z &= \frac{\partial w}{\partial z} \\
\gamma_{xy} &= \frac{\partial u}{\partial y} + \frac{\partial v}{\partial x} \\
\gamma_{yz} &= \frac{\partial v}{\partial z} + \frac{\partial w}{\partial y} \\
\gamma_{zx} &= \frac{\partial u}{\partial z} + \frac{\partial w}{\partial x}
\end{aligned} \tag{5.21}$$

If compactly defined, it will be described as follows:

$$\varepsilon = \partial u \tag{5.22}$$

where $\underline{\varepsilon} = \begin{Bmatrix} \varepsilon_x \\ \varepsilon_y \\ \varepsilon_z \\ \gamma_{xy} \\ \gamma_{yz} \\ \gamma_{zx} \end{Bmatrix}$ $\underline{\partial} = \begin{bmatrix} \frac{\partial}{\partial x} & 0 & 0 \\ 0 & \frac{\partial}{\partial y} & 0 \\ 0 & 0 & \frac{\partial}{\partial z} \\ \frac{\partial}{\partial y} & \frac{\partial}{\partial x} & 0 \\ 0 & \frac{\partial}{\partial z} & \frac{\partial}{\partial y} \\ \frac{\partial}{\partial z} & 0 & \frac{\partial}{\partial x} \end{bmatrix}$ $\underline{u} = \begin{Bmatrix} u \\ v \\ w \end{Bmatrix}$.

5.2.2.6 Stress–Strain Relationship

If General Hooke's law is applied, linear elastic material can be defined as follows:

$$\underline{\sigma} = \underline{D}\,\underline{\varepsilon} \tag{5.23}$$

where

$$\underline{D} = \frac{E}{(1+\nu)(1-2\nu)}\begin{bmatrix} 1-\nu & \nu & \nu & 0 & 0 & 0 \\ \nu & 1-\nu & \nu & 0 & 0 & 0 \\ \nu & \nu & 1-\nu & 0 & 0 & 0 \\ 0 & 0 & 0 & \frac{1-2\nu}{2} & 0 & 0 \\ 0 & 0 & 0 & 0 & \frac{1-2\nu}{2} & 0 \\ 0 & 0 & 0 & 0 & 0 & \frac{1-2\nu}{2} \end{bmatrix}$$

5.2.2.7 Principle of Minimum Potential Energy

For a linear elastic frame subjected to surface tractions $\underline{T}_{\underline{S}} = \left[p_x, p_y, p_z\right]^T$ and body forces $\underline{X} = [X_a, X_b, X_c]^T$, leading to displacements $u = [u, v, w]^T$, strains $\underline{e}$, and stresses $\underline{s}$, the possible energy P is expressed as the strain energy minus the potential energy of the loads requiring $\underline{X}$ and $\underline{T}_{\underline{S}}$:

$$\Pi = U - W \tag{5.24}$$

where $U = \frac{1}{2}\int_V \underline{\sigma}^T\underline{\varepsilon}\,dV$ and $W = \int_V \underline{u}^T\underline{X}\,dV + \int_{S_T} \underline{u}^T\underline{T}_{\underline{S}}\,dS$.

The one that fulfills the equilibrium equations amidst all allowable displacement fields also gives the potential energy P to a minimum. The 'allowable displacement field' has the following properties:

1. The first derivative of the displacement components prevails.
2. It fulfils the boundary states on S_u.

5.2.3 Beam

A beam in mechanical/civil engineering is a structural constituent piece whose length is longer compared to its cross-sectional area. In other words, it is a bar-like structural part whose main function is to bear transverse loading and convey it to the supports. The Euler–Bernoulli beam theory plays a critical part in structural examination because it helps mechanical engineers to compute the load-carrying and deflection characteristics of beams in a straightforward way.

When a beam is bent downward, the materials in the bottom part of the beam shrink, but those in the top part expand. These changes in the lengths of the materials produce stresses in the materials. Those that are enlarged possess tensile stresses applied on the materials in the orientation of the longitudinal axis of the beam, while those that are compressed are subjected to compressive stresses. There always exists a neutral surface of the beam containing materials that do not experience any compression or extension and thus are not subjected to any compressive or tensile stress. It, therefore, can be deduced that from the neutral axis, the value of the bending stress will change linearly with distance. Now, we start the analysis of the beam from the standpoint of elasticity.

5.2.3.1 Euler–Bernoulli Beam Theory

As shown in Fig. 5.9, before bending:

$$AB = CD = EF = \Delta x \tag{5.25}$$

As bending line segment *AB* is compressed, line segment *CD* is elongated, and line segment *EF* does not vary. Hence, line segment *EF* is referred to as the neutral surface. The junction of the neutral surface with the longitudinal plane of symmetry is defined as the neutral axis.

From the calculus and analytic geometry in Fig. 5.9a, we can find the curvature

$$\kappa = \frac{1}{\rho} = \frac{\Delta\theta}{\Delta x} \tag{5.26}$$

where κ is the curvature and ρ is the radius of the curvature.

5.2.3.2 Strain–Displacement Relationships

Because $\rho = \frac{\Delta x}{\Delta\theta}$ and $\Delta x = \rho\Delta\theta$, along line segment *AB*, the normal stain in the longitudinal direction can be expressed as follows:

$$\varepsilon_{AB} = \frac{\delta}{L} = \frac{L_f - L_i}{L_i} = \frac{(\rho - y)\Delta\theta - \rho\Delta\theta}{\rho\Delta\theta} = \frac{-y}{\rho} \tag{5.27}$$

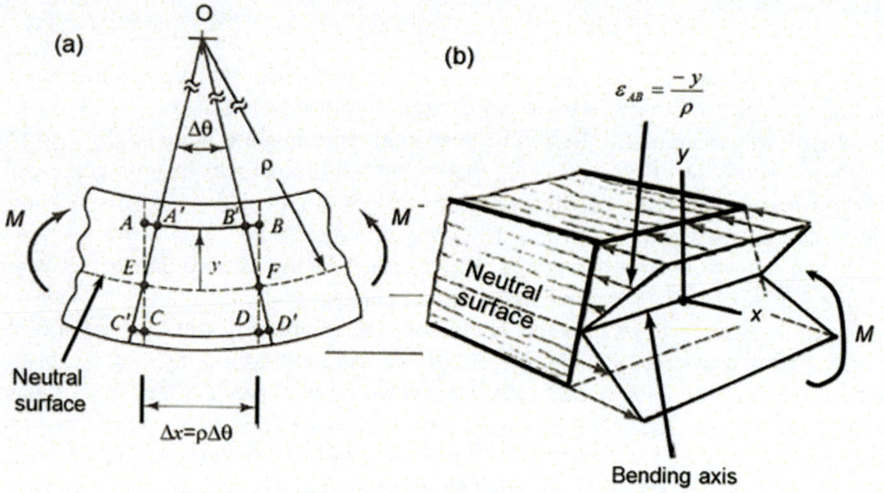

Fig. 5.9 A deformed section in the beam after bending

5.2.3.3 Stress–Strain Relationship

Since Hooke's law satisfies, the stresses are proportionate to the interval y from the neutral axis:

$$\sigma = E\varepsilon = \frac{-Ey}{\rho} \tag{5.28}$$

5.2.3.4 Equilibrium Equations

From Fig. 5.10, we find that internal stresses are statically identical to the external forces and moment. That is,

$$\sum F_{x(I)} = \sum F_{x(\mathrm{II})} \Rightarrow \int \sigma dA = 0 \tag{5.29}$$

$$\sum M_{z(I)} = \sum M_{z(\mathrm{II})} \Rightarrow -\int y\sigma dA = M \tag{5.30}$$

If Eq. (5.28) is substituted, we can obtain:

$$-\int y\frac{-Ey}{\rho}dA = M \tag{5.31}$$

$$M = \frac{E}{\rho}\int y^2 dA = \frac{EI}{\rho} \tag{5.32}$$

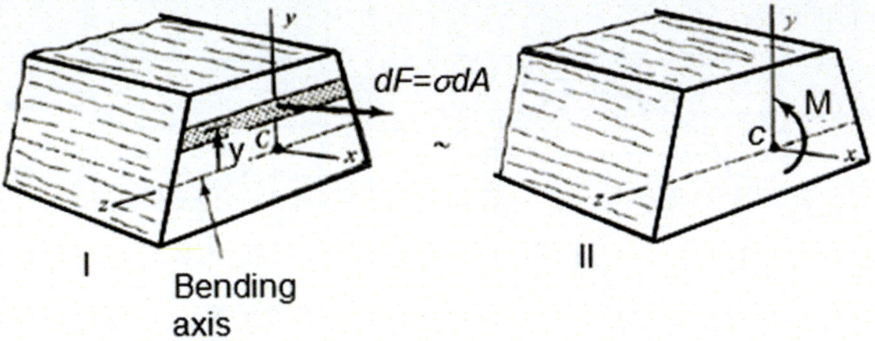

Fig. 5.10 Bending stresses on a beam cross section

where the moment of inertia is $I = \int y^2 dA$.

Because curvature, κ, equals $1/\rho$, the bending momentum can be redefined as follows:

$$M = \frac{EI}{\rho} = EI\kappa = EIy'' \tag{5.33}$$

Finally, if combining Eqs. (5.28) and (5.29), we can find the bending equation:

$$\sigma = \frac{My}{I} \tag{5.34}$$

5.2.4 *Flat Plate*

At first, plates are flat structural members possessing thicknesses much smaller than the other sizes. To discover the distribution of displacement and stress for a plate subjected to a specified set of forces, basic assumptions such as the Kirchhoff hypothesis are required as follows:

1. The deflecting action of the mid-surface is tiny compared with the thickness of the plate.
2. The mid-plane lasts unstrained following bending.
3. At first plane segments normal to the mid-surface and normal to that surface after bending.
4. The stress component normal to the mid-plane is tiny compared with the other stress components.

Based on these assumptions, a three-dimensional plate problem can be reduced to a two-dimensional one. Consider a plate with no load, in which the *xy* plane

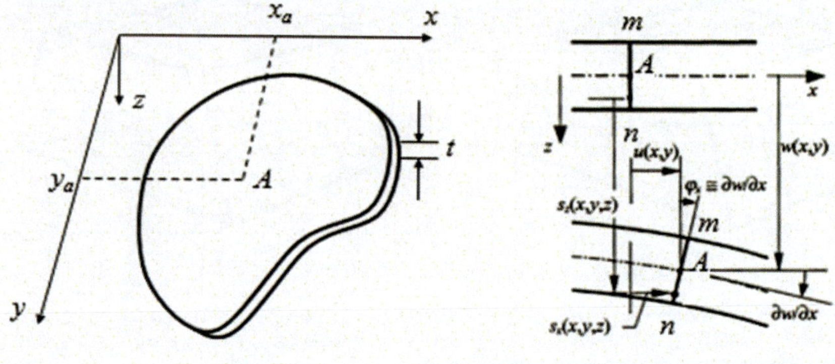

(a) Plate (b) Plate part before and after deflection

Fig. 5.11 A plate of constant thickness

corresponds with the midplane (Fig. 5.11). The components of displacement at a point are denoted by u, v, and w.

According to the preceding presumptions, the displacement field might be defined as follows:

$$u(x, y, z) = -z\frac{\partial w}{\partial x} \tag{5.35a}$$

$$v(x, y, z) = -z\frac{\partial w}{\partial y} \tag{5.35b}$$

$$w(x, y, z) = w(x, y) \tag{5.35c}$$

5.2.4.1 Strain–Displacement Relationships

The nonzero linear strains connected with the displacement field are defined as follows:

$$\varepsilon_x = \frac{\partial u}{\partial x} = -z\frac{\partial^2 w}{\partial x^2} \tag{5.36a}$$

$$\varepsilon_y = \frac{\partial v}{\partial y} = -z\frac{\partial^2 w}{\partial y^2} \tag{5.36b}$$

$$\gamma_{xy} = \frac{\partial u}{\partial y} + \frac{\partial v}{\partial x} = -2z\frac{\partial^2 w}{\partial x \partial y} \tag{5.36c}$$

where γ_{xy} is the shear strain and $(\varepsilon_x, \varepsilon_y)$ are the normal strain.

5.2.4.2 Stress–Strain Relationship

From Hooke's law, we can obtain stress because $\varepsilon_z = \gamma_{yz} = \gamma_{xz} = 0$. That is,

$$\sigma_x = \frac{E}{1-\nu^2}\left[\varepsilon_x + \nu\varepsilon_y\right] \tag{5.37a}$$

$$\sigma_y = \frac{E}{1-\nu^2}\left[\varepsilon_y + \nu\varepsilon_x\right] \tag{5.37b}$$

$$\tau_{xy} = G\gamma_{xy} \tag{5.37c}$$

From Eq. (5.36), we can simplify as follows:

$$\sigma_x = -\frac{E}{1-\nu^2}z\left[\frac{\partial^2 w}{\partial x^2} + \nu\frac{\partial^2 w}{\partial y^2}\right] \tag{5.38a}$$

$$\sigma_x = -\frac{E}{1-\nu^2}z\left[\frac{\partial^2 w}{\partial y^2} + \nu\frac{\partial^2 w}{\partial x^2}\right] \tag{5.38b}$$

$$\tau_{xy} = -\frac{E}{1+\nu}z\frac{\partial^2 w}{\partial x \partial y} \tag{5.38c}$$

5.2.4.3 Equilibrium Equations

The stresses dispersed over the thickness of the plate yield bending moments, twisting moments, and vertical shear forces. These moments and forces per unit length are also defined as stress resultants and described as follows:

$$\left\{\begin{matrix} M_x \\ M_y \\ M_{xy} \end{matrix}\right\} = \int_{\frac{-t}{2}}^{\frac{t}{2}} \left\{\begin{matrix} \sigma_x \\ \sigma_y \\ \tau_{xy} \end{matrix}\right\} zdz \tag{5.39}$$

Substituting Eq. (5.38) into Eq. (5.39), we can derive the following formulas for the bending and twisting moments:

$$\begin{aligned} M_x &= -D\left(\frac{\partial^2 w}{\partial x^2} + \nu\frac{\partial^2 w}{\partial y^2}\right) \\ M_y &= -D\left(\frac{\partial^2 w}{\partial y^2} + \nu\frac{\partial^2 w}{\partial x^2}\right) \\ M_{xy} &= -D(1-\nu)\frac{\partial^2 w}{\partial x \partial y} \end{aligned} \tag{5.40}$$

where $D = \frac{Et^3}{12(1-\nu^2)}$.

Therefore, for bending of a thin plate, the differential equation of equilibrium shall be described as follows:

$$\frac{\partial^2 M_x}{\partial x^2} + 2\frac{\partial^2 M_{xy}}{\partial x \partial y} + \frac{\partial^2 M_y}{\partial y^2} = -p \tag{5.41}$$

Exchanging Eq. (5.35) into Eq. (5.36), the governing equation for the plate can be derived as follows:

$$\frac{\partial^4 w}{\partial x^4} + 2\frac{\partial^4 w}{\partial x^2 \partial y^2} + \frac{\partial^4 w}{\partial y^4} = \frac{p}{D} \tag{5.42}$$

5.2.4.4 Boundary Conditions

The boundary conditions on the edge $x = a$ of the rectangular plate with edges parallel to the x- and y-axes are as follows (Fig. 5.12).

1. Clamped, fixed, or built-in edge: $w = 0,\ \frac{\partial w}{\partial x} = 0\ (x = a)$
2. Simply supported edge: $w = 0,\ M_x = -D\left(\frac{\partial^2 w}{\partial x^2} + \nu\frac{\partial^2 w}{\partial y^2}\right) = 0\ (x = a)$
3. Free edge: $\frac{\partial^2 w}{\partial x^2} + \nu\frac{\partial^2 w}{\partial y^2} = 0,\ \frac{\partial^3 w}{\partial x^3} + (2 - \nu)\frac{\partial^3 w}{\partial x \partial y^2} = 0\ (x = a)$

5.2.5 Torsion Member

Torsion is a distorted form of an object due to an exerted torque. Because the shear stresses for noncircular sections no longer denote the circumference of a curved geometry, the torsional theory of circular sections cannot be applied to the torsion of

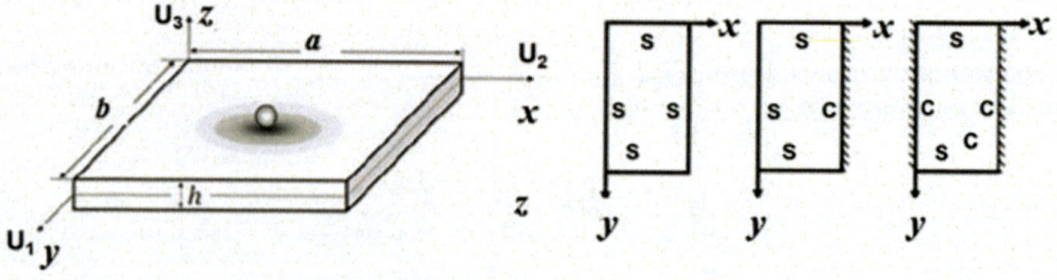

(S - simply supported edge, F - free edge, C - clamped or built-in edge)

Fig. 5.12 Various boundary conditions of a plate (S-simply supported edge, F-free edge, C-clamped or built-in edge)

noncircular sections. Moreover, plane cross-sections do not possess a plane and do not distort the implementation of torque, and warping of the cross section occurs.

As shown in Fig. 5.13a, assume a prismatic bar of isotropic homogeneous noncircular segments subjected to twisting action. Based on the mechanics convention, that is, the axis of the bar corresponds to the z-axis, the cross section is a set in the x, y plane. The bar is attached at $z = 0$ base, and the opposite base $z = l$ is twisted by angle $l\,\alpha$. We can acquire the following assumptions:

- The cross sections in the x, y plane revolve like a rigid body. In the case of a noncircular shape, the cross section is not planar, but it is diverted in the z direction.
- The deflection and twist rate α are constant along the whole length of the bar. Therefore, the problem is lessened to a two-dimensional problem.

Presume the following geometrical behavior: based on the Saint–Venant hypothesis, think about any point P in the section, which will rotate and warp owing to the implementation of T, as shown in Fig. 5.13b. The displacements u, v, and w in orientations x, y, and z under these assumptions can be expressed as follows:

$$u = -r\beta \sin\theta = -\beta y = -\alpha yz$$

$$v = -r\beta \cos\theta = \beta x = \alpha xz \tag{5.43}$$

$$w = \alpha\psi(x, y)$$

where $\beta = \alpha z$, and $\psi(x, y)$ is an unknown function defining the deflection.

Fig. 5.13 Noncircular section under twist

5.2.5.1 Strain–Displacement Relationships

It is assumed to be differentiable. A straightforward computation produces the equivalent strain (small deformation). Thus, we can assume the following:

$$\varepsilon_x = \varepsilon_y = \varepsilon_z = \gamma_{xy} = 0 \tag{5.44}$$

and, thus, the only shearing strains which exist are γ_{xz} and γ_{yz}, which are expressed as follows:

$$\gamma_{xz} = \frac{\partial w}{\partial x} + \frac{\partial u}{\partial z} = \alpha\left(\frac{\partial \psi}{\partial x} - y\right) \tag{5.45}$$

$$\gamma_{yz} = \frac{\partial w}{\partial y} + \frac{\partial v}{\partial z} = \alpha\left(\frac{\partial \psi}{\partial y} + x\right) \tag{5.46}$$

5.2.5.2 Stress–Strain Relationship

From Hooke's law, we can obtain stress from Eqs. (5.44) to (5.46). That is,

$$\sigma_x = \sigma_y = \sigma_z = \tau_{xy} = 0 \tag{5.47}$$

$$\tau_{xz} = G\alpha\left(\frac{\partial \psi}{\partial x} - y\right) \tag{5.48}$$

$$\tau_{yz} = G\alpha\left(\frac{\partial \psi}{\partial y} + x\right) \tag{5.49}$$

5.2.5.3 Equilibrium Equations

$$\frac{\partial \tau_{xz}}{\partial x} + \frac{\partial \tau_{yz}}{\partial y} = 0 \tag{5.50}$$

Let us introduce the Prandt stress function $\phi = \phi(x, y)$. We can define shear stress as $\tau_{xz} = \frac{\partial \phi}{\partial y}$, $\tau_{yz} = -\frac{\partial \phi}{\partial x}$. By differentiating Eqs. (5.48) and (5.49) in respect of y and x, separately, we can obtain the following equation:

$$\frac{\partial \tau_{xz}}{\partial y} - \frac{\partial \tau_{yz}}{\partial x} = \frac{\partial^2 \varphi}{\partial y^2} - \left(-\frac{\partial^2 \varphi}{\partial x^2}\right) = G\alpha\left(\frac{\partial^2 \psi}{\partial x \partial y} - 1\right)$$
$$- G\alpha\left(\frac{\partial^2 \psi}{\partial x \partial y} + 1\right) = -2G\alpha \tag{5.51}$$

The previous equality shall be restated into an inhomogeneous second-order partial differential equation, which is called the *Poisson equation*:

$$\nabla^2\phi = \frac{\partial^2\phi}{\partial y^2} + \frac{\partial^2\phi}{\partial x^2} = -2G\alpha \tag{5.52}$$

5.2.5.4 Boundary Conditions

Since zero surface forces are examined, the traction vector $T = \left(T_x, T_y, T_z\right)$ on the boundary has zero components. Inserting τ_{xy} and τ_{xz} from Eqs. (5.48) and (5.49) into equality $T_z = 0$, we obtain

$$T_z = \tau_{xz}n_x + \tau_{yz}n_y = \frac{\partial\phi}{\partial y}n_x - \frac{\partial\phi}{\partial x}n_y = \frac{\partial\phi}{\partial y}\frac{dy}{ds} + \frac{\partial\phi}{\partial x}\frac{dx}{ds} = \frac{\partial\phi}{\partial s} = 0 \tag{5.53}$$

Thus, Eq. (5.48) involves the tangent derivative of ϕ, which is equal to zero, and thus, ϕ is constant along each component of the boundary.

5.2.5.5 Torque, Section Moment, and Shear Stress

The torsional moment or torque, M, is calculated as the double integral over the cross section:

$$M = \iint\limits_{R^2} \left(-\tau_{xz}y + \tau_{yz}x\right)dxdy \tag{5.54}$$

where R^2 is the cross section of the bar.

$$M = -\iint\limits_{R^2} \left(\frac{\partial\phi}{\partial y}y + \frac{\partial\phi}{\partial x}x\right)dxdy \tag{5.55}$$

Utilizing the way of integration by parts, we can obtain the following results:

$$\iint\limits_{R^2} \frac{\partial\phi}{\partial y}ydxdy = \int\limits_{\Gamma} \phi y n_y ds - \iint\limits_{R^2} \phi(x, y)dxdy = -\iint\limits_{R^2} \phi(x, y)dxdy \tag{5.56}$$

$$\iint\limits_{R^2} \frac{\partial\phi}{\partial x}xdxdy = -\iint\limits_{R^2} \phi(x, y)dxdy \tag{5.57}$$

Therefore, we can obtain the torsional moment. That is,

$$M = 2 \iint_{R^2} \phi(x, y)\, dxdy \tag{5.58}$$

5.3 Stiffness of Mechanical Product—Vibration

5.3.1 Introduction

In this section, we will discuss vibrations of a mechanical system. We know that all human beings involve vibrations in one form or the other. For instance, we hear through eardrum vibration, see by light wave vibration, breath by lung vibration, and speak through tongues vibration. Vibration can be defined as a time-dependent displacement of a system of particles or a particle with respect to an equilibrium location.

Vibration in a mechanical system is a restoring force that continually pulls the system toward its equilibrium position by the elastic or springing property of the system. That is, when the spring is stretched, energy is stored as a potential energy, and a spring force is developed. When the block is released, the potential energy shall be changed to kinetic energy so that the block be pulled toward the equilibrium position, and it continues to oscillate around the equilibrium position.

A vibratory system is a dynamic system for which variables such as the input excitations and response outputs are time dependent. It relies on the external excitation as well as the initial conditions. Because most empirical vibrating systems are very complicated, it is not possible to examine all the details for a mathematical analysis. Only the major features are examined in the analysis to estimate the system attribute under defined input states. The analysis of a vibrating system includes (1) mathematical modeling, (2) derivation of the governing equation, (3) solution of the equation, and (4) interpretation of the result. That is, dynamic system can be analyzed as follows: (1) describe the motion, (2) apply physics $\sum_X F_{ext} = M\ddot{x}$, (3) draw forces in direction in which they act (free body diagram), assuming position values. After that, we can set up the N-DOF system linearized as follows: (1) $M\ddot{x} + C\dot{x} + Kx = F(t)$, (2) Find undamped frequencies ω_i and mode shapes $[u]$, (3) Invoke modal expansion theorem: $x = [u]q(t) = \sum_{i=1}^{N} \{\}^{(i)} q_i(t)$ and $M_i\ddot{q} + C_i\dot{q} + C_i q = Q_i(t)$.

There are basic terminologies to understand the vibration as follows: (1) *periodic motion* is a motion that repeats itself after equal intervals of time, (2) *time period* is a time that is required to finish one cycle, (3) *frequency* is the number of cycles per unit time, (4) *amplitude* is the greatest displacement of a vibrating body from its equilibrium position, (5) *natural frequency* is the frequency of free vibration without external excitation, and (6) *degree of freedom* is the least number of independent coordinates necessitated to state the motion of a system at any instant.

5.3.2 *Analysis of the Free Vibratory Motion in Mechanical Systems*

We will introduce the response of single degree-of-freedom systems. A single degree-of-freedom system is one whose motion is governed by a second-order differential equation. If vibration occurs. System has one natural frequency. Only two variables, position and velocity, are needed to describe the motion of the system. Many structures can be idealized as single degree-of-freedom systems. Understanding how a single degree-of-freedom system responds provides the engineer with important insights into the fundamental behavior of general systems. First of all, the response of a single degree-of-freedom system will be expressed with free vibration.

Motion of equation for vibration of the mechanical system (Fig. 5.14) is,

$$\sum F_x = M\ddot{x} = K_1\left(\frac{L}{2} - x\right) - K_2\left(\frac{L}{2} + x\right) \text{ or } M\ddot{x} + (K_1 + K_2)x = \frac{L}{2}(K_1 - K_2) \tag{5.59}$$

There is static x_s and dynamic x_d displacement in x. That is,

Let $x = x_s + x_d$ where $x_s = \left[\frac{\frac{L}{2}(K_1 - K_2)}{K_1 + K_2}\right]$. We know that $\dot{x} = \dot{x}_d$ or $\ddot{x} = \ddot{x}_d$.

So, Eq. (5.59) is redefined as follows:

$$M\ddot{x}_d + (K_1 + K_2)x_d + (K_1 + K_2)x_s = \frac{L}{2}(K_1 - K_2) \tag{5.60}$$

Because $(K_1 + K_2)x_s = \frac{L}{2}(K_1 - K_2)$, Eq. (5.60) is

$$M\ddot{x}_d + \left(K_{eq}\right)x_d = 0 \text{ where } K_{eq} = (K_1 + K_2) \tag{5.61}$$

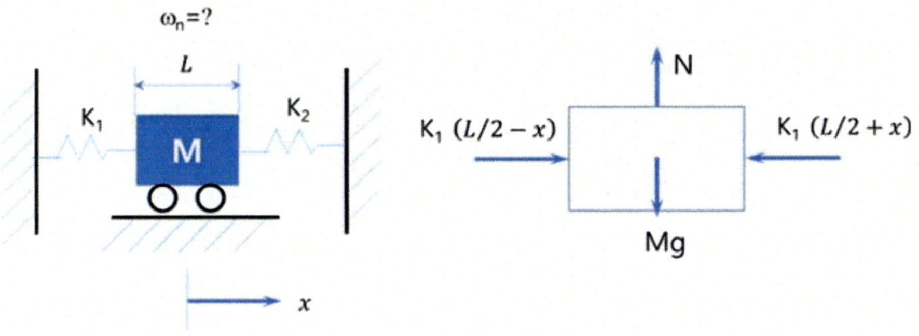

(a) Mechanical system (b) Free body diagram

Fig. 5.14 Vibration of the mechanical system with no damping

If $x_d(t) = A_1\cos(\omega t) + B_1\sin(\omega t)$ and substitute into Eq. (5.61), it is

$$\underbrace{\left(-M\omega^2 + K_{eq}\right)}_{=0}\underbrace{\left(A_1\cos(\omega t) + B_1\sin(\omega t)\right)}_{\neq 0} = 0 \tag{5.62}$$

The undamped natural frequency is

$$\omega_n = \sqrt{\frac{K_{eq}}{M}} \tag{5.63}$$

If $x_d(t) = x(t = 0)$ in Fig. 5.15, $A_1 = x_0$.
If $\dot{x} = -A_1\omega\sin(\omega t) + B_1\omega\sin(\omega t)$ and $v_0 = \dot{x}(t = 0) = B_1\omega_n$, $B_1 = \frac{v_0}{\omega_n}$

$$x(t) = x_0\cos(\omega t) + \frac{v_0}{\omega_n}\sin(\omega t) = A\cos(\omega_n t - \phi) \tag{5.64}$$

where $A = \sqrt{x_0^2 + \left(\frac{v_0}{\omega_n}\right)^2}$, $\phi = \tan^{-1}\left(\frac{B}{A}\right) = \tan^{-1}\left(\frac{v_0}{x_0\omega_n}\right)$
Motion of equation in Fig. 5.16 is

$$M\ddot{x} + C\dot{x} + Kx = 0 \tag{5.65}$$

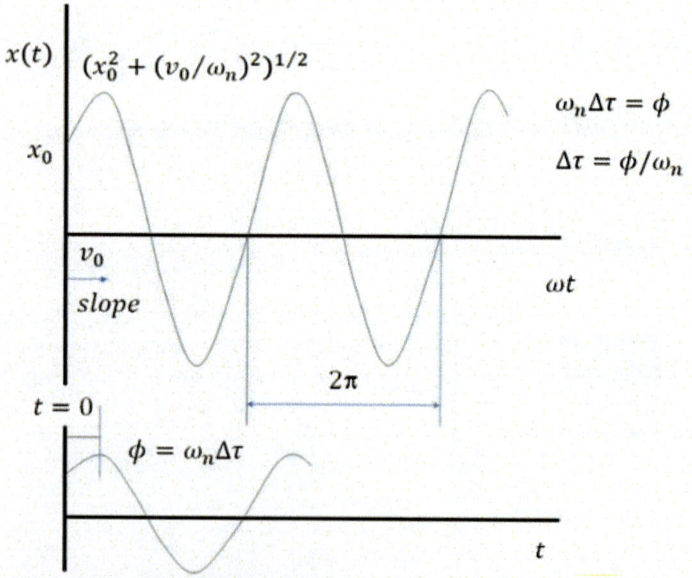

Fig. 5.15 Steady-state solution $x(t)$

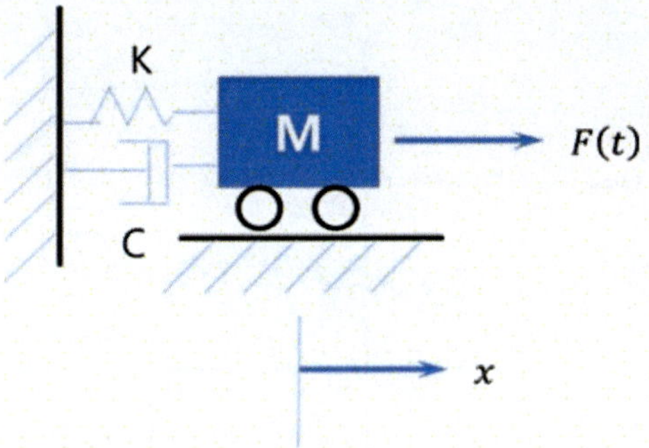

Fig. 5.16 Vibration of the mechanical system with damping and external force

If $C\dot{x} - F(t) = 0$ and Eq. (5.65) is solved, $\omega_n = \sqrt{\frac{K}{M}}$.

Because $A\cos(\omega_n t + \phi) =$ real part of $Ae^{i(\omega_n t - \phi)}$ in Eq. (5.64), $x(t) = e^{-st}$ can be assumed. That is, Eq. (5.65) can be transformed as follows:

$$\left(Ms^2 + Cs + K\right)e^{-st} = 0 \text{ or } s^2 + \frac{C}{M}s + \frac{K}{M} = 0 \tag{5.66}$$

The roots in Eq. (5.66) are

$$s_{1,2} = -\frac{C}{2M} \pm \sqrt{\left(\frac{C}{2M}\right)^2 - \frac{K}{M}} = -\left(\frac{C}{2M\omega_n}\right)\omega_n \pm \omega_n\sqrt{\left(\frac{C}{2M\omega_n}\right)^2 - 1}$$
$$= -\zeta\omega_n \pm \omega_n\sqrt{\zeta^2 - 1} \tag{5.67}$$

where $\zeta = \frac{C}{2M\omega_n}$

There are several cases in Eq. (5.67). That is,

$$\text{st} = \begin{cases} -\zeta\omega_n t \pm \omega_n\sqrt{\zeta^2 - 1t} & \text{for } \zeta > 1 \\ -\omega_n t & \text{for } \zeta = 1 \\ -\zeta\omega_n t \pm i\omega_n\sqrt{1 - \zeta^2}\, t & \text{for } \zeta < 1 \text{ (oscillation)} \end{cases} \tag{5.68}$$

The response $x(t)$ for $\zeta < 1$ (Fig. 5.17) is

$$x(t) = Ae^{-\zeta\omega_n t}\cos(\omega_d t - \phi) \tag{5.69}$$

where $\omega_d \equiv$ damped natural frequency $= \omega_n\sqrt{1 - \zeta^2}$, $\phi \equiv$ phased angle

Equation (5.69) is restated as follows:

$$x(t) = \left[\frac{x_0}{\sqrt{1 - \zeta^2}}\cos(\omega_d t - \Psi) + \frac{v_0}{\omega_d}\sin(\omega_d t)\right]e^{-\zeta\omega_n t} \tag{5.70}$$

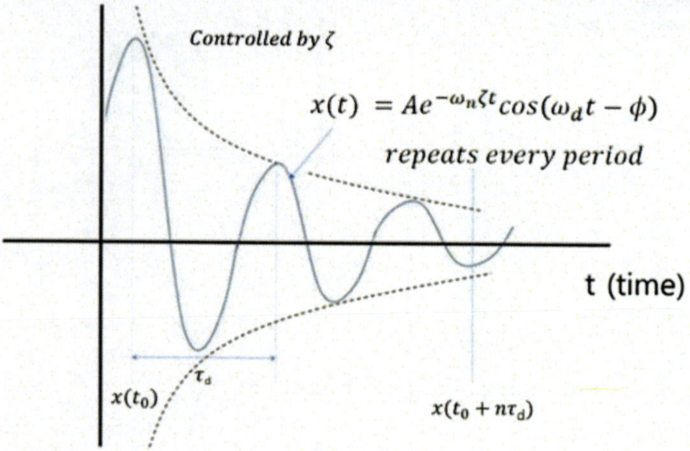

Fig. 5.17 Homogeneous solution x(t) for ζ < 1

$$x(t) = \left[x_0\cos(\omega_d t) + \frac{v_0}{\omega_d}\sin(\omega_d t)\right]e^{-\zeta\omega_n t}\, for\ small\ damping(\zeta < 0.1) \quad (5.71)$$

The ratio $x(t)/x(t + n\tau_d)$ from Eq. (5.69) is

$$\frac{x(t)}{x(t + n\tau_d)} = \frac{e^{-\omega_n\zeta t_0}}{e^{-\omega_n\zeta(t_0 + n\tau_d)}} = e^{\omega_n\zeta n\tau_d} \quad (5.72)$$

If Eq. (5.72) takes the logarithm, it is

$$\ln\frac{x(t)}{x(t + n\tau_d)} = \omega_n\zeta n\tau_d = \omega_n\zeta n\frac{2\pi}{\omega_d} = \omega_n\zeta n\frac{2\pi}{\omega_n\sqrt{1 - \zeta^2}} \quad (5.73)$$

If ζ is small, $\sqrt{1 - \zeta^2} \approx 1$. Equation (5.73) is

$$\ln\frac{x(t)}{x(t + n\tau_d)} = n2\pi\zeta \quad (5.74)$$

If Eq. (5.74) is rearranged, damping ζ is

$$\zeta = \frac{1}{2\pi n}ln\frac{x(t)}{x(t + n\tau_d)} \quad (5.75)$$

If n is such that $x(t + n\tau) = 50\% of\, x(t)$, Eq. (5.75) is

$$\zeta = \frac{1}{2\pi n_{50\%}}ln(2) = \frac{0.11}{n_{50\%}} \quad (5.76)$$

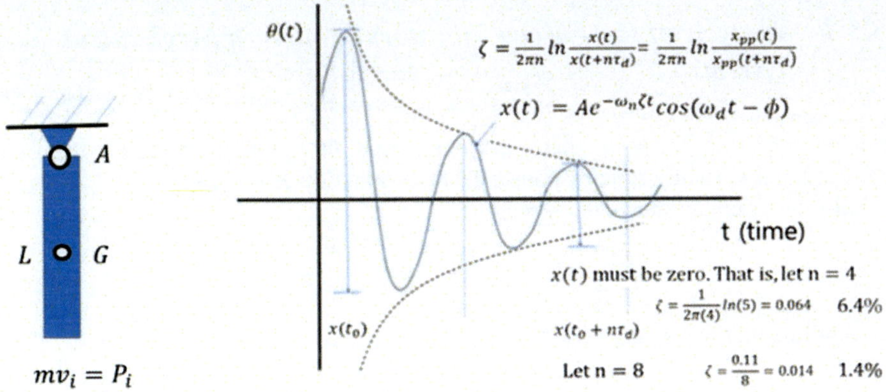

(a) Model (b) Response

Fig. 5.18 A simple bar

Example 5.1 Damping Analysis for a Simple Bar For a simple bar in Fig. 5.18, when 4 cycles or 8 cycles become half in amplitude, find the damping ratio.

Solution

$$\theta(t) \cong \left(\theta_0 \cos(\omega_d t) + \frac{\dot{\theta}_0}{\omega_d} \sin\omega_d t\right) e^{-\zeta \omega_n t}$$

$$\theta(t) = \frac{\dot{\theta}_0}{\omega_d} \sin\omega_d t$$

$$I_{zz} = \frac{ML^2}{3}$$

$$\zeta = \frac{c}{2m\omega_n} \Rightarrow \zeta = \frac{c_t}{2\left(I_{zz} + mL^2\right)\omega_n}$$

$$^{A}H_2 = {}^{A}H_1 = {}^{A}\left(I_{zz} + mL^2\right)\dot{\theta}_0 = mv_i L$$

$$\dot{\theta}_0 = \left(\frac{mv_i L}{I_{zz} + mL^2}\right)$$

$$^{A}H_i = r \times P_i = mv_i L\hat{i}_z$$

If n = 4, $\zeta = \frac{1}{2\pi n} ln(5) = 0.064\, or\, 6.4\%$. If n = 8, $\zeta = \frac{0.11}{8} = 0.014\, or\, 1.4\%$

5.3.3 *Analysis of the Forced Vibratory Motion in Mechanical Systems*

Forced vibration focus on harmonic excitation of the form $\cos(\omega t)$ (external excitation) and steady state response. Transfer function $H_{X/F}(\omega)$ is,

Motion of equation in Fig. 5.19 is

$$M\ddot{x} + C\dot{x} + Kx = F(t) = F_o \cos(\omega t) \tag{5.77}$$

Assume that the steady state solution $x_{s.s.} = x_0 \cos(\omega t - \phi)$. Plug in Eq. (5.77)

$$x_0\left[\left(k - M\omega^2\right)\cos(\omega t - \phi) - C\omega \sin(\omega t - \phi)\right] = F_0 \cos \omega t \tag{5.78}$$

By dividing K in Eq. (5.77), we obtain

$$x_0\left[\left(1 - \frac{\omega^2}{\omega_n^2}\right)\cos(\omega t - \phi) - 2\zeta\frac{\omega}{\omega_n}\sin(\omega t - \phi)\right] = \frac{F_0}{K}\cos \omega t \tag{5.79}$$

where $\omega_n = \sqrt{\frac{K}{M}}$, $\zeta = \frac{C}{2M\omega_n}$,

Because $\cos(\omega t - \phi) = \cos \omega t \cos \phi + \sin \omega t \sin \phi$ and $\sin(\omega t - \phi) = \sin \omega t \cos \phi - \cos \omega t \sin \phi$, we insert them into Eqs. (5.78) and (5.79). That is,

$$x_0\left[\left(1 - \frac{\omega^2}{\omega_n^2}\right)\cos \phi - 2\zeta\frac{\omega}{\omega_n}\sin \phi\right]\cos \omega t = \frac{F_0}{K}\cos \omega t \tag{5.80}$$

$$x_0\left[\left(1 - \frac{\omega^2}{\omega_n^2}\right)\sin \phi - 2\zeta\frac{\omega}{\omega_n}\cos \phi\right]\sin \omega t = 0 \tag{5.81}$$

The solutions are

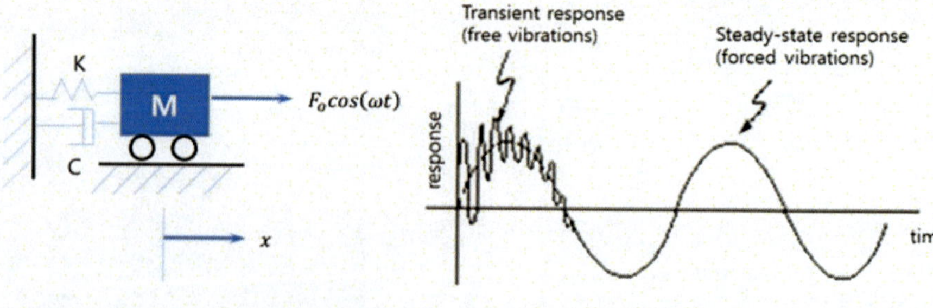

(a) Model with forced vibration (b) Response

Fig. 5.19 Vibration of the mechanical system with external force $F_o \cos(\omega t)$

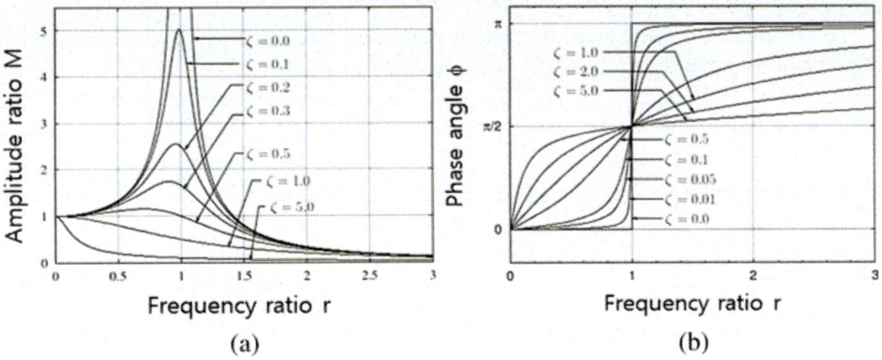

Fig. 5.20 Variation of M and ϕ with frequency ratio r

$$x_0 = \frac{\frac{F_0}{k}}{\left[\left(1-\frac{\omega^2}{\omega_n^2}\right)^2+\left(2\zeta\frac{\omega}{\omega_n}\right)^2\right]^{\frac{1}{2}}} = \left(\frac{F_0}{k}\right)\left|H_{X/F}(\omega)\right|e^{-i\phi} \text{ and } \phi = \tan^{-1}\left(\frac{2\zeta\frac{\omega}{\omega_n}}{1-\frac{\omega^2}{\omega_n^2}}\right) \tag{5.82}$$

If $x_{static} = \frac{F_0}{k}$, $\left|H_{X/F}(\omega)\right|(= M) = \frac{x_0}{x_{static}} = \frac{x_0}{\frac{F_0}{k}} = \frac{\frac{F_0}{k}}{\left[\left(1-\frac{\omega^2}{\omega_n^2}\right)^2+\left(2\zeta\frac{\omega}{\omega_n}\right)^2\right]^{\frac{1}{2}}}$ from Eq. (5.82). That is, variation of M and ϕ with frequency ratio $r = \frac{\omega}{\omega_n}$ are plotted as follows (Fig. 5.20).

5.3.4 Vibration Isolation of Mechanical Systems

The vibration design of a mechanical product is causally connected to the vibration isolation that lessens the unpleasant results of vibration. Therefore, action plans for a mechanical design require the introduction of a mounting rubber between the source of vibration and the mechanical product so that a lessening in the dynamic response of the system is attained under the stated circumstances of vibration excitation.

Vibration isolation shall be utilized when (1) the base of the vibrating machine is secured against its large unstable forces and (2) the product is secured from environmental circumstances such as transportation. The first pattern of isolation is utilized when a mechanical system is subjected to an excitation force. On the other hand, the second type of isolation can be found as follows: when a domestic refrigerator is transported, it shall be subjected to random vibration from the refrigerator base. Vibration should be isolated as the refrigerator uses mounting rubber. Vibration (or noise) can also bring uneasiness to customers.

The response of a mechanical system with mass M, damping C, and spring K in the motion of equation (Fig. 5.21) is expressed as follows:

$$M\ddot{x} + C\dot{x} + Kx = F(t) = F_o\cos(\omega t) = \text{Real part}\left[F_o e^{i\omega t}\right] \tag{5.83}$$

First, the steady state response is,

$$x(t) = x_0\cos(\omega t - \phi) = \text{Re}\left[x_o e^{i(\omega t - \phi)}\right] = \text{Re}\left[\underbrace{x_o e^{-i\phi}}_{X} \cdot e^{i\omega t}\right] = \text{Re}\left[X e^{i\omega t}\right] \tag{5.84}$$

By plugging Eq. (5.84) into Eq. (5.83), we arrive at

$$\left[\left(K - M\omega^2\right) + ic\omega\right]Xe^{i\omega t} = F_o e^{i\omega t} \tag{5.85}$$

From Fig. 5.21b, steady state response is

$$H_{X/F}(\omega) = \frac{X}{F_0} = \frac{1}{\left[\left(K - M\omega^2\right) + ic\omega\right]} = \frac{1/K}{\left[\left(1 - \frac{\omega^2}{\omega_n^2}\right) + 2i\zeta\frac{\omega}{\omega_n}\right]} = \left|H_{X/F}(\omega)\right|e^{-i\phi} \tag{5.86}$$

where $\left|H_{X/F}(\omega)\right| = \frac{\frac{1}{K}}{\left[\left(1-\frac{\omega^2}{\omega_n^2}\right)^2 + \left(2\zeta\frac{\omega}{\omega_n}\right)^2\right]^{\frac{1}{2}}}$, $\phi = \tan^{-1}\left(\frac{2\zeta\frac{\omega}{\omega_n}}{1-\frac{\omega^2}{\omega_n^2}}\right)$

Assume $y \ll x$, the transmitted force F_t from the bottom excitation in Fig. 5.22 is

$$F_t = C\dot{x} + Kx = [K + i\omega C]Xe^{i\omega t} \tag{5.87}$$

In this case, the response $\frac{F_t}{F_o}$ is

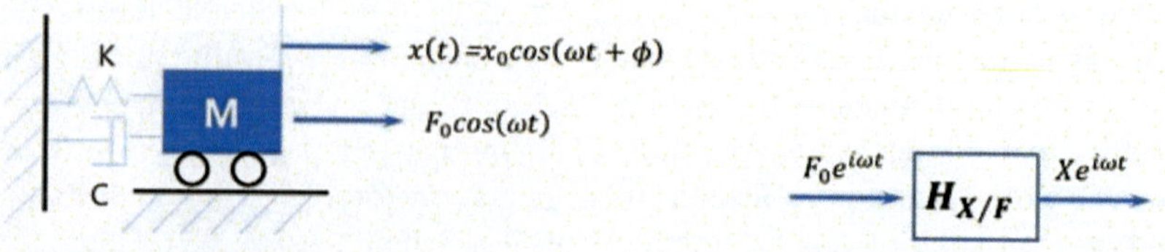

(a) Model (b) Steady state response (transfer function)

Fig. 5.21 Model and response of mechanical system

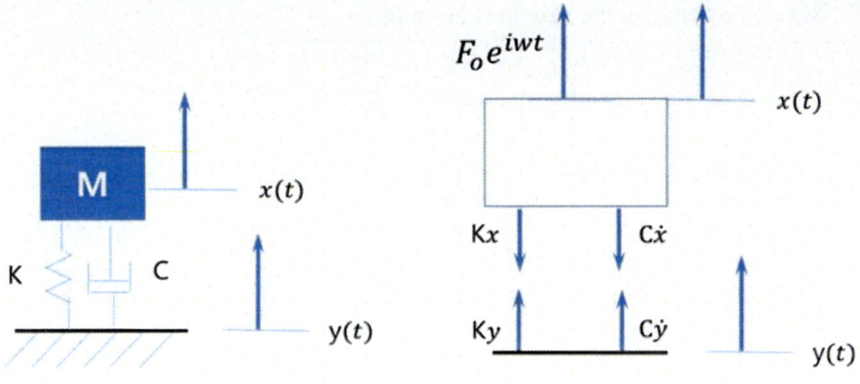

(a) Model (b) Free body diagram

Fig. 5.22 A vibratory mechanical system excited from the bottom

$$\frac{F_t}{F_o} = \frac{[K + i\omega C]Xe^{i\omega t}}{F_o e^{i\omega t}} = [K + i\omega C]H_{X/F}(\omega) \tag{5.88}$$

Transfer function is

$$\left|H_{\frac{F_t}{F_o}}\right| = |H_{X/Y}(\omega)| = \frac{\left[1 + \left(2\zeta\frac{\omega}{\omega_n}\right)^2\right]}{\left[\left(1 - \frac{\omega^2}{\omega_n^2}\right)^2 + \left(2\zeta\frac{\omega}{\omega_n}\right)^2\right]^{\frac{1}{2}}} \tag{5.89}$$

Example 5.2 Vibration Insulation When the motor (source @ 20 Hz) is attached on the end of the bar (total weight = 20 kg), some vibration is designed to reduce 90%. Find $\frac{\omega}{\omega_n}$ such that $|H_{X/y}| = 0.1$

Sol) For $\zeta = 0$, $|H_{X/y}| = \frac{1}{\left|1 - \frac{\omega^2}{\omega_n^2}\right|} = \frac{1}{\frac{\omega^2}{\omega_n^2} - 1} \Rightarrow 0.1\frac{\omega}{\omega_n} = x = \sqrt{11} = 3.31.$

If f = 20 Hz, $f_n = \frac{f}{3.31} = \frac{20\text{ Hz}}{3.31} = 6.04$ Hz, $\omega_n = 37.96\frac{\text{rad}}{s} = \sqrt{\frac{K}{M}}$.
If M = 20 kg, K = 28,827 N/m.

Example 5.3 An exhaust fan rotating at 1000 RPM is to be borne by four springs, each having a stiffness of K. If only 10% of the unbalanced force of the fan is to be transferred to the base, what should be the value of K? Presume the mass of the exhaust fan to be 40 kg.

As the transmissibility must be 0.1, the forcing frequency is given by

$$\omega = (1000 \times 2\pi)/60 = 104.72\text{ rad/s}$$

Natural frequency of the system is given by

$$\omega_n = \left(\frac{k}{m}\right)^{1/2} = \left(\frac{4K}{40}\right)^{1/2} = \frac{\sqrt{K}}{3.1623}.$$

By presuming the damping ratio to be $\zeta = 0$, we attain

$$T_f = 0.1 = \frac{1}{r^2 - 1} = \frac{1}{\left(\frac{104.72 \times 3.1623}{\sqrt{K}}\right)^2 - 1}$$

$$\frac{331.1561}{\sqrt{K}} = 3.3166 \qquad K = 9.97\ \text{N/m}$$

5.3.5 *Multiple Degree-of-Freedom (MDOF) Systems*

Multiple degree-of-freedom systems are defined as those requiring two or more coordinates to describe their motion. That is, MDOF systems require N equation of motion (N mode shapes) that consists of number of DOF ordinary differential equations. There are two groups in Multi-DOF systems as follows: (1) Simple lumped-parameter systems, consisting of only a few degrees of freedom and (2) Structures, which are now almost always set up by using the finite element method.

Motion of equation for Multi-DOF systems with free motion is

$$[M]\{\ddot{X}\} + [K]\{X\} = 0 \tag{5.90}$$

The solutions of MDOF are assumed as follows:

$$\{X\} = \begin{Bmatrix} x_1(t) \\ \vdots \\ x_N(t) \end{Bmatrix} = \begin{bmatrix} a_1 \\ \vdots \\ a_N \end{bmatrix} \cos(\omega_i t - \phi_i) \tag{5.91}$$

Differentiating Eq. (5.91), it is

$$\{\ddot{X}\} = \begin{Bmatrix} \ddot{x}_1(t) \\ \vdots \\ \ddot{x}_N(t) \end{Bmatrix} = -\omega^2 \begin{bmatrix} a_1 \\ \vdots \\ a_N \end{bmatrix} \cos(\omega t - \phi) \tag{5.92}$$

As plugging in Eq. (5.90), it is

$$-\omega^2[M]\begin{bmatrix} a_1 \\ \vdots \\ a_N \end{bmatrix}\cos(\omega t-\phi)+[K]\begin{bmatrix} a_1 \\ \vdots \\ a_N \end{bmatrix}\cos(\omega_i t-\phi_i)=0 \text{ or } \left[-\omega^2[M]+[K]\right]=0 \tag{5.93}$$

As Eq. (5.93) is multiplying by $[M]^{-1}$, it is

$$-\omega^2\begin{bmatrix} a_1 \\ \vdots \\ a_N \end{bmatrix}\cos(\omega t-\phi)+\underbrace{[M]^{-1}[K]}_{[A]}\begin{bmatrix} a_1 \\ \vdots \\ a_N \end{bmatrix}\cos(\omega_i t-\phi_i)=0 \tag{5.94}$$

$$[A]\begin{bmatrix} a_1 \\ \vdots \\ a_N \end{bmatrix}\cos(\omega_i t-\phi_i)=\omega^2\begin{bmatrix} a_1 \\ \vdots \\ a_N \end{bmatrix}\cos(\omega t-\phi) \text{ or } [A]X = \omega^2 X \Leftrightarrow \underbrace{\left[A-\omega^2 I\right]}_{det(A)=0}X=0 \tag{5.95}$$

Natural frequencies from characteristic Eq. (5.95) are

$$A-\omega^2 I=0(\text{characteristic equation }) \Rightarrow \omega^2=? \tag{5.96}$$

Equation of motion of a two degree-of-freedom mechanical system in Fig. 5.23 is

$$M_1\ddot{x}_1+(K_1+K_2)x_1-K_2x_2=F_1(t)=0 \tag{5.97}$$

$$M_2\ddot{x}_2-K_1x_1+K_2x_2=F_1(t)=0 \tag{5.98}$$

Equations (5.97) and (5.98) can be expressed as matrix form. That is,

$$\begin{bmatrix} M_1 & 0 \\ 0 & M_2 \end{bmatrix}\begin{Bmatrix} \ddot{x}_1 \\ \ddot{x}_2 \end{Bmatrix}+\begin{bmatrix} K_1+K_2 & -K_2 \\ -K_2 & K_2 \end{bmatrix}\begin{Bmatrix} x_1 \\ x_2 \end{Bmatrix}=\begin{Bmatrix} F_1(t) \\ F_2(t) \end{Bmatrix}=0 \text{ or } M\ddot{X}+KX=0 \tag{5.99}$$

To solve for $\omega_1\&\omega_2$(mode shape), consider a special case as follows: (1) $M_1=M_2=M$ and (2) $K_1=K_2=K$. Equation (5.99) is

$$\begin{bmatrix} M & 0 \\ 0 & M \end{bmatrix}\begin{Bmatrix} \ddot{x}_1 \\ \ddot{x}_2 \end{Bmatrix}+\begin{bmatrix} 2K & -K \\ -K & K \end{bmatrix}\begin{Bmatrix} x_1 \\ x_2 \end{Bmatrix}=\begin{bmatrix} 0 \\ 0 \end{bmatrix} \tag{5.100}$$

Fig. 5.23 A two degree-of-freedom mechanical system (example)

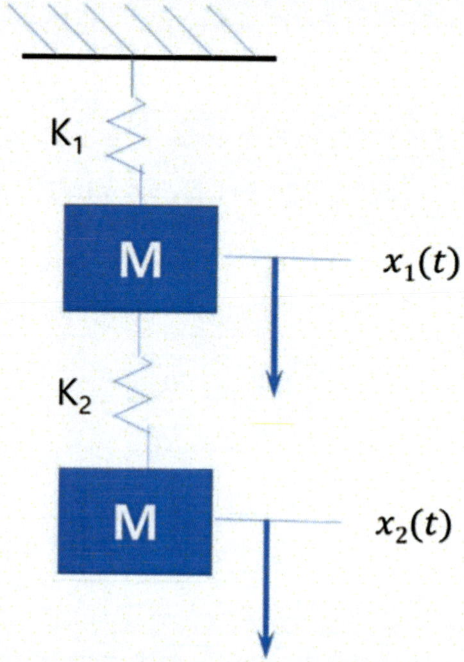

Assume masses oscillate with the same frequency ω and different amplitudes a_1 and a_2. The solutions are assumed as follows:

$$\begin{Bmatrix} x_1 \\ x_2 \end{Bmatrix} = \begin{bmatrix} a_1 \\ a_2 \end{bmatrix} \cos(\omega t - \phi) \tag{5.101}$$

Differentiating Eq. (5.101), it is

$$\begin{Bmatrix} \ddot{x}_1 \\ \ddot{x}_2 \end{Bmatrix} = -\omega^2 \begin{bmatrix} a_1 \\ a_2 \end{bmatrix} \cos(\omega t - \phi) \tag{5.102}$$

Substituting into Eq. (5.100), it is

$$\begin{bmatrix} -M\omega^2 & 0 \\ 0 & -M\omega^2 \end{bmatrix} \begin{bmatrix} a_1 \\ a_2 \end{bmatrix} \cos(\omega t - \phi) + \begin{bmatrix} 2K & -K \\ -K & K \end{bmatrix} \begin{bmatrix} a_1 \\ a_2 \end{bmatrix} \cos(\omega t - \phi) = \begin{bmatrix} 0 \\ 0 \end{bmatrix} \tag{5.103}$$

Dividing Eq. (5.103) by $\cos(\omega t - \phi)$, it is

$$\underbrace{\begin{bmatrix} -M\omega^2 + 2K & -K \\ -K & -M\omega^2 + K \end{bmatrix}}_{det[\mathrm{A}]=0} \begin{bmatrix} a_1 \\ a_2 \end{bmatrix} = \begin{bmatrix} 0 \\ 0 \end{bmatrix} \tag{5.104}$$

Solving det[A] for a_1, a_2, it is

$$\det\left[A\right] = \det\begin{pmatrix} -M\omega^2 + 2K & -K \\ -K & -M\omega^2 + K \end{pmatrix} = 0 \text{ or } M^2\omega^4 - 3KM\omega^2 + K^2 = 0 \tag{5.105}$$

From Eq. (5.105), natural frequencies are

$$\omega^2 = \left(\frac{3 \pm \sqrt{5}}{2}\right)\frac{K}{M} \tag{5.106}$$

From 1st row of Eq. (5.105), we can obtain the amplitude ratio as follows:

$$\left(-M\omega^2 + 2K\right)a_1 = Ka_2 \text{ or } \frac{a_1}{a_2} = \frac{K}{\left(-M\omega^2 + 2K\right)} \tag{5.107}$$

For $\omega^2 = \left(\frac{3-\sqrt{5}}{2}\right)\frac{K}{M} \Rightarrow \omega_1 = 0.618\sqrt{\frac{K}{M}}$, the amplitude ratio is

$$\frac{a_1}{a_2} = 0.618 \tag{5.108}$$

For $\omega^2 = \left(\frac{3+\sqrt{5}}{2}\right)\frac{K}{M} \Rightarrow \omega_2 = 1.618\sqrt{\frac{K}{M}}$, the amplitude ratio is

$$\frac{a_1}{a_2} = -1.618 \tag{5.109}$$

These are natural modes. We know the facts as follows: (1) ratios, not absolute magnitudes, (2) each natural mode is with a particular natural frequency ω_n, (3) these describe the situation in which the entire system is oscillating at one frequency, and (4) natural modes can be thought as special initial conditions that cause the system to decouple into independent subsystems.

$$\begin{Bmatrix} x_1 \\ x_2 \end{Bmatrix} = A_1\begin{Bmatrix} 0.618 \\ 1 \end{Bmatrix}\cos(\omega_1 t - \phi_1) + A_2\begin{Bmatrix} 0.618 \\ -1 \end{Bmatrix}\cos(\omega_2 t - \phi_2) \tag{5.110}$$

where A_1, A_2, ϕ_1, ϕ_2 are determined by initial conditions

5.3.6 Modal Analysis of Vibration Systems (Mode Superposition)

Modal analysis helps to determine the vibration characteristics (natural frequencies and mode shapes) of a mechanical structure or component, showing the movement of different parts of the structure under dynamic loading conditions.

The motion of equation is

$$M\ddot{X} + C\dot{X} + KX = F(t) \tag{5.111}$$

The solution of Eq. (5.111) is assumed as follows:

$$X = uq(t) \tag{5.112}$$

where X is generalized coordinate, $q(t)$ is modal coordinate (or natural coordinate)

By modal expansion theorem, Eq. (5.112) is restated as follows:

$$X = \{u\}^{(n)} q_n(t) = \left[\, \{u\}^{(1)} \cdots \{u\}^{(n)} \,\right] \begin{Bmatrix} q_1(t) \\ \cdots \\ q_n(t) \end{Bmatrix} \tag{5.113}$$

For two DOF system, it is

$$\{x(t)\} = \{u\}^{(1)} q_1(t) + \{u\}^{(2)} q_2(t) \tag{5.114}$$

Differentiating Eq. (5.114), it is

$$\dot{X} = u\dot{q}(t) \tag{5.115}$$

$$\ddot{X} = u\ddot{q}(t) \tag{5.116}$$

As plugging them into Eq. (5.111) for X, $\dot{X}$, $\ddot{X}$ and multiplying by u^T, it is

$$u^T Mu\ddot{q}(t) + u^T Cu\dot{q}(t) + u^T Kuq(t) = u^T F \text{ for ideal conditions} \tag{5.117}$$

We know that mode shapes are orthogonal to one another.

Because $u^T Mu = \begin{bmatrix} \ddots & & \\ & M_i & \\ & & \ddots \end{bmatrix}$, $u^T Ku = \begin{bmatrix} \ddots & & \\ & K_i & \\ & & \ddots \end{bmatrix}$, $u^T Cu = \begin{bmatrix} \ddots & & \\ & C_i & \\ & & \ddots \end{bmatrix}$, it is

$$[M]\ddot{q} + [C]\dot{q} + [K]q = u^T F = Q(t) \quad \text{(N independent DOF systems)} \tag{5.118}$$

Begin with response to initial conditions.

$$X = uq(t) \text{ or } X = \sum_i \{u\}^i q_i(t) \text{ (useful for critical modes)} \tag{5.119}$$

From the solution form of 1 DOF system, $q_i(t)$ can is expected as follows:

$$x(t) = \left[x_0 \cos(\omega_d t) + \frac{(v_0 + \zeta\omega_n x_0)}{\omega_d} \sin(\omega_d t)\right] e^{-\zeta\omega_n t} \tag{5.120}$$

Similarly, the forms of N modal solutions are

$$q_1(t) = \left[q_{10}\cos(\omega_{1d} t) + \frac{(\dot{q}_{10} + \zeta_1\omega_1 q_{10})}{\omega_d} \sin(\omega_{1d} t)\right] e^{-\zeta_1\omega_1 t} \tag{5.121}$$

$$q_2(t) = \left[q_{20}\cos(\omega_{2d} t) + \frac{(\dot{q}_{20} + \zeta_2\omega_2 q_{20})}{\omega_d} \sin(\omega_{2d} t)\right] e^{-\zeta_2\omega_2 t} \tag{5.122}$$

(Rayleigh) damping model is expressed as follows:

$$[C] = \alpha[M] + \beta[K] \tag{5.123}$$

$$u^T Cu = \alpha u^T Mu + \beta u^T Ku = \begin{bmatrix} \ddots & & \\ & C_i & \\ & & \ddots \end{bmatrix} = \alpha \begin{bmatrix} \ddots & & \\ & M_i & \\ & & \ddots \end{bmatrix} + \beta \begin{bmatrix} \ddots & & \\ & K_i & \\ & & \ddots \end{bmatrix} \tag{5.124}$$

For 2 DOF, model is

$$C_1 = \alpha M_1 + \beta K_1 \tag{5.125}$$

$$C_2 = \alpha M_2 + \beta K_2 \tag{5.126}$$

$$x(t) = \sum_{i=1}^{n} \{u\}^{(1)} q_1 + \{u\}^{(2)} q_2 + \cdots + \{u\}^{(n)} \tag{5.127}$$

5.3.7 *Response of 2-DOF Systems by the Use of Transfer Functions*

To characterize the response of N DOF system such as 2-DOF systems to forced vibrations, it is useful to define a transfer function or frequency response function between the input and output of the system.

The motion of equation in Fig. 5.24a is

$$(m_1 + m_2)\ddot{x} + b\dot{x} + kx = -\frac{1}{2} m_2 L\left(\ddot{\theta}\cos\theta - \dot{\theta}^2 \sin\theta\right) \tag{5.128}$$

If $\dot{\theta} = \omega = constant$, Eq. (5.128) is

$$(m_1 + m_2)\ddot{x} + b\dot{x} + kx = F(t) = \frac{1}{2} m_2 L \omega^2 \sin(\omega t) \tag{5.129}$$

On the other hand, the motion of equation in Fig. 5.24b is

$$M\ddot{x} + C\dot{x} + Kx = F(t) = \begin{Bmatrix} F_1 \\ F_2 \end{Bmatrix} e^{i\omega t} \tag{5.130}$$

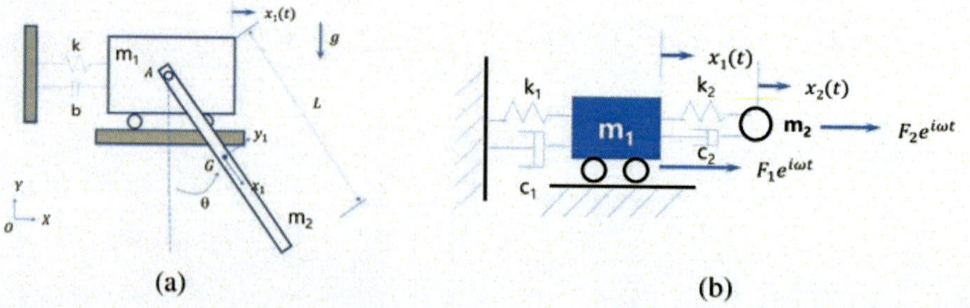

Fig. 5.24 2-DOF systems (example)

If $x = Xe^{i\omega t}$, $\dot{x} = i\omega Xe^{i\omega t}$ and $\ddot{x} = -\omega^2 Xe^{i\omega t}$.

If plugging them into Eq. (5.130), it is

$$\underbrace{\left[-\omega^2 M + i\omega C + K\right]}_{Z(\omega)\rightarrow impedence\ matrix} Xe^{i\omega t} = Fe^{i\omega t} \tag{5.131}$$

where $Z(\omega) = -\omega^2\begin{bmatrix} m_1 & 0 \\ 0 & m_2 \end{bmatrix} + i\omega\begin{bmatrix} c_1 + c_2 & -c_2 \\ -c_2 & c_2 \end{bmatrix} + \begin{bmatrix} k_1 + k_2 & -k_2 \\ -k_2 & k_2 \end{bmatrix} = \begin{bmatrix} Z_{11} & Z_{12} \\ Z_{21} & Z_{22} \end{bmatrix}$

$$ZX = F \text{ or } X = Z^{-1}F \equiv HF \tag{5.132}$$

where H is N × N matrix of transfer functions, $H_{ij} \equiv$ response at i per unit force j

For $N = 2(2 \times 2)$, inverse Z^{-1} is

$$Z^{-1} = \frac{\begin{bmatrix} Z_{22} & -Z_{12} \\ -Z_{21} & Z_{11} \end{bmatrix}}{det|Z|} = \frac{\begin{bmatrix} Z_{22} & -Z_{12} \\ -Z_{21} & Z_{11} \end{bmatrix}}{Z_{11}Z_{22} - Z_{12}^2} = \begin{bmatrix} H_{11} & H_{12} \\ H_{21} & H_{22} \end{bmatrix} \tag{5.133}$$

where $Z_{11} = -\omega^2 m_{11} + i\omega C_{11} + K_{11} = -\omega^2 m_1 + i\omega(c_1 + c_2) + k_1 + k_2$

$$\begin{Bmatrix} x_1 \\ x_2 \end{Bmatrix} = \begin{bmatrix} H_{11} & H_{12} \\ H_{21} & H_{22} \end{bmatrix}\begin{Bmatrix} F_1 \\ F_2 \end{Bmatrix} \tag{5.134}$$

If $F_2 = 0$ and $C = 0$, response x_1 is

$$x_1 = H_{11}F_1 + H_{12}F_2 = H_{11}F_1 \tag{5.135}$$

$$x_1 = \frac{Z_{22}}{Z_{11}Z_{22} - Z_{12}^2}F_1 = \frac{\left(-\omega^2 m_2 + k_2\right)}{\left[\left(-\omega^2 m_1 + k_1 + k_2\right)\left(-\omega^2 m_2 + k_2\right) - k_2^2\right]}F_1 \tag{5.136}$$

where $Z_{11} = -\omega^2 m_1 + k_1 + k_2$, $Z_{11} = -\omega^2 m_2 + k_2$, $Z_{12} = k_2$.

If $-\omega^2 m_2 + k_2 = 0$, $x_1 = 0$. That is,

$$\omega^2 = \frac{k_2}{m_2} \tag{5.137}$$

Because $\Omega_n = \sqrt{\frac{k_1}{m_1}} = \sqrt{\frac{k_2}{m_2}}$, we know that $x_1 = x_{1s}$

$$\left|\frac{x_1(\omega)}{x_{1s}}\right| = \frac{F_1|H_{11}(\omega)|}{F_1/k_1} = k_1|H_{11}(\omega)| \tag{5.138}$$

$x_2 = H_{21}F_1$ where $F_1 = 0$

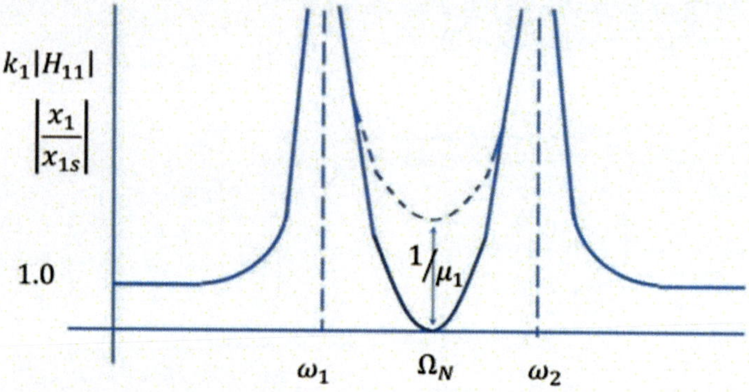

Fig. 5.25 Response of 2-DOF systems (example)

$$x_{2\omega=\Omega_N} = \frac{-Z_{12}}{Z_{11}Z_{22} - Z_{12}^2}F_1$$

$$= \frac{-k_2 F_1}{\left[(-\omega^2 m_1 + k_1 + k_2)\left(\underbrace{-\omega^2 m_2 + k_2}_{0}\right) - k_2^2\right]} = \frac{F_1}{k_2} \quad (5.139)$$

$$\frac{x_2}{x_{1s}} = \frac{F_1/k_2}{F_1/k_1} = \frac{k_1}{k_2} = \frac{m_1}{m_2} = \frac{1}{\mu} \quad (5.120)$$

where $\mu \equiv massratio$ (Fig. 5.25)

Example 5.4 Modal Analysis 1 If $m_1 = 0.3193$ kg, $m_2 = 0.6309$ kg, $k_1 = 55.87$ N/m, $k_2 = 36.047$ N/m in Fig. 5.20, carry out modal analysis of system in Fig. 5.26.

$$\begin{bmatrix} m_1 & 0 \\ 0 & m_2 \end{bmatrix}\begin{Bmatrix} \ddot{x}_1 \\ \ddot{x}_2 \end{Bmatrix} + \begin{bmatrix} C_1 & 0 \\ 0 & C_2 \end{bmatrix}\begin{Bmatrix} \dot{x}_1 \\ \dot{x}_2 \end{Bmatrix} + \begin{bmatrix} K_1 + K_2 & -K_2 \\ -K_2 & K_2 \end{bmatrix}\begin{Bmatrix} x_1 \\ x_2 \end{Bmatrix} = 0 \text{ or } M\ddot{x} + C\dot{x} + Kx = 0$$

$$M = \begin{bmatrix} 0.3193 & 0 \\ 0 & 0.630 \end{bmatrix}\text{kg,} = \begin{bmatrix} 91.97 & -36.047 \\ -36.047 & 36.047 \end{bmatrix}\text{N/M}$$

To find the undamped natural frequencies ω_n's and displacements $\{u\}^i$, it is

$$M\ddot{X} + KX = 0$$

$$X = \{u\}e^{i\omega t}$$

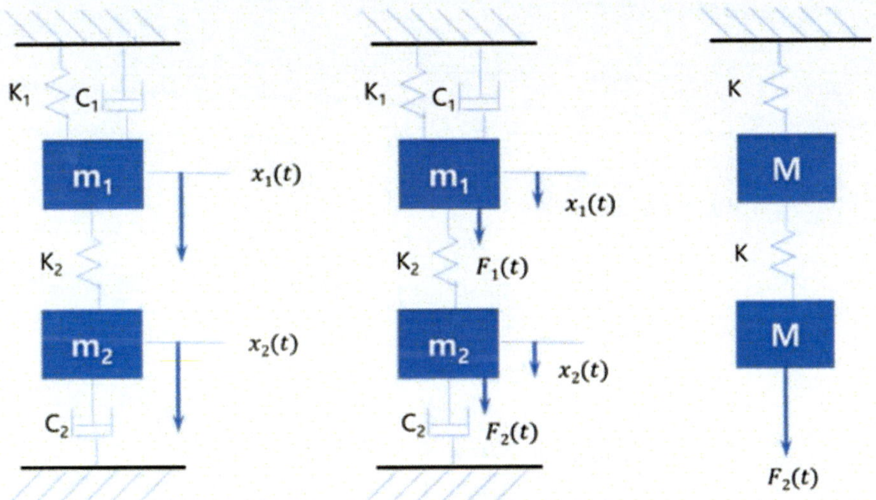

(a) No forced (b) Forced (case 1) (c) Forced (case 2)

Fig. 5.26 A two degree-of-freedom mechanical system with damping (example)

$$\underbrace{\left[-\omega^2 M + K\right]}_{\det(A)=0}\{u\}e^{i\omega t} = 0$$

The natural frequencies are

$$\omega_1 = 5.6546 \text{ rad/s}, \ \omega_2 = 17.695 \text{ rad/s}$$

$$u = \left[\begin{Bmatrix} 1.0 \\ 2.2667 \end{Bmatrix}^{(1)} \begin{Bmatrix} 1.0 \\ -0.2236 \end{Bmatrix}^{(2)}\right]$$

Verify $\omega_1 = \sqrt{\frac{K_1}{M_1}} = \sqrt{\frac{113.71}{3.556}} = 5.6546$ rad/s, $\omega_2 = \sqrt{\frac{K_2}{M_2}} = 17.695$ rad/s

$$u^T M u = \begin{bmatrix} M_1 & 0 \\ 0 & M_2 \end{bmatrix} = \begin{bmatrix} 3.5562 & 0 \\ 0 & 0.3508 \end{bmatrix}$$

$$u^T K u = \begin{bmatrix} K_1 & 0 \\ 0 & K_2 \end{bmatrix} = \begin{bmatrix} 113.71 & 0 \\ 0 & 109.839 \end{bmatrix}$$

By modal expansion theorem, modal equation is

$$M_1 \ddot{q}_1 + C_1 \dot{q}_1 + K_1 q_1 = Q_1(t)$$

$$u^T C u = \begin{bmatrix} C_1 & 0 \\ 0 & C_2 \end{bmatrix}$$

$$Q = u^T F$$

If $F = 0$, initial conditions are

$$x(t = 0) = \begin{Bmatrix} x_{10} \\ \vdots \\ x_{20} \end{Bmatrix}$$

$$\dot{x}(t = 0) = \begin{Bmatrix} v_{10} \\ \vdots \\ v_{20} \end{Bmatrix} = 0$$

Because $x = uq$, it is

$$x(0) = uq(0)$$

$$q(0) = u^{-1}x(0)$$

$$\dot{q}(0) = u^{-1}\dot{x}(0)$$

$$M_1\ddot{q}_1 + C_1\dot{q}_1 + K_1 q_1 = 0$$

Let $\begin{Bmatrix} x_{10} \\ x_{20} \end{Bmatrix} = \begin{Bmatrix} 1 \\ 2.260 \end{Bmatrix}^{(1)}, \begin{Bmatrix} \dot{x}_{10} \\ \dot{x}_{20} \end{Bmatrix} = 0$

$$\begin{Bmatrix} q_{10} \\ q_{20} \end{Bmatrix} = u^{-1}x(0) = \begin{bmatrix} 0.898 & 0.4016 \\ 0.9102 & -0.4016 \end{bmatrix} \begin{Bmatrix} 1 \\ 2.260 \end{Bmatrix}^{(1)} = \begin{Bmatrix} 1 \\ 0 \end{Bmatrix}$$

$$[C] = \alpha[M] + \beta[K]$$

$$u^T C u = \alpha[M] + \beta[K]$$

$$[C] = \begin{bmatrix} C_1 & 0 \\ 0 & C_2 \end{bmatrix}$$

$$C_1 = \alpha M_1 + \beta K_1$$

$$C_2 = \alpha M_2 + \beta K_2$$

$$\zeta_1 = \frac{C_1}{2\omega_1 M_1} = \frac{\alpha M_1}{2\omega_1 M_1} + \frac{\beta K_1}{2\omega_1 M_1} = \frac{\alpha}{2\omega_1} + \frac{\beta\omega_1}{2}$$

$$\zeta_2 = \frac{C_2}{2\omega_2 M_2} = \frac{\alpha}{2\omega_2} + \frac{\beta\omega_2}{2}$$

For test, it is

$$\zeta_1 \approx \frac{0.11}{n_{50\%}} = \frac{0.11}{2.5} = 0.044\, or\, 4.4\%$$

$$\zeta_2 \approx \frac{0.11}{n_{50\%}} = \frac{0.11}{10} = 0.011\, or\, 1.1\%$$

If solving them, $\alpha = 10.71$, $\beta = -0.033$

$$[C] = 10.71[M] - 0.033[K]$$

$$\begin{bmatrix} \ddots & & \\ & C_i & \\ & & \ddots \end{bmatrix} = u^T C u = \alpha \begin{bmatrix} \ddots & & \\ & M_i & \\ & & \ddots \end{bmatrix} + \beta \begin{bmatrix} \ddots & & \\ & K_i & \\ & & \ddots \end{bmatrix}$$

$$x(t) = \sum_{i=1}^{n} \{u\}^{(1)} q_1 + \{u\}^{(2)} q_2$$

$$q_1(t) = \left[q_{10} cos(\omega_{1d} t) + \frac{(\dot{q}_{10} + \zeta_1 \omega_1 q_{10})}{\omega_d} \sin(\omega_{1d} t) \right] \mathrm{e}^{-\zeta_1 \omega_1 t}$$

$$q_2(t) = \left[q_{20} cos(\omega_{2d} t) + \frac{(\dot{q}_{20} + \zeta_2 \omega_2 q_{20})}{\omega_d} \sin(\omega_{2d} t) \right] \mathrm{e}^{-\zeta_2 \omega_2 t}$$

For two degree-of-free system in Fig. 5.26c, it is

$$F = \begin{Bmatrix} 0 \\ F_2 \end{Bmatrix} \mathrm{e}^{i\omega t}$$

$$Q = u^T F = \begin{bmatrix} 1.0 & 2.2667 \\ 1.0 & -0.2236 \end{bmatrix} \begin{Bmatrix} 0 \\ F_2 \end{Bmatrix} \text{ or } Q_1 = 2.2667 F_2 e^{i\omega t}\, Q_2 = -0.2236 F_2 e^{i\omega t}$$

$$M_1 \ddot{q}_1 + C_1 \dot{q}_1 + K_1 q_1 = 2.2667 F_2 e^{i\omega t}$$

$$|q_1| = |Q_1|\left|Hq_{1/Q_1}(\omega)\right| = \frac{|Q_1|\left(1/K_1\right)}{\left[\left(1-\frac{\omega^2}{\omega_1^2}\right)^2+\left(2\zeta_1\frac{\omega}{\omega_1}\right)^2\right]^{\frac{1}{2}}}$$

If $\omega = \omega_1$, $|q_1| = \frac{2.2667F_2}{K_1}\cdot\frac{1}{2\zeta_1} = \frac{2.2667F_2}{113.71}\cdot 11.36 = 0.227F_2$

If $\frac{\omega}{\omega_2} = \frac{\omega_1}{\omega_2}$, $|q_2| = |Q_2|\left|Hq_{2/Q_2}(\omega)\right| = \frac{-0.2236F_2}{\underbrace{109.84}_{K_2}}\cdot$

$$\frac{1}{\underbrace{\left[\underbrace{\left(1-\frac{\omega^2}{\omega_1^2}\right)^2}_{0.806}+\underbrace{\left(2\zeta_1\frac{\omega}{\omega_1}\right)^2}_{5\times10^{-5}}\right]^{\frac{1}{2}}}_{0.898}} = 0.0023F_2$$

$$x = uq$$

$$q_1 = q_1\cos\left(\omega_1 t - \frac{\pi}{2}\right) = 0.227F_2\cos\left(\omega_1 t - \frac{\pi}{2}\right)$$

$$q_2 = q_2\cos(\omega_1 t - \phi_2) = 0.0023F_2\cos(\omega_1 t - \phi_2)$$

Example 5.5 Modal Analysis 2 If there is double pendulum system in Fig. 5.27, carry out modal analysis.

Motion of equation of double pendulum system in Fig. 5.20a is

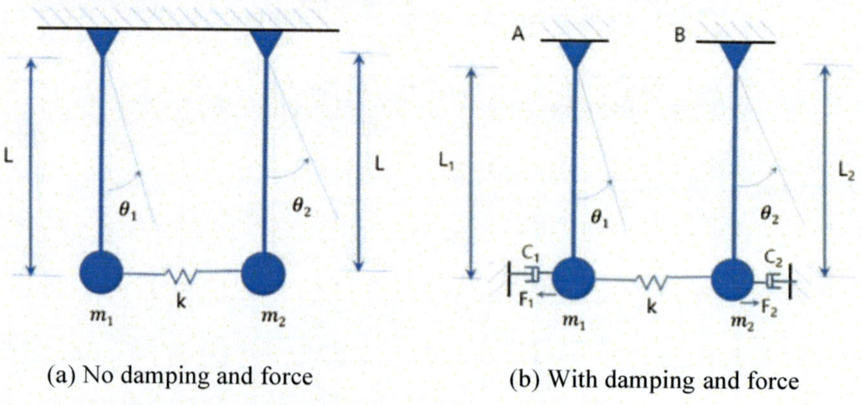

(a) No damping and force (b) With damping and force

Fig. 5.27 Double pendulum system (example)

$$\ddot{\theta}_1 + \left(\frac{g}{L} + \frac{k}{m_1}\right)\theta_1 - \frac{k}{m_1}\theta_2 = 0$$

$$\ddot{\theta}_2 - \frac{k}{m_2}\theta_1 + \left(\frac{g}{L} + \frac{k}{m_2}\right)\theta_2 = 0$$

Matrix form is

$$\begin{bmatrix} 1 & 0 \\ 0 & 1 \end{bmatrix}\begin{Bmatrix} \ddot{\theta}_1 \\ \ddot{\theta}_2 \end{Bmatrix} + \begin{bmatrix} \left(\frac{g}{L} + \frac{k}{m_1}\right) & -\frac{k}{m_1} \\ -\frac{k}{m_2} & \left(\frac{g}{L} + \frac{k}{m_2}\right) \end{bmatrix}\begin{Bmatrix} \theta_1 \\ \theta_2 \end{Bmatrix} = 0 \text{ or } M\ddot{\theta} + K\theta = 0$$

Let $\theta = \{u\}e^{i\omega t}$. Plug in motion of equation.

$$\underbrace{\left[-\omega^2 M + K\right]}_{\det(A)=0}\{u\}e^{i\omega t} = 0$$

Let $m_1 = m_2 = m$ and $h = \left(\frac{g}{L} + \frac{k}{m}\right)$. The characteristic equation is

$$\omega^2 - 2h\omega^2 + h^2 - \left(\frac{k}{m}\right)^2 = 0$$

$$\omega_1 = \sqrt{g/L},\ \omega_2 = \sqrt{g/L + 2k/m}$$

$$\left[\begin{bmatrix} -\omega^2 & 0 \\ 0 & -\omega^2 \end{bmatrix} + \begin{bmatrix} h & -k/m \\ -k/m & h \end{bmatrix}\right] = \begin{bmatrix} -\omega^2 + h & -k/m \\ -k/m & -\omega^2 + h \end{bmatrix}\begin{Bmatrix} u_1 \\ u_2 \end{Bmatrix} = 0$$

For $\omega = \omega_1$, $u_1 = u_2$

For $\omega = \omega_2$, $u_1 = -u_2$.

Initial conditions θ_{10}, θ_{20} are

$$\theta_{i0} = \begin{Bmatrix} 1 \\ 0 \end{Bmatrix} = C_1\begin{Bmatrix} 1 \\ 1 \end{Bmatrix} + C_2\begin{Bmatrix} 1 \\ -1 \end{Bmatrix} = 1/2\begin{Bmatrix} 1 \\ 1 \end{Bmatrix} + 1/2\begin{Bmatrix} 1 \\ -1 \end{Bmatrix}$$

$$1/2\begin{Bmatrix} 1 \\ 1 \end{Bmatrix}\cos\omega_1 t + 1/2\begin{Bmatrix} 1 \\ -1 \end{Bmatrix}\cos\omega_2 t = \begin{Bmatrix} \theta_1 \\ \theta_2 \end{Bmatrix}$$

$$1/2\cos\omega_1 t + 1/2\cos\omega_2 t = \theta_1(t) = \cos\left(\frac{\omega_1 - \omega_2}{2}\right)t\cos\left(\frac{\omega_1 + \omega_2}{2}\right)t$$

$$1/2\cos\omega_1 t - 1/2\cos\omega_2 t = \theta_1(t) = \sin\left(\frac{\omega_2 - \omega_1}{2}\right)t\sin\left(\frac{\omega_1 + \omega_2}{2}\right)t$$

Motion of equation of double pendulum system in Fig. 5.20b is

$$\begin{bmatrix} m_1L_1^2 & 0 \\ 0 & m_2L_2^2 \end{bmatrix}\begin{Bmatrix} \ddot{\theta}_1 \\ \ddot{\theta}_2 \end{Bmatrix} + \begin{bmatrix} C_1L_1^2 & 0 \\ 0 & C_2L_2^2 \end{bmatrix}\begin{Bmatrix} \dot{\theta}_1 \\ \dot{\theta}_2 \end{Bmatrix}$$
$$+ \begin{bmatrix} m_1gL_1 + kL_1^2 & -kL_1L_2 \\ -kL_1L_2 & m_2gL_2 + kL_2^2 \end{bmatrix}\begin{Bmatrix} \theta_1 \\ \theta_2 \end{Bmatrix} = \begin{Bmatrix} F_1L_1 \\ F_2L_2 \end{Bmatrix}$$

If $m_1 = m_2 = 0.5$ kg, $L_1 = 1.1$ m, $L_2 = 1.0$ m, $k = 0.5$ N/m,it is

$$M = \begin{bmatrix} 0.605 & 0 \\ 0 & 0.50 \end{bmatrix}, K = \begin{bmatrix} 6.005 & -0.55 \\ -0.55 & 5.405 \end{bmatrix}$$

The natural frequencies and displacements are

$$\begin{Bmatrix} \omega_1 \\ \omega_2 \end{Bmatrix} = \begin{Bmatrix} 3.0445 \\ 3.3851 \end{Bmatrix}, u = \begin{bmatrix} \begin{Bmatrix} 1.0 \\ 0.7139 \end{Bmatrix}^{(1)} & \begin{Bmatrix} 1.0 \\ -1.6949 \end{Bmatrix}^{(2)} \end{bmatrix}$$

By modal expansion theorem, $x = uq$. That is,

$$\begin{Bmatrix} \theta_1 \\ \theta_2 \end{Bmatrix} = uq = \sum_{i=1}^{N} \{u\}^{(i)} q_i(t) = \begin{Bmatrix} 1.0 \\ 0.7139 \end{Bmatrix}^{(1)} q_1(t) + \begin{Bmatrix} 1.0 \\ -1.6949 \end{Bmatrix}^{(2)} q_2(t)$$

Modal mass, stiffness matrices, and modal forces are,

$$[M] = u^T M u = \begin{bmatrix} 0.8599 & 0 \\ 0 & 2.0414 \end{bmatrix}\left(= \{ 1 \; -1.6949 \} \begin{bmatrix} 0.605 & 0 \\ 0 & 0.50 \end{bmatrix} \begin{Bmatrix} 1 \\ -1.6949 \end{Bmatrix} = m_2 \right)$$

$$[K] = u^T K u = \begin{bmatrix} 9.9699 & 0 \\ 0 & 23.3918 \end{bmatrix}$$

Let $F_1 = F_1 \cos\omega t$, $F_2 = 0 F_2 = 0$

$$Q = u^T F = \begin{bmatrix} 1 & 0.7139 \\ 1 & -1.6949 \end{bmatrix}\begin{Bmatrix} F_1L_1 \\ 0 \end{Bmatrix}\cos\omega t$$

Let $q_1 = q_1\cos(\omega t - \phi)$ and $\omega = \omega_1$, modal equation is

$$M_1\ddot{q}_i + C_1\dot{q}_1 + K_1q_1 = \left(\underbrace{-M_1\omega_1^2 q_1 + K_1 q_1}_{0} \right)$$
$$- C_1\omega_1 q_1 \sin(\omega t - \phi) = Q_1(t) = F_1L_1\cos\omega t$$

$$C_1\omega_1 q_1 = F_1 L_1$$

Let $\omega = \omega_2$, 2nd mode contribution is

$$\begin{Bmatrix} \theta_1 \\ \theta_2 \end{Bmatrix}^{(2)} = \begin{Bmatrix} \\ \end{Bmatrix}^{(2)} q_2(t)$$

$$q_2(t) = |Q_2|\left|Hq_2/_{Q_2}\right|\cos(\omega_2 t - \phi_2) = \frac{F_1 L_1\left(\frac{1}{K_2}\right)}{\left[\underbrace{\left(1 - \frac{\omega_2^2}{\omega_2^2}\right)^2}_{=0} + \left(2\zeta\frac{\omega_2}{\omega_2}\right)^2\right]^{\frac{1}{2}}} = \frac{F_1 L_1\left(\frac{1}{K_2}\right)}{2\zeta}$$

$$u^T C u = \alpha[M] + \beta[K] \approx \alpha[M] \text{ or } C_2 = \alpha M_2 = \alpha(2.04)$$

$$\zeta_2 = \frac{C_2}{2\omega_2 M_2} = \frac{\alpha(2.04)}{2\omega_2 M_2} = 0.02.$$

Chapter 6
Mechanical System Failure

6.1 Introduction

A typical failure mechanism of a mechanical system due to repetitive or varying loads is fatigue and fracture. As coined by the French engineer Jean-Victor Poncelet in the midst of the nineteenth century, the term ‘fatigue’ means that the matter becomes weak, which eventually leads to disintegration or fracture. It evolves in an imperfect part of a product as a consequence of cyclic plastic deformation in a limited area on the outside of the part where there is a tiny crack or a preexisting defect. Based on individualistic studies conducted by Battelle in 1982 [1], approximately 80–90% of all constructional unsuccessfulness happen by a fatigue mechanism. The cost due to these failures is approximated to be $1.5 billion per year. The required costs are crucial, and injury and hazards to human life due to many failures are infinitely more important. Hence, Battelle suggested that these can be lessened by 29% by the implementation of modern fatigue technology.

Materials can typically be divided into groups as follows: (1) metals, (2) polymers/plastics, and (3) ceramics. Metals are conductive (electrical & thermal), ductile, opaque, etc. Polymers/plastics are insulators, transparent, semitransparent, etc. Ceramics are compounds of metallic and nonmetallic matters (oxides, carbides, nitrides, sulfides), which are brittle insulators. Many materials are composed of many grains, each of which has grain boundaries and second-phase particles, and the material is structurally arranged at different length scales. Thus, it is uncomplicated to investigate plastic deformation in a sole crystal to erase the results of grain boundaries and second-phase particles. As a sole crystal in a metal is stressed in tension beyond its elastic limit, it elongates somewhat, which is defined as plastic deformation (Fig. 6.1).

Plastic deformation requires the disintegrating and remaking of chemical bonds. Plastic deformation shall occur by a slip, twinning, or a mixture of both. Plastic deformation may not be returned to its original state by changes. Under tensile stress, plastic deformation is distinguished by a strain-hardening region, necking region, and fracture (also called ‘rupture’).

S. Woo, *Design of Mechanical Systems*, Springer Series in Reliability Engineering,
https://doi.org/10.1007/978-3-031-28938-5_6

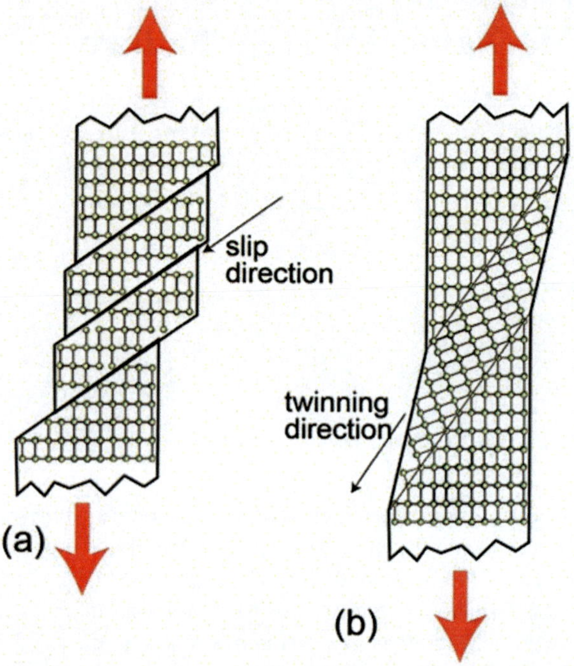

Fig. 6.1 Plastic deformation (metals) for axial force

Fracture is the breakdown of a sole member into pieces by an acted stress. Studies in the field of fracture mechanics cover the response in materials for (dynamic) loading and, in results, occurrence and propagation of its cracks. The process of crack formation in a material includes the subsequent fracture mechanisms: (1) fatigue, (2) brittle (cleavage) fracture, (3) ductile (shear) fracture, (4) deadhesion, and (5) crazing. The branch of fracture mechanics obviously depends on the material's behavior. Brittle fracture is the dominant failure mechanism of low-strength materials. At intermediate strength levels, there is a move between brittle ductile overload and fracture under linear elastic conditions. Fracture failure at very high strength values is determined by the flow properties of the matter.

6.2 Failure Mechanics and Design for Mechanical Products

6.2.1 Introduction

Critical design features in a newly designed mechanical structure should have enough stiffness and strength to endure a variety of loads. The product needs stiffness, and the opposition against reversible deformation can change over a broad span and rely on product implementation. Strength, the defense against irreversible deformation,

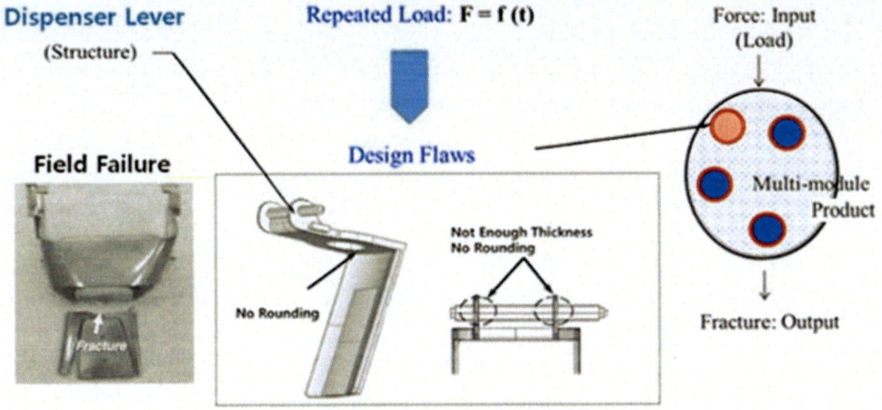

Fig. 6.2 Failure mechanics produced by a load on a part produced from a particular matter

is always necessitated to be big through the design of the product. As an insufficient strength and stiffness due to improper material/shape, this permanent deformation may cause cracks, fractures, and loss of the product's intended functions. That is, if there is a design fault in the structure where the loads are exerted, the product will fracture at that location in product lifetime because it does not have enough stiffness and strength to bear stress. Engineers would want to make products with better designs without faults to withstand repeated loads. Thus, the mechanical system shall be redesigned to be a good design in product life.

As seen in Fig. 6.2, the failure mechanics of mechanical parts shall be distinguished by two parameters: (1) the stress (or loads) on the structure, including the mechanism, and (2) the pattern of matters and form utilized in the structure. To stop failure in the field, mechanical engineers modify product designs by choosing a proper product material and shape before the products are released. Thus, the mechanical system of the products can withstand the loads (or stress) in its life with enough strength and stiffness.

Most mechanical systems are made up of multimodule structures. If the product structure, as a design result, is ideally designed in the developing process, it can robustly endure its own loading. Although the product is optimally designed, it might have design imperfections which shall suddenly appear in the field. In particular, fatigue arises due to the existence of a tiny crack or preexisting flaw on the surface of a part because it is almost unnoticeable using conventional finite element techniques. To prevent this problem, the structure should be experimentally confirmed if it has a superior quality—sufficient stiffness and strength. The design problems, therefore, should be revealed and altered by reliability testing, such as parametric ALTs.

6.2.2 *Quantum Mechanics*

Quantum mechanics is the investigation of physics on extremely little length scales, though there are macroscopic systems applicable. Particles in quantum mechanics have wavelike things. The Schrodinger equation, a specific wave equation, determines how these waves behave. Before discussing the Schrodinger wave equation, let us look at the historical timeline of how quantum mechanics happened.

Max Planck (1900) suggested that light with frequency ν is discharged in quantized lumps of energy, which originate from integral multiples of the quantity,

$$\mathrm{E} = h\nu = \hbar\omega \tag{6.1}$$

where $h \approx 6.63 \times 10^{-34}$ J s is Planck's constant, and $\hbar \equiv h/2\pi = 1.06 \times 10^{-34}$ J s.

Albert Einstein (1905) expressed that the quantization was actually intrinsic to the light and that the lumps shall be explained as particles, which we at the moment call "photons." This scheme was a consequence of his study on the photoelectric effect, which addresses the absorption of light and the emission of elections from a matter. That is, E = pc for a light wave. Thus, Planck's $\mathrm{E} = \hbar\omega$ relation is

$$\mathrm{E} = \hbar\omega \Rightarrow \mathrm{pc} = \hbar(ck) \Rightarrow p = \hbar\mathrm{k} \tag{6.2}$$

Niels Bohr (1913) expressed that electrons in atoms have wavelike behaviors. It accurately stated a few objects about hydrogen, particularly the quantized energy levels which were familiar. Louis de Broglie (1924) expressed that all particles are related to waves, where the frequency and wavenumber of the wave are specified by the same connections we discovered above for photons, that is, $\mathrm{E} = \hbar\omega$ and $p = \hbar\mathrm{k}$. The larger E and p are, the larger ω and k are. Even for little E and p, which are typical of a photon, ω and k are very large because $\hbar$ is so little. Therefore, any everyday-sized particle with big (in comparison) energy and momentum values shall have very large ω and k values.

Werner Heisenberg (1925) and Erwin Schrodinger (1926) defined a edition of quantum mechanics. In particular, Schrodinger wrote down a wave equation which determines how the waves develop in space and time. Max Born (1926) explained Schrodinger's wave as a probability amplitude. By "amplitude", we signify that the wave shall be squared to attain the required probability. This yields the probability of discovering a particle at a specified position. Paul Dirac (1926) explained that Heisenberg's and Schrodinger's types of quantum mechanics were identical in that they might both be derived from a more widespread kind of quantum mechanics.

6.2.3 The Schrodinger Equation

We will derive the Schrodinger equation. The energy of a particle is the total of the kinetic and potential energies. We presume that the potential is a function of just x.

$$\mathrm{E} = \mathrm{K} + \mathrm{V} = \frac{1}{2}mv^2 + V(x) = \frac{p^2}{2m} + V(x) \tag{6.3}$$

De Broglie asserted that all particles shall be expressed as waves with frequency ω and wavenumber k and that $\mathrm{E} = \hbar\omega$ and $p = \hbar \mathrm{k}$. That is,

$$\hbar\omega = \frac{\hbar^2 k^2}{2m} + V(x) \tag{6.4}$$

A wave with frequency ω and wavenumber k shall be expressed as $\psi(x, t) = Ae^{i(kx-\omega t)}$. In 3-D, we have $\psi(r, t) = Ae^{i(kr-\omega t)}$. Let us just consider as 1-D. We define it as follows:

$$\frac{\partial\psi}{\partial t} = -i\omega\psi \Rightarrow \omega\psi = i\frac{\partial\psi}{\partial t} \text{ and } \frac{\partial^2\psi}{\partial x^2} = -k^2\psi \Rightarrow k^2\psi = -\frac{\partial^2\psi}{\partial x^2} \tag{6.5}$$

As we increase the energy equation in Eq. (6.4) by ψ and thus is connected in these relationships, we attain it as follows:

$$\hbar(\omega\psi) = \frac{\hbar^2}{2m}(k^2\psi) + V(x)\psi \Rightarrow i\hbar\frac{\partial\psi}{\partial t} = \frac{-\hbar^2}{2m}\frac{\partial^2\psi}{\partial x^2} + V\psi \tag{6.6}$$

This is the time-dependent Schrodinger equation. If we set the x and t arguments back in, the equation has the form

$$i\hbar\frac{\partial\psi(x, t)}{\partial t} = \frac{-\hbar^2}{2m}\frac{\partial^2\psi(x, t)}{\partial x^2} + V(x)\psi(x, t) \tag{6.7}$$

In 3-D, the x dependence becomes dependent on all three coordinates (x, y, z), and the $\frac{\partial^2\psi}{\partial x^2}$ term becomes $\nabla^2\psi$ (the total of the second derivatives). Born interprets that $|\psi|^2$ takes the probability of discovering the particle at location x.

6.2.4 Solving the Schrodinger Equation—Infinite Square Well

Consider an (electric) particle that is overly controlled to proceed only in the x way from x = 0 to x = a. The potential energy, V(x), will be defined as follows:

$$\mathrm{V}(x) = \begin{cases} 0(0 \le x \le a) \\ \infty(x < 0 \text{ or } x > a) \end{cases} \tag{6.8}$$

The time-independent Schrodinger wave equation in operator form shall be defined as follows:

$$\hat{H}\psi = E\psi \tag{6.9}$$

where $\hat{H}$ is the Hamiltonian operator in the x direction, ψ is the wave function, and E is the (electron) energy.

If $\hat{H} = -\frac{h^2}{8\pi^2 m}\frac{d^2}{dx^2} + V$, we can put this in Eq. (6.9). That is,

$$\frac{d^2\psi}{dx^2} + \frac{8\pi^2 m}{h^2}(E - 0)\psi = 0 \text{ or } \frac{d^2\psi}{dx^2} + \frac{8\pi^2 m}{h^2}(E - V)\psi = 0 \tag{6.10}$$

where m is the electron mass, h is the Planck constant, and V is the potential energy.

Because $V = \infty$ outside the walls, this is possible only when $\psi = 0$. That is, particles are not outside the walls. Because $V = 0$ inside the walls, Eq. (6.10) can be stated as follows:

$$\frac{d^2\psi}{dx^2} + \frac{8\pi^2 m}{h^2}(E - 0)\psi = 0 \text{ or } \frac{d^2\psi}{dx^2} + K^2\psi = 0 \tag{6.11}$$

where $K^2 = \frac{8\pi^2 mE}{h^2}$

We can assume the solution of Eq. (6.11) as follows:

$$\psi(x) = A\sin Kx + B\cos Kx \tag{6.12}$$

where A, B = constants

Because $x = 0$ or $x = a$ at walls, $\psi(0) = \psi(a) = 0, B = 0, K = \frac{n\pi}{a}$, and $E = \frac{n^2h^2}{8ma^2}$, $n = 1,2,3,4$

Therefore, we can state Eq. (6.12) as follows:

$$\psi(x) = A\sin\left(\frac{n\pi}{a}\right)x \tag{6.13}$$

The probability of discovering the particle in a little region between x and $x + \mathrm{d}x$ is given as follows:

$$\int_0^a \psi^2(x)dx = 1 \text{ or } \int_0^a \left(A\sin\left(\frac{n\pi}{a}\right)x\right)^2 dx = 1 \tag{6.14}$$

Therefore, we can obtain the solution of Eq. (6.14) as follows:

$$\psi(x) = \sqrt{\frac{2}{a}} \sin\left(\frac{n\pi}{a}\right)x \tag{6.15}$$

where $\psi(x + a) = \psi(x)$, a is the (periodic) interval, and n is the principal quantum number

6.2.5 *Flux*

The flux, F, of any species is defined as the $F \equiv$ number going through unit area in unit time. To attain a formula for the flux F, think about the instance of the movement of charged impurities in a positive way in a crystal. The atoms of the crystal establish some potential hills which hinder "the movement of the charged impurities. The limit of the potential barrier, W, is usually on the order of electron volts in nearly all matters. The interval between consecutive potential barriers, a, is on the disposition of the lattice spacing, which is typically several angstroms.

As indicated in Fig. 6.3, if a continual electric field is exerted, the potential distribution as a function of distance shall be caused to move into a sloping position, as shown in Fig. 6.3b. It shall migrate positively charged particles to the right more uncomplicated and make their passage to the left harder. Let us now compute the flux F at position x. This flux shall be the mean of the fluxes at position $(x - a/2)$ and at $(x + a/2)$. Successively, these two fluxes are specified by $F_1 - F_2$ and $F_3 - F_4$, separately,

Think about the component F_1. It shall be specified by the multiplication of (i) the density per unit area of impurities at the potential hill at $(x - a)$; (ii) the probability of a leap of any of these impurities to the valley at x; and (iii) the frequency of attempted jumps v. Therefore, we shall express

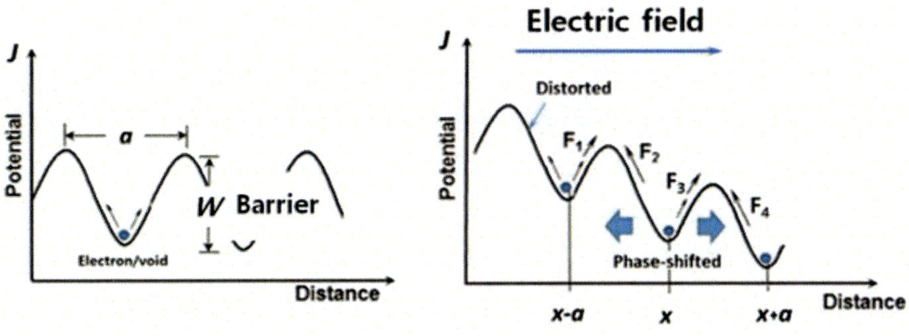

(a) Initial state (b) Final state

Fig. 6.3 Model of ionic motion in a crystal

$$F_1 = [aC(x-a)] \cdot \exp\left[-\frac{q}{kT}\left(W - \frac{1}{2}a\xi\right)\right] \cdot v \tag{6.16}$$

where $[aC(x-a)]$ is the density per unit area of particles located in the valley at $(x - a)$, and the exponential factor is the probability of a successful leap from the hill at $(x-a)$ to the hill at x. Bear in mind the lowering of the obstacle due to the electric field ξ.

Alike formulas shall be expressed for F_1, F_2, and F_3. As these are combined to take a formula for the flux F at position x, with the concentrations $C(x \pm a)$ closed to $C(x) \pm a(\partial C/\partial x)$, we attain

$$F_1 = -\left[a^2 ve^{-qW/kT}\right] \cdot \cosh \frac{qa\xi}{2kT}\frac{\partial C}{\partial x} + \left[2ave^{-qW/kT}\right] C \sinh \frac{qa\xi}{2kT} \tag{6.17}$$

A very main limiting form of this equation is attained for the occasion when the electric field is comparatively small, i.e., $\xi = kT/qa$. In this occasion, we can enlarge the cos*h* and sin*h* terms in the above equation. Bear in mind that $\cosh(x) = 1$ and $\sinh(x) = x$ for $x \to 0$, which results in the limiting shape of the flux equation for a positively charged species,

$$F(x) = -D\frac{\partial C}{\partial x} + \mu\xi C \tag{6.18}$$

where $D \equiv a^2 ve^{-qW/kT}$ and $\mu \equiv a^2 ve^{-qW/kT}/kT/q$

Bear in mind that the mobility μ and the diffusivity D are connected by

$$D = \frac{kT}{q}\mu \tag{6.19}$$

It is usual to recognize the contribution to the flux that is proportionate to the concentration gradient as the *diffusion* term, while the contribution which is proportionate to the concentration itself is mentioned as the drift expression.

6.2.6 *Diffusion*

Mass transport in a liquid or gas requires the move of fluid (e.g., convection currents), though atoms also diffuse. Solids, on the other hand, shall bear shear stresses and thus do not move except by diffusion requiring the leaping of atoms on a fastened network of locations.

Presume that such jumps shall in a way be attained in the solid state, with a frequency ν with each jump over a distance λ.

For random jumps, the root mean square distance is as follows:

$$\overline{x} = \lambda\sqrt{n} = \lambda\sqrt{\nu t} \text{ where } n \text{ is the number of jumps and } t \text{ is the time} \tag{6.20}$$

Concentration of solute, C, number/m^3. Each plane has $C\lambda$ atoms/m^2 (Fig. 6.4)

$$\delta C = \lambda \frac{\partial C}{\partial x} \tag{6.21}$$

Atomic flux, J, atoms/(m^2 s)

$$J_{L\rightarrow R} = \frac{1}{6}\nu C\lambda \tag{6.22}$$

$$J_{R\rightarrow L} = \frac{1}{6}\nu(C + \delta C)\lambda \tag{6.23}$$

Thus, the net flux along x is specified by

$$J_{net} = -\frac{1}{6}\nu\delta C\lambda = -\frac{1}{6}\nu\lambda^2\frac{\partial C}{\partial x} \equiv -D\frac{\partial C}{\partial x} \tag{6.24}$$

This is Fick's first law, where the constant of proportionality is specified as the diffusion coefficient in m^2/s. Fick's first law is applicable to steady state flux in a constant concentration gradient. Therefore, our equation for the mean diffusion

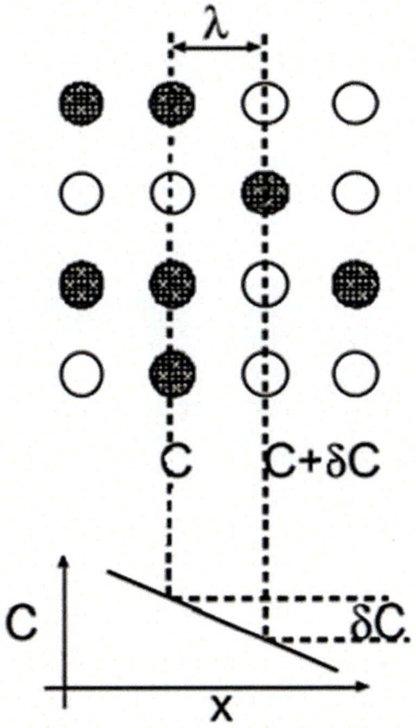

Fig. 6.4 Diffusion Gradient

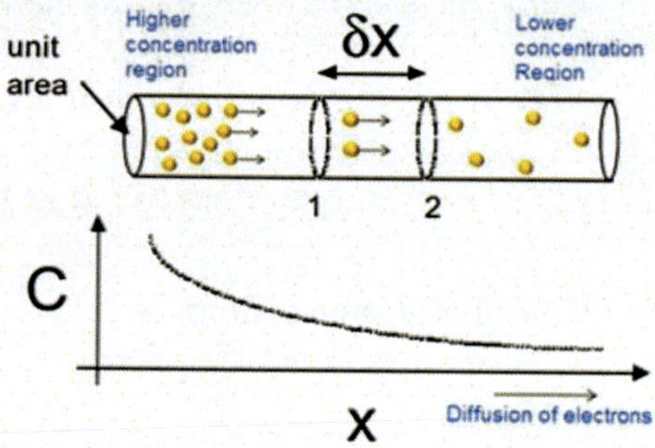

Fig. 6.5 Non–uniform concentration gradient

distance shall be defined in terms of the diffusivity as follows:

$$\overline{x} = \lambda\sqrt{\upsilon t} \text{ with } D = \frac{1}{6}\upsilon\lambda^2 \text{ giving } \overline{x} = \sqrt{6Dt} \simeq \sqrt{Dt} \tag{6.25}$$

Assume that the concentration gradient is not constant (see Fig. 6.5)

$$Flux\ in = -D\left(\frac{\partial C}{\partial x}\right)_1 \tag{6.26}$$

$$Flux\ out = -D\left(\frac{\partial C}{\partial x}\right)_2 = -D\left[\left(\frac{\partial C}{\partial x}\right)_1 + \delta x\frac{\partial^2 C}{\partial x^2}\right] \tag{6.27}$$

In the time interval δt, the concentration changes δC

$$\delta C\delta x = (Flux\ in - Flux\ out)\delta t \tag{6.28}$$

$$\frac{\partial \boldsymbol{C}}{\partial \boldsymbol{t}} = \boldsymbol{D}\frac{\partial^2 \boldsymbol{C}}{\partial \boldsymbol{x}^2} \tag{6.29}$$

Presume that the diffusivity is unrelated of the concentration. This is Fick's second law of diffusion.

This is compliant to numerical results for the common occasion, but there are a couple of interesting systematic answers for specific boundary circumstances. For an occasion where a fastened amount of solution is plated onto a semi–infinite bar (Fig. 6.6).

The boundary circumstances can be described as follows:

$$\int_0^\infty C(x, t)dx = B \text{ and } C(x, t = 0) = 0 \tag{6.30}$$

$$C(x, t) = \frac{B}{\sqrt{\pi Dt}}\exp\left(\frac{-x^2}{4Dt}\right) \tag{6.31}$$

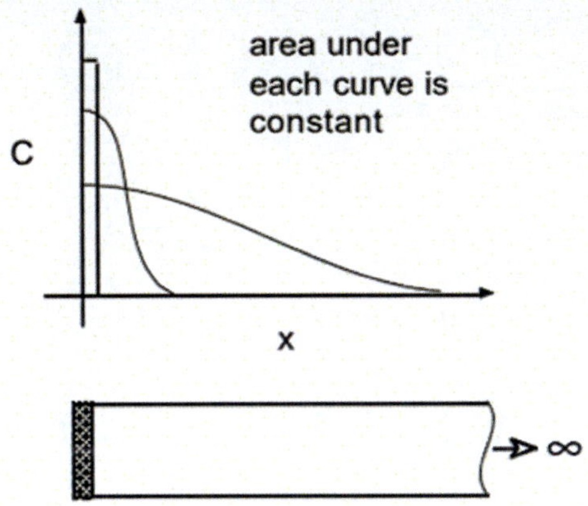

Fig. 6.6 Non–uniform concentration on a fastened amount of solute plated on a semi–infinite bar

Assume that we produce the diffusion couple by heaping an infinite group of narrow sources on the ending of one of the bars. Diffusion shall therefore be served by taking an entire group of the exponential functions attained above, each rather displaced along the x-axis, and summing (integrating) up their individual results. The integral is in fact the error function.

$$erf(x) = \frac{2}{\sqrt{\pi}} \int_0^x \exp(-u^2)du \tag{6.32}$$

Therefore, the result to the diffusion equation, based on boundary conditions, is

$$C(x = 0, t) = C_s \quad \text{and} \quad C(x, t = 0) = C_0 \tag{6.33}$$

$$C(x, t) = C_s - (C_s - C_0)erf\left(\frac{x}{2\sqrt{Dt}}\right) \tag{6.34}$$

This solution shall be utilized in numerous situations where the surface concentration is kept constant, for instance, in carburization or decarburization processes. The results expressed here also exert to the diffusion of heat.

Atoms in the solid-state drift by leaping into vacancies. The vacancies shall be interstitial or in substitutional sites. Nevertheless, there is an obstacle to the movement of the atoms because the movement is related to a transient distortion of the lattice (Figure 6.7).

Presuming that the atom tries to jump at a frequency ν_0, the frequency of doing well jumps is specified as follows:

$$\nu = \upsilon_0 \exp\left(\frac{-G^*}{kT}\right) \equiv \upsilon_0 \exp\left(\frac{S^*}{k}\right) \exp\left(\frac{-H^*}{kT}\right) \tag{6.35}$$

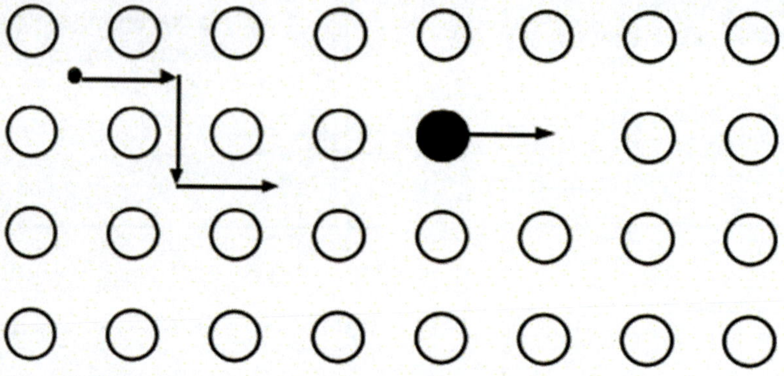

Fig. 6.7 The movement is related to a transient distortion of the lattice

where k is the Boltzmann constant, T is the absolute temperature, separately, and H^* and S^* are the activation enthalpy and activation entropy, separately. Since $D \propto \nu$, we discover

$$D = D_0 \exp\left(\frac{-H^*}{kT}\right) \tag{6.36}$$

6.2.7 *Mechanism of Slip*

If a mechanical system (or a part) is subjected to a repeated stress (or loads), plastic deformation (or failure) occurs at an early stage. In material science, a dislocation is a linear crystallographic defect that possesses a sudden exchange in the positioning of atoms. The motion of dislocations (or slip) brings atoms to tilt over each other at low stress levels. Slip happens on planes which have the highest planar density of atoms and in the orientation with the highest linear density of atoms. In other words, slip happens in the orientations in which the atoms are packed with little space since this necessitates the lowest quantity of energy. Thus, they shall slide past each other with force. The slide flow relies upon the repeated structure of the crystal, which brings the atoms to shear away from their first neighbors. It, thus, passes along the face and joins up with the atom of new crystals.

A slip happens as a consequence of a straightforward shearing stress. The resolve of the axial tensile load F takes two loads. That is, one F_s is the shear load along the slip plane, and the other F_N is a normal tensile load perpendicular to the plane. According to the examination and test, the greatest shear stress happens at 45°. Figure 6.8 represents the packing of atoms on a slip plane. We recognize that there are three ways in which the atoms are close-packed, and these shall be the uncomplicated slip orientations.

Parts of the crystal on either side of a particular slip plane go in opposing orientations and become to rest with the atoms in almost equilibrium locations, so that there is minute change in the lattice positioning. Therefore, the outer form of the crystal is altered without terminating it. Simply, slip shall be described in a face-centered cubic (FCC) lattice. The (111) plane is the slip plane possessing the largest number of atoms. It bisects the (001) plane in the line AC, (110) direction possessing the greatest number of atoms on it. A slip is regarded as a motion along the (111) planes in the close-packed (110) orientation (Fig. 6.9a).

From the simplified illustration of slip in an FCC crystal, one can presume that the atoms slip one after another without interruption, beginning at one location or at a few locations in the slip plane, and thus moving external over the rest of the plane. For example, if one attempts to slide the whole rug as one piece, the resistance is too

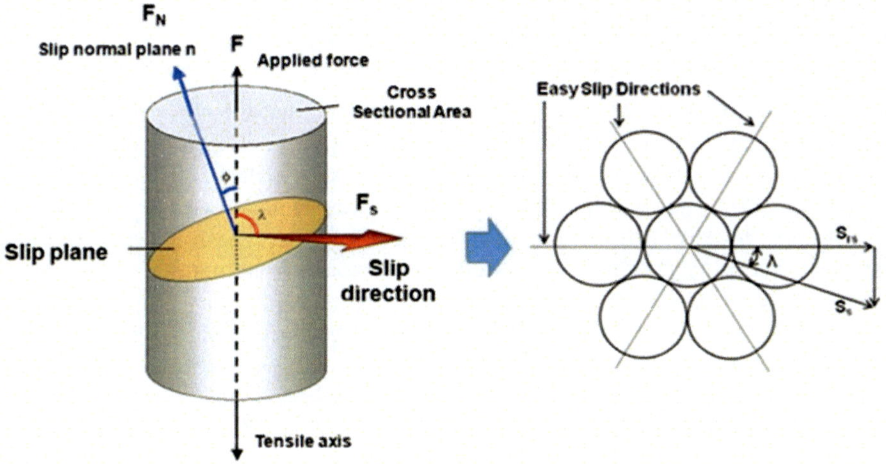

Fig. 6.8 Components of force on a slip plane

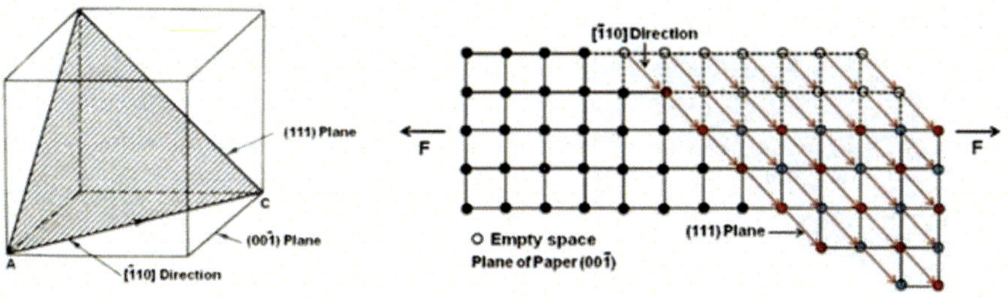

(a) Slip plane in the FCC lattice

(b) Simplified illustration

Fig. 6.9 Simplified illustration of the slip plane in the FCC lattice much. What one shall do is to make a fold in the rug and thus slide the entire rug a little at a time by thrusting the fold along. An alike analogy to the fold in the rug is the motion of an earthworm. It moves in an orientation by proceeding a component of its body at a time.

By implementing the shear force, first, an additional plane of atoms (specified as a dislocation) is created above the slip plane. Thus, the bond between atoms breaks, which leads to the production of a new bond between atoms and a dislocation. On continued implementation of force, this dislocation proceeds by disintegrating old bonds and remaking new bonds. In the following move, as the bond between atoms is disintegrated, a new bond is re-made between atoms repeatedly, resulting in a dislocation. As a result, this dislocation goes across the slip plane and leaves a step when it appears at the exterior of the crystal. Each time the dislocation goes across the slip plane, the crystal goes one atom spacing (Fig. 6.9b).

6.2.8 Stress Concentration at the Crack Tip

Though most mechanical products are designed such that the nominal stress lasts elastic ($S_n < \sigma_{ys}$), stress concentrations in a mechanical system often give rise to plastic strains that occur in the neighborhood of design faults or stress raisers such as holes, grooves, notches, and fillets where the stress is increased.

The fracture strength of a matter is connected to the cohesive forces between atoms. One shall approximate that the conceptual cohesive strength of a matter must be one-tenth of the elastic modulus (E). However, the exploratory fracture strength for a brittle material is usually $E/100$–$E/10{,}000$ below this conceptual quantity. This much lower fracture strength comes from the stress concentration due to the existence of microscopic defects or cracks discovered on the exterior of the matter. The stress profile along the x-axis is concentrated on an elliptically formed inner crack (Fig. 6.10).

Stress has a greatest value at the crack tip, and it decreases to the nominal exerted stress with growing distance away from the crack. Defects such stress raisers or stress concentrators have the capacity to enlarge the stress at a designated position. The value of amplification relies on the crack orientation and geometry.

Inglis's solution (1913) not only used elliptical coordinates to resolve the elliptical hole issue but also utilized complex numbers [2]. He obtained the consequences that the confined stresses around a corner or hole in a stressed plate could be many times larger than the mean acted stress. The greatest stress at the end of the ellipse is connected to its dimension and form by

$$\sigma_{\max} = \sigma_{\infty}\left[1 + 2\frac{a}{b}\right] \tag{6.37}$$

Obviously, Inglis's elliptical consequence lessens to the known $\sigma_{\max} = 3\sigma_{\infty}$ for the particular occasion of a hole when $a = b$. That is, a familiar result might be

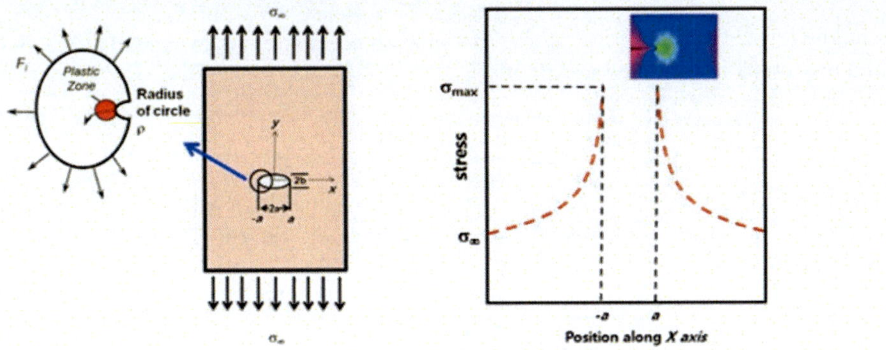

(a) Shape of cracks in a part (b) Simplified stress outline along the X axis at the crack tip

Fig. 6.10 Stress concentration at crack tip locations

seen in most strength-of-materials books. On the other hand, the maximum stress is forecasted to approach infinity as the ellipse makes flat to generate a crack ($b \rightarrow 0$).

The radius of curvature, ρ, at the end of an ellipse is connected to its length and width by

$$\rho = \frac{b^2}{a} \tag{6.38}$$

Resolving this for b and exchanging into the a/b ratio in Eq. (6.37), it can be estimated by Eq. (6.39):

$$\sigma_{\max} = \sigma_\infty \left[1 + 2\sqrt{\frac{a}{\rho}} \right] \tag{6.39}$$

where ρ is the radius of curvature, σ_∞ is the applied stress, $\sigma_{\max}$ is the stress at the crack tip, and a is the half-length of the interior crack or the comprehensive length for a surface defect.

As the crack is alike to an elliptical hole through plate and is aligned perpendicular to the exerted stress, the greatest stress $\sigma_{\max}$ happens at a crack tip. The quantity of the theoretical acted tensile stress is σ_∞; the radius of the curvature of the crack tip is ρ; and a depicts the length of a surface crack, or half-length of an interior crack. For a comparatively lengthy microcrack, the factor $(a/\rho)^{1/2}$ can be very large. Therefore, Eq. (6.39) can be restated as follows:

$$\sigma_{\max} \cong 2\sigma_\infty \left(\frac{a}{\rho} \right)^{1/2} \tag{6.40}$$

The proportion between the greatest stress and the nominal acted tensile stress is defined as the stress concentration factor K_t. The stress concentration factor is a straightforward potion of the degree to which an outer stress is expanded at the end of a tiny crack and defined as follows:

$$K_t = \frac{\sigma_{\max}}{\sigma_o} \approx 2\left(\frac{a}{\rho}\right)^{1/2} \tag{6.41}$$

As an outer stress is expanded at the end of a crack, Eq. (6.41) shall be restated as follows:

$$\sigma_{\max} = 2\sigma_\infty\left(\frac{a}{\rho}\right)^{1/2} = K_t\sigma_\infty \tag{6.42}$$

Stress amplification not only happens at tiny defects or cracks on a microscopic range of matter but shall also happen in stress concentrations such as sharp corners, fillets, holes, and notches on the macroscopic quantity. Cracks with sharp tips spread more easily than cracks with blunt tips. Because of an expanding acted stress, stress concentration can happen at microscopic flaws, sharp corners, interior discontinuities (voids/inclusions), notches, and scratches which are usually defined as stress raisers. Stress raisers are usually more devastating in brittle matters. Ductile materials possess the capacity to plastically deform in the area neighboring the stress raisers, which in succession uniformly disperses the stress load near the defect. The greatest stress concentration factor results in a value less than which discovered for the theoretical value. Since brittle materials shall not plastically deform, the stress raisers shall produce the conceptual stress concentration in circumstance.

The quantity of this amplification relies on microcrack positioning, dimensions, and geometry. For instance, the stress concentration at sharp-edged corners relies on the fillet radius (Fig. 6.11).

6.2.9 *Fracture Toughness and Crack Propagation*

Cracks with sharp-edged tips grow and propagate more easily than cracks possessing blunt-edged tips. In ductile materials, plastic deformation at a crack tip 'blunt' evolves to the crack. Elastic strain energy is kept in a matter as elastically deformed. This energy is freed when the crack grows. The formation of new surfaces necessitates energy.

As a crack has expanded into a solid to a depth a, an area of matter adjoining the free surfaces is unloaded, and its strain energy is released. A straightforward manner of seeing this energy release is to consider two triangular regions near the crack flanks, of width a and height πa, as totally unloaded, as the existing matter remains the full stress σ. The whole strain energy U released is thus the strain energy per unit volume in both triangular areas:

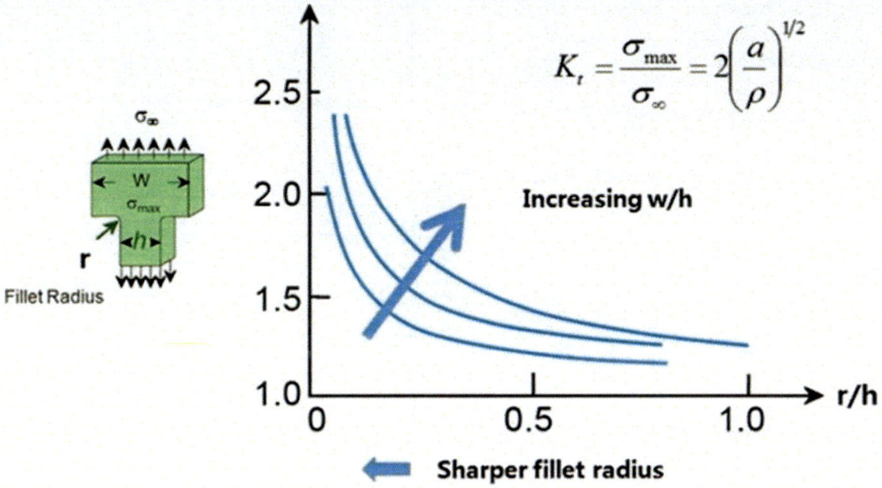

Fig. 6.11 Stress concentration at sharp corners in accordance with fillet radius [3]

$$U^* = -\frac{\sigma^2}{2E} \cdot \pi a^2 \tag{6.43}$$

At this time, the area normal to the x–y plane is taken to be unity, so U is the strain energy released per unit thickness of specimen. This strain energy is released by crack growth. However, in shaping the crack, bonds shall be destroyed, and the necessary bond energy is effectively used by the matter. The surface energy S related to a crack of length a (and unit depth) is

$$S = 2\gamma a \tag{6.44}$$

where γ is the surface energy and the factor 2 is necessitated since two free surfaces have been found.

The whole energy related to the crack is thus the total of the (positive) energy used to produce the new surfaces, plus the (negative) strain energy released by allowing the regions near the crack flanks to be unloaded.

When the crack propagates, the strain energy depends on the surface energy. Beyond a critical crack length a_c, the system shall become its lowest energy by putting the crack grow longer. Up to the point where $a = a_c$, the crack shall produce only if the stress increases. Beyond the position, crack growth is unforced and catastrophic (Fig. 6.12).

The quantity of the crucial crack length shall be obtained by putting the derivative of the whole energy $S + U^*$ to zero:

$$\frac{\partial(S + U^*)}{\partial a} = 2\gamma - \frac{\sigma_f^2}{E} \cdot \pi a = 0 \tag{6.45}$$

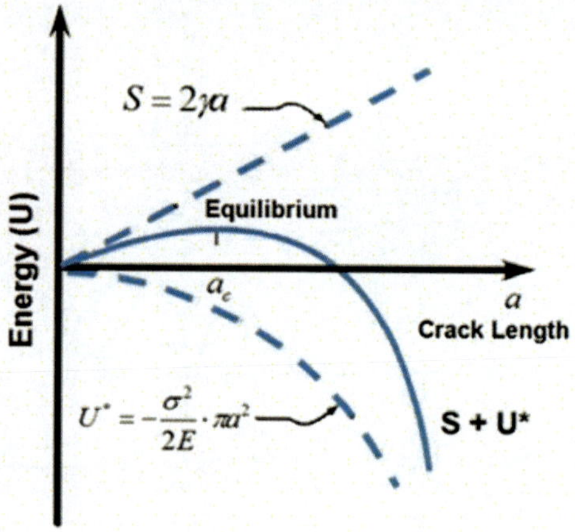

Fig. 6.12 Total energy related to the crack

Because fast fracture is approaching when this circumstance is fulfilled, we can solve Eq. (6.45). The required crucial stress for crack propagation is expressed as follows:

$$\sigma_c = \left(\frac{2E\gamma_s}{\pi a}\right)^{1/2} \tag{6.46}$$

where γ_s is the specific surface energy.

As the tensile stress at the tip of the crack surpasses the crucial stress quantity, the crack grows and evolves in fracture. Most metals and polymers have plastic deformation. For ductile materials, the specific surface energy γ_s shall be substituted into $\gamma_s + \gamma_p$, where γ_p is the plastic deformation energy. Therefore, Eq. (6.46) shall be expressed as follows:

$$\sigma_c = \left(\frac{2E(\gamma_s + \gamma_p)}{\pi a}\right)^{1/2} \tag{6.47}$$

For highly ductile materials, $\gamma_p \gg \gamma_s$ is correct. Therefore, Eq. (6.47) shall be restated as follows:

$$\sigma_c = \left(\frac{2E\gamma_p}{\pi a}\right)^{1/2} \tag{6.48}$$

Most brittle materials possess a population of tiny defects which are a variety of sizes. As the size of the tensile stress at the tip of the crack surpasses the crucial stress quantity, the crack grows and evolves in fracture. Only minute and almost defect-free metallic and ceramic matters have been generated with facture strength which approximates their conceptual quantities.

Example 6.1 There is a long plate of glass subjected to a tensile stress of 30 MPa. As the modulus of elasticity and particular surface energy for this glass are 70 GPa and 0.4 J/m^2, respectively, discover the crucial length of a surface flaw which shall have no fracture.

From Eq. (6.18), $E = 70$ GPa, $\gamma_s = 0.4$ J/m^2, and $\sigma = 40$ MPa. Therefore, the crucial length shall be attained as

$$a_c = \left(\frac{2E\gamma_s}{\pi\sigma^2}\right) = \left(\frac{2 \cdot 70\ \text{GPa} \cdot 0.4\ \text{J/m}^2}{\pi \cdot (30\ \text{MPa})^2}\right) = 2.0 \times 10^{-6}\ \text{m}$$

Because of the catastrophic nature of fractures, many attempts have been attempted to determine fracture mechanics. From a collaboration of basic and observed causes, brittle fracture shall happen when the fracture toughness (K_c) of a matter is surpassed. That is, fracture toughness K_c is a matter's resistance to fracture when a crack exists. It, thus, denotes the quantity of stress necessitated to grow a defect. It shall be expressed as follows:

$$K_c = Y(a/w)\sigma_c\sqrt{\pi a} \tag{6.49}$$

Now, $Y(a/w)$ is a geometrical element which relies on the crack sizes, where a is the crack length, w is the sample thickness, σ_c is the crucial stress for the crack to grow, and a is the length of a surface crack of half the length of an interior crack.

If $a \to 0$ or $w \to \infty$, $Y \to 1$. As the sample thickness increases, the fracture toughness decreases till the plane strain area is attained. Fracture toughness relies on temperature, strain rate, and microstructure. Its magnitude decreases with growing strain rate and lessening temperature. If the yield strength is improved by alloying and strain hardening, the fracture toughness shall enhance with a lessening in grain size.

6.2.10 Crack Growth Rates

Metal fatigue starts at a surface (or an interior) defect by concentrated stresses and progresses at the start of shear flow along slip planes. Slip can occur in the (111) plane in an FCC lattice because the atoms are most compactly packed. Over a number of (random) loading cycles in the field, this slip produces intrusions and extrusions which start to resemble a crack. A true crack running inward from an intrusion region shall grow at first along one of the first slip planes but in the end turns to move transversely to the principal normal stress.

As repetitive loads, the slip bands shall develop into minute shear-driven microcracks. These Stage I cracks shall be explained as a back and forth slip on a series of adjacent crystallographic planes to produce a band. In these slip bands, pore nucleation and coalescence occur. It finally guides to microcrack formation. Often, extrusion and intrusions shall also emerge, which, being a much-localized discontinuity, the outcomes in much faster microcrack formation.

Microcracks join to emerge a macrocrack in Stage II of fatigue. Here, the crack is long enough to get away shearing stress control and be operated by normal stress that generates a continuous growth, cycle by cycle, on a plane which is not any more crystallographic but rather normal to outer loads. Ahead of this macrocrack, two plastic lobes are produced by stress concentration. The cracks increase perpendicular to the dominant stress and grow strikingly due to plastic stresses at the crack tip, as depicted in Fig. 6.13.

It is crucial that designers can forecast the rate of crack growth during load cycling in engineering structures so that the troublesome components can be modified before the crack extends a critical length. A great amount of test verification assists the view that the crack growth rate shall be altered with the cycle variation in the stress intensity factor [4]:

$$\frac{da}{dN} = A\Delta K^m \tag{6.50}$$

where da/dN is the fatigue crack rate per cycle, $\Delta K = K_{\min} - K_{\max}$ is the stress intensity factor range during the cycle, and A and m are parameters which rely on the matter, environment, frequency, temperature, and stress ratio.

The rate of fatigue crack propagation during Stage II relies on the stress level, crack size, and materials. It is well familiar as the 'Paris law', which brings to charts alike to those depicted in Fig. 6.14.

Some given values of constants m and A for some alloys are specified in Table 6.1. The exponent m is often near 4 for metallic systems, which could be justified as the damage accumulation being connected to the volume V_p of the plastic zone: since the volume V_p of the zone scales with r_p^2 and $r_p \propto K_I^2$, then $da/dn \propto \Delta K^4$.

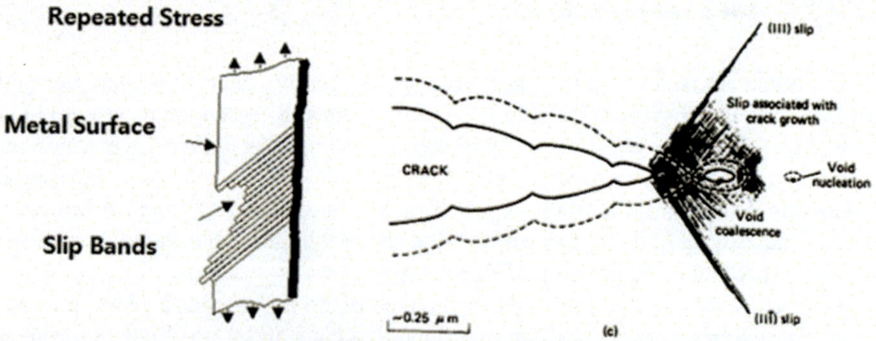

Fig. 6.13 A simplified illustration of common slip generating nucleation and growth of voids

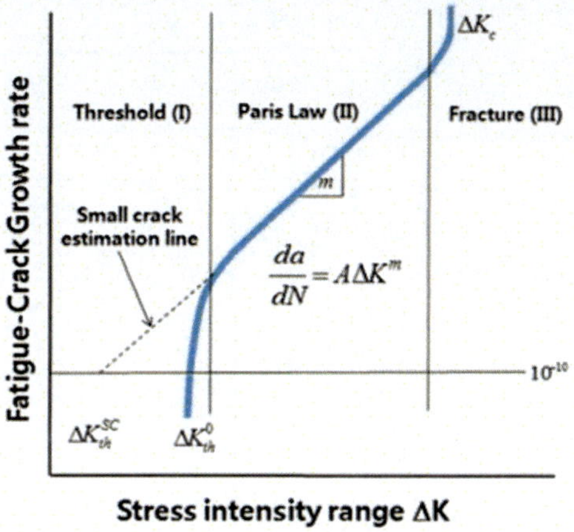

Fig. 6.14 Paris law

Table 6.1 Numerical parameters in the Paris equation

Alloy	m	A
Steel	3	10^{-11}
Aluminum	3	10^{-12}
Nickel	3.3	4×10^{-12}
Titanium	5	10^{-11}

6.3 Fatigue Failure

6.3.1 Introduction

A. Wöhler initially started the modern study of fatigue. He, a German engineer on the Lower Silesia-Brandenberg Railroad, worked for the railroad system in the mid-nineteenth century and was the head manager of rolling stock. Wöhler worried about the sources of fracture in railcar axles after lengthened use. A railcar axle is basically a circular beam with four-point bending, which yields a compressive stress along the top area and a tensile stress along the bottom. If the axle is revolved a half turn, the bottom becomes the top and vice versa, so the stresses on a specific area of the matter at the surface change repetitively from compression to tension. Although the metal becomes weak, fatigue was specified to explain this pattern of damage. This is just now known as completely reversed fatigue loading. Some of Wöhler's data are for Krupp axle steel and are plotted in terms of nominal stress (S) versus the number of cycles to failure (N), which is known as the S–N diagram. Each curve on such a illustration is still mentioned as a Wöhler line (Fig. 6.15).

Since 1830, it has been acknowledged that metal under a fluctuating or repeated load shall be unsuccessful at a stress level lower than that necessitated to lead failure

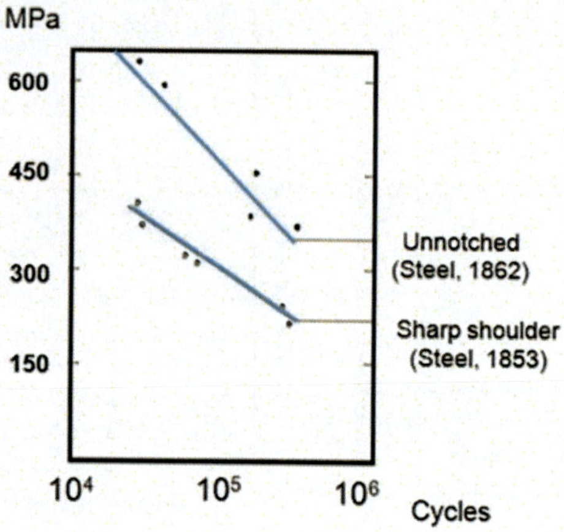

Fig. 6.15 Some of Wöhler's data for rail car axle steel on the S–N curve [5],

under an individual implementation of the same load. Figure 6.16 represents a bar-shaped part subjected to a constant sinusoidal changing force. As a period of time passes, a crack might start to form on the perimeter of the hole. This crack then grows throughout the part till the whole segment is unable to endure the applied stresses, and finally, the part is unsuccessful.

The physical growth of a crack shall usually be split into two different phases. These are the crack beginning phase (Phase I) and the crack growth phase (Phase II). Fatigue cracks start by the release of shear strain energy. The above illustration represents how a tiny crack or preexisting defect starts and the shear stresses develop in internal plastic deformation along slip planes. When the sinusoidal loading is

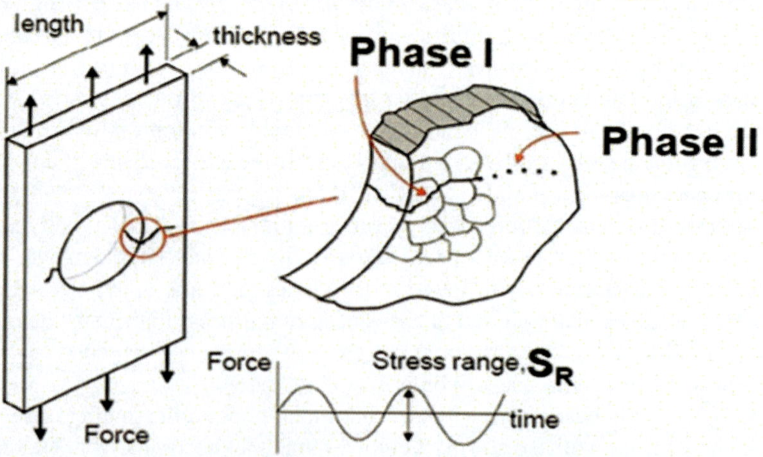

Fig. 6.16 Bar-shaped part subjected to a constant sinusoidal changing force

cycled, the slip planes pass back and forth, resulting in little extrusions and intrusions on the crystal exterior. These surface disturbances are roughly 1–10 μm in height and add up to embryonic cracks.

Phase I growth pursuits the orientation of the greatest shear plane, or 45° to the orientation of loading. A crack starts in this way and extends to the grain boundary. The mechanism at this spot is slowly moved to the adjoining grain. Once the crack has grown through roughly three grains, it is seen to alter its orientation of propagation. The physical mechanism for fatigue moves to Phase II. The crack is adequately large to generate a geometrical stress concentration. A tensile plastic zone is produced at the crack tip, as depicted in Fig. 6.19. After this stage, the crack moves perpendicular to the orientation of the acted load.

As previously mentioned, approximately 90% of all constructional failures happen through a fatigue mechanism. Fatigue occurs after a member is subjected to repeated cyclic loadings and repetitive deformations. It depends on the form, matter, and how close to the elastic limit it deforms. It also represents itself in the formation of cracks, which are developing at specific positions in diverse types of mechanical structures that might include mechanisms such as airplanes, boats, cranes, overhead cranes, turbines, machine parts, reactor vessels, bridges, offshore platforms, canal lock doors, transmission towers, chimneys, and masts.

The crack tip travels a very small distance in each loading cycle; if supplied, the stress is tall enough but not too high to give rise to unexpected comprehensive fracture. It is not almost possible to discover any growing variations in matter manner during the fatigue process, so brittle failures usually happen with no caution. Rest time, with the fatigue stress eliminated, does not cause any noticeable recovery or healing. With the naked eye, we can observe a 'clam shell' structure in the crack plane. Under a microscope, 'striations' shall be observed, which distinguish the positions of the crack tip after each single loading cycle. Because crack growth is very small in each load cycle, a large number of cycles are necessitated before the whole failure happens. Fatigue failure was first found in the nineteenth century by noticing the destitute service lifetime of railroad axles designed based on static design limits (Fig. 6.17).

The fatigue lifetime of a structure subjected to repetitive cyclic loads is expressed as the number of stress cycles it shall rise before failure. The bodily result of a repetitive load on a matter is dissimilar from that of a static load. Failure every time shall be brittle fracture regardless of whether the matter is ductile or brittle. Fatigue mainly happens at stresses well below the static elastic strength of the matter. Relying on the structural particular shape, its fabrication or the matter utilized, four crucial factors shall affect the fatigue strength: (1) the stress difference called the stress range, (2) the material, (3) the structural shape, and (4) the environment.

Fatigue in ductile metals represents the formation of cracks with high stress concentrations, such as holes and grooves, and propagates it [6, 7]. Fatigue stresses in mechanical products take the shape of a sinusoidal pattern. In periodic patterns, the peaks on both the low side (minimum) and the high side (maximum) are crucial. The altered Goodman diagram explains the fatigue of part with alternating stress on the y-axis and average stress on the x-axis. Therefore, it shall be expressed as follows:

Fig. 6.17 Fracture of train wrecks due to metal fatigue (Wikipedia)

$$\frac{S_a}{S_e} + \frac{S_m}{S_{ut}} = 1 \tag{6.50}$$

where S_a is the alternating stress, S_m is the mean stress, S_e is the fatigue limit for completely reversed loading, and S_{ut} is the ultimate tensile strength of the material.

The cycles to failure are represented as a function of mean stress and range along the lines of constant R-values [8]. The altered Goodman relation expresses a failure envelope such that any alternating stress which drops in the diagram shall not give rise to failure (Fig. 6.18).

However, fatigue failure standards, containing the typical Goodman diagram [10–13], have difficulty approximating the life cycles of multimodule products because

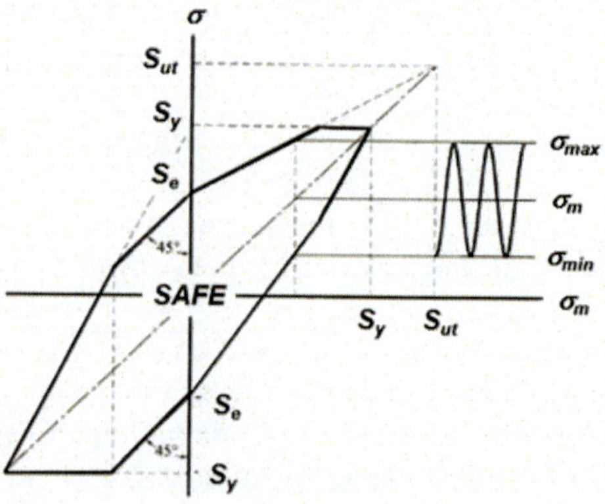

Fig. 6.18 The "Complete" modified Goodman diagram [9]

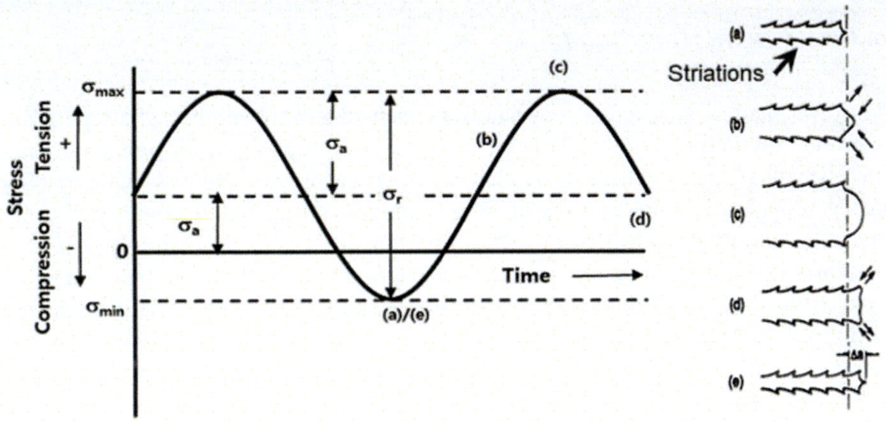

Fig. 6.19 Fatigue with striations under fluctuating and asymmetrical cyclic stresses at approximately zero stress

little samples of parts, not modules, are tested and part failures due to design faults rarely happen in the market. These systematic methods fail to reproduce the field failure and have been controversial.

As there are design flaws where a product is subjected to repeated loads, the product shall fail in its lifetime. In other words, fatigue occurs. It is the enfeebling of a matter brought by cyclic loading. In particular, the low-cycle fatigue (LCF) regime is distinguished by high cyclic stress levels in excess of the endurance limit of the material and is generally comprehended to be in the area of 10^4–10^5 cycles. A great amount of awareness is now being given to the LCF performance of superalloys, especially in the area of turbine-engine designs [14–18].

Many papers have been published concerning the LCFs of nickel-base polycrystalline [19–21] and monocrystalline alloys [22]. There are some fatigue tests on dissimilar patterns of samples according to stated fatigue testing standards [23–25]. However, it is hard to attain experimental data of multimodule products available to straightly build S–N curves for LCFs because comprehensive testing might have been too expensive due to time and sample size.

In Chaps. 7 and 8, we will discuss a structured method for reliability tests for enhancing the LCF failure of a mechanical system subjected to repeated loading. It covers (1) an ALT strategy based on the product BX life, (2) a load investigation, (3) a tailored sample of ALT with the design alternations, and (4) an assessment of whether the last designs of the mechanical system fulfill the objective BX life.

6.3.2 Fluctuating Load

Most mechanical structures—especially mechanisms—are subjected to repeatedly variable, fluctuating loading. This kind of repeated loading is very threatening not only because limited stresses are much lower than those accepted for static loading but also because of the essence of fatigue failure, products fail abruptly with no signs of yielding in their lifetime. There are three dissimilar fluctuating stress–time patterns in fatigue loading: (1) symmetrical about zero stress, (2) asymmetrical about zero stress, and (3) random stress cycle. For asymmetrical stress of approximately zero, a fluctuating load (stress) which results in fatigue is distinguished by the mean stress σ_m, the range of stress $\Delta\sigma$, the amplitude stress σ_a, the proportion of the mean stress over the amplitude stress χ, and the stress ratio R (Fig. 6.19).

The fatigue cycle begins when the nominal stress is at position 'a'. When the stress increases in tension through position 'b', the crack tip is open to give rise to confined plastic deformation, while it expands into the virginal metal at position 'c'. As the tensile stress just now lessens through 'd', the crack tip closes, and the lasting plastic deformation generates a distinguishing saw tooth profile familiar as a 'striation'. On realization of the cycle at position 'e', the crack has proceeded through length Δa and has established an extra striation.

As seen in Figs. 6.20 and 6.21, additional cyclic stresses which result in fatigue are (1) periodic and symmetrical about zero stress and (2) random stress fluctuations. In mechanical systems such as aircraft, automobiles, refrigerators, machine components, and bridges, fatigue failures under fluctuating/cyclic stresses are necessary:

- A maximum tensile stress of very high value
- A sufficiently large variation or fluctuation in the exerted stress
- A very large number of exerted stress cycles.

They are represented as the following equations:

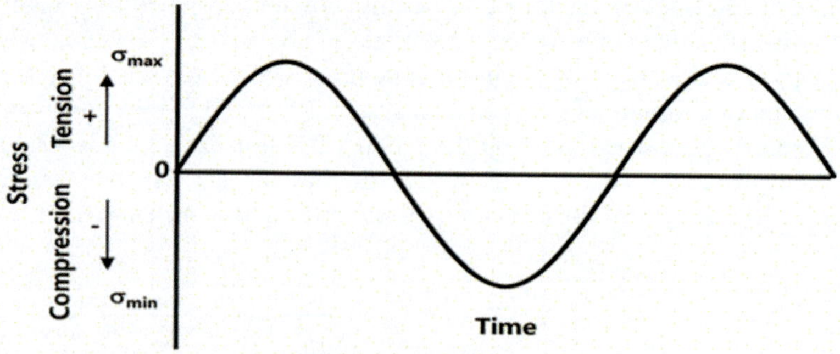

Fig. 6.20 Periodic and symmetrical about zero stress in fatigue

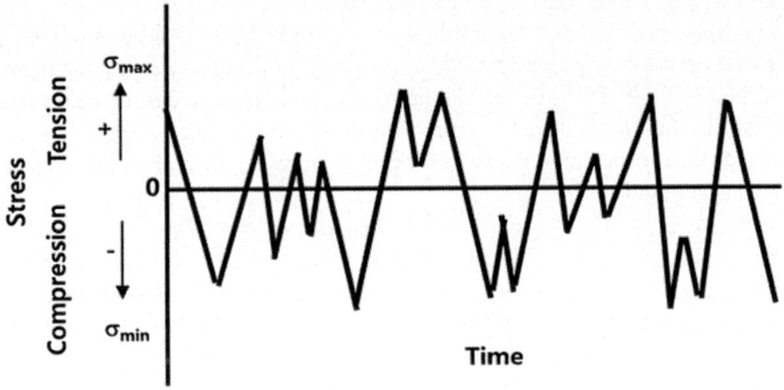

Fig. 6.21 Random stress fluctuations in fatigue

$$\sigma_m = \frac{(\sigma_{\max} + \sigma_{\min})}{2} \tag{6.51}$$

$$\Delta\sigma = (\sigma_{\max} - \sigma_{\min}) \tag{6.52}$$

$$\sigma_a = \frac{(\sigma_{\max} - \sigma_{\min})}{2} \tag{6.53}$$

$$\chi = \frac{\sigma_m}{\sigma_a} \tag{6.54}$$

$$R = \sigma_{\min} \big/ \sigma_{\max} \tag{6.55}$$

There are two typical load patterns: one with $\chi = 1$, the load is pulsing, and the other with $\chi = 0$, the load is reversed. The latter is the most threatening shape of load variation. The stress bound for fluctuating loading (the endurance limit) is expressed as a value of stress which is secure for a specified specimen regardless of the number of load repetitions. It generally remains in a compact relation to the ultimate stress limit.

There are two methods to decide when a part is in danger of metal fatigue: predicting when a failure shall happen due to the shape/iteration/matter/force mix and substituting the unsafe matters before this happens or carrying out examinations to identify the microscopic cracks and performing replacements once they happen. Choosing materials that are unlikely to be hurt by metal fatigue during the product life is the optimal result, but it is not always possible. Avoiding shapes with sharp corners restricts metal fatigue by lessening stress concentrations, but it does not eradicate it.

Fatigue in product designs is a crucial failure mechanism to be contemplated. It is a process in which damage accumulates due to the repeated loads below the yield point, which is brittle-like even in ductile materials. Fatigue cracks start at very small and initially grow very slowly until the crack length come close to the crucial length.

Therefore, it is risky because it is difficult to originally identify accumulative fatigue damage without a telescope, microscope, or other device until the crack has evolved to near-crucial length. The general fracture surface is perpendicular to the direction of the acted stress. Fatigue failure has three clear phases: (1) crack starting in the regions of stress concentration (near stress raisers), (2) incremental crack propagation, and (3) last catastrophic failure.

An instance of 'fatigue' for a multitude of causes has been investigated as the Versailles rail accident, and the disaster of Comet aircraft happened when they became big enough to propagate catastrophically (see Chap. 1). Fatigue failure happens in both metallic and nonmetallic matters, and it accounts for approximately 80–90% of all constructional failures—aircraft landing, motor shaft, gear machine parts, automobile crank-shaft, bridges, etc. Consequently, designing for a greatest stress shall not ensure proper product life. Most fractures produced are associated with this classification.

Engineering stress is asymmetrical near stress raisers such as notches, fillets, or holes which concentrate on the stress. In compound drawings, engineers often omit these design defects, which may produce reliability disasters. For example, the vibration of an aircraft wing during a long flight shall result in tens of thousands of load cycles. If not properly designed, these structures shall collapse. It is significant to discover the design defects and correct them. In Chaps. 7 and 8, we shall address how to search the design faults by utilizing the ALT.

The main problem in designing against fracture in high-strength materials is that the preexistence of cracks shall alter the local stresses to such an extent that the elastic stress analyses done so deliberately by the engineers are inadequate. As a crack extends to a definite crucial length, it shall propagate to an extremely unsuccessful extent through the structure, even though the gross stress is much less than would usually bring yield or failure in a tensile specimen. The term 'fracture mechanics' mentions to an essential specialization in solid mechanics in which the existence of a crack is presumed, and we wish to search out quantitative relationships between the crack length, the matter's intrinsic resistance to crack growth, and the stress at which the crack propagates at high speed to give rise to a structural failure.

A fast fracture shall happen in a few loading cycles. For instance, fatigue failures in 1200 rpm motor shafts took less than 12 h from installation to last fracture, approximately 830,000 cycles. On the other hand, crack growth in steadily rotating process equipment shafts has taken many months and more than 10,000,000 cycles to be unsuccessful.

6.4 Fracture Failure

6.4.1 Introduction

Fracture is the breakdown of a structure into pieces when subjected to a (repetitive) stress. It occurs whenever the acted loads (or stresses) are more than the resisting strength of the body. It begins with a crack which breaks with no making completely apart. Fracture due to overstress is as likely as not the most widespread failure mechanism in mechanical systems and can be categorized as brittle (cleavage) fracture, ductile (shear) fracture, fatigue fracture, deadhesion, and crazing.

As a crack propagates, a new free surface is produced, possessing a particular surface energy γ. It is supplied by the exterior load and is also obtainable as accumulated elastic energy. Not all available energy, however, is utilized for the production of new crack surfaces. It is also transferred into other energies, such as dissipative heat or kinetic energy. As much available energy is utilized for crack growth, the fracture is said to be brittle. As a large amount of energy is transferred into other energies, mostly due to dissipative mechanisms, the fracture is designated ductile.

As seen in Fig. 6.22, brittle fracture propagates quickly on a crack with minimum energy absorption and plastic deformation. Brittle fracture happens along characteristic crystallographic planes defined as cleavage planes. The mechanism of brittle fracture was originally described by the Griffith theory [26]. Griffith suggested that there are microcracks in a brittle matter that act on the concentrated stress at their tips. The crack might originate from a number of sources as flow happened during a surface scratch or solidification. As plastic deformation at the crack tip is forbidden, the crack shall move through grains by breaking atom bonds in lattice planes. This cleavage fracture shall exist in matters with little or no close-packed planes, possessing BCC or HCP crystal structures.

Brittle materials are ceramics, glasses, metals, and some polymers. They have the next attributes:

- There is no considerable plastic deformation, so crack propagation is too fast.
- The crack propagates almost perpendicular to the orientation of the acted stress.

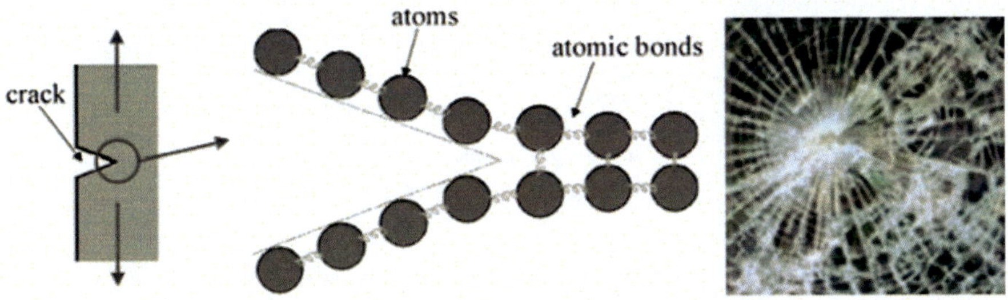

Fig. 6.22 Brittle (cleavage) fracture mechanism and its example (glass)

- Cracks frequently propagate by cleavage—breaking atomic bonds along particular crystallographic planes (cleavage planes).

For ductile fracture, as a crystalline matter is filled, dislocations begin to go through the lattice due to local shear stresses. Additionally, many dislocations shall grow. Because the interior structure is altered irreversibly, the macroscopic deformation is permanent (plastic). The dislocations unite at grain boundaries and accumulate to form a void. These voids will grow, and one or more of them shall move in a macroscopic crack.

Because the start and growth of cracks are aroused by shear stresses, this mechanism is mentioned as shearing. Plastic deformation is crucial, so this mechanism is usually noticed in FCC crystals, which have many close-packed planes. The fracture surface has a 'dough-like' structure with dimples, the shape of which designates the loading of the crack (Figs. 6.23 and 6.24).

- Brittle fracture: Separation along crystallographic planes due to breakdown of atomic bonds (V-shaped chevron, cleavage, intergranular)

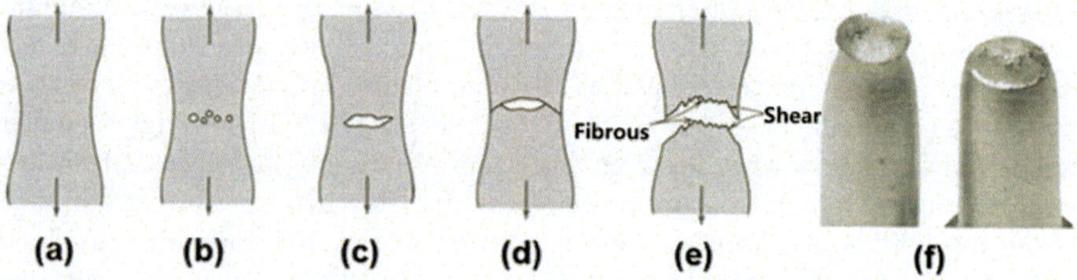

Fig. 6.23 Ductile fracture failure mechanism. **a** Necking. **b** Formation of micro-voids. **c** Coalescence of micro-voids. **d** Crack propagation by shear. **e** Fracture. **f** Cup and cone

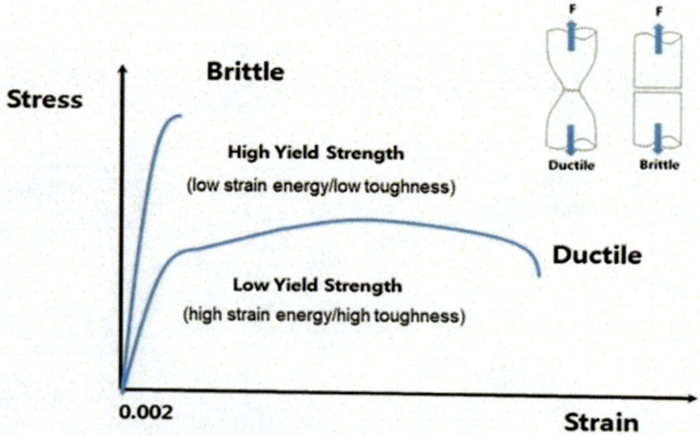

Fig. 6.24 Brittle versus ductile fracture in matter

- Ductile fracture: Initiation, growth, and coalescence of micro-voids (cup-and-cone, dimple).

However, matter in sale is polycrystalline, whose crystal axes are situated at random. As a polycrystalline matter is subjected to stress, slip begins first in those grains in which the slip system is most positively located with respect to the acted stress. As contact at the grain boundaries is continued, it can be requisite for more than one slip system to run. The rotation into the axis of tension causes other grains, initially less positively situated, into a position where they can now deform. As deformation and rotation move, the individual grains tend to elongate in the orientation of flow.

As a crystal distorts the shape, there is some deformation of the lattice structure. This distortion is maximum on the slip planes and grain boundaries and grows with deformation. This is obvious by a grow in resistance to further deformation. The matter undergoes strain hardening or work hardening. As dislocations pile up at grain boundaries, metals shall be hardened by lessening the size of the grains.

6.4.2 Ductile–Brittle Transition Temperature (DBTT)

The ductile-to-brittle transition temperature (DBTT) is generally noticed in metals which are reliant on their composition. For some steels, the transition temperature shall be approximately 0 °C, and in winter, the temperature in some areas of the world might be below this. Consequently, some steel structures are very likely to be unsuccessful in winter. That is, at high temperatures, where the impact energy taken in before failure is large, ductile failure happens with comprehensive elastic and plastic deformation before failure. At low temperatures, where the impact energy absorbed before failure is low, brittle fracture happens with little deformation before failure.

The governing mechanism of this change stays uncertain despite the many efforts made in theoretical and experimental research. All ferrous materials (except the austenitic grades) show a transition from ductile to brittle when tested above and below a definite temperature, defined as DBTT. FCC metals such as Cu and Ni remain ductile down to very low temperatures. For ceramics, this type of transition happens at much higher temperatures than for metals (Fig. 6.25).

Steels possessing DBTTs just below room temperature were utilized. At low temperatures, steels may be seriously brittle. At higher temperatures, the impact energy is large, which is consistent with a ductile mode of fracture. When the temperature drops, the impact energy decreases abruptly over a comparatively narrow temperature span, which is consistent with the mode of brittle fracture. Fatigue cracks form a nucleus at the corners of square hatches and propagate quickly by brittle fracture.

As the well-known weld fractures in some US army ships (Liberty Ships, tankers) in World War II are explored, the ductile-to-brittle transition shall be computed by

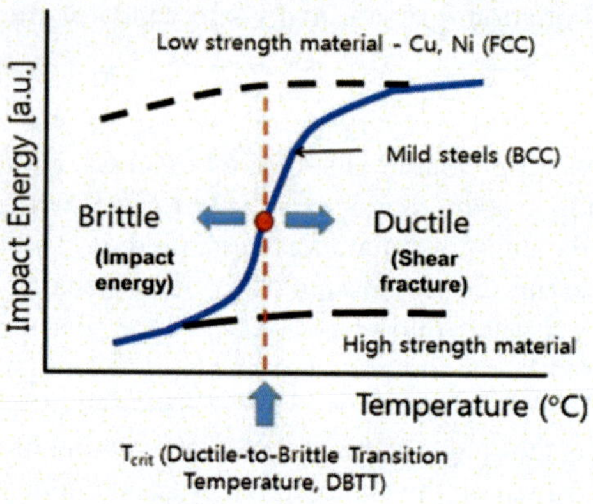

Fig. 6.25 Ductile-to-brittle transition temperature

impact testing such as Charpy V-notch testing (Fig. 6.26). Though the tensile stress–strain curve supplies a sign for ductile/brittle failure, the archetype experiment to look over this is the Charpy V-notch test. The principal benefit of this test is that it supplies a straightforward measure for the dissipated energy in fast crack propagation.

The sample is a beam with a 2 mm deep V-shaped notch, which possesses a 90° angle and a 0.25 mm root radius. It is carried and loaded as in a three-point bending test. The load is supplied by the impact of a weight at the end of a pendulum. A crack shall begin at the tip of the V-notch and run through the sample. The matter distorts at a strain rate of generally 103 s^{-1}. The energy dissipated during fracture shall be computed effortlessly from the height of the pendulum weight before and after impact. The dissipated energy is the impact toughness C_v.

The impact toughness shall be decided for diverse specimen temperatures T. For inherent brittle materials such as high-strength steel, the dissipated energy shall be

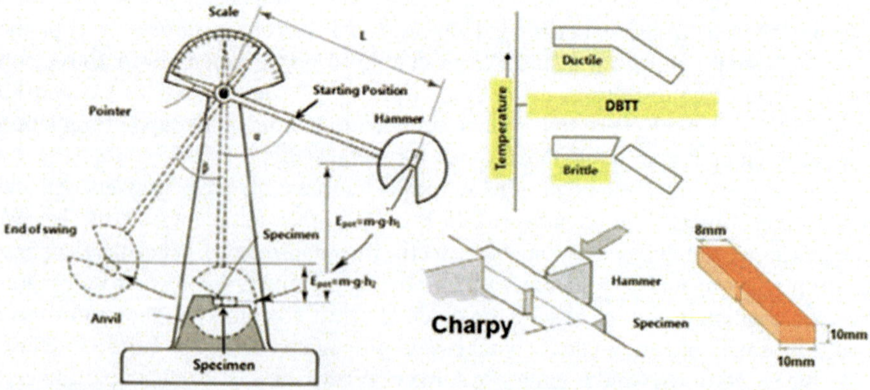

Fig. 6.26 Schematic of conventional Charpy V-notch testing

low for all temperature T. For inherent ductile materials such as FCC metals, C_v shall be high for all T. Many materials represent a change from brittle to ductile fracture with growing temperature.

References

1. Duga JJ, Fisher WH, Buxaum RW, Rosenfield AR, Buhr AR, Honton EJ, McMillan SC (1982) The economic effects of fracture in the United States. Final Report, September 30, 1982, Battelle Laboratories, Columbus, OH. Available as NBS Special Publication 647-2
2. Inglis CE (1913) Stresses in a plate due to the presence of cracks and sharp corners. Trans R Inst Nav Architectes 60:219–241
3. Neugebauer GH (1943) Stress concentration factors and their effect in design. Prod Eng 14:82–87
4. Paris PC, Gomez MP, Anderson WE (1961) A rational analytic theory of fatigue. Trend Eng 13:9–14
5. Wöhler A (1870) Über die Festigkeitsversuche mit Eisen und Stahl. Z für Bauwesen 20:73–106
6. Kim WH, Laird C (1978) Crack nucleation and stage I propagation in high strain fatigue—II Mechanism. Acta Metall 26(5):789–799
7. Woo S, O'Neal D, Pecht M (2010) Reliability design of a reciprocating compressor suction reed valve in a common refrigerator subjected to repetitive pressure loads. Eng Fail Anal 17(4):979–991
8. Sutherland HJ (1999) On the fatigue analysis of wind turbines, SAND99–0089. Sandia National Laboratories, Albuquerque, p 132
9. Nisbett RG, Budynas JK (2014) Shigley's mechanical engineering design, 10th edn. McGraw-Hill Higher Education, Boston, pp 308–324
10. Miner MA (1945) Cumulative damage in fatigue. J Appl Mech 12:149–164
11. Palmgren AG (1924) Die Lebensdauer von Kugellagern Zeitschrift des Vereines Deutscher Ingenieure. Sci Res 68(14):339–341
12. Burhan I, Kim H (2018) S-N curve models for composite materials characterization—an evaluative review. J Compos Sci 2(3):38–66
13. Sutherland HJ, Mandell JF (2005) Optimized Goodman diagram for the analysis of fiberglass composites used in wind turbines blades. In: 43rd AIAA aerospace sciences meeting and exhibit, Reno, Nevada
14. Murakami Y, Miller KJ (2005) What is fatigue damage? A view point from the observation of low cycle fatigue process. Int J Fatigue 27(8):991–1005
15. Qiu B, Kan Q, Kang G, Yu C, Xie X (2019) Rate-dependent transformation ratcheting-fatigue interaction of superelastic NiTi alloy under uniaxial and torsional loadings—Experimental observation. Int J Fatigue 127:470–478
16. Sun J, Su A, Wang T, Chen W, Guo W (2019) Effect of laser shock processing with post-machining and deep cryogenic treatment on fatigue life of GH4169 super alloy. Int J Fatigue 119:261–267
17. Kumar N, Goel S, Jayaganthan R, Owolabi GM (2018) The influence of metallurgical factors on low cycle fatigue behavior of ultrafine grained 6082 Al alloy. Int J Fatigue 110:130–143
18. Maier G, Riedel H, Somsen C (2013) Cyclic deformation and lifetime of Alloy617B during isothermal low cycle fatigue. Int J Fatigue 55:126–135
19. Sola JF, Kelton R, Meletis EI, Huang H (2019) Predicting crack initiation site in polycrystalline nickel through surface topography changes. Int J Fatigue 124:70–81
20. Liu Y, Kang M, Wu Y, Wang M, Wang J (2018) Crack formation and microstructure-sensitive propagation in low cycle fatigue of a polycrystalline nickel-based superalloy with different heat treatments. Int J Fatigue 108:79–89

21. Wahi RP, Auerswald J (1997) Damage mechanisms of single and polycrystalline nickel base superalloys SC16 and IN738LC under high temperature LCF loading. Int J Fatigue 19(93):89–94
22. Wilhelm F, Affeldt E, Fleischmann E, Glatzel U, Hammer J (2017) Modeling of the deformation behavior of single crystalline Nickel-based superalloys under thermal mechanical loading. Int J Fatigue 97:1–8
23. American Society for Testing and Materials (1999a) Standard practice for strain-controlled fatigue testing. In Annual Book of ASTM Standard, vol. 03.01. American Society for Testing and Materials, West Conshohocken, PA, no. E606.
24. American Society for Testing and Materials (1999b) Standard practice for cycle counting in fatigue analysis. In: Annual book of ASTM standard, vol. 03.01. American Society for Testing and Materials, West Conshohocken, PA, no. E1049.
25. American Welding Society (1996) Structural welding code—steel, 15th edn. American Welding Society, prepared by AWS Structural Welding Committee, Miami
26. Griffith AA (1921) The phenomena of rupture and flow in solids. Philos Trans R Soc Lond A 221:163–198

Chapter 7
Design Methodology—Parametric Accelerated Life Testing

7.1 Introduction

Due to hostile needs in the marketplace, mechanical products might be designed to achieve desired functioning and tall reliability. As the system designs are assessed before launching, new attributes are swiftly integrated into a product and brought to the market. The application of all these newly developed features affects a wide range of customer sectors where structural safety is a major concern: automobiles, refrigerators, airplanes, nuclear power plants, civil or naval structures, etc. With either restricted trial or no evident apprehension of how introduced design traits shall be employed by the end user, system introductions with design flaws can badly influence the manufacturer's brand [1].

To transmit torque and speed, a gear system of auger motor utilized in a refrigerator ice-maker is a rotating circular mechanical part having cut teeth, which mesh with another (compatible) toothed component. The advantages of gear drives are their exact velocity ratio, large powers, lofty efficiency, reliable service, and compact layout. On the other hand, manufacturing gears requires particular tools and apparatuses. If the mistake in cutting teeth has, it causes vibration and noise in functioning. The basic operation of gears is analogous to that of levers. That is, the auger motor assembly, including the gear system, is designed to crush the ice through a greater rotational torque by decelerating speed through a plurality of driven gears interlocked with a drive pinion gear installed in the input shaft. To stop a gear system from being unsuccessful in the field before its expected lifetime, a company should prove the new gear combined with appropriate ISO Standards [2, 3] and information sheet [4] and or properly perform reliability testing before the product is released.

The Boeing 737 MAX passenger aircraft from March 2019 to December 2020 was grounded after 346 persons lost their life in pair crashes. The airplanes adopted the CFM International LEAP-1B engines using the optimized 68-in. fan design; these engines consumed 12% less fuel and were 7% lighter than other engines [5]. Investigators, including the Ethiopian Civil Aviation Authority, tentatively deduced that the crash was created by the aircraft's design engine. To secure a product that

S. Woo, *Design of Mechanical Systems*, Springer Series in Reliability Engineering,
https://doi.org/10.1007/978-3-031-28938-5_7

is successful in the field, problematic parts need to be recognized and adjusted by employing a systematic testing way—parametric ALT, which may produce reliability quantitative (RQ) statements [6].

Product material flaws, such as cracks, notches, local thin areas, etc., when subjected to repeated (impact) loads can cause fatigue to occur. Fatigue is the leading cause of lack of success in metallized components, expressing roughly 80–95% of all constructural failures or economic losses [7]. That is, fatigue in metals becomes visible in the form of cracks, which develop in regions that stress may collect, such as grooves, sharp-edged, holes, etc., and propagate it. Structural failures in any of these sectors may have evident serious consequences in terms of human lives and environmental disasters. It is necessary to understand the different mechanisms generating critical and subcritical processes in structural materials and to develop assessment techniques and management procedures for the corresponding structures.

To avoid these structural failures, the ALT integrated with the reliability block diagram has been carried out as an alternative method for solving product problems [8]. It included a test scheme for the product, identifyng failure mechanics—fatigue and fracture, and utilizing sample size formulation, elevated loads, etc. Elsayed [9] grouped statistical, statistics/physics, and experimental/physics–accepted mock-ups for the failure model. Hahn and Meeker [10] proposed some empirical procedures to make arrangements for an ALT. Performing a parametric ALT [11, 12] involves several idea, such as the BX lifetime for the systematic tests, a simple life-stress model, sample size equation, and fracture mechanics, because failure can instantly appear due to vulnerable parts in a module system [13–15]. The present test approach [16–19] cannot easily reproduce the design imperfections of parts in a multimodule product because the techniques evaluate too few samples and insufficient testing time.

To robustly acquire the design of a mechanical product, engineers have used established approaches such as the strength of materials [20]. Awareness is being paid to the development of engineering procedures defining the structural integrity conditions of a given component. Engineers have also utilized quantum mechanics to recognize which fatigue failures started from the atomic and microstructural scales in numerous metallic alloys or many engineering plastics [21, 22]. To recognize the fatigue origin of a mechanical product, a (generalized) life-stress formulation may be utilized as an established design method and associated approaches to recognize the lack of success of electronic components due to matter defects or tiny cracks. Finite element methods (FEMs) and strength of material cannot recognize the origin of failure because field failure stochastically occurs in the areas of locally high stress concentration, not continuum in material [23–26]. Alternatively, there are other ways—structural health monitoring (SHM)—which permit the observation of the failure origin [27]. However, it is hard to get test data for multimodule systems because comprehensive tests might have been necessary, which may be too expensive due to a period of time and required samples.

To show the success of identifying and changing the design flaws of a mechanical system, a parametric ALT may be utilized as a structured way to produce mission cycles such as RQ specifications. It covers (1) a product BX lifetime produced on the ALT procedure, (2) a load test, (3) adapted ALTs with alternations, and (4) an evaluation of whether the system design(s) attains the targeted BX life. The time-to-failure formulation, the sample size formulation, and BX lifetime are proposed. To confirm the validity of ALT, it would be necessitated to notice the initiation of the new design in the market to ensure that it passed the objective reliability.

7.2 Parametric ALT for Mechanical System

7.2.1 BX Lifetime in a Product

In products, power may be employed to a mechanical advantage to supply a job that necessitates forces and motion by properly adapting mechanisms. In the process, they will be subjected to repetitive loading. They should have a robust design for the expected stresses. An example is a domestic refrigerator that is planned to give chilled air from an evaporator and preserve the freshness of foods. One of the refrigerator modules is the ice-maker, including the gear system, which decreases the speed and increases the torque. As a result, the ice-maker can obtain sufficient torque to crush the ice. A refrigerator can also include a range of subsystems (or modules)—the door, cabinet, shelves and drawers, controlled instruments, motor and electronics, compressor, evaporator and condenser, water supplying equipment, and numerous unlike components. The lifetime of the refrigerator is determined by the wear-out (or random) failure of a newly designed module such as the ice-maker, including an auger motor with a gear system that has design defects (Figs. 7.1 and 7.2).

To carry out a parametric ALT, the BX lifetime (or "Bearing Life"), L_B, shall be determined as an index of the product life. The BX lifetime metric originated from bearing manufacturing but has become an index for product lifetime, which can be employed in various industries today. That is, BX life is a proceeded time when X% of a group of products under consideration is unsuccessful. A "BX life Y years" is a manner for showing the product life. If a system is a B20 life 10 years, it signifies that 20% of a collection of samples under consideration will fail during 10 years of working. By using this measure, the ALT may recognize the cumulative failure rate and satisfy the field demands for the life needs of the system.

For instance, a refrigerator has approximately 2000 parts, including ice-makers. If an ice-maker has design defects, it may affect the possible life of the refrigerator. If a domestic refrigerator's life target is set to have a B20 life 10 years, the life of each unit shall be a B1 life 10 years because the refrigerator is made up to 20 units and each unit has 100 parts.

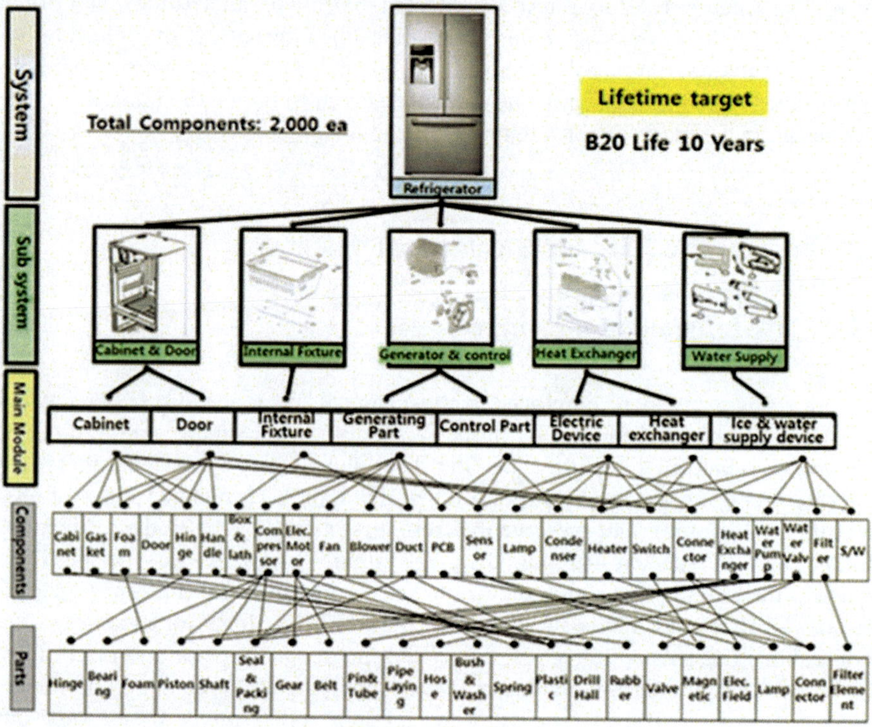

Fig. 7.1 A systematic class of multimodule refrigerators

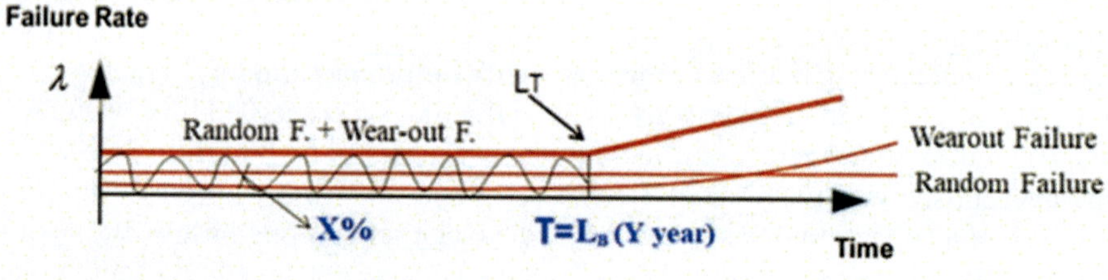

Fig. 7.2 System life resolved by new subsystem (or module)

7.2.2 *Positioning a Total Parametric ALT Procedure*

The reliability of a mechanical product may be expressed as the capability required to continually carry out the planned function under the described operational/environmental circumstances for a needed period of time [28]. If system reliability is explained by utilizing a “bathtub curve”, there are three sections: (1) in the first section, during the premature system lifetime, there is some lessening in the failure rate, (2) in the second section, during its center lifetime, there is a comparatively continual failure rate, and (3) in the third section, there is an increasing failure rate until the last lifetime of the system is reached.

If T is a random variable designating the time to failure, the proportion of outliving at time t may be stated as:

$$R(t) = P(T > t) \tag{7.1}$$

The cumulative distribution function (CDF), $F(t)$, is $F(t)(or\ X) = 1 - R(t)$. The failure rate, λ, on the slanted bathtub curve in Fig. 7.3 can be expressed as:

$$\lambda(t) = \frac{f(t)}{R(t)} = \frac{\mathrm{d}F(t)/\mathrm{d}t}{R(t)} = \frac{-R'(t)}{R(t)} \tag{7.2}$$

where f is the failure density function.

If the failure rate in Eq. (7.2) takes the integral, the $X\%$ cumulative failure, $F(L_B)$, at $T = L_B$ can be attained as follows:

$$F(= X) = \int_0^T \lambda(t) \cdot \mathrm{d}t = -\ln R(L_B) \cong \langle\lambda\rangle \cdot L_B \tag{7.3}$$

Assuming that T_I is the period of the earliest failure in the second section of the bathtub, the reliability function $R(t)$ shall also be obtained as:

$$R(t) = P(T_1 > t) = P(\text{no failure in } (0, t]) = \frac{(m)^0 \mathrm{e}^{-m}}{0!} = \mathrm{e}^{-m} = \mathrm{e}^{-\lambda t} \tag{7.4}$$

As product life is increased, the failure rate in the field shall drop down. Instead, the system life extends. For such situations, the reliability of a mechanical product can be declared as follows:

$$R(L_B) = 1 - F(L_B) = \mathrm{e}^{-\lambda L_B} \cong 1 - \lambda L_B \tag{7.5}$$

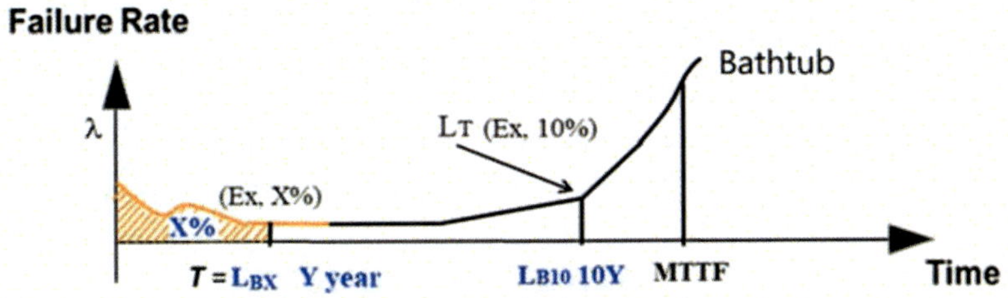

Fig. 7.3 Lifetime index; BX Life (L_{BX}) on the bathtub

The equation can be proven to be below approximately 20% of the cumulative failure [28]. We also know that the product reliability (or unreliability) can be obtained from the multiplication of failure rate and lifetime.

As a case study in this study, an ice-maker, including an auger motor with a gear system, repeatedly requires a straightforward mechanical operation: (1) water provides the ice tray; (2) water will freeze into ice by blowing chilled air through the evaporator (heat exchanger); and (3) the ice harvests until the ice bucket is in full. As the consumer exerts force on the lever to move it away, crushed (or cubed) ice will dispense. The unsuccessful products returned from the marketplace were crucial for comprehending and distinguishing the repeated usage methods of consumers and recognizing design defects equipped in the structural system. Based on the market statistics, the basic cause(s) of the problematic ice-maker, including the gear system, were identified. After putting the targeted lifetime by using a parametric ALT, L_B, the mechanical system might be changed by pinpointing the troublesome parts and modifying them (Fig. 7.4).

To put the targeted life of a mechanical product—an ice-maker—by parametric ALT, which is made up of up to eight modules (or subsystems) (see Fig. 7.1), there are the following potential modules: (1) altered individualistic units, (2) new individualistic units, and (3) individualistic units similar to the preceding design entrenched on a requirement in the market. The altered ice-maker in the refrigerator investigated here is utilized as a case investigation. It was initially altered to increase the performance of the ice-maker, including the auger motor with gear system, and make the product more good quality in the market. However, the modified ice-maker had design defects that were in need of being altered because of field failures.

The modified module E from the market data had a failure rate of 0.31% per year and a B1 life of 3.2 years (Table 7.1). Based on field details, the lifetime of the ice-maker, including the auger motor with gear system, was an anticipated B1 life of 1.61 years because they had a failure rate of 0.3% per year. To fulfil customer

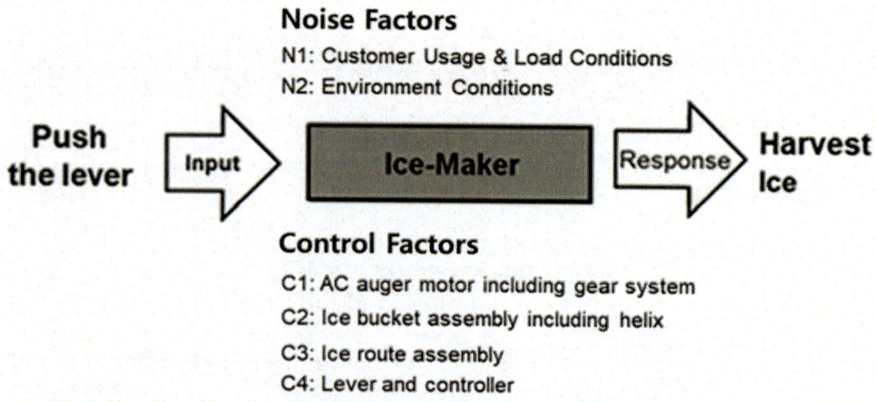

Fig. 7.4 Intended function of ice-maker adding gear system (instance)

Table 7.1 Comprehensive ALT strategy of mechanical modules (or subsystems) in a product such as a refrigerator

Modules	Market data		Predicted reliability				Intended reliability	
	Failure rate per year, λ (%/Year)	BX life, L_B (Year)	Failure rate per year, λ (%/Year)			BX life, L_B (Year)	Failure rate per year, λ (%/Year)	BX life, L_B (Year)
A	0.30	3.3	The same	× 1	0.30	3.33	0.10	10(BX = 1.0)
B	0.35	2.9	The same	× 1	0.35	2.9	0.10	10(BX = 1.0)
C	0.24	4.2	New	× 5	1.20	0.83	0.10	10(BX = 1.0)
D	0.15	6.7	Adjusted	× 2	0.30	3.33	0.10	10(BX = 1.0)
E	0.31	3.2	Adjusted	× 2	0.62	1.61	0.10	10(BX = 1.0)
Others (F/G/H)	0.50	10.0	The same	× 1	0.50	10.0	0.50	10(BX = 5.0)
System	1.9	2.9	–	–	3.27	0.83	1.00	10(BX = 10)

requirements, a new life of the product, such as an ice-maker, including an auger motor with a gear system, was aimed at achieving a B1 life of 10 years.

7.2.3 *Failure Model and Sample Size Formulation*

Mechanical products generally transport power from one position to another by choosing a suitable mechanism, such as the gear system in an ice-maker auger motor. As the consumer wants ice, the ice-maker in a refrigerator as this newly designed function is added. The main parts in an ice-maker are made up of the auger motor, helix support, bucket case, blade dispenser, helix dispenser clamp, helix upper dispenser, blade, etc.

An auger motor as a gear train is two or more gears functioning jointly by engaging their teeth and rotating each other in a system to cause torque and speed. That is, as electric motors are utilized, the gear systems decrease speed and increase torque. That is, an auger motor driven by an alternating current (AC) increases the torque through the gearbox to deliver ice and crush it at the end of an ice-maker. In the process, the system shall be subjected to repeated stresses because of (impact) loading due to crushing ice and a severely cold temperature. If there is a structural flaw that brings a strength when the (impact) loads are employed, the system shall unexpectedly fail in its expected life. Therefore, the components with field failure should be required to be fixed or replaced (Fig. 7.5).

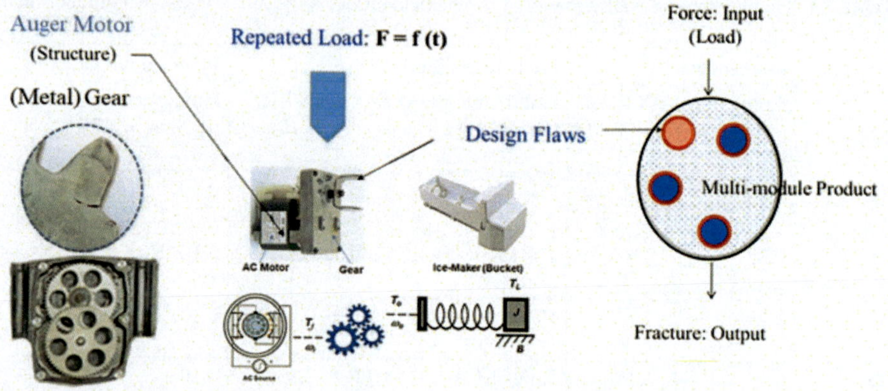

Fig. 7.5 Fatigue unsuccessfulness on the structural component generated by repeated (impact) load and design defects

7.2.3.1 The Schrodinger Equation

We shall derive the Schrodinger equation. Our beginning point is the classical nonrelativistic formulation for the energy of a particle, which is the total of the kinetic and potential energies. We presume usually that the potential is a function of only x. We have

$$E = K + V = \frac{1}{2}mv^2 + V(x) = \frac{p^2}{2m} + V(x) \tag{7.6}$$

We shall cite de Broglie's assertion that all particles is expressed as waves with frequency ω and wavenumber k and that E = $\hbar\omega$ and p = $\hbar$k. It does the expression for the energy into

$$\hbar\omega = \frac{\hbar^2 k^2}{2m} + V(x) \tag{7.7}$$

A wave with frequency ω and wavenumber k shall be expressed as usual as $\psi(x,t) = Ae^{i(kx-\omega t)}$ (the convention is to take a minus sign in front of ωt). In 3-D, we would have $\psi(x,t) = Ae^{i(kx-\omega t)}$, but let us just talk about 1-D. We bear in mind that

$$\frac{\partial \psi}{\partial t} = -i\omega\psi \Rightarrow \omega\psi = i\frac{\partial \psi}{\partial t} \text{ and } \frac{\partial^2 \psi}{\partial x^2} = -k^2\psi \Rightarrow k^2\psi = -\frac{\partial^2 \psi}{\partial x^2} \tag{7.8}$$

If we multiply the energy formulation in Eq. (7.7) by ψ and insert these relations, we attain

$$\hbar(\omega\psi) = \frac{\hbar^2}{2m}(k^2\psi) + V(x)\psi \Rightarrow i\hbar\frac{\partial\psi}{\partial t} = \frac{-\hbar^2}{2m}\frac{\partial^2\psi}{\partial x^2} + V\psi \tag{7.9}$$

In 3-D, the x dependence becomes dependent on whole three coordinates (x, y, z), and the $\frac{\partial^2\psi}{\partial x^2}$ term becomes $\nabla^2\psi$ (the total of the second derivatives). Recall that Born's (correct) explanation of $\psi(x)$ is that $|\psi(x)|^2$ takes the probability of discovering the particle at position x.

7.2.3.2 Infinite Square Well

The principal issue for parametric ALT is to determine how speedy the possible failure manner may be pinpointed. To accomplish this purpose, a simple failure illustration is methodically prepared, and the accurate coefficients for the life model are determined. That is, a life-stress (LS) prototype (or time-to-failure) that has stresses and reaction parameters should be developed. It therefore incorporates numerous failures—fatigue (or fracture). Fatigue failures on the exterior of a component can occur not only due to partial stresses but also due to flaws such as cracks or thin surfaces.

That is, fatigue may initiate from material defects—electron/void—that arise on a macro, microscopic or nano-range. From such a conceptual standpoint, it may be stated as transport processes such as the diffusion of shallow level dopants of silicon in semiconductors.

First, think about an electric particle that is restricted to transport only in the x orientation from $x = 0$ to $x = a$. The time-independent Schrodinger wave equation in operator form shall be defined as follows:

$$\hat{H}\psi = \mathrm{E}\psi \tag{7.10}$$

If $\hat{H} = -\frac{h^2}{8\pi^2 m}\frac{d^2}{dx^2} + V$, we can put this in Eq. (7.10).

$$-\frac{h^2}{8\pi^2 m}\frac{d^2\psi}{dx^2} + V\psi = E\psi \quad \text{or} \quad \frac{d^2\psi}{dx^2} + \frac{8\pi^2 m}{h^2}(E - V)\psi = 0 \tag{7.11}$$

where m is the electron mass, h is the Planck constant, and V is the potential energy.

Because $V = \infty$ outside the walls, this is possible only when $\psi = 0$. That is, the particle is not outside the walls. Because $V = 0$ inside the walls, Eq. (7.11) can be stated as follows:

$$\frac{d^2\psi}{dx^2} + \frac{8\pi^2 m}{h^2}(E - 0)\psi = 0 \quad \text{or} \quad \frac{d^2\psi}{dx^2} + K^2\psi = 0 \tag{7.12}$$

where $K^2 = \frac{8\pi^2 mE}{h^2}$

We can assume the solution of Eq. (7.12) as follows:

$$\psi(x) = A \sin Kx + B \cos Kx \tag{7.13}$$

where A, B = constants.

Because $x = 0$ or $x = a$ at walls, $\psi(0) = \psi(a) = 0$, B = 0, $K = \frac{n\pi}{a}$, and $E = \frac{n^2h^2}{8ma^2}$, n = 1, 2, 3, 4.

Therefore, we can state Eq. (7.13) as follows:

$$\psi(x) = A \sin\left(\frac{n}{a}\right)x \tag{7.14}$$

The probability of finding the particle in a small space between x and $x + \mathrm{d}x$ is given as follows:

$$\int_0^a \psi^2(x)\mathrm{d}x = 1 \quad \text{or} \quad \int_0^a \left(A \sin\left(\frac{n\pi}{a}\right)x\right)^2 \mathrm{d}x = 1 \tag{7.15}$$

Therefore, we can obtain the solution of Eq. (7.13) as follows:

$$\psi(x) = \sqrt{\frac{2}{a}} \sin\left(\frac{n\pi}{a}\right)x \tag{7.16}$$

where $\psi(x + a) = \psi(x)$, a is the (periodic) distance, and n is the principal quantum number.

7.2.3.3 Transport Processes

The atoms of the crystal establish a series of potential barriers that hinder the movement of the charged impurities. As an electromagnetic field, ξ, is applied, the barriers of potential junction energy as a function of distance will be reduced and distorted/phase-shifted. The impurities in matter, produced through electronic movement, are effortlessly migrated to the right because the passage to the left becomes difficult (Fig. 7.6).

Transport processes are therefore stated as follows (see Table 7.2):

$$J = LX \tag{7.17}$$

where J is a flux vector that is identified as the transport attribute. X is stated as a driving force, which is identified as the slope of the electrical potential, fluid velocity, concentration, temperature, etc. L is a transport numerical quantity.

For example, solid-state diffusion in semiconductor technology is the most practical procedure that controls the type and concentration of impurities in specific regions of a crystal. It includes electromigration-induced voiding, growth of chloride ions, and catching of electrons or holes.

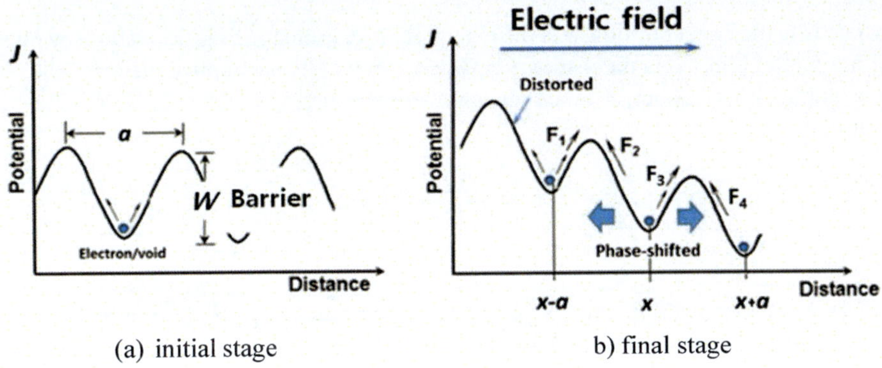

Fig. 7.6 Potential variation in material—silicon—when an electromagnetic field is applied

Table 7.2 Linear transport phenomena

Ohm's law of electrical conduction: $j = -\sigma \nabla V$		
$J =$ electric current density, j (units: A/cm^2)	$X =$ electric field, $-\nabla V$ (units: V/cm, V = electrical potential)	$L =$ conductivity, $\sigma = 1/\rho$ [units: $\rho =$ resistivity (Ω cm)]
Fourier's law of heat transport: $q = -\kappa \nabla T$		
$J =$ heat flux, q (units: W/cm^2)	$X =$ thermal force, $-\nabla T$ (units: °K/cm, T = temperature)	$L =$ thermal conductivity, κ (units: W/°K cm)
Fick's law of diffusion: $F = -D\nabla C$		
$J =$ material flux, F (units:/s cm^2)	$X =$ diffusion force, $-\nabla C$ (units:/cm^4, $C =$ concentration)	$L =$ diffusivity, D (units: cm^2/s)
Newton's law of viscous fluid flow: $F_u = -\mu \nabla u$		
$J =$ fluid velocity flux, F_u (units:/s^2 cm)	$X =$ viscous force, $-\nabla u$ (units:/s, $u =$ fluid velocity)	$L =$ viscosity, μ (units:/s cm)

The solid-state diffusion of impurities doped in silicon J is the amount/area per time. It can be formulated as [30, 31]:

$$J = [aC(x-a)] \cdot \exp\left[-\frac{q}{kT}\left(W - \frac{1}{2}a\xi\right)\right] \cdot v \tag{7.18}$$

[Density/Area] · [Jump Probability] · [Jump Frequency]

$$= -\left[a^2 ve^{-qw/kT}\right] \cdot \cosh \frac{qa\xi}{2kT} \frac{\partial C}{\partial x} + \left[2ave^{-qw/kT}\right] C \ \sinh \frac{qa\xi}{2kT}$$
$$= \Phi(x, t, T) \sinh(a\xi) \exp\left(-\frac{Q}{kT}\right) = B \sinh(a\xi) \exp\left(-\frac{Q}{kT}\right)$$

where C is the concentration, q is the extent of electric charge, ν is the frequency rate of attempted jump, a is the distance between atoms, ξ is the applied electric field, k is Boltzmann's constant, T is the absolute temperature, Q is the energy, and B is a constant.

$$K = K^{+} - K^{-} = a\frac{kT}{h}\mathrm{e}^{-\frac{\Delta E - aS}{kT}} - a\frac{kT}{h}\mathrm{e}^{-\frac{\Delta E + aS}{kT}} = B\ \sin h(aS)\exp\left(-\frac{\Delta E}{kT}\right) \tag{7.19}$$

where K is the reaction rate, S is the (chemical) field effect, T is the absolute temperature, k is Boltzmann's parameter, E is the activation energy, and Δ is the difference.

The junction function, J, from Eqs. (7.18) and (7.19) can be stated as:

$$J = B \sin \mathrm{h}(aS) \exp\left(-\frac{E_a}{kT}\right). \tag{7.20}$$

7.2.3.4 Life-Stress (LS) Prototype

If Eq. (7.20) puts a converse function, the life-stress (LS) prototype could be restated as:

$$TF = A[\sin h(aS)]^{-1} \exp\left(\frac{E_a}{kT}\right) \tag{7.21}$$

The sine hyperbolic form $[\sin \mathrm{h}(aS)]^{-1}$ in Eq. (7.21) has the following characteristics: (1) $(S)^{-1}$ at the beginning has some linear effect; (2) $(S)^{-n}$ has what is formed as a middle effect; and (3) $\left(\mathrm{e}^{aS}\right)^{-1}$ in the end is high (Fig. 7.7).

As a parametric ALT in the medium range is generally carried out, Eq. (7.21) might be expressed as:

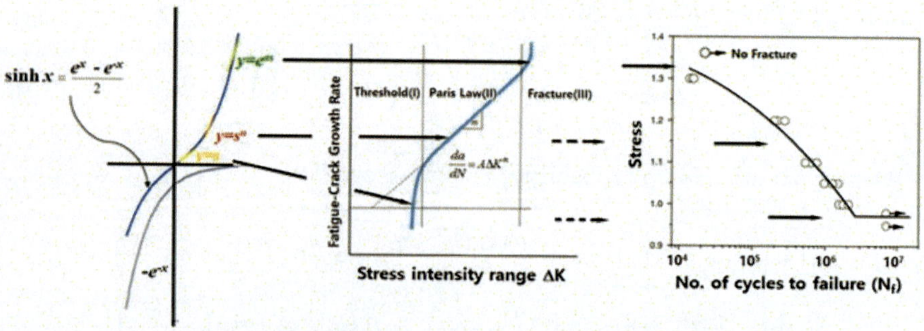

Fig. 7.7 Definition of hyperbolic sine stress expression on Paris law and S–N curve

Table 7.3 Energy flow in the multiport physical system

Modules	Effort, $e(t)$	Flow, $f(t)$
Mechanical translation	Force, $F(t)$	Velocity, $V(t)$
Mechanical rotation	Torque, $\tau(t)$	Angular velocity, $\omega(t)$
Compressor, pump	Pressure difference, $\Delta P(t)$	Volume flow rate, $Q(t)$
Electric	Voltage, $V(t)$	Current, $i(t)$
Magnetic	Magneto-motive force, e_m	Magnetic flux, φ

$$TF = A(S)^{-n} \exp\left(\frac{E_a}{kT}\right) \tag{7.22}$$

As the stress quantity in the structural components is difficult to calculate in parametric ALT, Eq. (7.22) is required to state again. Because the power for numerous energy areas is defined as the process of combining effort and flows, stresses start from effort in an energy transport system (Table 7.3) [32].

Stress is a material extent which defines the inner forces joining a minute portion of a continuum matter to exert to bear on each other. Because stress comes from effort in a mechanical system, Eq. (7.22) might be restated as:

$$TF = A(S)^{-n} \exp\left(\frac{E_a}{kT}\right) = B(\mathrm{e})^{-\lambda} \exp\left(\frac{E_a}{kT}\right) \tag{7.23}$$

where A and B are quantities that do not alter their values.

To fulfill the acceleration factor (AF), which could be expressed as the correlation between the elevated stress and common operation situations, it might be modified to merge with the effort idea:

$$AF = \left(\frac{S_1}{S_0}\right)^n \left[\frac{E_a}{k}\left(\frac{1}{T_0} - \frac{1}{T_1}\right)\right] = \left(\frac{e_1}{e_0}\right)^\lambda \left[\frac{E_a}{k}\left(\frac{1}{T_0} - \frac{1}{T_1}\right)\right] \tag{7.24}$$

As the majority of elevated testing is performed at usual (room) temperatures, Eq. (7.24) may be expressed as:

$$AF = \left(\frac{S_1}{S_0}\right)^n = \left(\frac{e_1}{e_0}\right)^\lambda. \tag{7.25}$$

7.2.4 Derivation of Sample Size Equation—Version I

Because of the cost and period limitation, it is not easy to perform a test with large samples for reliability testing of a system. If fewer parts are performed for test, the reliability testing in statistical analysis shall be more unknown. For a more exact consequence, enough samples should be tested. However, this testing will require considerable time and cost. Thus, utilizing the sample size formulation with AF in Eq. (7.21) and combining the fundamentals of statistics such as confidence level, it is critical to evolve a way for elevated testing.

In statistical testing, the first step of parametric ALT requires deciding how the sample size might be drawn from the population. The test samples are selected randomly. It is also associated with the confidence levels and the statistical span of the calculated failure values. Thus, the confidence levels for the life evaluation are critical because it is impossible to collect the life of some limited sample sizes. In statistics, the failure functioning of the limited sample shall strongly deviate from the real failure behavior of the population itself. The central idea further helps through the confidence levels, which may state the confidence of the test consequences and assess the failure behavior of the population.

Various ways have evolved to decide the sample size equation. The Weibull examination is a famous and extensively received way of examining reliability data. The Weibayes prototype based on Weibull examination is utilized to determine the sample size. However, because it is difficult to directly apply because of its mathematical complication, we shall differentiate the occasion of failures ($r \geq 1$) from the occasion of no failures ($r = 0$). A clarified equation therefore shall be evolved.

In choosing the set of values of the model parameters which make the likelihood function as large as possible, maximum likelihood estimation (MLE) in statistics is a method of approximating the parameters of a statistical prototype from a specified data set. The characteristic life η_{MLE} shall be expressed as follows:

$$\eta_{\mathrm{MLE}}^{\beta} = \sum_{i=1}^{n} \frac{t_i^{\beta}}{r} \tag{7.26}$$

As the number of failures is $r \geq 1$ and the confidence level is $100(1-\alpha)$, the characteristic life, η_α, would be closed from Eq. (7.26):

$$\eta_{\alpha}^{\beta} = \frac{2r}{\chi_{\alpha}^{2}(2r+2)} \cdot \eta_{\mathrm{MLE}}^{\beta} = \frac{2}{\chi_{\alpha}^{2}(2r+2)} \cdot \sum_{i=1}^{n} t_i^{\beta} \quad \text{for } r \geq 1 \tag{7.27}$$

Assuming there are no failures, the p value is α, and $\ln(1/\alpha)$ is mathematically identical to the chi-squared value, $\frac{\chi_{\alpha}^{2}(2)}{2}$ [33]. That is,

$$p\text{-value: } \alpha = \int_{\chi_\alpha^2(2)}^{\infty} \left(\frac{e^{-\frac{x}{2}} x^{\frac{\nu}{2}-1}}{2^{\frac{\nu}{2}} \Gamma\left(\frac{\nu}{2}\right)} \right) dx = \int_{2\ln\alpha^{-1}}^{\infty} \left(\frac{e^{-\frac{x}{2}} x^{\frac{\nu}{2}-1}}{2^{\frac{\nu}{2}} \Gamma\left(\frac{\nu}{2}\right)} \right) dx \quad \text{for } x \geq 0 \tag{7.28}$$

where ν is the shape parameter and Γ is the gamma function.

For $r = 0$, the characteristic life η_α from Eq. (7.27) is expressed as follows:

$$\eta_\alpha^\beta == \frac{2}{\chi_\alpha^2(2)} \cdot \sum_{i=1}^{n} t_i^\beta = \frac{1}{\ln\frac{1}{\alpha}} \cdot \sum_{i=1}^{n} t_i^\beta, \quad \text{for } r = 0 \tag{7.29}$$

Because Eq. (7.27) is valid for all occasions $r \geq 0$, the characteristic life η_α may be expressed as follows:

$$\eta_\alpha^\beta = \frac{2}{\chi_\alpha^2(2r+2)} \cdot \sum_{i=1}^{n} t_i^\beta \quad \text{for } r \geq 0 \tag{7.30}$$

From CDF in Weibull, the relation between BX life and characteristic life shall be defined as follows:

$$L_{BX}^\beta = \left(\ln \frac{1}{1-x} \right) \cdot \eta^\beta \tag{7.31}$$

As the approximated characteristic life of the p value α, η_α, in Eq. (7.27), is exchanged into Eq. (7.31), we attain the BX life equation:

$$L_{BX}^\beta = \left(\ln \frac{1}{1-x} \right) \cdot \frac{2}{\chi_\alpha^2(2r+2)} \cdot \sum_{i=1}^{n} t_i^\beta \tag{7.32}$$

Because most life testing usually has inadequate samples to approximate life and the allowed number of failures would be less than that of the sample size, the planned testing time will begin as follows:

$$n \cdot h^\beta \geq \sum t_i^\beta \geq (n-r) \cdot h^\beta \tag{7.33}$$

If Eq. (7.33) is exchanged into Eq. (7.32), the BX life formulation can be expressed as follows:

$$\begin{aligned} L_{BX}^\beta &\cong \left(\ln \frac{1}{1-x} \right) \cdot \frac{2}{\chi_\alpha^2(2r+2)} \cdot n h^\beta \\ &\geq \left(\ln \frac{1}{1-x} \right) \cdot \frac{2}{\chi_\alpha^2(2r+2)} \cdot (n-r) h^\beta \geq L_{BX}^{*\beta} \end{aligned} \tag{7.34}$$

Table 7.4 Characteristics of $\frac{\chi^2_{0.4}(2r+2)}{2}$ at the α = 60% confidence level

r	$1-\alpha$	$\frac{\chi^2_{0.4}(2r+2)}{2}$	$\frac{\chi^2_{\alpha}(2r+2)}{2} \approx r+1$	$1-\alpha$
0	0.4	0.92	1	0.63
1	0.4	2.02	2	0.59
2	0.4	3.11	3	0.58
3	0.4	4.18	4	0.57

The sample size formulation with the number of failures shall also be obtained as follows:

$$n \geq \frac{\chi^2_{\alpha}(2r+2)}{2} \cdot \frac{1}{\left(\ln \frac{1}{1-x}\right)} \cdot \left(\frac{L^*_{BX}}{h}\right)^{\beta} + r \tag{7.35}$$

For a 60% confidence level, the first term $\frac{\chi^2_{\alpha}(2r+2)}{2}$ shall be closed to $(r+1)$. Additionally, if the cumulative failure rate, x, is below approximately 20%, the denominator of the second term $\ln \frac{1}{1-x}$ is close to x by the Taylor expansion (Table 7.4).

Therefore, the sample size Eq. (7.35) should be close as follows:

$$n \geq (r+1) \cdot \frac{1}{x} \cdot \left(\frac{L^*_{BX}}{h}\right)^{\beta} + r \tag{7.36}$$

Therefore, we recognize that the sample size equation $n \approx$ (number of failures + 1) · (1/cumulative failure rate)((target lifetime/(test hours))$^{\beta+r}$.

As the *AF*s in Eq. (7.24) are attached into the planned testing time h, Eq. (7.36) turns out to be

$$n \geq (r+1) \cdot \frac{1}{x} \cdot \left(\frac{L^*_{BX}}{AF \cdot h_a}\right)^{\beta} + r \tag{7.37}$$

If the life target of a mechanical system such as a refrigerator compressor is set to have a B1 life of 10 years, the assigned mission cycles can be attained for a specified set of samples subjected to elevated conditions. Because $\frac{\chi^2_{\alpha}(2r+2)}{2} \cong (r+1)$, $\ln(1-x)^{-1} \cong x$, and $h = h_a \cdot AF$ in Eq. (7.37), the approximated life L_B in each ALT is also close as follows:

$$L_B^{\beta} \cong x \cdot \frac{n \cdot (h_a \cdot AF)^{\beta}}{r+1} \tag{7.38}$$

Let $x = \lambda \cdot L_B$. The approximated failure rate of the design samples λ shall be expressed as

Table 7.5 The computed sample size with h = 1080 h testing time

β	Failure number	Sample size	
		Equation (7.31) by Minitab	Equation (7.32)
2	0	3	3
2	1	7	7
3	0	1	1
3	1	3	3

$$\lambda \cong \frac{1}{L_{BX}} \cdot (r+1) \cdot \frac{L_{BX}^{\beta}}{n \cdot (h_a \cdot AF)^{\beta}} \tag{7.39}$$

In every ALT stage, we shall express the quantity of the reliability from the product of the assessed L_B life and failure rate λ. The usual working cycles of a product in its lifetime are computed under the anticipated consumer usage circumstances. If the failed number, objective lifetime, AF, and accumulative failure rate are decided, the necessitated real testing cycles under the elevated circumstances shall be attained from Eq. (7.37).

To apply the accelerated effort to samples, ALT equipment shall be inventively designed entrenched on the load examination of the product and its working mechanism. Utilizing ALT with a close sample size of an AF, we shall attain the actual mission cycles, $h_{a,}$ from Eq. (7.37). After obtaining the failed samples in mission cycles, we can decide whether the reliability target is fulfilled. To prove the accuracy of the approximated sample size in Eq. (7.32), we present in Table 7.5 without considering the AF.

If the approximated failure rate from the reliability tests is not larger than the objective failure rate (λ^*), the number of sample sizes (n) could also be attained. The approximated failure rate with a common-sense level of confidence (λ) shall be expressed as follows:

$$\lambda^* \geq \lambda \cong \frac{r+1}{n \cdot (AF \cdot h_a)} \tag{7.40}$$

By resolving Eq. (7.40), we shall also attain the sample size

$$n \geq (r+1) \cdot \frac{1}{\lambda^*} \cdot \frac{1}{AF \cdot h_a} \tag{7.41}$$

By multiplying the targeted life (L_B^*) into the numerator and denominator of Eq. (7.41), we shall obtain another sample size formulation:

$$n \geq (r+1) \cdot \frac{1}{\lambda^* \cdot L_B^*} \cdot \frac{L_B^*}{AF \cdot h_a} = (r+1) \cdot \frac{1}{x} \cdot \left(\frac{L_B^*}{AF \cdot h_a}\right)^1 \tag{7.42}$$

Thus, we recognize that $\lambda^{*} \cdot L_{B}^{*}$ is transferred into the accumulative failure rate x.

We may find two formulations for sample size which have a close form—Eqs. (7.37) and (7.42). For wear-out failure, it is fascinating that the quantity of the third term for the two equations is 1 or β, which is greater than 1. As the sample size formulation for the failure rate is added and the allowed failed number r is 0, the sample size formulation for the life could be a generalized equation to attain the reliability objective.

As the testing period of an item (h) is more than the objective lifetime (L_{B}^{*}), the reduction factor R is approximated to 1. The sample size formulation in Eq. (7.37) can be restated as follows:

$$n \geq (r+1) \cdot \frac{1}{x} \tag{7.43}$$

If the objective life, L_{B}^{*}, for a module is assigned to have B1 life 10 years, it is effortlessly attained from the computation by hand. For a refrigerator, the number of working cycles for one day was five; the worst case was nine. Thus, the objective lifetime for ten years could be 32,850 cycles.

The other type of sample size equation, which is derived by Wasserman [33], is defined as follows:

$$\begin{aligned} n &= -\frac{\chi_{\alpha}^{2}(2r+2)}{2m^{\beta} \ln R_{L}} = \frac{\chi_{\alpha}^{2}(2r+2)}{2m^{\beta} \ln R_{L}^{-1}} = \frac{\chi_{\alpha}^{2}(2r+2)}{2m^{\beta} \ln\left(1-F_{L}^{-1}\right)} \\ &= \frac{\chi_{\alpha}^{2}(2r+2)}{2} \cdot \frac{1}{\ln\left(1-F_{L}^{-1}\right)} \cdot \left(\frac{L_{BX}}{h}\right)^{\beta} \end{aligned} \tag{7.44}$$

where $m \cong h/L_{BX}$.

When $r = 0$, the sample size formulation shall be obtained as follows:

$$\begin{aligned} n &= \frac{\ln(1-C)}{m^{\beta} \ln R_{L}} = \frac{-\ln(1-C)}{-m^{\beta} \ln R_{L}} = \frac{\ln(1-C)^{-1}}{m^{\beta} \ln R_{L}^{-1}} = \frac{\ln \alpha^{-1}}{m^{\beta} \ln R_{L}^{-1}} \\ &= \frac{\chi_{\alpha}^{2}(2)}{2} \cdot \frac{1}{\ln\left(1-F_{L}^{-1}\right)} \cdot \left(\frac{L_{BX}}{h}\right)^{\beta} \end{aligned} \tag{7.45}$$

Thus, we recognize that Wasserman's sample size formulation (7.44) is identical to Eq. (7.32). In particular, the proportion between the product life versus the testing period in Eq. (7.37) shall be expressed as a reduction factor. It shall be utilized to decide whether ALT is proper. That is,

$$R = \left(\frac{L_{BX}^{*}}{h}\right)^{\beta} = \left(\frac{L_{BX}^{*}}{AF \cdot h_{a}}\right)^{\beta} \tag{7.46}$$

To successfully begin the parametric ALT, we should obtain the acute states which shall increase the AF and the shape factor β. Therefore, the position and form of the unsuccessful product in both field and parameter ALT consequences are alike. If the real test time h_a is lengthy than the planned testing time, which is stated in the reliability objective, the reduction fraction shall be less than one. Thus, we shall attain elevated circumstances which will lessen the testing time and sample size number.

7.2.5 Simplified Sample Size Equation—Version II

Typically, the characteristic life is expressed as follows:

$$\eta^\beta = \frac{\sum t_i^\beta}{r} \cong \frac{n \cdot h^\beta}{r} \tag{7.47}$$

where β is the shape parameter in a Weibull distribution.

As product (or part) reliability improves, the failure becomes less recurrent in lab tests. It will be harder to assess the characteristic life in Eq. (7.47). We shall take a feasible statement for characteristic life as follows:

$$\eta^\beta \cong \frac{n \cdot h^\beta}{r+1} \tag{7.48}$$

As a random variable T is depicted to the life of a product, the Weibull CDF of T, designated $F(t)$, shall be expressed as all individuals possessing a $T \le t$, divided by the whole number of individuals. It also takes the probability P of randomly selecting an individual having a value of T equal to or less than t, and therefore, we have

$$P(T \le t) = F(t) \tag{7.49}$$

If the system pursuits a Weibull distribution, the cumulative failure rate, $F(t)$, is defined as follows:

$$F(t) = 1 - \mathrm{e}^{-\left(\frac{t}{\eta}\right)^\beta} \tag{7.50}$$

As $t = L_B$ in Eq. (7.50), the relation between BX life (L_B) and characteristic life η shall be expressed as follows:

$$L_B^\beta \cong x \cdot \eta^\beta \tag{7.51}$$

where $x = 0.01F(t)$.

If Eq. (7.51) is inserted into (7.52), the BX life shall be redefined as follows:

$$L_B^\beta \cong x \cdot \eta^\beta \cong x \cdot \frac{n \cdot h^\beta}{r+1} \ge L_B^{*\beta} \tag{7.52}$$

The straightforward sample size equation shall be discovered from Eq. (7.52):

$$n \geq (r+1) \cdot \frac{1}{x} \cdot \left(\frac{L_B}{h}\right)^{\beta} \tag{7.53}$$

As the accelerated factor in Eq. (7.24) is attached to the planned testing time h, Eq. (7.53) is

$$n \geq (r+1)\frac{1}{x}\left(\frac{L_{BX}}{AF \cdot h_a}\right)^{\beta} \tag{7.54}$$

7.2.6 Derivation of Sample Size Equation—Version III

To obtain the number of assigned mission cycles of a parametric ALT from the targeted BX life on the test scheme, the sample size equation integrated with AF must be decided.

First, each Bernoulli trial has one of the two results, success or failure. The cumulative probability that follows a binomial distribution for failures can be expressed as follows:

$$L(p) = \sum_{r=0}^{c} \binom{n}{r} p^r \cdot (1-p)^{n-r} \leq \alpha \tag{7.55}$$

where n is the number of test samples and c is the assumed number of failures.

When probability p is tiny and n is not small, Eq. (7.55), which follows a Poisson distribution for failures, should be redefined as follows:

$$L(n \cdot p) = \sum_{r=0}^{c} \frac{1}{r!}(n \cdot p)^r \cdot \mathrm{e}^{-(np)} = \sum_{r=0}^{c} \frac{1}{r!} m^r \cdot \mathrm{e}^{-m} \leq \alpha \tag{7.56}$$

where $m = \text{parameter} = n \cdot p$.

From Eq. (7.56), we can assume that parameter m follows the chi-square distribution, $\chi_\alpha^2()$, when the p value is α. That is,

$$m = n \cdot p \sim \frac{\chi_\alpha^2(2r+2)}{2} \tag{7.57}$$

The Weibull distribution is widely used because it shall be straightforwardly expressed as a shape parameter and characteristic life. That is, if the product follows the Weibull distribution, the cumulated failure rate, $F(t)$, is expressed as

$$F(t) = 1 - e^{-\left(\frac{t}{\eta}\right)^{\beta}} \tag{7.58}$$

where t is time, η is characteristic life, and β is shape parameter.

In the case of unreliability, $p = F(t)$, and reliability, $1 - p = R(t)$, we can insert Eq. (7.58) into Eq. (7.56). That is,

$$L(p) = \sum_{r=0}^{c} \binom{n}{r} \left(1 - e^{-\left(\frac{t}{\eta}\right)^{\beta}}\right)^{r} \cdot \left(e^{-\left(\frac{t}{\eta}\right)^{\beta}}\right)^{n-r} \leq \alpha \tag{7.59}$$

Because $e^{-\left(\frac{t}{\eta}\right)^{\beta}} \cong 1 - \left(\frac{t}{\eta}\right)^{\beta}$, Eq. (7.59) can be approximated as follows:

$$L(p) \cong \sum_{r=0}^{c} \frac{1}{r!} \left(\frac{t}{\eta}\right)^{\beta r} \cdot \left(1 - \left(\frac{t}{\eta}\right)^{\beta}\right)^{n-r} \leq \alpha \tag{7.60}$$

Because Eqs. (7.59) and (7.60) have similar forms, the characteristic life with the confidence level 100 $(1 - \alpha)$ shall be stated as follows:

$$m = n \cdot p = n \cdot \left(\frac{t}{\eta}\right)^{\beta} \sim \frac{\chi_{\alpha}^{2}(2r + 2)}{2} \quad or \quad \eta_{\alpha}^{\beta} = \frac{2}{\chi_{\alpha}^{2}(2r + 2)} \cdot n \cdot t^{\beta} \tag{7.61}$$

At BX life, L_B, in Eq. (7.58), testing time, t, is h.

$$L_{B}^{\beta} \cong x \cdot \eta_{\alpha}^{\beta} = x \cdot \frac{2}{\chi_{\alpha}^{2}(2r + 2)} \cdot n \cdot t^{\beta} = x \cdot \frac{2}{\chi_{\alpha}^{2}(2r + 2)} \cdot n \cdot h^{\beta} \geq L_{B}^{*\beta} \quad \text{for } x \leq 0.2 \tag{7.62}$$

where $x = 0.01\ F(t)$.

If Eq. (7.62) is rearranged, the sample size equation shall be defined as follows:

$$n \geq \frac{\chi_{\alpha}^{2}(2r + 2)}{2} \times \frac{1}{x} \times \left(\frac{L_{B}^{*}}{h}\right)^{\beta} \tag{7.63}$$

For a 60% confidence level, because the 1st term $\frac{\chi_{\alpha}^{2}(2r+2)}{2}$ in Eq. (7.63) will be estimated as $(r + 1)$, Eq. (7.63) can be defined as follows:

$$n \geq (r + 1) \times \frac{1}{x} \times \left(\frac{L_{B}^{*}}{h}\right)^{\beta} \tag{7.64}$$

As the acceleration factors in Eq. (7.24) are added to the planned testing time h, Eq. (7.64) is

$$n \geq (r+1) \times \frac{1}{x} \times \left(\frac{L_B^*}{AF \cdot h_a}\right)^{\beta} \tag{7.65}$$

References

1. Bigg G, Billings S (2014) The iceberg risk in the *Titanic* year of 1912: was it exceptional? Significance 11(3):6–10
2. ISO 1328-2:2020 (2020) Cylindrical gears—ISO system of flank tolerance classification—Part 2: definitions and allowable values of double flank radial composite deviations: general procedures. ISO, Geneva, Switzerland
3. ISO 17804:2020 (2020) Founding—Ausferritic spheroidal graphite cast irons—classification cylindrical gears. ISO, Geneva, Switzerland
4. AGMA 939-A07 (2018) Austempered ductile iron for gears—AGMA Information Sheet
5. DVB Bank SE Aviation Research (AR) (2014) An overview of commercial jet aircraft 2013, p 20
6. Woo S, Pecht M, O'Neal D (2019) Reliability design and case study of the domestic compressor subjected to repetitive internal stresses. Reliab Eng Syst Saf 193:106604
7. Duga JJ, Fisher WH, Buxaum RW, Rosenfield AR, Buhr AR, Honton EJ, McMillan SC (1982) The economic effects of fracture in the United States. Final report. Available as NBS Special Publication 647-2. Battelle Laboratories, Columbus, OH, USA
8. Modarres M, Kaminskiy M, Krivtsov V (2016) Reliability engineering and risk analysis: a practical guide, 3rd edn. CRC Press, Boca Raton, FL, USA
9. Elsayed EA (2012) Reliability engineering. Wiley, Hoboken, NJ, USA
10. Hahn GJ, Meeker WQ (2004) How to plan an accelerated life test (e-book). ASQ Quality Press, Milwaukee, WI, USA
11. Moura EC (2004) How to determine sample size and estimate failure rate in life testing. ASQ Quality Press, Milwaukee, WI, USA
12. McPherson J (2010) Reliability physics and engineering: time-to-failure modelling. Springer, New York, NY, USA
13. Griffith AA (1921) The phenomena of rupture and flow in solids. Philos Trans R Soc Lond A 221:163–198
14. Wang Y, He Z (2022) Experimental and statistical study of the fracture mechanism of Sn96.5Ag3Cu0.5 solder joints via ball shear test. Materials 15:2455
15. Anderson TL (2017) Fracture mechanics—fundamentals and applications, 3rd edn. CRC, Boca Raton, FL, USA
16. ASTM E606/E606 M (2019) Standard test method for strain-controlled fatigue testing. ASTM International, West Conshohocken, PA, USA
17. ASTM E399 (2020) Standard test method for linear-elastic plane-strain fracture toughness of metallic materials. ASTM International, West Conshohocken, PA, USA
18. ASTM E647 (2015) Standard test method for measurement of fatigue crack growth rates. ASTM International, West Conshohocken, PA, USA
19. ASTM E739-10 (2015) Standard practice for statistical analysis of linear or linearized stress-life (S–N) and strain-life (ε-N) fatigue data. ASTM International, West Conshohocken, PA, USA
20. Elishakoff I (2019) Stepan Prokofievich Timoshenko and America. ZAMM J Appl Math Mech 99(3)

21. David JG (2018) Introduction to quantum mechanics, 3rd edn. Cambridge University Press, Cambridge, UK
22. Cemal B (2021) Introduction to unified mechanics theory with applications, 1st edn. Springer Nature, Berlin, Germany
23. Weingart RG, Stephen P (2007) Timoshenko: father of engineering mechanics in the U.S. Structure Magazine, 1 Aug 2007
24. Reddy JN (2020) An introduction to the finite element method, 4th edn. McGraw-Hill, New York, NY, USA
25. Dyniewicz B, Bajkowski JM, Bajer CI (2022) Effective viscoplastic-softening model suitable for brain impact modeling. Materials 15:2270
26. Vijaya Kumar SD, Karuppanan S, Ovinis M (2022) Artificial neural network-based failure pressure prediction of API 5 L X80 pipeline with circumferentially aligned interacting corrosion defects subjected to combined loadings. Materials 15:2259
27. Vega MA, Todd MD (2022) A variational Bayesian neural network for structural health monitoring and cost-informed decision-making in miter gates. Struct Health Monit 21(1):4–18
28. IEEE Standard Glossary of Software Engineering Terminology. IEEE STD 610.12-1990 (2020) Standards Coordinating Committee of the Computer Society of IEEE. (reaffirmed September 2002). Available online: https://ieeexplore.ieee.org/document/159342. Accessed on 31 Dec 2020
29. Kreyszig E (2011) Advanced engineering mathematics, 10th edn. Wiley, Hoboken, NJ, USA, p 683
30. Grove A (1967) Physics and technology of semiconductor device, 1st edn. Wiley International Edition, New York, NY, USA, p 37
31. Minges ML (1989) Electronic materials handbook, vol 1. ASM International, Cleveland, OH, USA, p 888
32. Karnopp DC, Margolis DL, Rosenberg RC (2012) System dynamics: modeling, simulation, and control of mechatronic systems, 6th edn. Wiley, New York, NY, USA
33. Wasserman G (2003) Reliability verification, testing, and analysis in engineering design. Marcel Dekker, New York, NY, USA, p 228
34. Tang LC (2004) Multiple-steps step-stress accelerated life tests: a model and its spreadsheet analysis. Int J Mater Prod Technol 21(5):423–434
35. William EL, David M, Christoper NM, Stephen WB, Richard JF, Timothy F, Thomas AS, Frank WG (2005) Mechanical properties of structural steels. NIST NCSTAR 1-3D

Chapter 8
Case Studies of Parametric Accelerated Life Testing (ALT)

8.1 Improving the Lifetime of a Localized Ice-Maker

As the customer wants to have an ice function in a refrigerator, an ice-maker system is equipped to produce ice. As the consumer exerts force on the lever, crushed (or cubed) ice is distributed through the route of ice. The main components in an ice-maker are the auger motor, including the gear system, helix support, helix upper dispenser, helix dispenser clamp, blade, blade dispenser, and bucket case, as manifested in Fig. 8.1. They require high-strength fatigue under low temperatures because of the repeated impact stress.

In the ice-making operation, the components in an ice-maker experience various mechanical load. Domestic refrigerators in the United States are planned to harvest from 10 cubes per usage to 200 cubes per day (20 times). As the ice-maker is repeatedly utilized in both crushed and cubed ice types, it is repeatedly subjected to (impact) loads in the ice-maker system, including auger motors with gear systems. Ice production can additionally be affected by consumer use conditions, such as ice consumption, water pressure, notch settings in refrigerators, and the number of open doors.

In the marketplace, some ice-maker components in a refrigerator failed under unidentified customer operations. Market data demonstrated that the returned products could have had design defects—improper material such as cast iron that must not be utilized under severe cold temperature (below − 20 °C) in the freezer section. Therefore, the rotating gear system made of cast iron repetitively impacts each other while crushed (or cubed) ice is produced. A crack (or fracture) at the root fillet and tooth end of the gear may suddenly occur in the auger motor and no longer work. Engineers should find the root causes by failure analysis or reliability testing and correct them (Fig. 8.2).

By employing failure analysis (and laboratory tests) for returned field products, cracks that started in the fillet (or ends) of teeth in mechanical parts, such as gear, propagated it to the end. To keep it functioning for its anticipated lifetime, the manufacturer was required to redesign the product failure, such as gear cracks, in the auger

S. Woo, *Design of Mechanical Systems*, Springer Series in Reliability Engineering,
https://doi.org/10.1007/978-3-031-28938-5_8

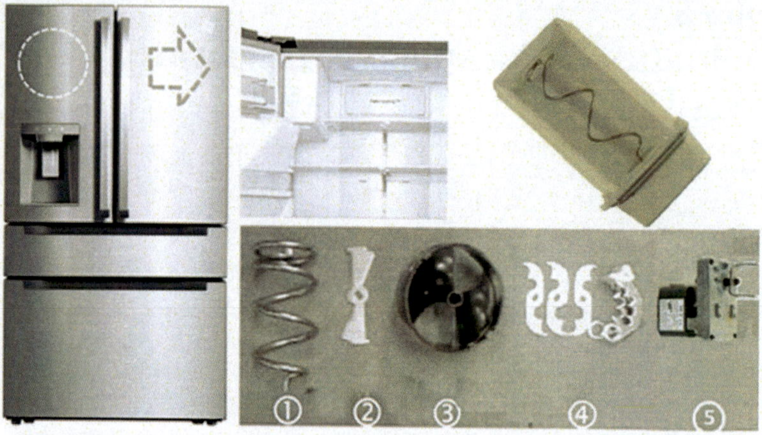

(a) refrigerator (b) mechanical components of an ice-maker assembly: helix support ①, blade dispenser ②, helix upper dispenser ③, blade ④, and auger motor ⑤.

Fig. 8.1 A domestic refrigerator with ice-maker

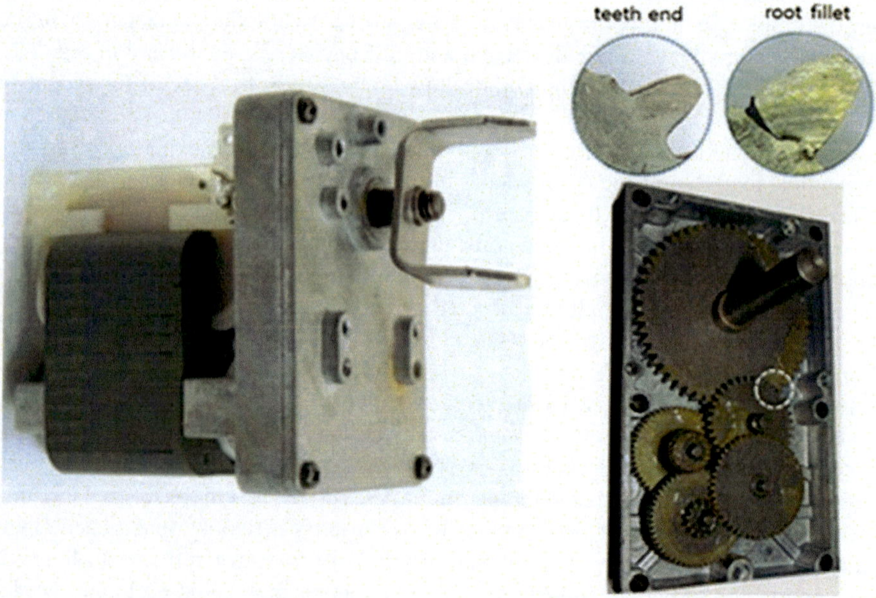

Fig. 8.2 Damaged auger motor after use

motor. That is, if there are design flaws—the gear system in the ice-maker's auger motor—where repetitive loads are applied under severe temperature conditions, the structure will be unsuccessful in its anticipated lifetime. To reproduce the problematic part(s) and alter them, an engineer was necessitated to carry out parametric ALT for a newly designed product. It was made up to (1) a load examination for the problematic product, (2) the action of making practical and effective use of ALTs with design modifications, and (3) the judgement of whether the lifetime objective of last designs had been fulfilled.

Figure 8.3 shows a schematic outline of the power transfer in an ice-making process by utilizing a bond graph formulation. To produce sufficient torque to compress forcefully to break the ice at the end of the ice-maker, an AC auger motor supplies power by the gear system that is additionally moved to the bucket and ice crusher blade assembly. Therefore, the ice-maker system in the bucket will increase torque that has enough force and be subjected to different loads.

To attain the governing equations, the bond graph prototype in Fig. 8.3b shall be resolved at each node as follows:

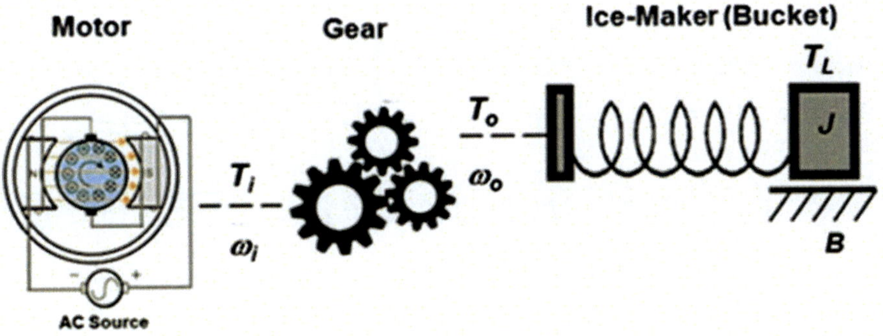

(a) Illustrative drawing of auger motor, ice crusher, ice bucket, etc.

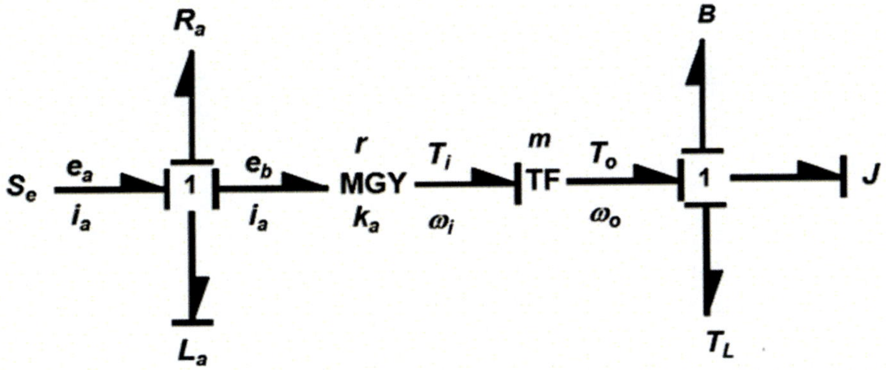

(b) Bond graph formulation of ice-maker.

Fig. 8.3 Design idea of an ice-maker

$$\mathrm{d}f \times E_2/\mathrm{d}t = 1/L_a \times eE_2 \tag{8.1}$$

$$\mathrm{d}fM_2/\mathrm{d}t = 1/J \times eM_2 \tag{8.2}$$

The junction from Eq. (8.1) is

$$eE_2 = e_a - eE_3 \tag{8.3}$$

$$eE_3 = R_a \times fE_3 \tag{8.4}$$

The junction from Eq. (8.2) is

$$eM_2 = eM_1 - eM_3 \tag{8.5}$$

$$eM_1 = (K_a \times i) - T_{\mathrm{Pulse}} \tag{8.6}$$

$$eM_3 = B \times fM_3 \tag{8.7}$$

Because $fM_1 = fM_2 = fM_3 = \omega$ and $i = fE_1 = fE_2 = fE_3 = i_a$ from Eqs. (8.3 and 8.4),

$$eE_2 = e_a - R_a \times fE_3 \tag{8.8}$$

$$fE_2 = fE_3 = i_a \tag{8.9}$$

If substituting Eqs. (8.8) and (8.9) into Eq. (8.1), then

$$\mathrm{d}i_a/\mathrm{d}t = 1/L_a \times (e_a - R_a \times i_a) \tag{8.10}$$

In addition, from Eqs. (8.5–8.7), we can obtain

$$eM_2 = [(K_a \times i) - T_L] - B \times fM_3 \tag{8.11}$$

$$i = i_a \tag{8.12}$$

$$fM_3 = fM_2 = \omega \tag{8.13}$$

If substituting Eqs. (8.11–8.13) into (8.2), then

$$\mathrm{d}\omega/\mathrm{d}t = 1/J \times [(K_a \times i) - T_L] - B \times \omega \tag{8.14}$$

We shall attain the state formulation from Eqs. (8.10) and (8.14) as follows:

$$\begin{bmatrix} \mathrm{d}i_a/\mathrm{d}t \\ \mathrm{d}\omega/\mathrm{d}t \end{bmatrix} = \begin{bmatrix} -R_a/L_a & 0 \\ mk_a & -B/J \end{bmatrix} \begin{bmatrix} i_a \\ \omega \end{bmatrix} + \begin{bmatrix} 1/L_a \\ 0 \end{bmatrix} e_a + \begin{bmatrix} 1 \\ -1/J \end{bmatrix} T_L \tag{8.15}$$

When Eq. (8.15) is integrated, the output of the AC motor and ice bucket assembly is attained as

$$y_p = \begin{bmatrix} 0 & 1 \end{bmatrix} \begin{bmatrix} i_a \\ \omega \end{bmatrix} \tag{8.16}$$

From Eq. (8.15), we realize that the life of the ice bucket fabrication relies on the stress (or torque) due to the forces necessitated to crush the ice. The life-stress (LS) prototype in Eq. (8.16) shall thus be altered as

$$TF = A(S)^{-n} = AT_L^{-\lambda} = A(F_c \times R)^{-\lambda} = B(F_c)^{-\lambda} \tag{8.17}$$

Thus, the AF in Eq. (7.20) shall be modified as

$$AF = \left(\frac{S_1}{S_0}\right)^n = \left(\frac{T_1}{T_0}\right)^\lambda = \left(\frac{F_1 \times R}{F_0 \times R}\right)^\lambda = \left(\frac{F_1}{F_0}\right)^\lambda \tag{8.18}$$

We can perform ALT from Eq. (7.29) till the assigned cycles that supply the life objective—B1 life 10 years—are fulfilled.

The environmental working circumstances of the ice bucket fabrication in a refrigerator icemaker shall change from roughly – 15 to – 30 °C with a relative humidity varying from 0 to 20%. Relying on consumer usage, an ice dispenser is utilized on a mean of roughly 3–18 times per day. Under greatest usage for 10 years, the dispenser happens approximately 65,700 use cycles. To determine the stress level for parametric ALT, based on the permitted use span of the Auger motor company in bench-marked data, which were attained from other chief manufacturers, we employed the step-stress life test which hall assess the life under a constant used-circumstance for many accelerated loads, such as 0.8, 1.0, and 1.47 kN-cm. As the dissimilar stress level was altered because the usual torque is 0.69 kN-cm, we might notice the failure cycles of the auger motor at specific stress levels.

Technical data from the auger motor manufacturer showed that the usual torque was 0.69 kN-cm and the greatest torque was 1.47 kN-cm. Presuming the accumulative damage factor $\lambda = 2$, the AF was roughly 5 in Eq. (8.18).

For a B1 life of 10 years, the mission cycles for 10 samples (computed utilizing Eq. 7.29) were roughly 42,000 cycles if the shape parameter was assumed to be 2.0. The ALT was planned to secure a lifetime target—B1 life 10 years—if it might be unsuccessful less than once for 42,000 cycles. Figure 8.4 manifests the test framework of a parametric ALT for reproducing the unsuccessful auger motor, including the gear

(a) ALT Apparatus (b) controller

Fig. 8.4 Apparatus utilized in parametric ALT

Fig. 8.5 Duty cycles of disturbance load T_L on the band clamper

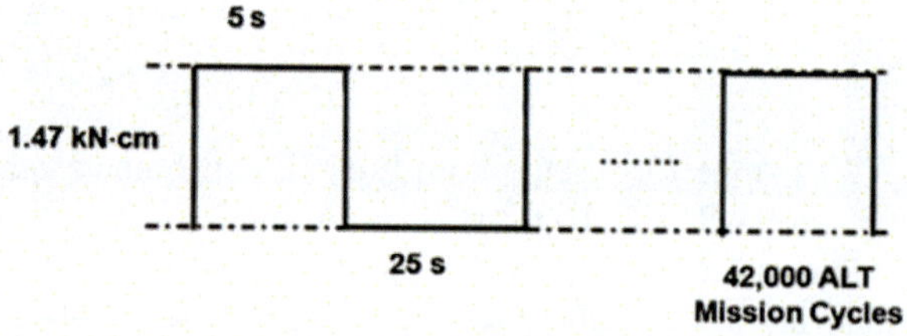

system in the market. Figure 8.5 shows the duty cycles for the ice-crushing torque T_L.

The chamber apparatus was cooled to a temperature of approximately – 30 °C. The control console located outside may begin or cease the apparatus and may display the whole test cycles and time periods, such as test sample on/off time. To utilize the greatest ice-crushing torque T_L, the helix upper dispensers with the blade dispenser were fastened jointly by a band clamper. When the controller applies the beginning signal, the equipment, including the auger motor, revolves. In the process, the ice-maker system will be applied with the greatest ice-crushing torque (1.47 kN-cm).

The ice-maker is generally made up to (cast, carbon, stainless, alloy, etc.) steel. The permissible stresses are expressed as a variable quantity of the tensile stress (F_u) or yield stress (F_y) of the part material. For steel, the scope of yield strength, F_y, and ultimate or tensile strength, F_u, usually utilized are 248–345 MPa and 400–483 MPa, respectively.

To place a scale of stress quantity through the step-stress life test, we examined the failure cycles at the subsequent stress quantity—0.8, 1.0, and 1.47 kN-cm, which may be attained from the bolted force of a band clamper with the helix upper dispenser. For 0.8 kN-cm, the ice-maker stopped nearly 12,000 cycles. For 1.0 kN-cm, the ice-maker stopped near 10,000 cycles and 12,000 cycles. On the other hand, for 1.47 kN-cm, the ice-maker stopped near 6000 cycles and 7000 cycles. Thus, we resolved the stress level as 1.47 kN-cm for ALT because it had comparatively fine data—linearity on the Weibull plot—contrasted with other stress levels.

In the first ALT, the teeth of the gear system in the auger motor fractured near 6000 cycles, 6900 cycles, 8500 cycles, and 8700 cycles when ice-makers were broken down in the failed samples. Figure 8.6 manifests a photo contrasting the product returned from the market and that from the 1st ALT, separately. Using a stereomicroscope, we also observed the fractured surface at the 1st ALT. It showed fatigue cracks and mechanical fractures. As they were alike in form, through parametric ALT, we might reproduce the fractured gear system in the marketplace. There was a material design flaw—cast iron that cannot be endured under the cold temperatures (– 20 °C below) in the freezer section. As the gear teeth (cast iron) repeatedly struck each other, they started to crack and finally fractured because this material was brittle under these conditions. Figure 8.7 manifests the graphic examination of the ALT consequences and market data on a Weibull chart. The shape parameter in the 1st ALT was approximated to be 2.0. For the last design, it was confirmed to be 4.38 on the Weibull plot.

To endure repeated impact loads, the material of the troublesome gear system utilized in the market was altered from cast-iron (Carbon, 3 wt% and silicon 2 wt%) to a sinter-hardened powder metallurgy nickel steel.

In the second ALT, near 9900 cycles and 12,000 cycles, the fracturing and cracking of helices made of polycarbonates (PC) occurred in the contact region of the blade dispenser (Fig. 8.8). To identify the root cause of the unsuccessful system, we checked

(a) Unsuccessful product in the market (b) Product with crack after the 1st ALT

Fig. 8.6 Unsuccessful gear systems in the market and during the first ALT

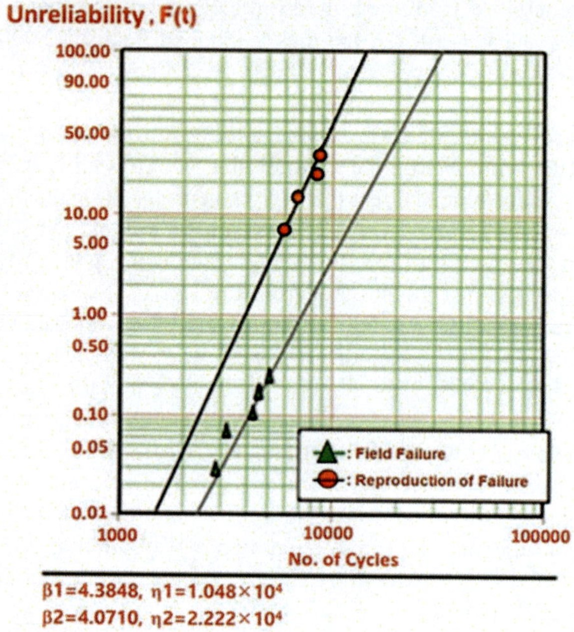

Fig. 8.7 Field statistics and outcomes of ALT on the Weibull plot

the failed product. We found that there was a structural design defect—the weld line between the blade dispenser and the helix upper dispenser, which had numerous micro-voids that were produced in the process of plastic injection. As the blade dispenser made of stainless steel hit the helix upper dispenser made of plastic under severe cold conditions, it cracked and fractured near the weld line. As a modification, we added an enforced rib on the side and front of the helix. Then, finite element analysis (FEA), which shall be united with ALT, was performed. As the helix upper dispenser was fixed against the wall, straightforward impact loads (1.47 kN-cm), as manifested in Fig. 14, were exerted. Utilizing matters and processing circumstances alike to those of the helix upper dispenser, the constitutive properties of the matters, such as polycarbonates (helix structure), were decided. As a result, the mechanical concentrated stress of the samples through finite element examination was decreased from 36.9 to 21.3 kPa.

As the gear material was altered and reinforced ribs on the front and side of the helix upper dispenser were added, the life of the ice-maker, including the auger motor with gear teeth, was extended. However, because the ice-maker system has insufficient fatigue strength for repeated impact stress, 42,000 mission cycles in the 2nd ALT have not yet been satisfied. Therefore, we carried out the 3rd ALT to confirm the design of the ice-maker.

In the 3rd ALT, there were no matters till 42,000 cycles. Over the route of three ALTs with modifications, the auger motor, including the gear system, was established to be a B1 life of 10 years. Figure 8.9 and Table 8.1 represent an abridged result of the ALTs.

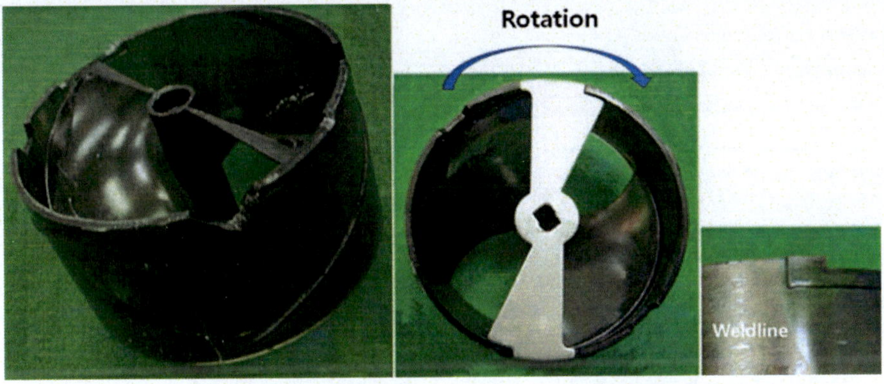

(a) unsuccessful products in the 2nd ALT (b) the root cause of 2nd ALT failure

Fig. 8.8 Unsuccessful products after the 2nd ALT

Fig. 8.9 Unsuccessful products after the 2nd ALT

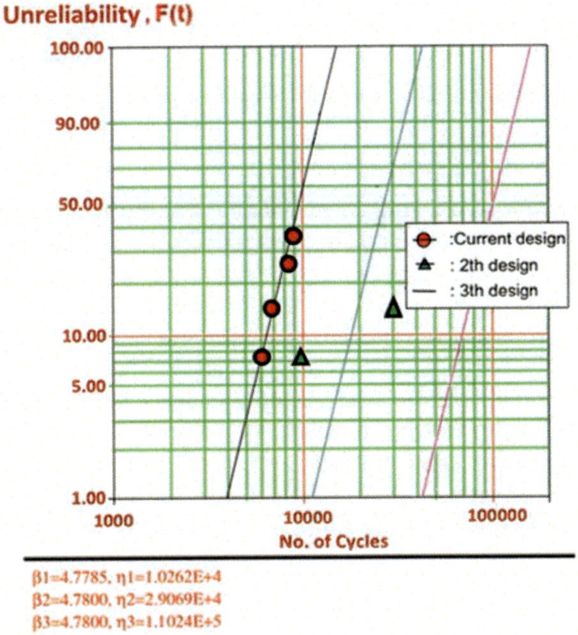

8.2 Residential-Sized Refrigerators During Transportation

Rail transportation of products is a leading way of moving products from cargo ships to the product's eventual destination—which could be a distributor, warehouse, or end user. We discuss two cases: (1) imported refrigerators were shipped from London in the United Kingdom to customers who lived in Skopje, the capital and largest city

Table 8.1 Consequences of ALT

Parametric ALT	1st ALT	2nd ALT	3rd ALT
	Draft design	–	Final design
Over the course of 42,000 cycles, the gear system has no problems	6000 cycles: 1/10 fracture 6900 cycles: 1/10 fracture 8500 cycles: 1/10 fracture 8700 cycles: 1/10 fracture (Failed gear samples)	9900 cycles: 1/10 fracture 12,000 cycles: 1/10 fracture (Failed helix samples)	42,000 cycles: 10/10 OK
Structure			
Action plans	C1: material: from cast-iron to a sinter-hardened powder steel	C2: added reinforced rib on side and front of helix	

of North Macedonia in southeastern Europe—a total distance of 2520 km over the course of two days, and (2) refrigerators were delivered from the West Coast of the United States to customers (or end-users) who lived on the East Coast. This trip was 7200 km for a total travel time of seven days.

To assess the ride attribute of a product installed on a vehicle, one of the most insightful mathematical representations of a vehicle suspension system is a quarter car prototype that shall be simply analyzed for vibration modes. The quarter car, two-degrees-of-freedom model is used to study wheel-hop as well as the bounce mode. It is set up using mutual connections of masses, springs, and dampers. Although a quarter car prototype has two degrees of freedom (DOF) and four state variables, it fulfills the aim of deciding the vehicle movement in transit. In these diagrams, the presumed prototype of the vehicle includes the sprung mass and the un-sprung mass respectively. The sprung mass, m_s, denotes one-fourth of the principal part of the vehicle; the un-sprung mass, m_u, denotes one wheel of the vehicle. The principal suspension is formulated as a spring k_s and a damper b_s aligned, which attaches the un-sprung mass to the sprung mass. The tire (or rail) is modeled as a spring constant of the tire (or rail), k_t, and denotes the transfer of the road force to the un-sprung mass (Fig. 8.10).

After the quarter car prototype is decomposed, the governing equations of movement can be represented as

$$m_s \ddot{x}_3 = -k_s(x_3 - x_2) - b_s(\dot{x}_3 - \dot{x}_2) \tag{8.19}$$

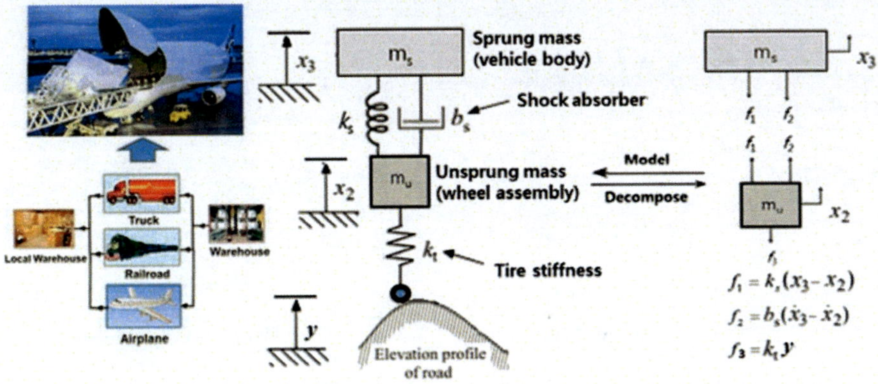

(a) A different mode of transportation (b) A dynamic model

Fig. 8.10 A quarter car model subjected to random loads from the base (or road)

$$m_u\ddot{x}_2 = k_s(x_3 - x_2) + b_s(\dot{x}_3 - \dot{x}_2) - k_t(x_2 - y) \tag{8.20}$$

Therefore, Eqs. (8.19) and (8.20) might be simply defined as

$$\begin{bmatrix} m_s & 0 \\ 0 & m_u \end{bmatrix}\begin{bmatrix} \ddot{x}_3 \\ \ddot{x}_2 \end{bmatrix} + \begin{bmatrix} b_s & -b_s \\ -b_s & b_s \end{bmatrix}\begin{bmatrix} \dot{x}_3 \\ \dot{x}_2 \end{bmatrix} + \begin{bmatrix} k_s & -k_s \\ -k_s & k_t + k_s \end{bmatrix}\begin{bmatrix} x_3 \\ x_2 \end{bmatrix} = \begin{bmatrix} 0 \\ k_t y \end{bmatrix} \tag{8.21}$$

As a consequence, Eq. (8.21) might be expressed in a matrix form:

$$[M]\ddot{X} + [C]\dot{X} + [K]X = F_T e^{i\omega t} \tag{8.22}$$

For Eq. (8.22), the (random) system response of an equivalent single degree of freedom (SDOF) is described as follows (see Fig. 8.11a):

$$m_e\ddot{X}(t) + c_e\dot{X}(t) + k_e X(t) = c_e\dot{Y}(t) + kY(t) \tag{8.23}$$

As the variables in the time domain are changed into the frequency domain, we can obtain a frequency domain function specified as the fatigue damage spectrum (FDS). Therefore, Eq. (8.23) can be stated as follows:

$$\ddot{X}(f) + 2j\varsigma\omega_n\dot{X}(f) + \omega_n^2 X(f) = \left[-2j\varsigma\frac{f_n}{f} - \left(\frac{f_n}{f}\right)^2\right]\ddot{Y}(f) \tag{8.24}$$

where ζ is the damping ratio, $\omega_n (= 2\pi f_n)$ is the natural frequency, and $\omega (= 2\pi f)$ is the excitation frequency.

The acceleration $\ddot{X}(f)$ can be expressed in the enforced acceleration $\ddot{Y}(f)$. That is,

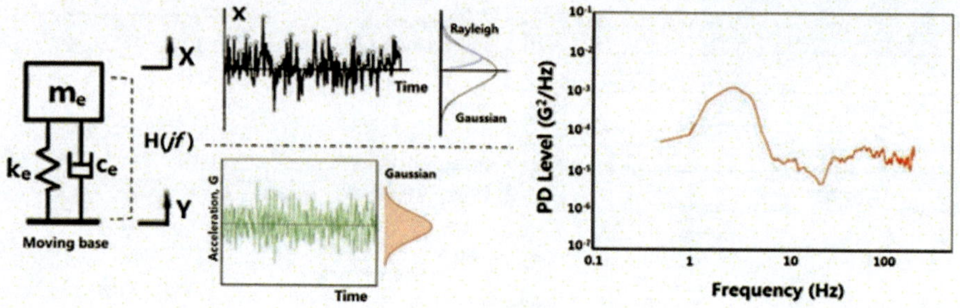

(a) Time domain (b) Frequency domain

Fig. 8.11 (Intermodal) vehicle responses for base random vibrations

$$\ddot{X}(f) = \left[\frac{2j\varsigma\left(\frac{f_n}{f}\right) + 1}{1 - \left(\frac{f_n}{f}\right)^2 + 2j\varsigma\left(\frac{f_n}{f}\right)}\right]\ddot{Y}(f) = H(jf)\ddot{Y}(f) \tag{8.25}$$

where *H(jf)* is the frequency transfer function (FRF).

From Eq. (8.25), the power spectral density (PSD) as the strength of the variations (energy) in the frequency domain can be attained. The PSD is used extensively to analyze the mechanical system characteristics for random loads such as forces and moments. The PSD of the acceleration $W_{\ddot{X}}$ can be stated in the PSD of the transmitted acceleration $W_{\ddot{Y}}$. That is,

$$W_{\ddot{X}}(f) = |H(f)|^2 W_{\ddot{Y}}(f) \tag{8.26}$$

where

$$|H(f)| = \left[\frac{\left(2\varsigma\frac{f_n}{f}\right)^2 + 1}{\left(1 - \left(\frac{f_n}{f}\right)^2\right)^2 + \left(2\varsigma\frac{f_n}{f}\right)^2}\right]^{1/2} \tag{8.27}$$

The root mean square (rms) value of the acceleration $\ddot{X}$ shall be computed as

$$a = \ddot{X}_{rms} = \int_0^\infty W_{\ddot{X}}(f)\,\mathrm{d}f \tag{8.28}$$

The integral in Eq. (8.28) can also be approximated (numerical quadrature) by the trapezoidal method to calculate $\ddot{X}_{rms}$ numerically.

$$\ddot{X}_{rms}(f_n) = \sqrt{\sum_{k=1}^{N} |H(f_n, f_k)|^2 W_{\ddot{X}}(f_k) \Delta f_k} \tag{8.29}$$

If the input from the base is a random vibration characterized as (1) a Gaussian distribution, (2) a constant PSD, and (3) a damping ratio less than 0.1 ($\zeta < 0.1$), it is possible to assume that the amplitude versus occurrence in the system response pursues the normal distribution (i.e., a Gaussian distribution) and that the peak versus occurrence manifests a Rayleigh distribution (Fig. A2a). However, in some cases, the market data may not pursue a Gaussian distribution because of the base random excitation. If the response of the equivalent SDOF in Eq. (A4) is displayed in the frequency domain, a plot (Fig. 8.11b) of the PSD can be obtained.

If the (accelerated) PSD response of the product is applied, it is possible to evaluate the design robustness due to the amplitude of the product at its natural frequency during its transit. That is, the resonance-induced impact of weak material parts due to repeated large displacement, strains, and stresses near natural frequency can be assessed on the potential for fracture starts at voids and then propagates it to the end.

In other words, from Eqs. (8.21) and (8.22), the stresses come from the transmitted vibration loads (F_T) that can be expressed as the PSD level of acceleration for a certain frequency band. Eq. (7.19) can be defined as

$$TF = A(S)^{-n} \exp\left(\frac{E_a}{kT}\right) = A(\mathrm{e})^{-\lambda} \exp\left(\frac{E_a}{kT}\right) = B(F_T)^{-\lambda} \exp\left(\frac{E_a}{kT}\right) \tag{8.30}$$

where A and B are constants.

The force transmissibility, Q, due to the base load, F_T, from Eq. (8.27) can be modeled as follows:

$$Q = \frac{F_T}{kY} = \left[\frac{1 + (2\zeta r)^2}{\left(1 - r^2\right)^2 + (2\zeta r)^2}\right]^{1/2} = |H(f)| \tag{8.31}$$

where $r = \omega_n/\omega = f_n/f$, ζ is the damping ratio ($c/c_c = c/2\omega_n$), k is the spring constant, and Y is the range of sinusoidal base excitations.

We also know that the PSD level for the given frequency band, a, is calculated from Eqs. (8.28) and (8.29). Thus, the acceleration factor (AF) is defined as the proportion between the typical circumstance and the elevated circumstance. It shall be defined as the multiplication of the amplitude proportion of gravitational acceleration R and force transmissibility Q. In other words,

$$\begin{aligned} \mathrm{AF} &= \left(\frac{\mathrm{S_1}}{\mathrm{S_0}}\right)^{\mathrm{n}} \left[\frac{\mathrm{E_a}}{\mathrm{k}}\left(\frac{1}{\mathrm{T_0}} - \frac{1}{\mathrm{T_1}}\right)\right] \\ &= \left(\frac{a_1}{a_0}\frac{\mathrm{F_T}}{\mathrm{kY}}\right)^{\lambda} \left[\frac{\mathrm{E_a}}{\mathrm{k}}\left(\frac{1}{\mathrm{T_0}} - \frac{1}{\mathrm{T_1}}\right)\right] \end{aligned}$$

$$= (R \times Q)^{\lambda}\left[\frac{E_a}{k}\left(\frac{1}{T_0} - \frac{1}{T_1}\right)\right] \tag{8.32}$$

where a_1 is the accelerated PSD level for the determined frequency band, a_0 is the normal PSD level for the determined frequency band, T_o is the normal temperature, and T_1 is the accelerated temperature.

Because accelerated testing is conducted at normal (room) temperature, Eq. (8.32) is redefined as

$$AF = \left(\frac{S_1}{S_0}\right)^n = \left(\frac{F_1}{F_0}\right)^{\lambda} = \left(\frac{a_1}{a_0}\frac{F_T}{kY}\right)^{\lambda} = (R \times Q)^{\lambda} \tag{8.33}$$

According to market data, after both shipping projects, the refrigerator compressor rubber in the mechanical compartments was torn out, and the joined tubes to the compressor fractured under unknown stress conditions during railway transport. For Europe, the distance at which failure first occurred during rail transportation was approximately 2400 km in Nis, Serbia. Over two days, when the refrigerators traveled a distance of 2520 km from London to Skopje, 10% of the products failed. On the other hand, in the United States, the distance at which failure first occurred during rail transportation was approximately 2500 km over two days. In Chicago, 27% of the transported products failed. Over seven days, when the refrigerators traveled a distance of 7200 km from Los Angeles to Boston, 67% of the products failed (Fig. 8.12).

Data from the field indicated that the unsuccessful refrigerators had design flaws. After identifying the problematic refrigerator designs in laboratory tests, the manufacturer could modify the problematic desi To attain the PSD obtained along rail routes in Fig. 4b that incorporated the mainline, side-line, and industrial line in the United States, it was measured in vertical and horizontal directions. That is, a SAVER 3X90 shock and vibration field data recorder (Lansmont Corp., CA, USA) was utilized to gather the spectral data for whole travel. The SAVER was installed immediately to the floor situated to the middle of the storage region where the door-frame hole is. After analyzing them, we obtained the vibration environment spectra, which showed a graphical plot of the PSD levels versus frequency. In this investigation, the PSD spectra were represented from 0.3 to 400 Hz to assess the design of the refrigerator.

After analyzing the measured vibration spectra, we calculated the acceleration factor that multiplies the amplitude proportion of acceleration R and force transmissibility Q from Eq. (19) if we know the damping ratio, natural frequency, normal PSD level, and accelerated PSD level. A programmed shaker table was used to apply the elevated external random vibrations, expressed by PSD, to each refrigerator sample. A table was set with refrigerator test samples that could be driven by actuators. As random waves and periodic waves such as sine, rectangular, and triangular shall be selected as the input motion, random shaking using observed strong PSD motions could particularly simulate the robustness of the product and the design weaknesses for RQ specification. The amplitude of the powerful random vibrations, expressed

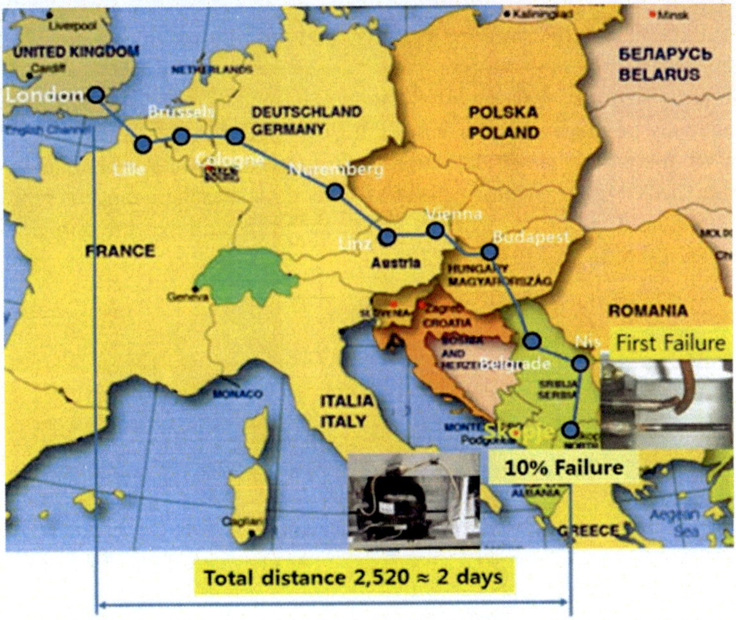

(a) Failures in Europe

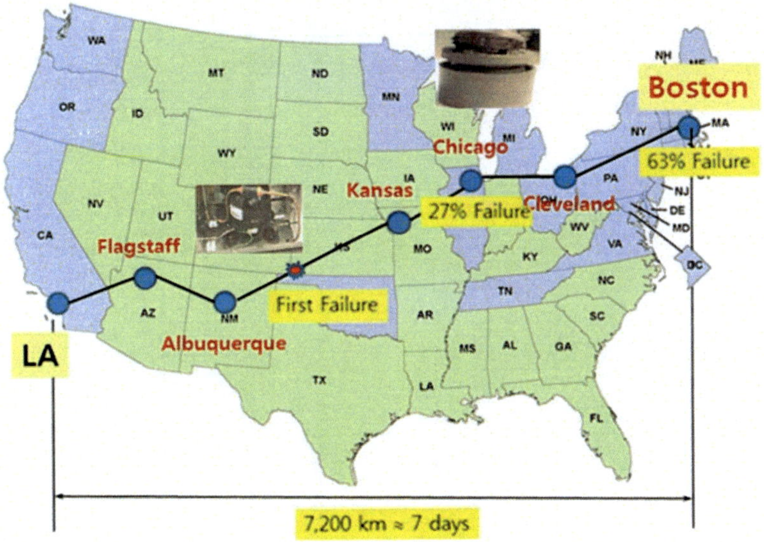

(b) Failures in the United States

Fig. 8.12 Failed places from the marketplace

as G_{rms}, could be adjusted according to the capacity of the specimen for accelerated loads.

In applying harmonic waves with various frequencies on the refrigerator, the natural frequency of the horizontal vibration (left ↔ right) was found to be 5 Hz. The natural frequency of the vertical vibration (up ↔ down) was 9 Hz in the vibration test. To attain the AF for horizontal or vertical directions, based on random vibrations observed in the field, amplified PSD loads for each orientation were applied to the refrigerator on a shaking table (Fig. 8.13).

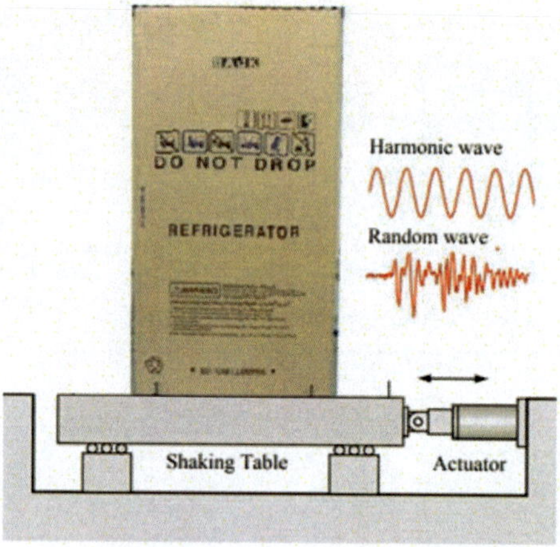

(a) Programmed shaking table for product testing

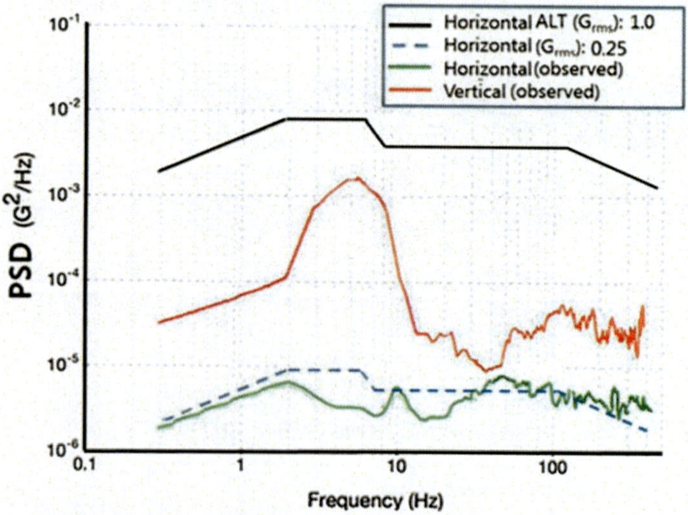

(b) Applied PSD loads

Fig. 8.13 Vibration test for the parametric ALT

Table 8.2 Conditions for parametric ALT of the refrigerator

System conditions	Worst case	ALT	AF
Transmissibility, Q ($r = 1.0, \zeta = 0.096 \approx 0.1$)	–	5.30 (from Eq. (8.31)	28.1①
Amplitude ratio of acceleration, R (a_1/a_0)	0.25 G_{rms}	1 G_{rms}	16.0②
Total AF (= ① × ②)			450

To decide the stress quantity for ALT and assess the lifetime, we employed the step-stress life test under constant used conditions for diverse amplitudes of the random vibrations, such as 0.40 G_{rms}, 0.60 G_{rms}, 0.80 G_{rms}, and 1.00 G_{rms}, at the natural frequency ($r = 1.0, \zeta \approx 0.1$). Because the damping ratio ($\zeta = 0.096 \approx 0.1$) with a settling time of 2 s and an overshoot of approximately 5 due to direct contact between the train/refrigerator and frequency ratio was expected, $r (= \omega/\omega_n) = 1$ at the natural frequency ω_n was applied to the refrigerator by rail, and the force transmissibility, Q, had a value of roughly 5.3 from Eq. (17). The AF due to gravitational acceleration for 1 G_{rms} was 4.0 because the refrigerator reached an acceleration of 1 G_{rms} on a shaker table, compared to that of worst-case 0.25 G_{rms}. Using a cumulative damage factor, λ, of 2.0, the entire AF in Eq. (8.33) was determined to be 450.0 (Table 8.2).

Based on the calculated AF, the needed testing time for a given sample size was acquired if the lifetime target was assigned. That is, suppose that the shape parameter in the Weibull plot was 2.0, and the life target was put to have a B1 life for seven days. The test time acquired from Eq. (7.29) was approximately 130 min for three sample pieces. If the refrigerator fails less than once in 130 min, the refrigerator design was suitable for a whole travel interval of 7200 km (seven days) to endure the fatigue damage by random vibration and have a B1 life with approximately a 60% level of confidence.

In the 1st ALT, we discovered the failure time of the next stress levels—0.40 G_{rms}, 0.60 G_{rms}, 0.80 G_{rms}, and 1.00 G_{rms} at the natural frequency ($r = 1.0, \zeta \approx 0.1$), we found refrigerator samples with torn rubber mounts and broken tubes. For 0.40 G_{rms}, it fractured at 70 min. For 0.60 G_{rms}, it fractured at 60 min. For 0.80 G_{rms}, it fractured at 45 min. For 1.00 G_{rms}, it fractured at 20 min. Therefore, we chose 1.00 G_{rms} because it has a good result with rapidity and linearity on the Weibull plot (see Fig. 8.15).

As 1.00 G_{rms} at the natural frequency ($r = 1.0, \zeta \approx 0.1$) was applied, and three refrigerators in the first parametric ALT fractured as follows: one sample at 20 min and two samples at 40 min from the horizontal vibration (left ↔ right). As the rubber mounts tore, the connecting refrigerant tubing in the machine compartment breaks, the refrigerant gas leaks out of the system, and the three samples no longer work. Figure 8.14 depicts the fractured products from the marketplace and the failed samples from the ALT. The forms and places of the failures in the ALT were alike to those shown in the marketplace. To explore the fracture surfaces, they were also captured by SEM. We discovered some irregular/coarser voids generated because of falling out rubber particles.

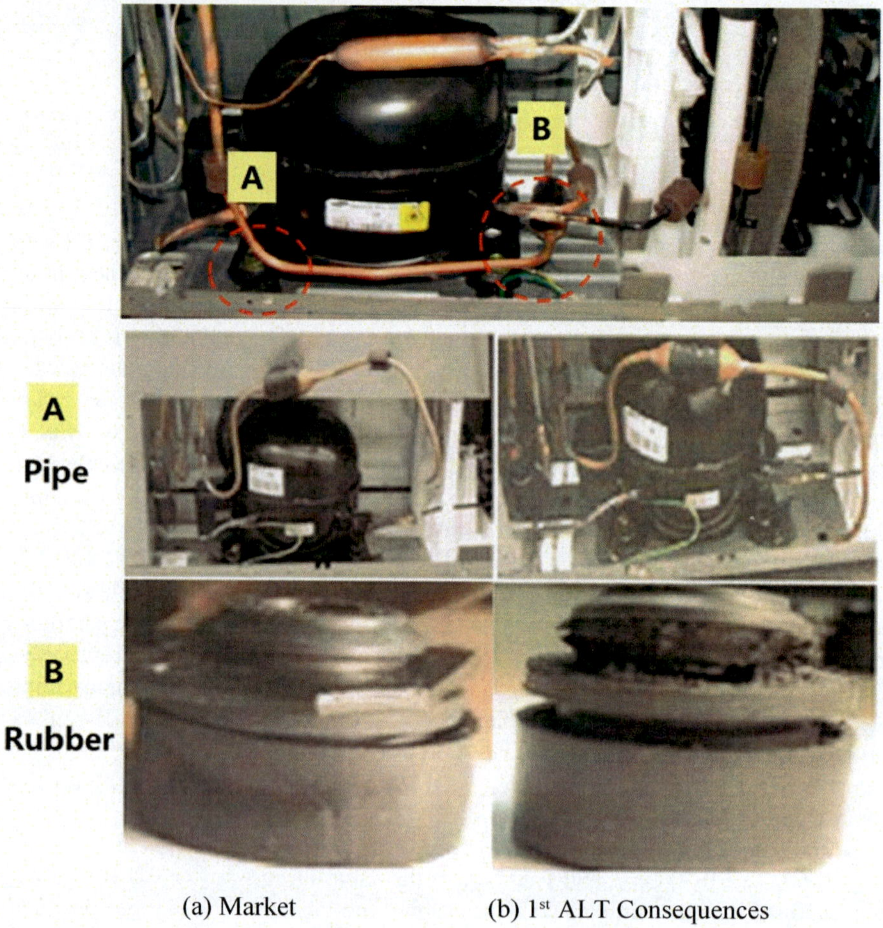

Fig. 8.14 Unsuccessful refrigerator tubes and rubber from the marketplace and 1st ALT outcome

Figure 8.15 depicts a graphical examination of the market data and ALT outcomes on a Weibull chart. The shape parameters of the ALT (β_1) and field data (β_2) were alike. That is, under the close repeated stresses near resonance, we knew that the failure patterns displayed in the first ALT and market were close to the failure patterns in the refrigerator from the field. Therefore, the shape parameter, β, was affirmed to be 6.13. Based on both test results and the Weibull chart, the parametric ALT was well-founded because it pinpointed the design fragilities which were answerable for the unsuccessfulness from the market.

The rubber tearing in the first ALT happened because there was no support for the rubber mount to withstand the horizontal vibration (left ↔ right). Due to the design imperfections—such as no rubber support in the high-stress regions—the repetitive random loading may have produced rubber tearing and broken connection pipes.

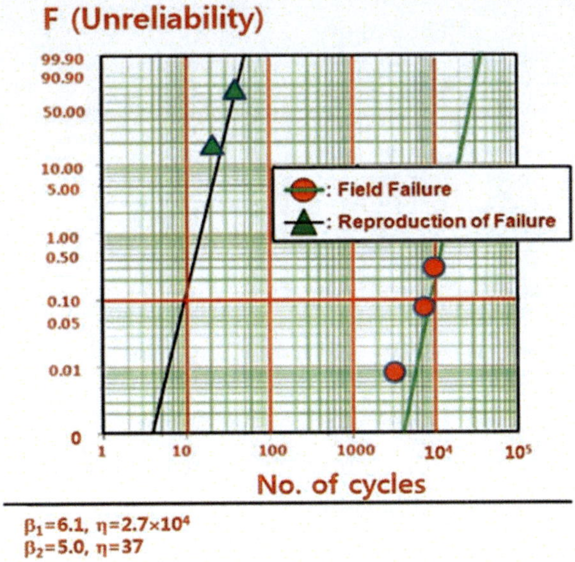

Fig. 8.15 Market statistics and outcomes of ALT on the Weibull plot

As action plans, the refrigerator was redesigned: (1) a reshaped compressor rubber mount shape, C1 (Fig. 8.16a); and (2) a reshaped joined tube shape, C2 (Fig. 8.16b). Second ALTs were then conducted using these modified designs.

The quantities of AF and β in Table 8.2 and Fig. 8.16 were affirmed to be 452.0 and 6.41, respectively. Based on the test data, because the lifetime objectives of newly designed samples were less than a B1 life for the whole travel distance (7 days), the recomputed test time in Eq. (20) for the three sample refrigerators was 40 min, which would be the specification of the parametric ALT. During the second ALT, the altered designs were successful in protecting the refrigerator from the random induced vibrations. As a result, the refrigerators were not fractured until 60 min. When a refrigerator reached an acceleration of 1 G_{rms} on a shaking table, we found that the natural frequency of the horizontal vibration (left ↔ right) moved from 5 to 8 Hz due to the increase in damping in the system (Fig. 8.17).

Through two rounds of ALTs, to endure fatigue damage by random vibration, the newly designed refrigerator was assured to be a B1 life for the whole travel distance of 7200 km (7 days) (Table 8.3).

8.3 Hinge Kit System (HKS) in a Kimchi Refrigerator

A refrigerator is devised to provide (chilled) air from the evaporator to the freezer (or refrigerator) compartment to store fresh food. To shut the door in a refrigerator comfortably, an HKS with a spring-damper mechanism was devised. Before releasing the newly designed HKS, it was required to discover possible design defects and make

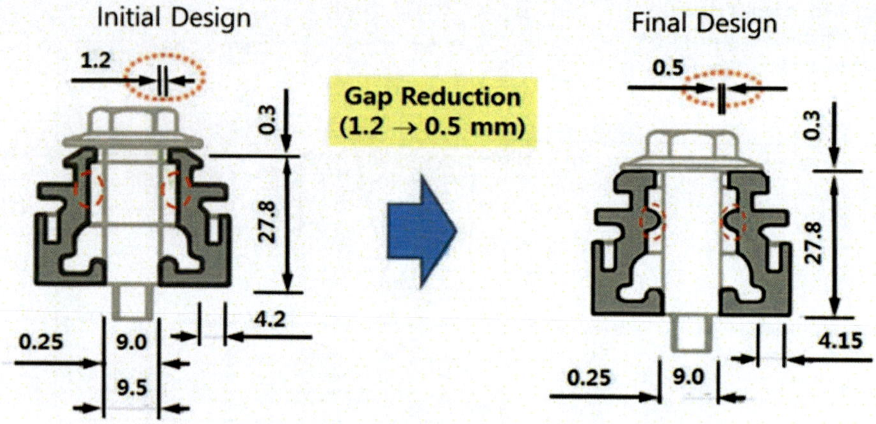

(a) Problematic rubber shape in refrigerator

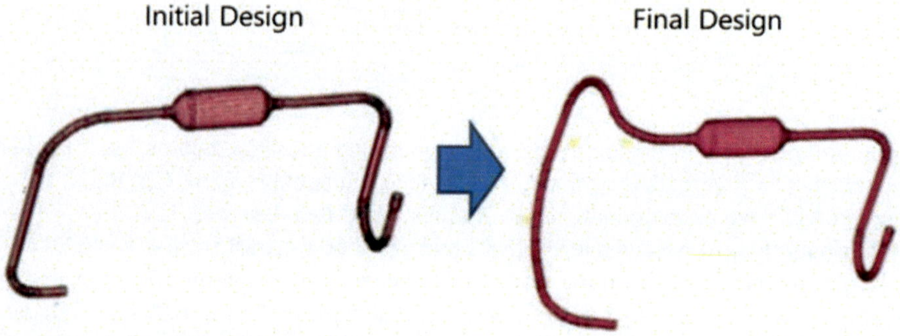

(b) Problematic connecting tube shape in refrigerator

Fig. 8.16 Design modifications

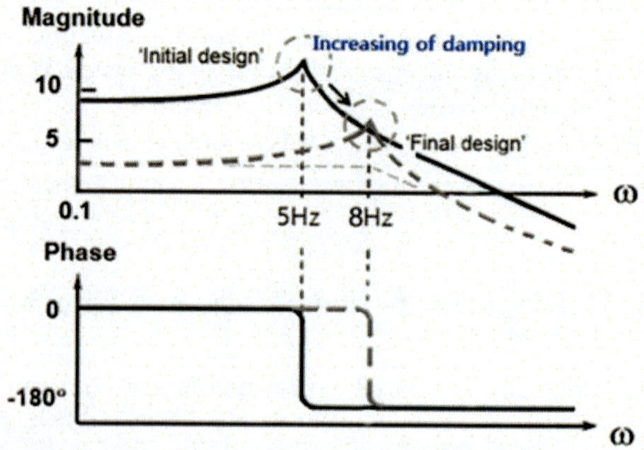

Fig. 8.17 Change in the natural frequency of horizontal vibration due to the modifications of product design

Table 8.3 Results of ALT

Parametric ALT	1st ALT	2nd ALT
	Initial design	Final design
In 40 min, there is no fracture of the connecting tube in refrigerator	20 min: 1/3 fracture 40 min: 2/3 fracture	40 min: 3/3 OK 60 min: 3/3 OK
Machine room in refrigerator	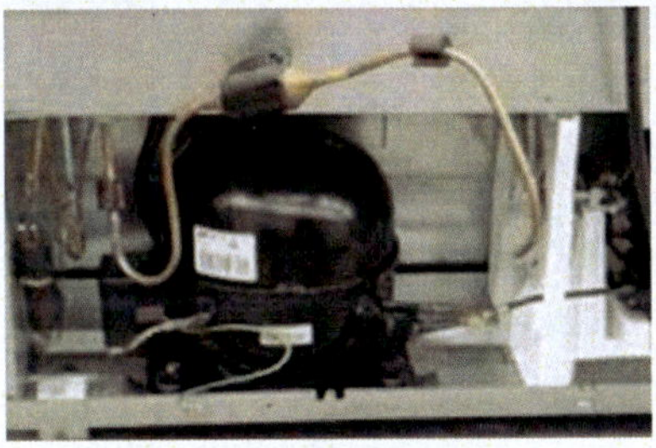	
Specification	C1: Shape of the compressor rubber C2: Connecting tube shape	

sure its reliability. The main parts in the HKS was made up to a kit cover, shaft, cam, spring, and an oil damper, as depicted in Fig. 8.18.

In the market, the HKS components in a refrigerator were cracked and fractured due to design failures. Engineers did not realize which design tests were required to reproduce the real consumer usage and load circumstances. To secure the product working for its anticipated life, a manufacturer might be designed to have a mechanical system robustly and optimally. If there are design faults when the product is subjected to repetitive loads, it shall be unsuccessful before its wanted life. Therefore, an HKS product's real life relies on the problematic components. To recognize and alter them in the design, a design engineer necessitates a reliable methodology.

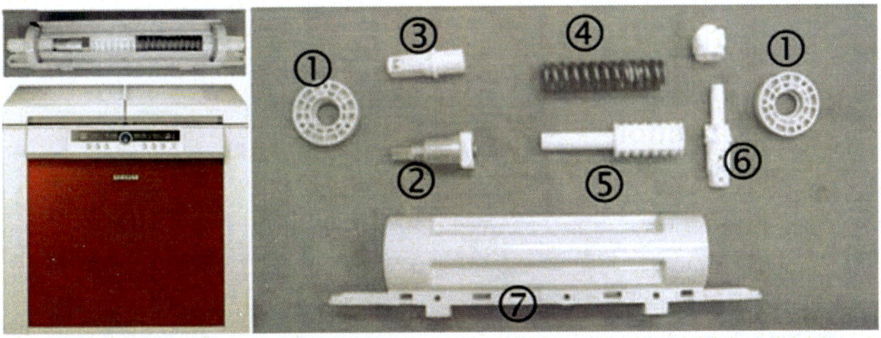

(a) Kimchi Refrigerator

(b) Mechanical components of the hinge kit system: kit cover ①, oil damper ②, spring ④, shaft ⑥, and oil damper ⑦

Fig. 8.18 Refrigerator and mechanical components of the HKS

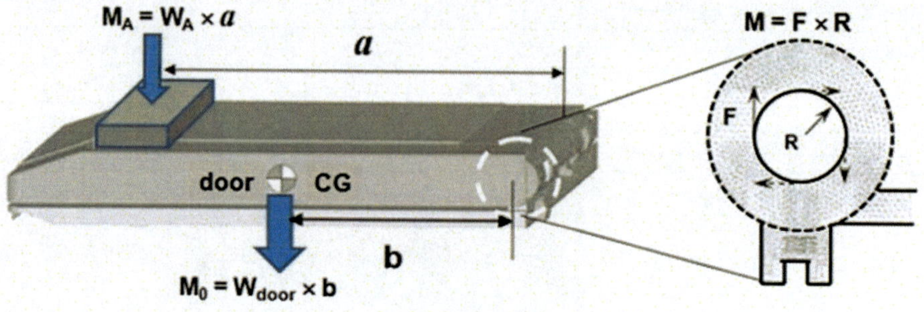

Fig. 8.19 Design idea of the HKS

It covers as follows: (1) a failure (or load) examination for the returned product, (2) employing ALTs with design alternations, and (3) demonstrating if the reliability objective of designs is achieved.

Based on the unidentified customer use circumstances in the market, the HKS was subjected to dissimilar loads during the functioning of the refrigerator door (Fig. 8.19).

The moment balance around the HKS shall be expressed as

$$M_0 = W_{\text{door}} \times b = T_0 = F_0 \times R \tag{8.34}$$

As the elevated weight on the refrigerator door was attached, the moment balance around HKS shall be altered as

$$\sum M = M_1 = M_0 + M_A = W_{\text{door}} \times b + W_A \times a = T_1 = F_1 \times R \tag{8.35}$$

Under the same environmental circumstances, the LS model in Eq. (7.19) shall be altered as follows:

$$TF = A(S)^{-n} = AT^{-\lambda} = A(F \times R)^{-\lambda} = B(F)^{-\lambda} \tag{8.36}$$

where A and B are constants.

We recognize that the product life depends on the exerted impact force. Thus, the AF from Eq. (7.21) shall be expressed as

$$AF = \left(\frac{S_1}{S_0}\right)^n = \left(\frac{T_1}{T_0}\right)^\lambda = \left(\frac{F_1 \times R}{F_0 \times R}\right)^\lambda = \left(\frac{F_1}{F_0}\right)^\lambda \tag{8.37}$$

The environmental operating circumstances of an HKS in a refrigerator changed from roughly 0–43 °C with a relative humidity varying from 0 to 95%. The HKS shall be subjected to between 0.2 and 0.24 g of acceleration. The number of door closing cycles is affected by particular customer use patterns. Customer statistics manifested that the door system of the refrigerator was usually opened and closed between three and ten times per day in the Korean domestic marketplace. With a life cycle design criterion of 10 years, the life of the HKS L_B^* happened approximately 36,500 use cycles for the worst occasion.

For this worst occasion, the impact force around the HKS was 1.10 kN, which was the anticipated greatest force exerted by the usual customer. For the ALT with an elevated weight, the impact force on the HKS was 2.76 kN. Utilizing a cumulative exponent, λ, of 2.0, the AF was discovered to be roughly 6.3 in Eq. (8.37).

For the 10-year life with a cumulated failure rate of 1%, the test cycles for six samples computed in Eq. (7.29) were 24,000 cycles if the shape parameter was assumed to be 2.0. The parametric ALT was designed to assure a 10-year life with a cumulated failure rate of 1% in roughly a 60% level of confidence that it might be unsuccessful less than once during 23,000 cycles. Figure 8.20 manifests the test arrangement of the ALT with labeled equipment for the reliability design of the HKS. Repeated stress shall be defined as the duty effect due to the on/off cycles, and HKS reduces part lifetime.

In the first ALT, the housing of the HKS fractured at 3000 and 15,000 cycles. Figure 8.21 manifests a photo comparing the unsuccessful product from the market and that from the first ALT. Figure 8.22 presents the graphical investigation of the ALT outcomes and market data on a Weibull chart. The shape parameter in the first ALT was approximated to be 2.0. For the final design, the shape parameter from the Weibull chart was affirmed to be 2.0. As seen in the market and the 1st ALT, they were very alike. By ALT, we reproduced the problematic housing structure of the HKS.

If there are design imperfections in the system where the loads are exerted, the HKS failure might fracture in its life. Thus, to have enough strength against repeated

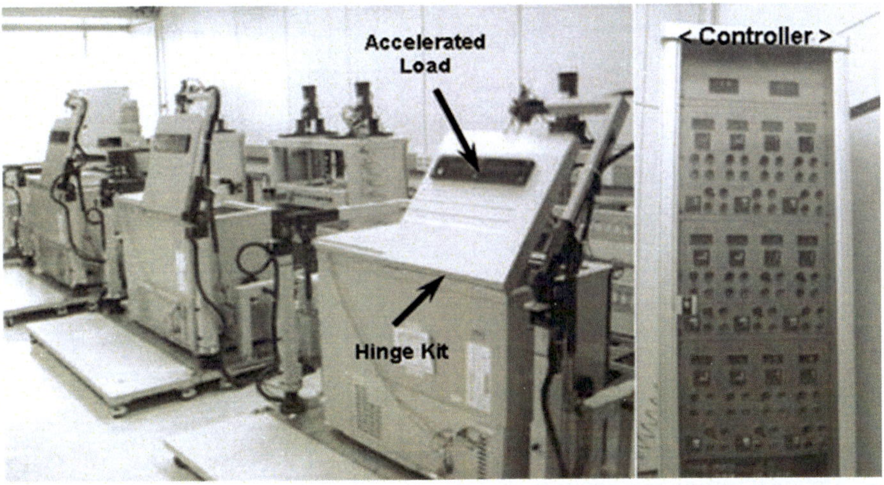

(a) ALT Apparatus (b) Controller

Fig. 8.20 Apparatus employed in ALT and controller

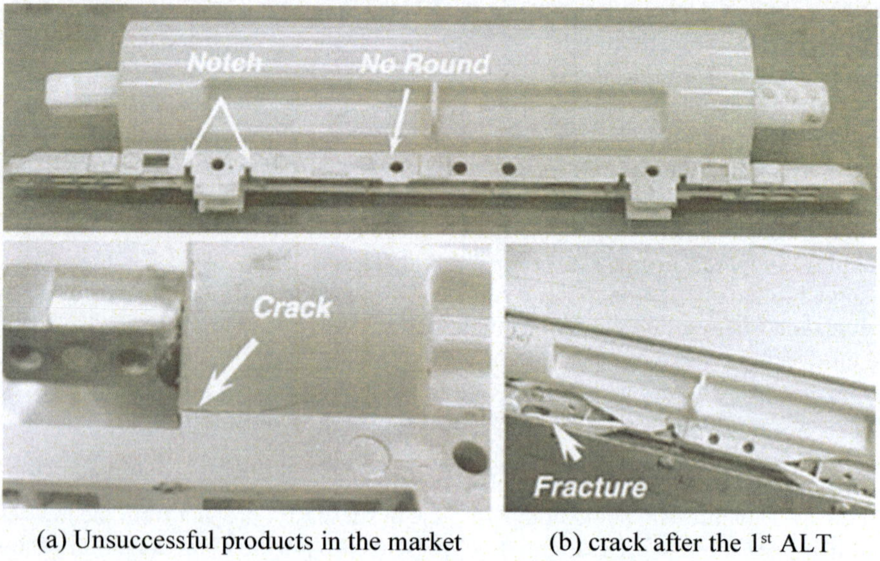

Fig. 8.21 Unsuccessful products in the market and crack after the 1st ALT

Fig. 8.22 Field data and consequences of ALT on the Weibull chart

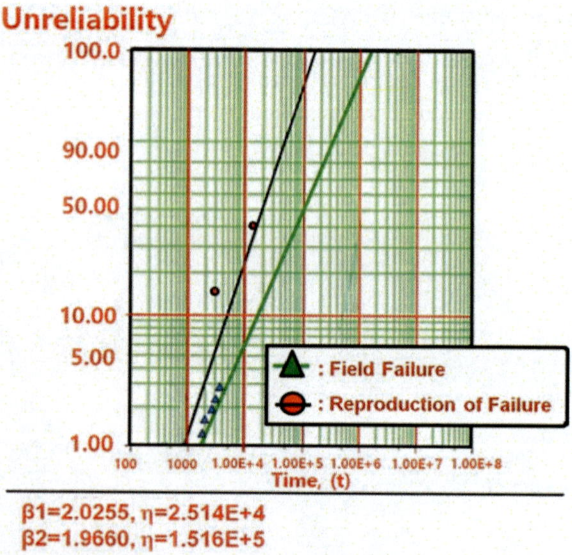

loads, the breakable structure of the hinge kit housing system was altered. The notch was removed, and the design was rounded outside and inside. Enforced ribs were also attached to the housing and decks (Fig. 8.23).

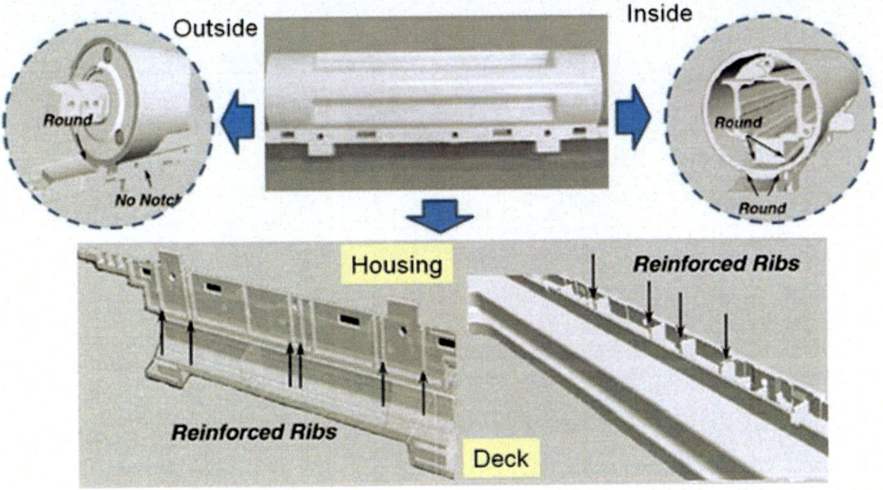

Fig. 8.23 Redesigned HKS housing structure

The maximum concentrated stresses of the housing hinge kit were roughly 21.2 MPa when finite element analysis was carried out. The high-stress risers came from the design defects such as sharp corners/angles, housing notches, and poorly reinforced ribs. The modifications were to apply fillets, attach the reinforced ribs, and eliminate the notching on the housing of the HKS. Implementing the modified designs and analyzing them by finite element analysis, we found that the stress concentrations in the housing of the HKS lessened from 21.2 to 18.9 MPa.

As disassembling the unsuccessful HKS samples, the damper oil in the HKS fabrication leaked at 15,000 cycles (Fig. 8.24). The basic cause for this failure originated from the oil damper sealing design which had added an O-ring, Teflon, and an O-ring with a gap of 0.5 mm. It was decided that there might be intervention between the O-ring and Teflon. To have the O-ring firmly clasped by the Teflon and have sufficient strength against impact, the sealing design was altered, as manifested in Fig. 8.25.

In the 2nd ALT, the HKS cover fractured at 8000, 9000, and 14,000 cycles (Fig. 8.26). The basic cause originated from the selection of the matter for the HKS cover. As working the HKS, the oil damper support made of aluminum die casting was hitting the kit cover made of plastic.

As a consequence, the support began to crack and propagate it to the end. The HKS failure for the second ALTs originated from the kind of matters utilized in the structure. As a corrective action, to have enough matter strength for its own loading, the matter of the HKS cover was altered from plastic to Al die casting (Fig. 8.27). By parametric ALT, the problematic cover of the HKS was altered.

To withstand the repetitive impact loads, the troublesome HKS system in the market was altered as follows: (1) the housing design of HKS was enforced, C1 (Fig. 8.23); (2) the sealing design in the oil damper was altered, C2 (Fig. 8.25); and (3) the kit cover matter, C3, was altered from plastic to aluminum die casting

Fig. 8.24 Leaked oil damper in the 1st ALT

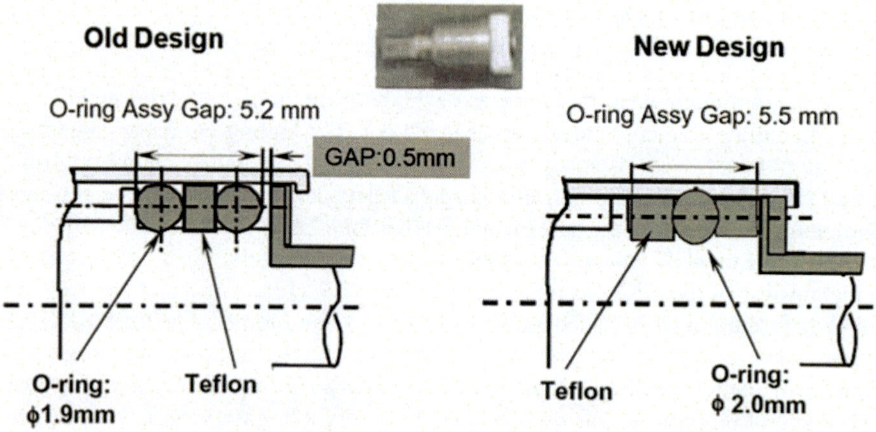

Fig. 8.25 Redesigned oil damper

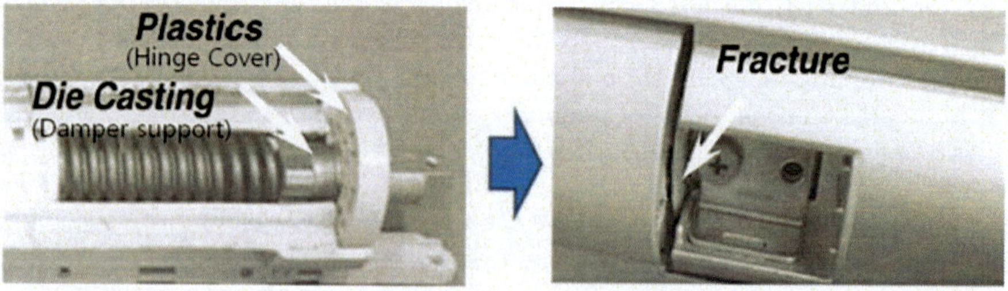

Fig. 8.26 Design of troublesome products at the 2nd ALT

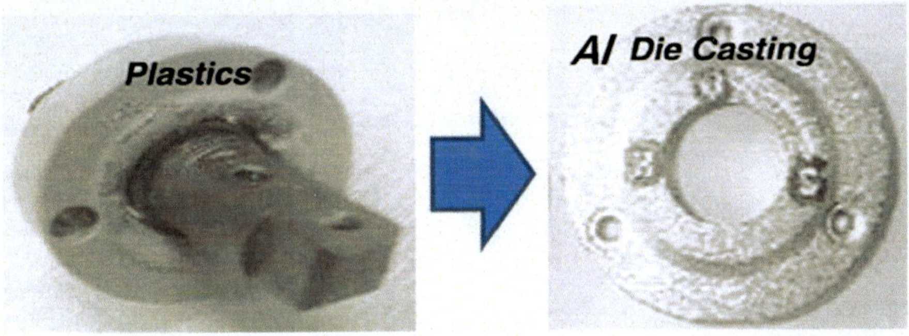

(a) Old design (b) New design

Fig. 8.27 Altered kit cover

(Fig. 8.27). With these design modifications, the refrigerator might be functioning as designed to satisfy the need of the product life because there were no issues till 23,000 mission cycles. Table 8.4 manifests the abridge of the outcomes of the ALTs. Throughout three ALTs, the samples were ensured to be B1 life 10.0 years with a cumulated failure rate of 0.1% per year.

8.4 Freezer Drawer System in a Refrigerator

To keep food freshly, a refrigerator provides chilly air from the evaporator to the refrigerator and freezer compartments through the conventional VCR. As consumers want to have suitable use to stock food, a freezer drawer in a refrigerator is devised to manipulate the necessitated food storage loads under anticipated customer use circumstances over a refrigerator's life. Stocking food in a freezer drawer has the following repeated handling procedures: (1) opening the drawer to stock the food, (2) taking food out of the drawer, and (3) closing the drawer. Figure 8.28 depicts a French door refrigerator with a new freezer drawer system.

French door refrigerators were being returned from the market because the handle of the freezer drawer fractured. As a consequence, customers were necessitated to exchange the refrigerators because they did not work anymore. Analysis of the troublesome refrigerators manifested that the drawer had significant design defects in the structure. Therefore, the freezer drawer had to be altered so that it might withstand repeated loading under consumer working circumstances and enhance its lifetime (Fig. 8.29).

As inspecting the customer use pattern of the freezer drawer, it was subjected to repeated food loads due to the functioning of the drawer. Because the troublesome drawer had serious design defects, engineers had to reproduce the circumstances which generated the failures in the freezer drawer and alter them.

Table 8.4 Results of ALT

	1st ALT	2nd ALT	3rd ALT
	Initial design	Second design	Final design
In 23,000 mission cycles, there is no problem in HKS	3000 cycles: 2/6 fracture 15,000 cycles: 4/6 fracture	7800 cycles: 1/6 fracture 9200 cycles: 3/6 facture 14,000: 1/6 fracture 26,200: 1/6 fracture	23,000 cycles: 6/6 OK 41,000 cycles: 6/6 OK
	HKS structure		
Photo	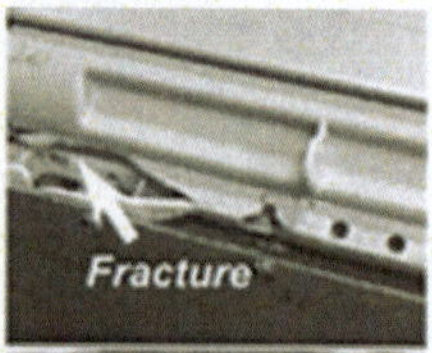 	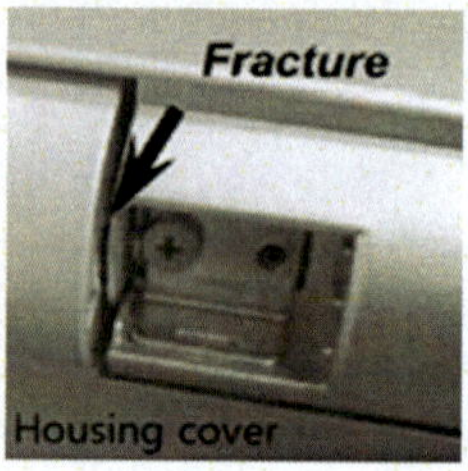	
Action plans	C1: Reinforced housing of HKS C2: Modified sealing structure of oil damper	C3: Kit cover material (Plastic →Al die casting)	

As seen in Fig. 8.30, the force balance in the free-body diagram of the freezer drawer shall be defined as follows:

$$F_{\text{draw}} = \mu W_{\text{load}} \tag{8.38}$$

Because the exerted stress of the freezer drawer system depended on the force proportional to the food weights, the time to failure from Eq. (7.19) shall be expressed as follows:

$$TF = A(S)^{-n} = A(F_{\text{draw}})^{-\lambda} = A(\mu W_{\text{load}})^{-\lambda} \tag{8.39}$$

The AF in Eq. (7.21) shall be represented,

$$AF = \left(\frac{S_1}{S_0}\right)^n = \left(\frac{F_1}{F_0}\right)^\lambda = \left(\frac{\mu W_1}{\mu W_0}\right)^\lambda = \left(\frac{W_1}{W_0}\right)^\lambda \tag{8.40}$$

(a) French door refrigerant

(b) Mechanical components of the drawer: 1) handle, 2) drawer, 3) slide rail, and 4) pocket box

Fig. 8.28 French door refrigerated and freezer drawer assembly

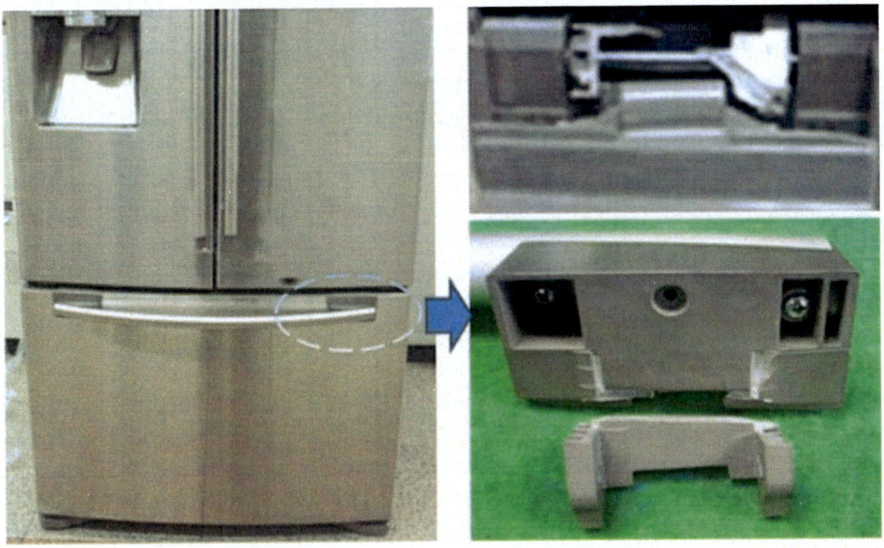

Fig. 8.29 A damaged product after use

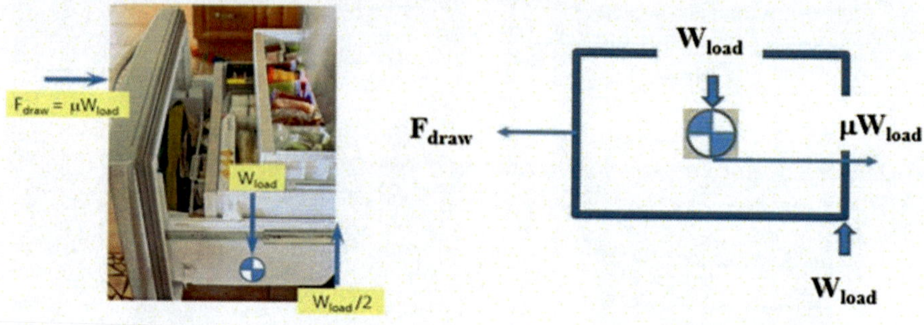

(a) Freezer drawer (b) Free body diagram

Fig. 8.30 Functional design concept of the freezer drawer system

For the freezer drawer system in the French door refrigerator, the usual working circumstances for a customer span from 0 to 43 °C with a relative humidity varying from 0 to 95%. The design circumstances for transportation or operation presumed that the freezer drawer was subjected to 0.2–0.24 g of acceleration. In the United States, the functioning cycles of the freezer drawer relied on the customer use profile. Field data manifested that customers employ the drawer system of a French door refrigerator between five and nine times per day. With a design life cycle of 10 years, the freezer drawer happened approximately 36,500 usage cycles.

Presuming the worst-case circumstance for the food weight in the drawer, the necessitated force on the handle of the freezer drawer was 0.34 kN (35 kg_f). The exerted food weight force for the ALT was 0.68 kN (70 kg_f). The load proportion was 2 (= 0.68 kN/0.34 kN). Utilizing Eq. (8.40) with a quotient λ of 2, the whole *AF* was roughly 4.0.

For the life objective of a B1 life of 10 years, the mission cycles for three sample units computed from Eq. (7.29) were 67,000 cycles if the shape parameter β was assumed to be 2.0. This parametric ALT was designed to secure a B1 life of 10 years with approximately a 60% level of confidence that it might be unsuccessful less than once during 67,000 cycles. Figure 8.31 manifests the experimental structure of the ALT with tagged apparatus for the robust design of the freezer drawer. Under room temperature and humidity conditions, the ALT as a fatigue test was carried out.

French door refrigerators returned from the market had a main failure mode with a fractured handle due to repeated food loads in the process of functioning the freezer drawer. Market data designated that the troublesome products might have had a design defect. Due to this problematic design, the repeated pressure loads might produce excessive stresses on the drawer handle, bringing it to fracture and producing a failure of the drawer. To reproduce the failure mode of the drawer, parametric ALT was performed.

In the first ALT, the handle of the drawer fractured at 7000 and 8000 cycles. Figure 8.32 manifests a photo comparing the unsuccessful product from the market and the first ALT. The unsuccessful form of the 1st ALT was very alike to those

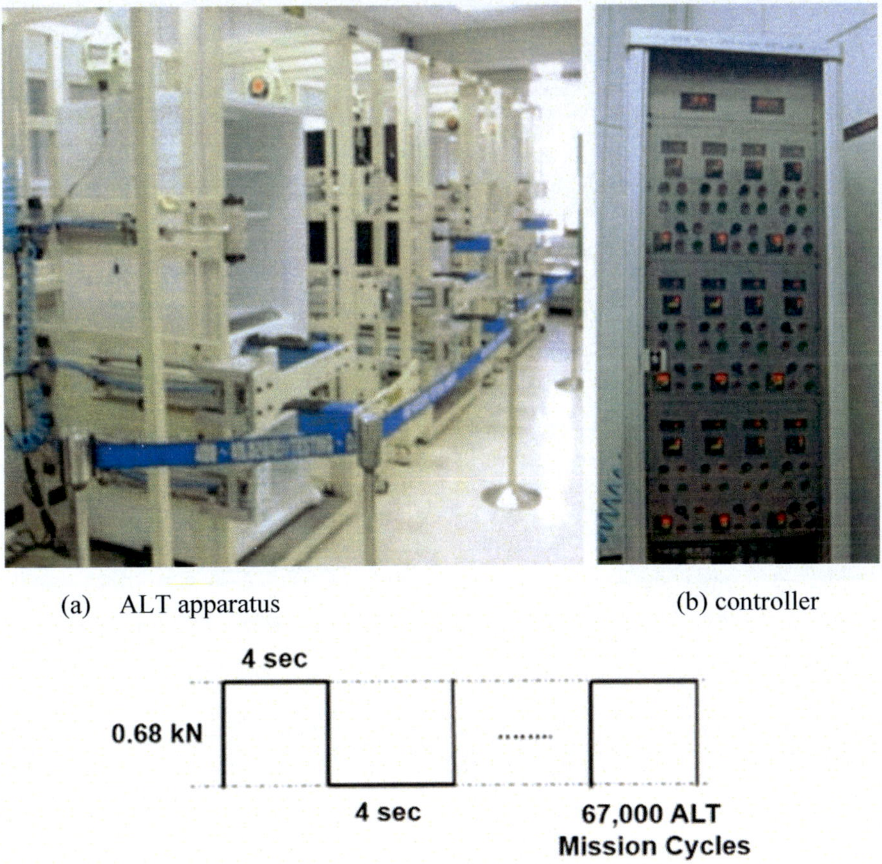

(a) ALT apparatus (b) controller

(c) Duty cycles of repeated food weight on the drawer

Fig. 8.31 ALT apparatus and duty cycles

from the market. The test consequences affirmed that the freezer drawer was not well designed for the functioning of its door. By parametric ALT, the problematic handle structure of the freezer drawer was identified and altered.

The basic causes of the fractured drawer handle originated from the inadequate added region of the handle to the drawer. This design defect might bring the drawer handle to snap off immediately when subjected to repeated food weights. To stop the drawer handle from fracturing due to the repeated usages, the handle was altered as follows: (1) enhancing the width of the enforced handle, C1, Width 1 from 90 to 122 mm and (2) enhancing the handle hooker size, C2, Width 2 from 8 to 19 mm (Fig. 8.33).

In the second ALT, the slide rails in the drawer failed at 15,000 and 16,000 cycles (Fig. 8.34). As the food weight for the ALT was exerted repeatedly, the tests uncovered

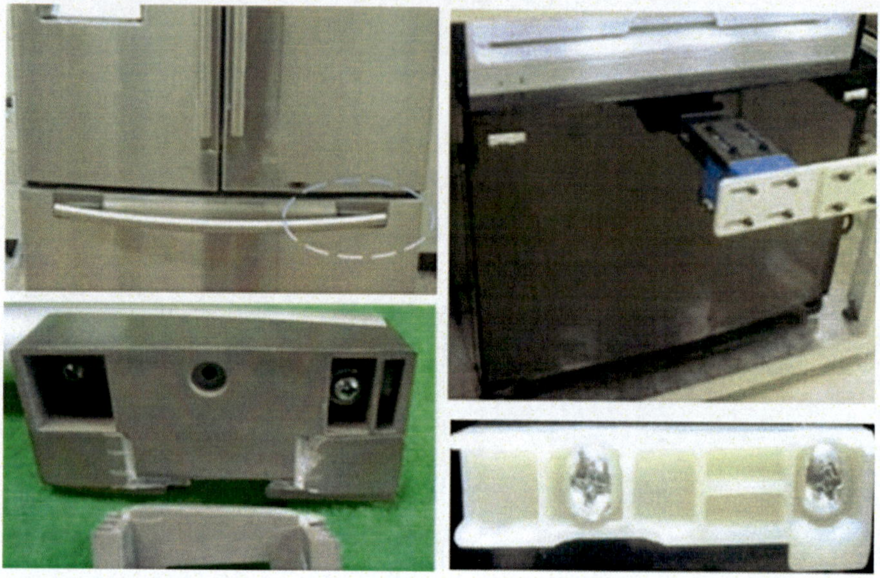

(a) Field (b) The 1st ALT consequences

Fig. 8.32 Failure of freezer drawer handles in the market and the 1st ALT result

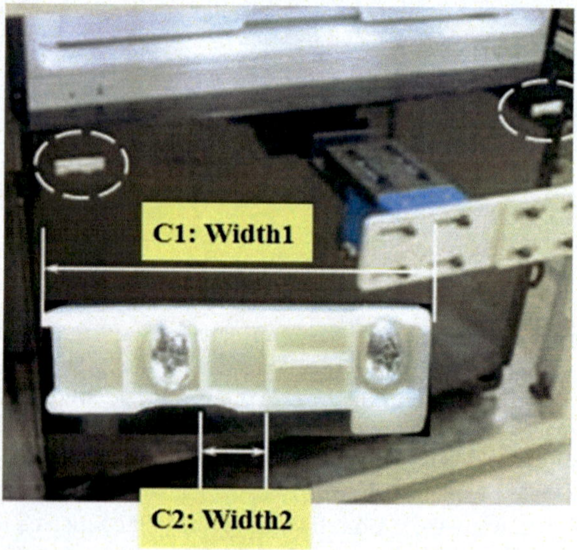

Fig. 8.33 Altered freezer drawer handle

the slide rails as a breakable part. The basic cause of this failure started from the form of the corner of the slide rails. As a consequence, the rail began to crack and fractured in its end. Corrective actions on the slide rail covered (1) enhancing the rail fastening screw number, C3, from 1 to 2; (2) attaching an inner chamber and plastic material, C4, from high impact polystyrene (HIPS) to acrylonitrile butadiene styrene (ABS); (3) thickening the boss, C5, from 2.0 to 3.0 mm; and (4) attaching a new support rib, C6. By exerting a parametric ALT, the problematic slide rail of the freezer drawer was altered (Fig. 8.35).

In the third ALT, there were no design issues with the freezer drawer system till the test was performed for 67,000 cycles. It was understood that the alternations to the design pinpointed from the 1st and 2nd ALT were effective in enhancing the life of the drawer system. Table 8.3 abridges the parametric ALT consequences. With the

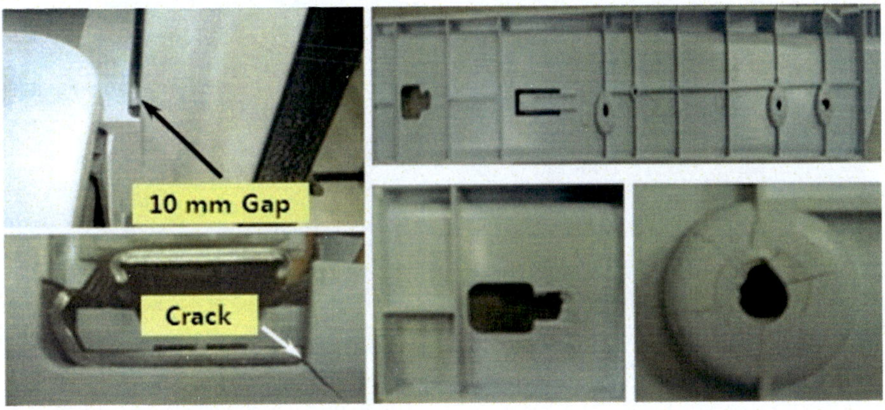

(a) Fractured slide rail in Freezer drawer (b) Detailed problem on the back of rail

Fig. 8.34 Unsuccessful slide rails in the 2nd ALT

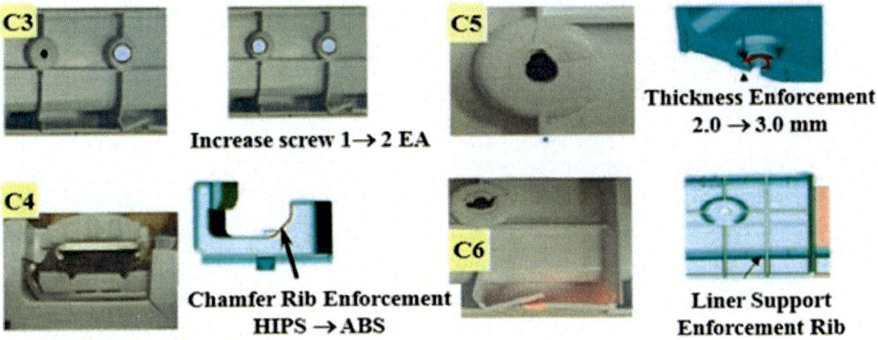

Fig. 8.35 Redesigned slide rail

Table 8.5 Outcomes of ALT

Parametric ALT	1st ALT	2nd ALT	3rd ALT
	Initial design	Second design	Final design
In 32,000 cycles, there is no problem in drawer	7500 cycles: 2/6 fracture 12,000 cycles: 1/3 OK	16,000 cycles: 2/3 fracture (rail system)	32,000 cycles: 3/3 OK 67,000 cycles: 3/3 OK
Photo	<Handle> 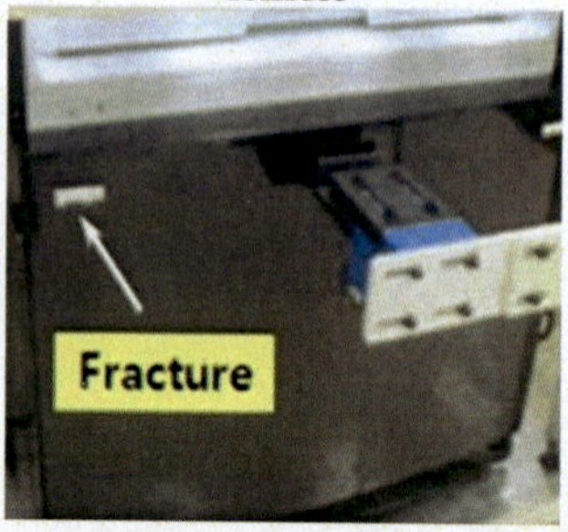 	<Rail> 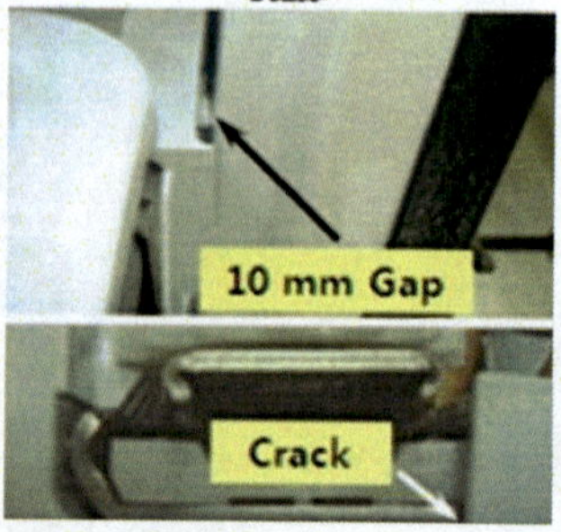 	
Action plans	C1: Width1: 90 mm → 122 mm C2: Width2: L8 → L19.0	C3: Screw number: 1.0 → 2.0 C4: Corner chamfer and material (HIPS →ABS) C5: Thickening the boss: 2.0 → 3.0 mm C6: Adding a new support rib	

altered designs, the final freezer drawer system was ensured to reach its life objective of a B1 life of 10 years (Table 8.5).

8.5 Compressor Suction Reed Valve

Working with unidentified consumer use circumstances, the suction reed valves in domestic compressors utilized in the market were in vain. The fractured suction reed valve brought the domestic compressor to lock and stop working. As the main aim of the refrigerator, involving the compressor, was lost, customers would request to have the product changed. To deal with the issue, it was critical to reproduce the failure mode(s) of the compressor in simulated circumstances in a space or building equipped for engineering tests. Apparently, the problematic compressors which originated from the market had two obvious design defects: (1) the suction reed valve had an amount of overlap with the valve plate, and (2) the valve plate had a sharp edge. When the suction reed valve repeatedly impacted the valve plate, it might be unsuccessful before its anticipated life (Fig. 8.36).

To cool the stocked foods in a refrigerator, the refrigerator provides chilled air from the evaporator heat exchanger to the freezer (or refrigerator) sections. The VCR cycle in a refrigerator includes a compressor, condenser, capillary tube, and evaporator. Electrical energy in a compressor motor is transformed to work energy in the compressor which is utilized to raise the refrigerant pressure. With refrigerant streaming in the system, heat energy absorbed by the evaporator is proceeded to the condenser, where it is discarded to the neighboring air. A capillary tube drops the refrigerant pressure from the high-pressure in the condenser to the low-pressure in the evaporator. In a perfect vapor-compression refrigeration (VCR) cycle, the refrigerant enters the compressor inlet as a saturated vapor and leaves its outlet as the superheat vapor (Processes 1–2) and is cooled down to the saturated liquid state

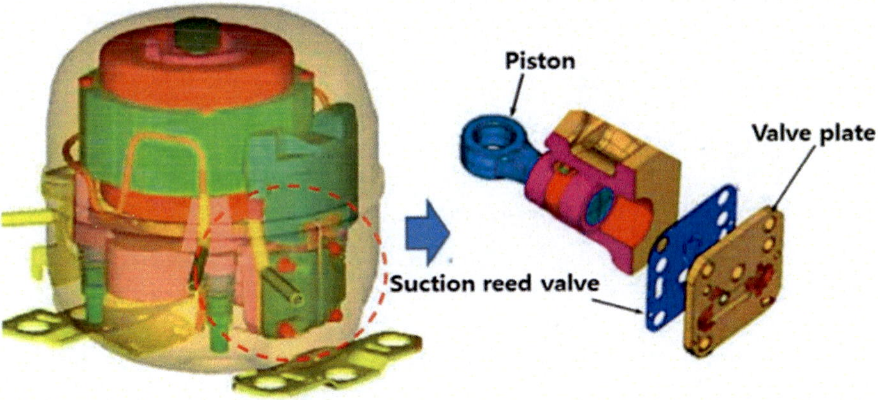

(a) Suction reed valve and valve plate in a domestic compressor.

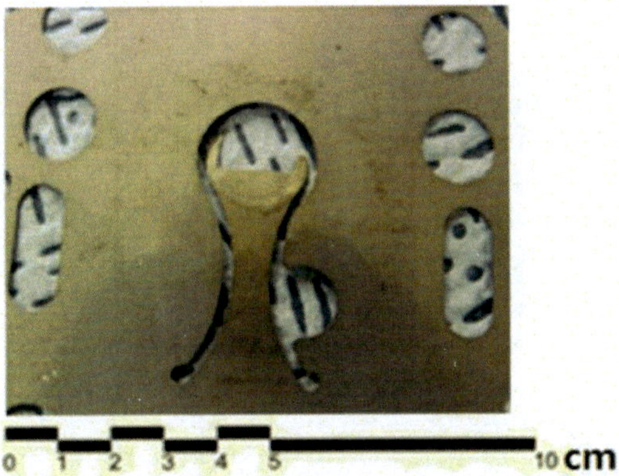

(b) Cracked suction reed valve from the field or ALT

Fig. 8.36 Fractured suction reed valve

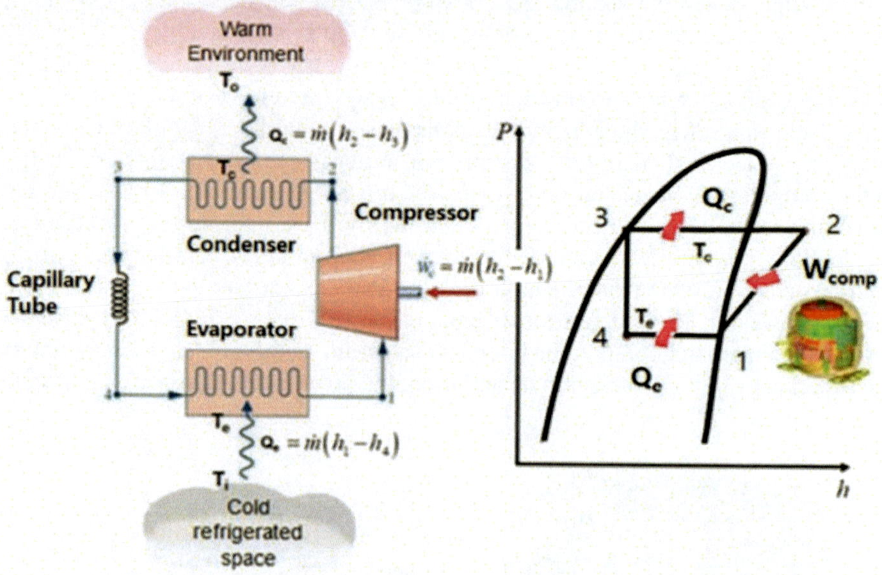

(a) Refrigeration cycle (b) T-s diagram

Fig. 8.37 Functional design idea of the compressor in the VCR cycle

in the condenser (Processes 2–3). It thus drops in pressure as it goes through the expansion device (capillary tube) to the lower pressure evaporator (Processes 3–4). In the evaporator, the refrigerant is vaporized as it soaks up heat from the refrigerated room (Processes 4–1) (Fig. 8.37).

During usual operation, refrigerator compressors are subjected to repeated stresses due to pressure load differences in the compressor. If there is a design defect in a component, such as an inadequacy of strength, when the loads are exerted in the compressor, the component shall suddenly be unsuccessful and not satisfy its anticipated life. By recognizing the failure by an ALT, an engineer shall choose adequate matter and reshape the component utilizing the best or most favorable manner so that the compressor shall withstand repetitive pressure loads and its lifetime shall be enhanced. It is required to examine the pressure loads in the compressor.

In a typical VCR cycle, to evaluate the design, it was necessary to decide both the condensing temperature, T_c, and evaporating temperature, T_e. The mass flow rate of refrigerant in a compressor shall be defined as

$$\dot{m} = PD \times \frac{\eta_v}{v_{\text{suc}}} \tag{8.41}$$

where PD is the piston displacement, η_v is the compressor's volumetric efficiency, and v_{suc} is the specific volume at compressor suction port.

The mass flow rate of a refrigerant in a capillary tube shall be defined as follows:

$$\dot{m}_{\text{cap}} = A\left[\frac{-\int_{P_3}^{P_4} \rho \mathrm{d}P}{\frac{2}{D} f_m \Delta L + \ln\left(\frac{\rho_3}{\rho_4}\right)}\right]^{0.5} \tag{8.42}$$

By balance of mass, the mass flow rate shall be expressed as:

$$\dot{m} = \dot{m}_{\text{cap}} \tag{8.43}$$

The energy balance at the condenser shall be expressed as

$$Q_c = \dot{m}(h_2 - h_3) = (T_c - T_o)/R_c \tag{8.44}$$

The energy balance at the evaporator shall be expressed as

$$Q_e = \dot{m}(h_1 - h_4) = (T_i - T_e)/R_e \tag{8.45}$$

Utilizing Eqs. (8.43) through (8.45), it is feasible to attain the estimates of the mass flow rate $\dot{m}$, evaporator temperature T_e, and condenser temperature T_c. Because the saturation pressure, P_{sat}, is a function of temperature, the evaporator pressure, P_e (or condenser pressure P_c), can be obtained as follows:

$$P_e = f(T_e) \text{ or } P_c = f(T_c) \tag{8.46}$$

Both the condensing pressure, P_c, and evaporating pressure, P_e, are critical when inspecting the load on the compressor. These pressures rely on the environmental circumstances, heat exchanger size, and customer usage circumstances in the design phase.

During usual compressor functioning, it shall be subjected to repetitive stresses due to the pressure differences between the suction and discharge. The internal stress of the compressor depends on the pressure difference between suction pressure between suction pressure, P_{suc}, and discharge pressure, P_{dis}.

$$\Delta P = P_{\text{dis}} - P_{\text{suc}} \cong P_c - P_e \tag{8.47}$$

For a compressor system, the time-to-failure *TF* from Eq. (7.19) shall be defined as

$$TF = A(S)^{-n} \exp\left(\frac{E_a}{kT}\right) = A(\Delta P)^{-\lambda} \exp\left(\frac{E_a}{kT}\right) \tag{8.48}$$

where A is a constant, k is Boltzmann's constant, E_a is the activation energy, T is the absolute temperature, n is the quotient, and λ is the cumulative damage exponent in Palmgren–Miner's rule.

The *AF* from Eq. (7.20) shall be redefined as

$$AF = \left(\frac{S_1}{S_0}\right)^n \left[\frac{E_a}{k}\left(\frac{1}{T_0} - \frac{1}{T_1}\right)\right] = \left(\frac{\Delta P_1}{\Delta P_0}\right)^\lambda \left[\frac{E_a}{k}\left(\frac{1}{T_0} - \frac{1}{T_1}\right)\right] \tag{8.49}$$

where S_1 (or P_1) is the mechanical stress (or pressure difference) under accelerated conditions and S_0 (or P_0) is the mechanical stress (or pressure difference) under representative conditions.

For a reciprocating compressor in a French door refrigerator, the usual functioning circumstances for a consumer span from 0 to 43 °C with a relative humidity varying from 0 to 95%. The design circumstances for transportation or operation presumed that the compressor was subjected to 0.2–0.24 g of acceleration.

As the compressor works, the suction reed valve opens to lead the refrigerant to stream into the compressor. A compressor is anticipated to cycle on and off 22 cycles per day. A worst-case scenario was also identical, with on and off 98 cycles per day. Under the worst occasions, the compressor functioning for 10 years shall happen roughly 357,700 usage cycles.

The compressor is made of carbon, cast, stainless, alloy, etc., steel. The allowable stresses are expressed as a function of the yield stress (F_y) or tensile stress (F_u) of the structural matter. For steel, the ranges of yield strength, F_y, and ultimate or tensile strength, F_u, usually utilized are 248–345 MPa and 400–483 MPa, respectively. R134a is the refrigerant employed in the VCR cycle. It utilizes synthetic refrigeration compressor oils which have a high viscosity index (VI). They have slight viscosity changes in relation to temperature changes. Thus, the gradient of the viscosity of a synthetic lubricant with a high VI is flatter with respect to temperature. The viscosity thus remains stable across a wide temperature usage range.

In the worst occasion, the pressure difference was 1.27 MPa, and the compressor dome temperature was 90 °C. For an ALT, the pressure difference was raised to 2.94 MPa, and the temperature of compressor dome was also elevated to 120 °C, with an accumulative damage exponent, λ, of 2. The total AF computed from Eq. (8.49) was 7.3 (Table 8.6).

The test cycles of the ALTs computed from Eq. (7.29) were 49,000 cycles for 100 samples if the shape parameter, β, was assumed to be 2.0. This ALT was designed to assure a life objective—B1 life of 10 years—with an approximate 60% level of confidence that it might be unsuccessful less than once during 49,000 cycles. The duty cycles of the pressure difference between suction pressure, P_{suc}, and discharge pressure, P_{dis} were applied.

Table 8.6 Compressor ALT circumstances

System conditions		Worst case	ALT	AF
Pressure (MPa)	High-side	1.27	2.94	5.36①
	Low-side	0.0	0.0	
	ΔP	1.27	2.94	
Temperature (°C)	Dome	90	120	1.37②
Total AF (=(① × ②))				7.32

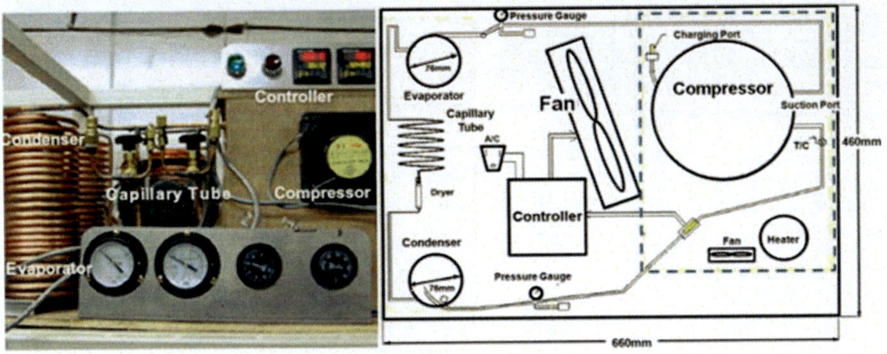

Fig. 8.38 Apparatus used in accelerated life testing

To evaluate the design of the compressor, a simplified refrigeration cycle was formed. It consisted of a compressor, condenser, capillary tube, and evaporator. The temperature in the enveloped fiberglass package is controlled by two 60 W lamps and a fan. A thermal switch added to the cover of the compressor managed a 51 m^3/h axial fan. The test circumstances and their boundaries were put up on the control board. When the test began, the high-side and low-side pressures were monitored on the pressure gauges or apparatus screen (Fig. 8.38).

A sample in the 1st ALT ($n = 100$) locked at 3500 cycles. Another sample remained working but had a partially broken suction reed at 7500 cycles (Fig. 8.39). The shape parameter, β, established from the 1st ALT was 2.0 (see Fig. 8.39). The shapes and locations of the failure in the samples attained from the 1st ALT and the market were similar (Fig. 10). The fractured suction reed valve started from the design defects: (1) it had an amount of overlap on the valve plate, (2) it had a sharp edge on the valve plate, and (3) it utilized an inadequately strong matter.

As the suction reed valve often stroked the valve plate, it finally fractured before its anticipated life. The major failure mode of the compressor was locking due to the fractured/cracked suction reed valve. The structured ALT method was well-found for pinpointing the fatigue failure obtained from the samples in the market. First, the shapes of the unsuccessful suction reed valves from the field and those in the 1st ALT were very alike in shape.

The 1st ALT failure and field failure data showed a similar pattern on a Weibull chart (Fig. 8.40). Because the data for the two prototypes had alike slopes on the chart, each load circumstance of the 1st ALT and that from the market over the product life were alike. Thus, it would be expected that the test samples in the laboratory might break similarly to those in the market. For the shape parameter, β, the final shape parameter from the plot was affirmed to be 2.0, compared with the approximated value—2.0. Based on both test results in the Weibull chart, the ALT was successful because it pinpointed the design flaws that were accountable for the field failures. In other words, as manifested by two factors—the visual likeness in both photos and similar slopes in the Weibull chart—these ALTs were warrant in pinpointing

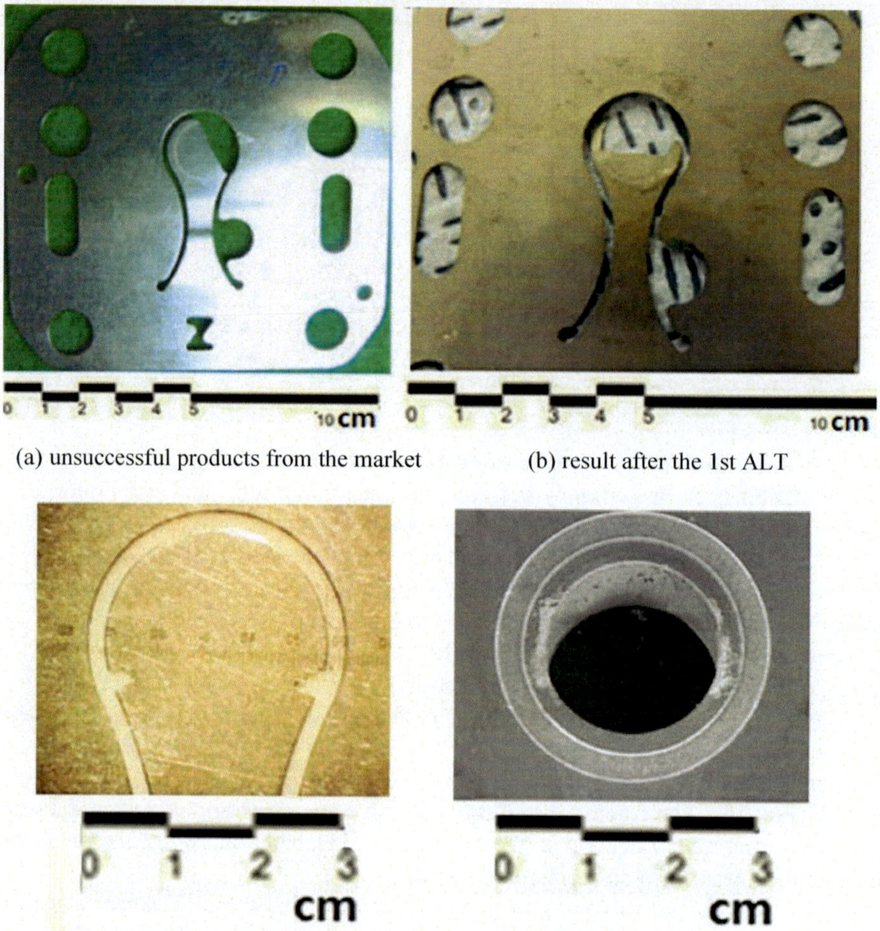

(a) unsuccessful products from the market (b) result after the 1st ALT

(c) Partially smashed suction reed valve (d) Shattered valve plate

Fig. 8.39 Unsuccessful suction reed valves from the market and 1st ALT

the design defects which were concluded as failures from the field. These failures decided the product life.

Refrigerators which originated from the market had a main failure mode without cooling because the compressor did not work. Market data proposed that the problematic compressors may have had design defects. Due to these flaws, the repeated pressure loads might generate unexpected stresses on the suction reed valve, bringing it to crack and propagate it to its end. To reproduce the failure mode of the compressor, ALTs were performed. Based on the 1st ALT and field data, we recognized that the AF and β values were 7.3 and 2.0, respectively (see Fig. 8.40). For designated test samples, the test cycles were calculated in Eq. (7.29) if the product life was assured to be a B1 life 10 years.

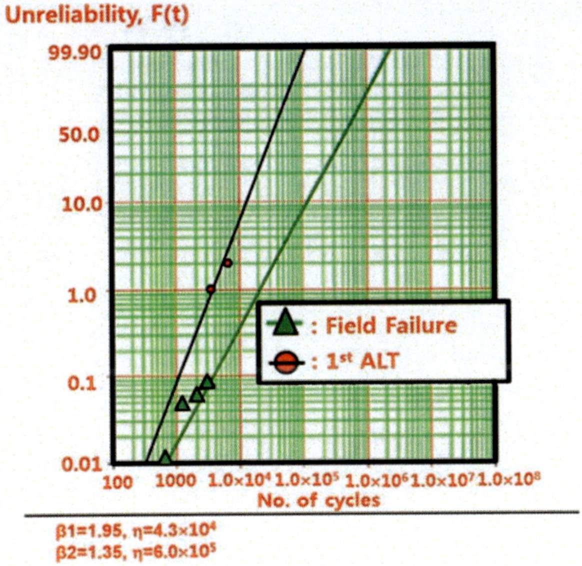

Fig. 8.40 Market data and outcomes of ALT on the Weibull plot

To check a compressor which was unsuccessful at 3500 cycles in the 1st ALT, the problematic compressors from the field and the 1st ALT were contrasted to decide the possible design flaws. The mode of compressor failure in the 1st ALT was very alike to those from the field. The suction reed valves failed in regions where they were incompletely overlapped by the valve plate. The tests explained that the compressor wrongly designed to function with this suction reed valve.

The unsuccessful suction reed valve started from the inappropriate design defects: (1) an amount of overlap with the valve plate; (2) a sharp edge on the valve plate; and (3) inadequately strong matter (0.178 t) used in the design of the suction reed valve. These flaws would bring the compressor system to fracture suddenly when subjected to repeated pressure loads (see Figs. 8.36 and 8.41).

To secure the unsuccessful suction reed valve failure which was due to the repeated pressure stresses in the compressor's life, the valve plate and suction reed valve were altered: (1) trespan size from 0.73 to 1.25 mm; (2) attaching ball peening and brush process; (3) thickening the suction reed valve from 0.178 t to 0.203 t; and (4) adding tumbling process (see Fig. 8.42).

For the 2nd ALT, three samples were locked near 17,000 cycles. The troublesome compressor system originated as follows: (1) abrasion of the crankshaft and (2) intrusion between the crank shaft and thrust washer. The design alternation was supplied to the heat treatment on the outside of the crank shaft which shall alter the physical, and sometimes chemical, properties of a matter to fulfill the wanted outcome, such as hardening of a weak crankshaft material (Ductile Iron FCD450—C ≥ 2.5 wt%, S ≤ 0.02 wt%, Mg ≤ 0.09 wt%).

In the 3rd ALT, there were no design problems in the compressor till the ALT was carried out for 49,000 cycles. We therefore inferred that the design alternations

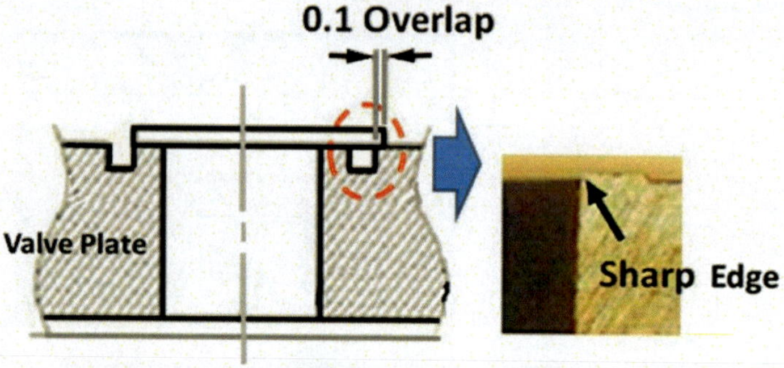

(a) overlapped suction reed valve and valve plate with sharp edge

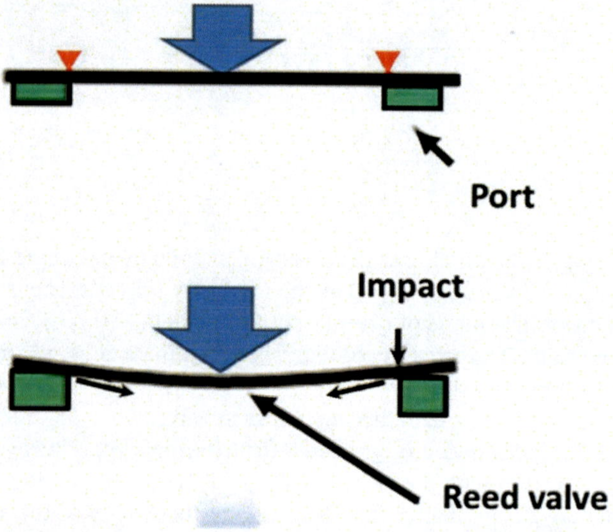

(b) impact load in combination with design flaws when the compressor is functioned.

Fig. 8.41 Design flaws of suction reed and valve plate in a compressor

attained from the 1st and 2nd ALTs were effective. Table 8.7 abridges the ALT results. With the altered designs, the compressor samples were assured to fulfill the life objective—B1 life of 10 years—with an approximately 60% confidence level.

8.6 Failure Analysis and Redesign of the Evaporator Tubing

Figure 8.43 manifests the Kimchi refrigerator with the aluminum cooling evaporator tubing proposed for cost savings. As a customer stocks the food in the refrigerator, the

(a) Trespan in valve plate (b) Ball peening process

Fig. 8.42 Altered valve plate

Table 8.7 Results of ALTs

Parametric ALT	First ALT	Second ALT	Third ALT
	Original design	Design	Final design
In 49,000 cycles, there are no issues in the compressor	3500 cycles: 1/100 locking 7500 cycles: 1/100 broken suction reed valve	17,000 cycles: 3/100 locking	49,000 cycles: 100/100 OK
Structure	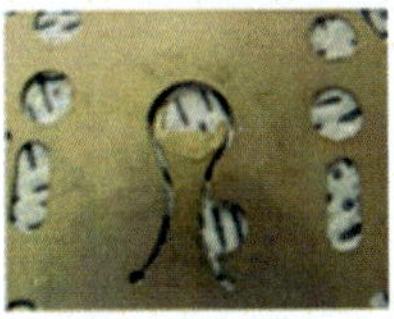	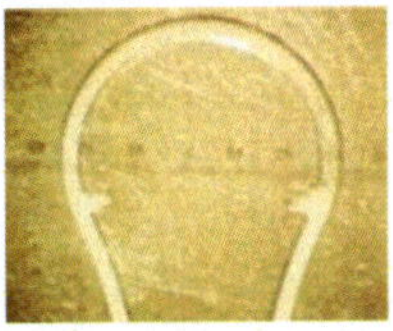	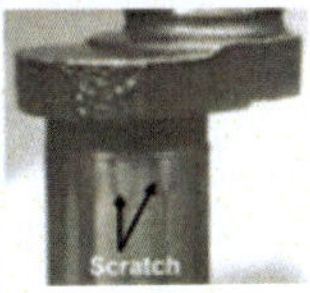
Action plans	C1: Trepan size: 0.73 mm → 1.25 mm C2: attaching ball peening and brush process C3: SANDVIK 20C: 0.178 t → 0.203 t C4: expanding tumbling: 4 h → 14 h	C5: FCD500 + no heat treatment → FCD500 + heat treatment on the crankshaft	

refrigerant streams through the evaporator tubing in the cooling enclosure to continue a constant temperature and conserve the food fresh. To carry out this purpose, the evaporator tube needs to be designed to function under the functioning circumstances it is subjected to by the customers who buy and utilize the Kimchi refrigerator. The evaporator tube fabrication in the cooling enclosure is made up to an inner case (1), evaporator tubing (2), lockring (3), and adhesive tape (4), as depicted in Fig. 8.43b.

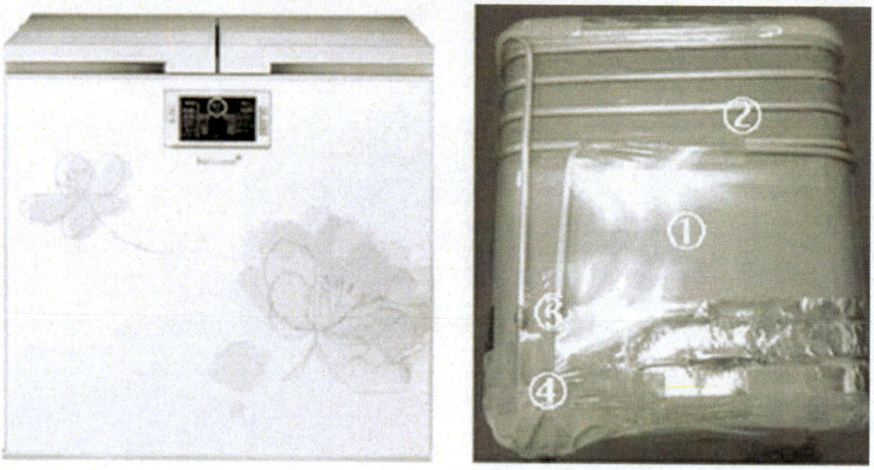

(a) Kimchi Refrigerator

(b) Mechanical components of the cooling closure: Inner Case (1), Evaporator tubing (2), Lokring (3), and Cotton adhesive tape (4)

Fig. 8.43 Kimchi refrigerator (**a**) and cooling evaporator assembly (**b**)

In the market, the evaporator tubing had been pitting, bringing the leakage of the refrigerant in the system and outcome in the losing of cooling in the refrigerator. The statistics on the unsuccessful products in the market were critical for comprehending the use environment of customers and identifying design alternations which necessitated to be made to the product (Fig. 8.44).

Market data designated that the failed products would have had some design defects. The design defects integrated with the repeated loads might bring failure. The pitted outsides of an unsuccessful sample from the market were attributed by SEM and EDX spectroscopy (Fig. 8.45; Table 8.8). We discovered a concentration of chlorine in the pitted outside. As ion liquid chromatography was utilized to calculate the chlorine concentration, the outcome for the tubing that had cotton adhesive tape was 14 ppm. In contrast, the chlorine concentration for tubing that had generic transparent tape was 1.33 ppm. It was speculated that the high chlorine concentration discovered on the outside should have originate from the cotton adhesive tape.

The evaporator tubing fabrication in the Kimchi refrigerator is made up to some mechanical components. Relying on the customer use circumstances, the evaporator tubing encountered repeated thermal duty loads due to the typical on/off cycling of the compressor to fulfill the thermal load in the refrigerator. As the refrigerant temperatures are usually below the dew point temperature of the air, condensation shall occur on the outside of the tubing.

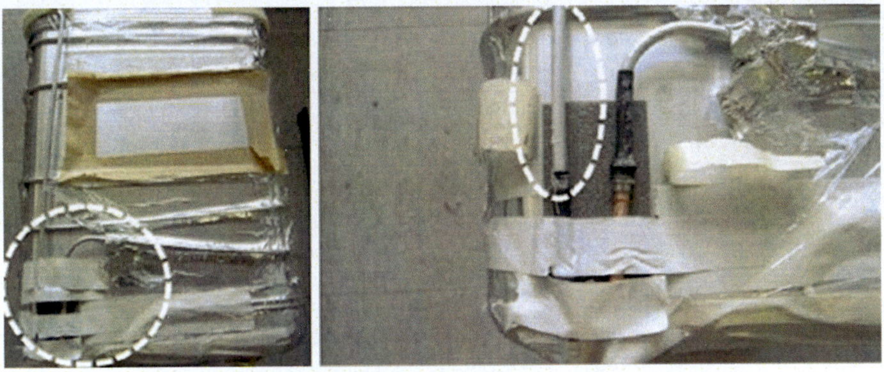

(a) Pitted evaporator tube in the market

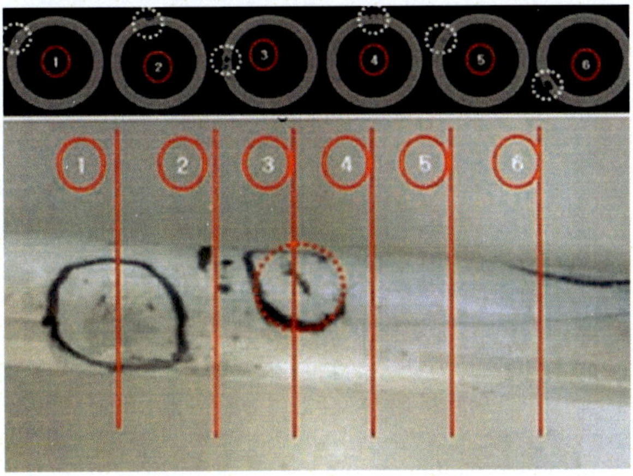

(b) X-ray photograph manifesting pitting corrosion on the evaporator tube

Fig. 8.44 A failed refrigerator after usage

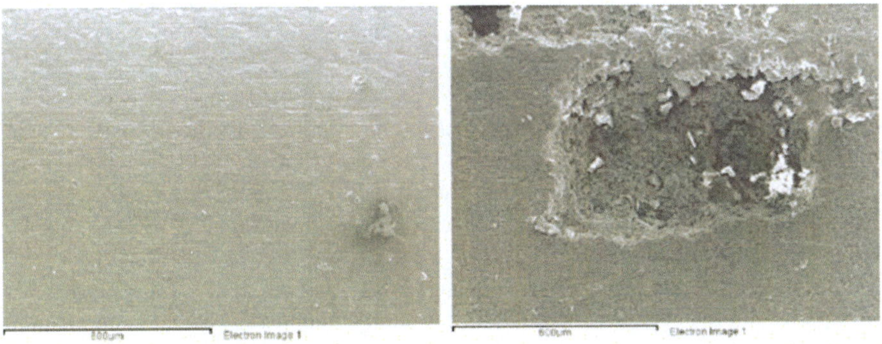

(a) No Pitting (b) Pitting

Fig. 8.45 SEM fractography manifesting pitting corrosion on the evaporator tube

Table 8.8 Chemical composition of the no pitting and pitting surfaces

	No pitting		Pitting	
	Weight	Atomic	Weight	Atomic
O	11.95	19.65	25.92	37.39
Al	97.29	90.74	69.29	59.61
Cl	0.33	0.23	3.69	2.41
Si	0.42	0.39	0.66	0.55
Ca			0.70	0.40
K			0.50	0.30
Na			0.34	0.34
Totals	100.00		100.00	

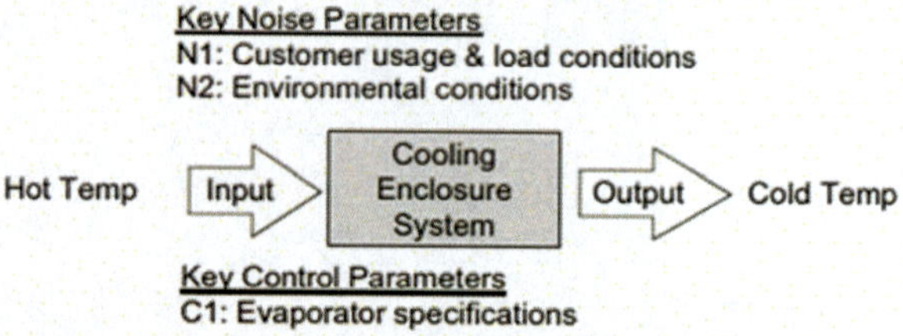

Fig. 8.46 Robust design illustration of a cooling enclosure system

Figure 8.46 manifests a robust design simplified illustration of the cooling evaporator system. Figure 8.47 manifests the mechanism of the crevice (or pitting) corrosion which happens because of the chemical reaction between the aluminum evaporator tubing and the cotton adhesive tape. As a Kimchi refrigerator works, water behaves as an electrolyte and condenses between the aluminum tubing and the cotton adhesive tape. Crevice (or pitting) corrosion shall start.

It shall be abridged as (1) passive film breakdown by Cl^- attack; (2) rapid metal dissolution: $Al \rightarrow Al^{+3} + 3e^-$; (3) electromigration of Cl into the pit; (4) acidification by the hydrolysis reaction: $Al^{+3} + 3H_2O \rightarrow Al(OH)_3 \downarrow + 3H^+$; (5) big cathode: external surface, small anode area: pit; (6) and the big voltage drop (i.e., "IR" drop, according to Ohm's law $V = I \times R$, where R is the equivalent path resistance and I is the average current) between the pit and the outer surface is the driving force for the propagation of pitting.

The number of Kimchi refrigerator working cycles is affected by particular customer use circumstances. In the Korean domestic marketplace, the compressor shall be anticipated to cycle on and off 22–99 times per day to keep the actual temperature inside the refrigerator. Because the corrosion stress of the evaporator tubing relies on the corrosive load (F) which shall be defined as the concentration of chlorine, the LS prototype from Eq. (7.19) shall be altered as

$$TF = A(S)^{-n} = A(F)^{-\lambda} = A(\text{Cl}\%)^{-\lambda} \tag{8.50}$$

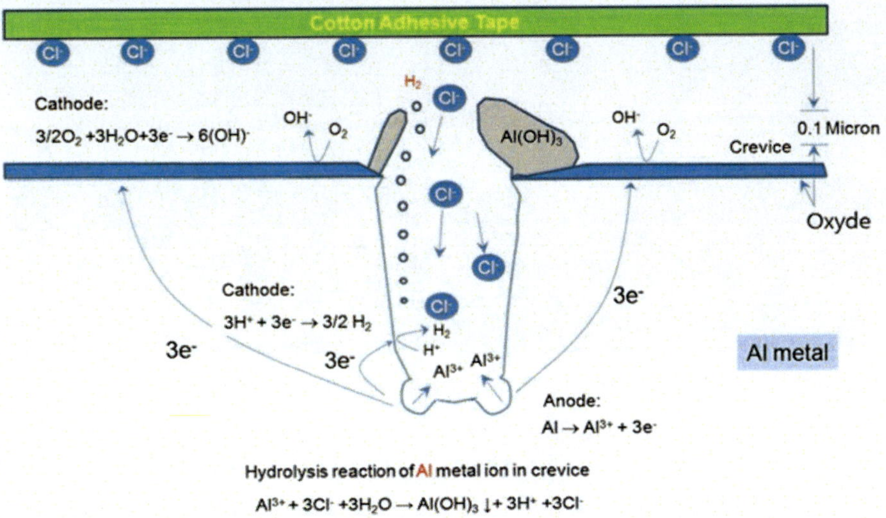

Fig. 8.47 Elevating corrosion in the crevice due to low pH, high Cl^- concentration, depassivation and IR drop

The *AF* from Eq. (7.21) shall be defined as

$$AF = \left(\frac{S_1}{S_0}\right)^n = \left(\frac{F_1}{F_0}\right)^\lambda = \left(\frac{Cl_1\%}{Cl_0\%}\right)^\lambda \tag{8.51}$$

The compressor in a Kimchi refrigerator is anticipated to cycle on 22–99 times per day. With a life cycle design point of 10 years, the Kimchi refrigerator happens 359,000 cycles. The chlorine concentration of the cotton adhesive tape was 14 ppm. To speed up the pitting of the evaporator tubing, the chlorine concentration of the cotton tape was modified to roughly 140 ppm by attaching some salt. Utilizing a stress dependence of 2.0, the *AF* was discovered to be roughly 100 in Eq. (8.51).

For a B1 life of 10 years, the test cycles and test sample numbers with the shape parameter $\beta = 6.41$ computed in Eq. (7.29) were 4700 cycles and 19 pieces, separately. The ALT was designed to assure a B1 10 years of life with approximately a 60% level of confidence that it might be unsuccessful less than once during 4700 cycles. Figure 8.47a manifests the Kimchi refrigerators in the ALT and evaporator tubing in the enclosure containing a 0.2 M NaCl water solution. Figure 8.48b manifests the duty cycles for the corrosive force (*F*) due to the chlorine concentration.

Figure 8.49 manifests the unsuccessful product from the market and from the ALT. In the both pictures, the form and position of the failure in the ALT were alike to those manifested in the market. Figure 8.50 manifests a graphical investigation of the ALT outcomes and market data on a Weibull chart. These structured methods were well founded in recognizing the weak designs responsible for failures in the market and were supported by two discoveries in the data. From the Weibull chart,

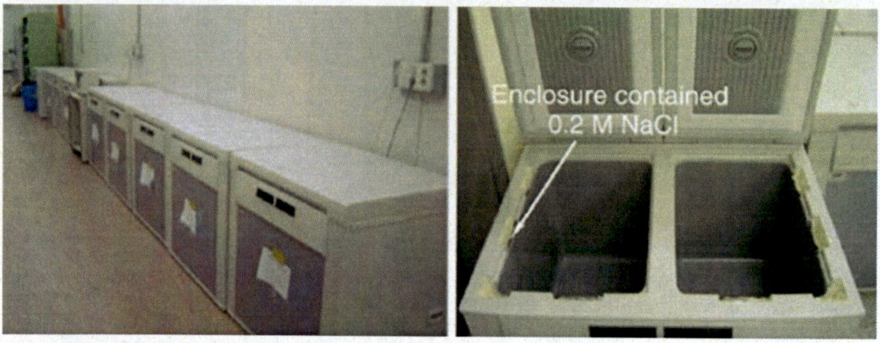

(a) Kimchi refrigerators in testing with 0.2 M NaCl water solution on an evaporator

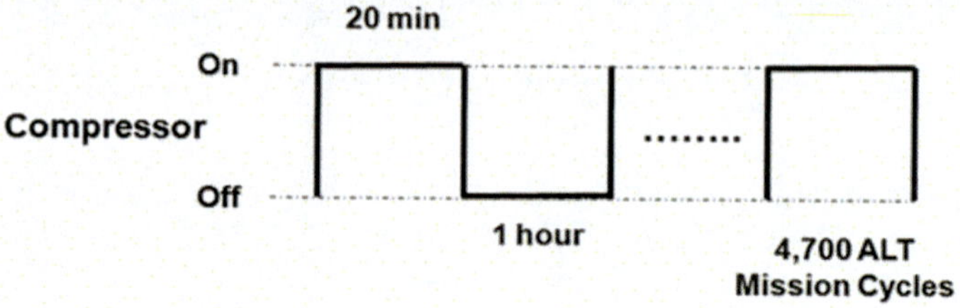

(b) Duty cycles of repeated corrosive load F

Fig. 8.48 Kimchi refrigerators in ALT and duty cycles of repeated corrosive load F

the shape parameters of the ALT, ($\beta 1$), and market data, ($\beta 2$) were discovered to be alike.

The pitting of the evaporator tubing in both the market products and the ALT test specimens happened in the inlet/outlet of the evaporator tubing (Fig. 8.51). Based on the altered designs, corrective measures put to enhance the life cycle of the evaporator tubing system comprised (1) expanding the length of the contraction tube (C1) from

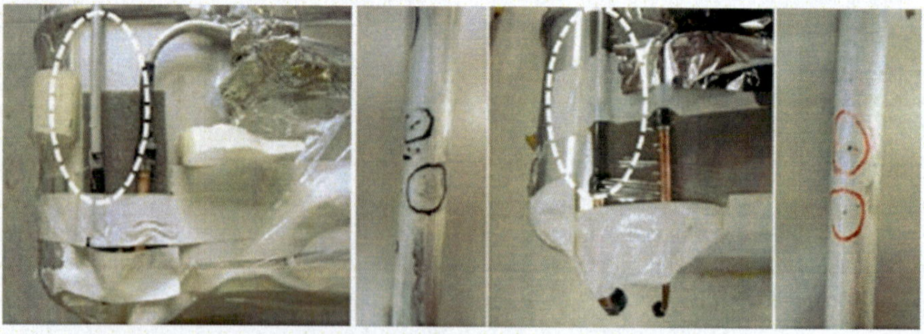

(a) Failed product from the market (b) ALT

Fig. 8.49 Unsuccessful products in the market and ALT

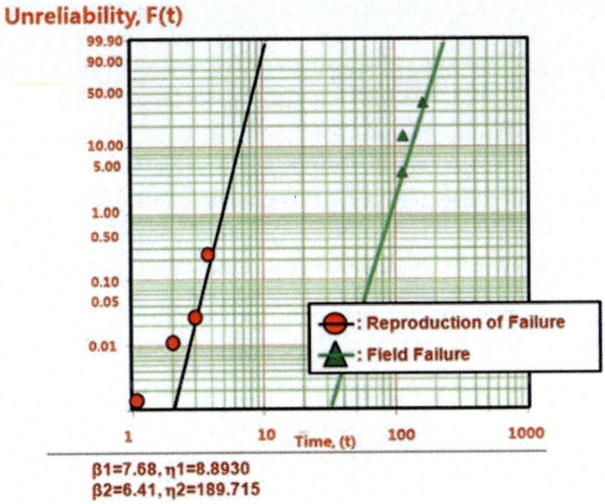

Fig. 8.50 Market data and outcomes of ALT on the Weibull plot

50.0 to 200.0 mm and (2) substituting the cotton adhesive tape (C2) with generic transparent tape.

Figure 8.52 manifests an altered evaporator tubing with high corrosive fatigue strength. The affirmed values of AF and β in Fig. 8.50 are 100.0 and 6.41, separately. The test cycles and sample size recomputed in Eq. (7.29) were 5300 and 9 EA, separately. Based on the objective BX life, two ALTs were carried out to recognize the design defects and alter them. In the two ALTs, the outlet of the evaporator tubing

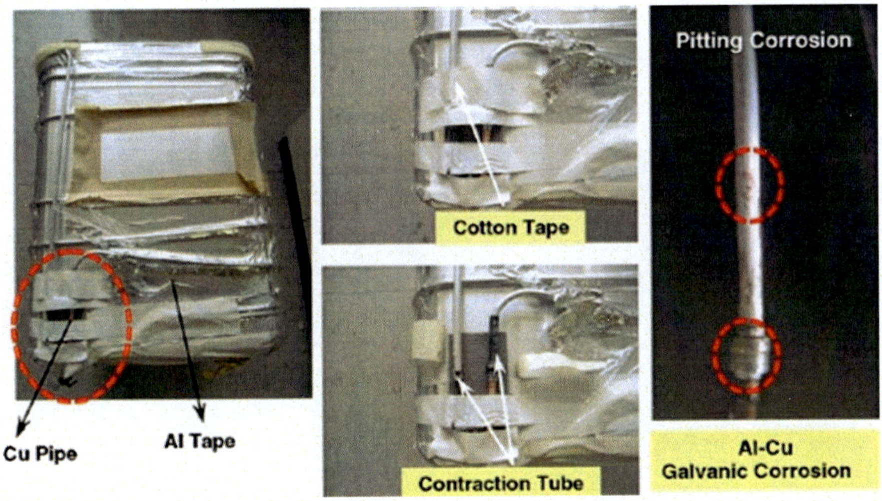

Fig. 8.51 Structure of pitting the corrosion tubing in the market and the ALT test samples

was pitted in the 1st ALT and was not pitted in the 2nd ALT. The repeated corrosive force in combination with the high chlorine concentration of the cotton tape and the crevice between the cotton adhesive tape and the evaporator tubing containing the condensed water as an electrolyte might have been pitting.

With these altered designs, the Kimchi refrigerator shall preserve the food for a longer time with no failure. Figure 8.53 and Table 8.9 manifest the graphical outcomes of ALT represented in a Weibull chart and an abridged of the ALT outcomes, respectively. Throughout the two ALTs, the B1 life of the samples enhanced by over 10.0 years.

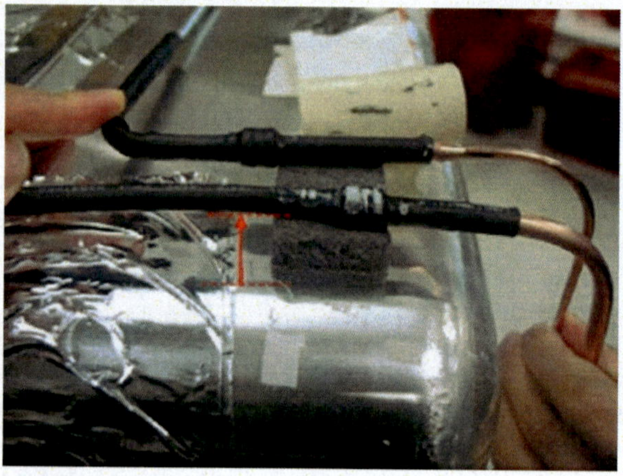

Fig. 8.52 A redesigned evaporator tubing

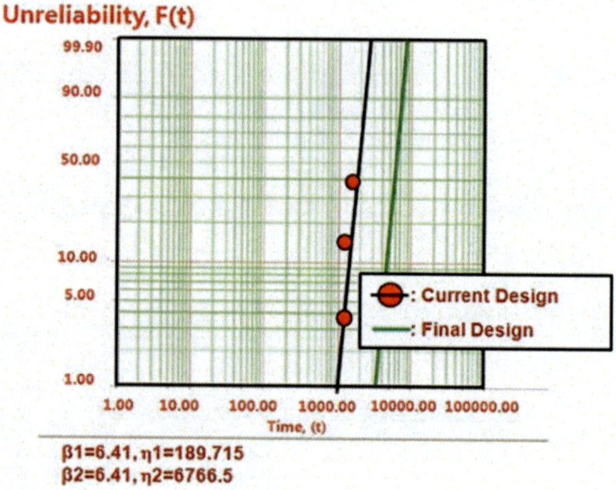

Fig. 8.53 Outcomes of ALT in the Weibull plot

Table 8.9 Results of ALT

	1st ALT	2nd ALT
	Initial design	Second design
In 5300 cycles, corrosion of evaporator pipe is less than 1	1130 cycles: 1/19 pitting 1160 cycles: 2/19 pitting 1690 cycles: 4/19 pitting 1690 cycles: 11/19 OK	5300 cycles: 9/9 OK
Evaporator pipe structure	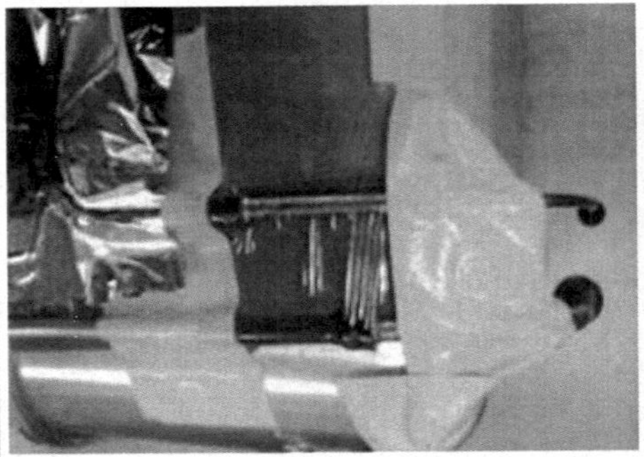	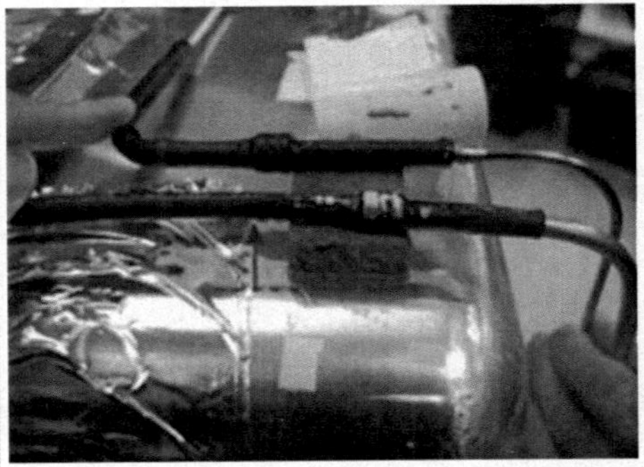
Material and specification	Length of the contraction tube C1: 50.0 mm → 200.0 mm Adhesive tape type C2: cotton type → generic transparent tape	

8.7 Improving the Noise of a Mechanical Compressor

A reciprocating compressor is a positive-displacement device which utilizes a piston to compress a gas and bring it at high pressure through a slider-crank mechanism. A refrigerator system, which works utilizing the fundamental concepts of thermodynamics, made up to a compressor, a condenser, a capillary tube, and an evaporator. The VCR cycle gets work from the compressor and moves heat from the evaporator to the condenser. The chief function of the refrigerator is to supply chilled air from the evaporator to the freezer and refrigerator sections. As a consequence, it maintains the stocked food fresh.

To enhance its energy efficiency, the engineer would select the fine functioning of the compressor. Figure 8.54 manifests a reciprocating compressor with an altered rotor and stator. The redesign was evolved to enhance the energy efficiency and lessen the noise from the compressors in a side-by-side refrigerator. For these implementations, the compressor necessitated to be designed robustly to work under a wide span of customer use circumstances (Fig. 8.55).

The reciprocating compressor in the refrigerators made noise in the market, bringing the customer to appeal a replacing of their refrigerator. One of the particular causes of compressor failure during the functioning was the compressor suspension spring. As the sound level during compressor shutdown of troublesome refrigerators in the market was recorded, the consequence was roughly 46 dB (6.2 sones). The design defects of the suspension spring in the troublesome compressor were the number of turns and the mounting spring diameter. As the compressor stops abruptly, the spring occasionally does not seize the stator frame firmly and causes noise (Fig. 8.56).

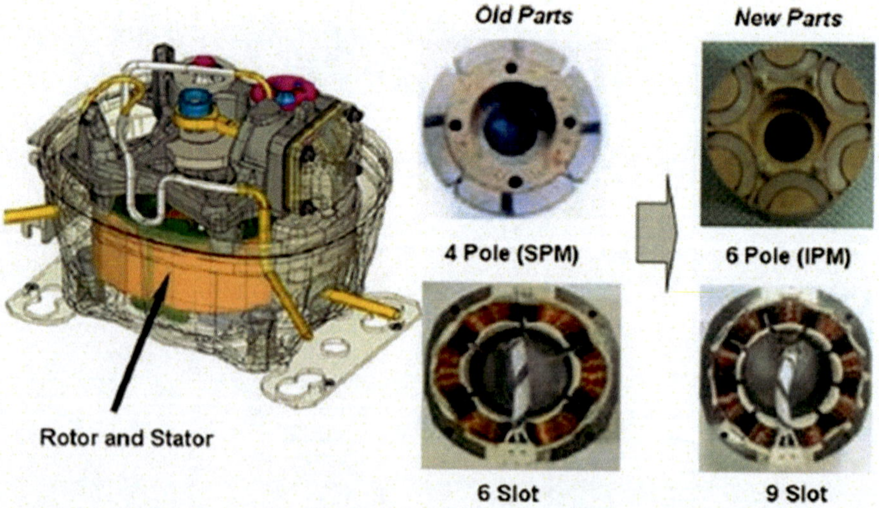

Fig. 8.54 Reciprocating compressors with an altered rotor and stator

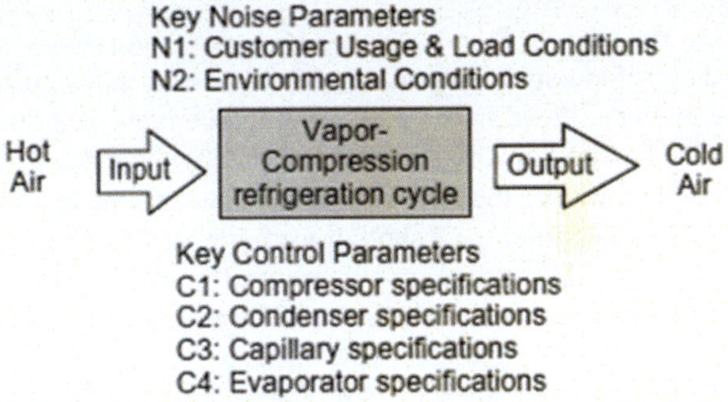

Fig. 8.55 Parameter diagram of the refrigeration cycle

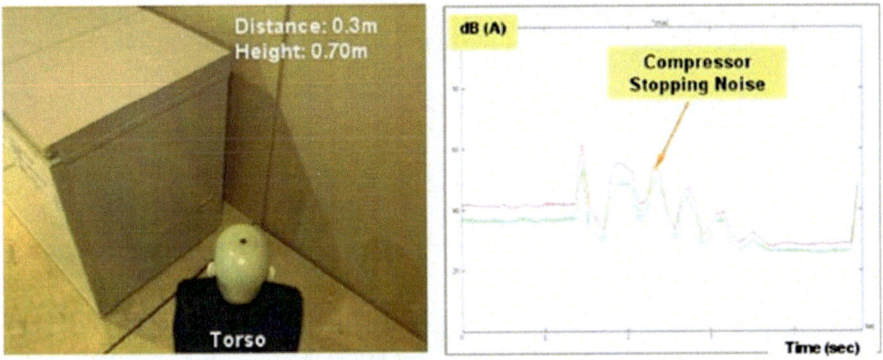

(a) Compressor stopping noise recorded with torso

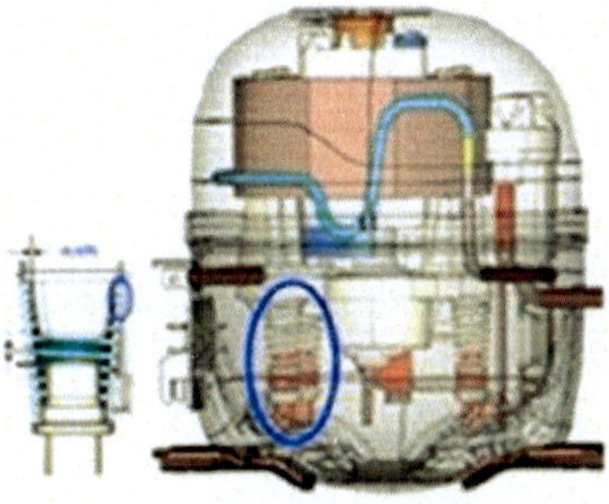

	Current	Modified
Suspension Spring		
Height, mm	30.2	31.2
Turn	7/ Upper 2 / Lower 2	7/ Upper 3 / Lower 2
Diameter, ϕ	15.55	15.35
Weight	6.9 kg	

(b) Reciprocating compressor and the design defects of the suspension spring

Fig. 8.56 Stopping noise of the reciprocating compressor

As recognizing the design defects related to a new compressor system, it was critical to alter the problematic compressor either through redesign of components or change of the matter utilized in the parts. Failure analysis of field data and ALT shall help affirm the design flaws in a newly designed compressor system.

To assess the ride feature of the newly designed stator frame and piston–cylinder assembly installed on the compressor shell, the compressor shall be modeled to two degrees of freedom (Fig. 8.57). Although it is straightforwardly four state variables in its prototype, it fulfills the reason of deciding the compressor movement in operation. The presumed prototype of the vehicle is made up to the sprung mass and the unsprung mass. The sprung mass m_s depicts the stator frame and piston–cylinder assembly, and the unsprung mass m_{us} depicts the rotor–stator assembly. The chief suspension is modeled as a spring k_s and a damper c_s in parallel, which connects the unspring to the sprung mass. The compressor suspension spring on its shell is modeled as a spring k_{us} and depicts the transfer of the road force to the unsprung mass.

$$m_s\ddot{x}_s + c_s(\dot{x}_s - \dot{x}_{us}) + k_s(x_s - x_{us}) = F\sin\omega t \tag{8.52}$$

$$m_{us}\ddot{x}_{us} + c_s(\dot{x}_{us} - \dot{x}_s) + (k_{us} + k_s)x_{us} - k_s x_s = 0 \tag{8.53}$$

Therefore, the above equations of motion can be concisely represented as follows:

$$\begin{bmatrix} m_s & 0 \\ 0 & m_{us} \end{bmatrix}\begin{bmatrix} \ddot{x}_s \\ \ddot{x}_{us} \end{bmatrix} + \begin{bmatrix} c_s & -c_s \\ -c_s & c_s \end{bmatrix}\begin{bmatrix} \dot{x}_s \\ \dot{x}_{us} \end{bmatrix} + \begin{bmatrix} k_s & -k_s \\ -k_s & k_{us} + k_s \end{bmatrix}\begin{bmatrix} x_s \\ x_{us} \end{bmatrix} = \begin{bmatrix} F\sin\omega t \\ 0 \end{bmatrix} \tag{8.54}$$

As a result, Eq. (8.54) shall be defined in a matrix form:

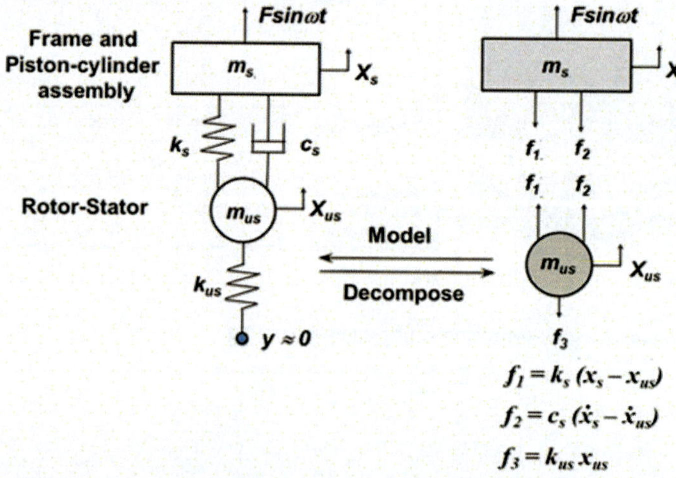

Fig. 8.57 A compressor assembly model and its decomposition

$$[M]\ddot{X} + [C]\dot{X} + [K]X = F \tag{8.55}$$

As Eq. (8.55) is integrated, the time response of the state variables shall be attained. At this time, the excited force due to compressor operation have to be comprehended. In a refrigeration cycle design, it is required to decide both the condensing pressure P_c and the evaporating pressure P_e. One indicator of the internal stresses on components in a compressor relies on the pressure difference between suction pressure, P_{suc}, and discharge pressure, P_{dis}.

Because an excited force due to the reciprocating movement of the piston originates from the pressure difference between the condenser and evaporator, the general LS prototype from Eq. (7.19) shall be altered as follows:

$$TF = A(S)^{-n} = A(F)^{-\lambda} = A(\Delta P)^{-\lambda} \tag{8.56}$$

The *AF* shall be expressed as follows:

$$AF = \left(\frac{S_1}{S_0}\right)^n = \left(\frac{F_1}{F_0}\right)^\lambda = \left(\frac{\Delta P_1}{\Delta P_0}\right)^\lambda \tag{8.57}$$

The usual number of functioning cycles per day was roughly 24; the worst occasion was 74. Under the worst occasion, the compressor life cycles for 10 years would be 270,100 cycles. From the ASHRAE Handbook test data for R600a, the typical pressure was 0.40 MPa at 42 °C, and the compressor dome temperature was 64 °C. As pressure had 1.39 MPa for the elevated testing, the *AF* was 12.6 with a quotient, λ, of 2. The whole *AF* was roughly 12.6 (Table 8.10).

The design standard of the newly designed compressor shall be more than the lifetime target—B1 10 years. Presuming the shape parameter β was 2.0, the test cycles for 100 test samples computed in Eq. (7.29) were 21,400 cycles. The ALT was designed to ensure a B1 life of 10 years with approximately a 60% level of confidence that it would fail less than once during 21,400 cycles. For the ALT apparatus, a straightforward vapor compression refrigeration system was manufactured (Fig. 8.58a). Figure 8.58b manifests the duty cycles for the repeated pressure difference ΔP.

Figure 8.59 manifests the stopping noise and vibration of a compressor from the ALT. In the plot, the peak noise level and vibration of a normal sample in the compressor were 52 dB and 0.09 g, respectively, when it stopped. On the other hand, for unsuccessful sample #1, the peak noise levels and vibration were 65 dB and

Table 8.10 ALT circumstances in a vapor compression cycle for R600a

System circumstances		Worst case	ALT	*AF*
Pressure (MPa)	High side	0.40	1.39	12.6
	Low side	0.02	0.04	
	ΔP	0.38	1.35	

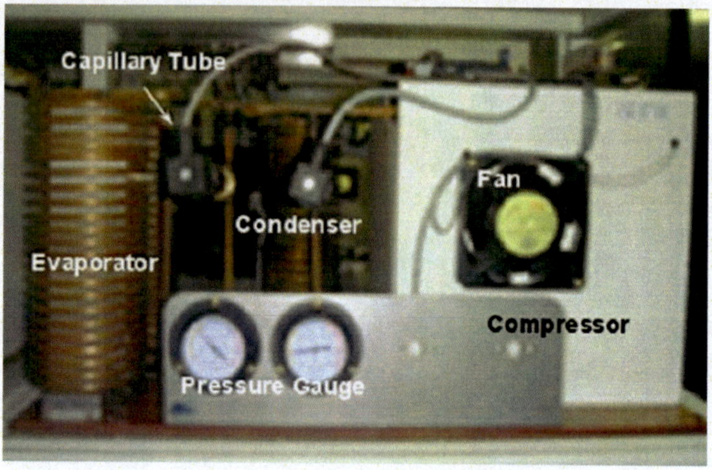

(a) Apparatus utilized in ALT.

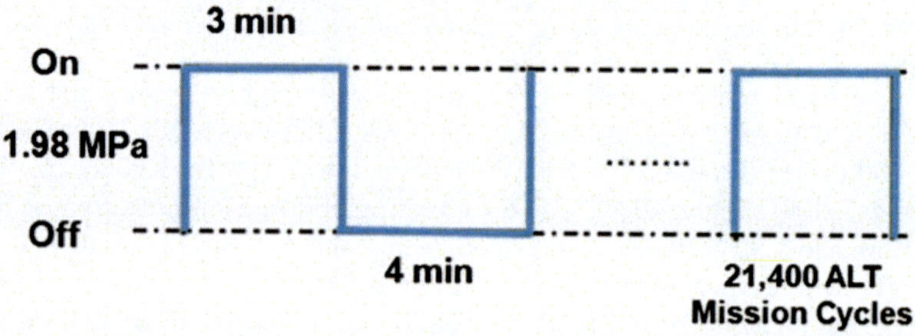

(b) Duty cycles of the repeated pressure difference on the compressor.

Fig. 8.58 Duty cycles and apparatus utilized in ALT

0.52 g, respectively. For unsuccessful sample #2, the peak noise levels and vibration were 70 dB and 0.60 g, respectively. Contemplating that the vibration specifications were less than 0.2 g, the failed sample vibrations did not satisfy the specification. As the troublesome samples in ALT equipment were mounted on the test refrigerator, the vibration was also reproduced with 0.25 g and violated the specification. In the market, the customer would appeal that the unsuccessful samples be substituted. Figure 8.60 presents the graphical investigation of the ALT outcomes and market data on a Weibull chart. For the shape parameter, the approximated value on the plot was 1.9.

As the unsuccessful samples were broken apart, a scratch was discovered inside the upper shell of the compressor where the stator frame had struck the shell. The gap between the frame and the shell was calculated to be 2.9 mm. The design gap specification might have been more than 6 mm to stop the compressor from striking the shell for the worst occasion. It was deduced that the stopping noise originated from

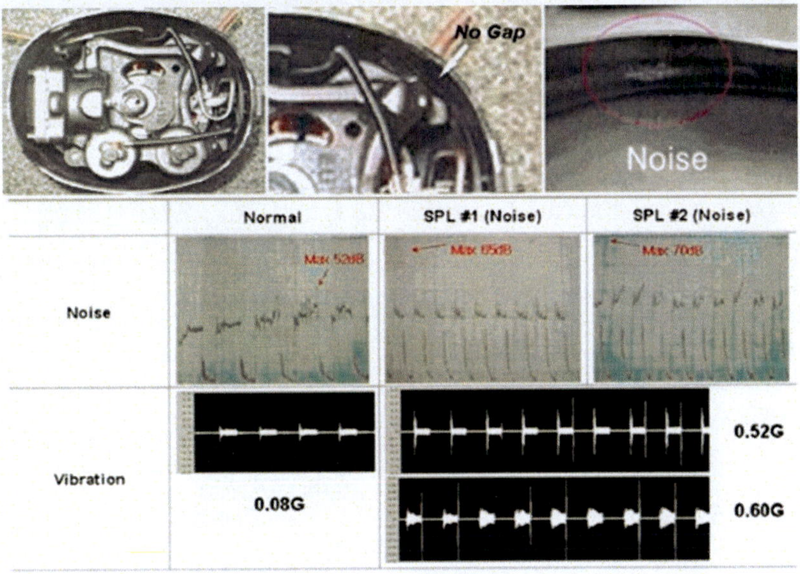

(a) Noise in ALT equipment

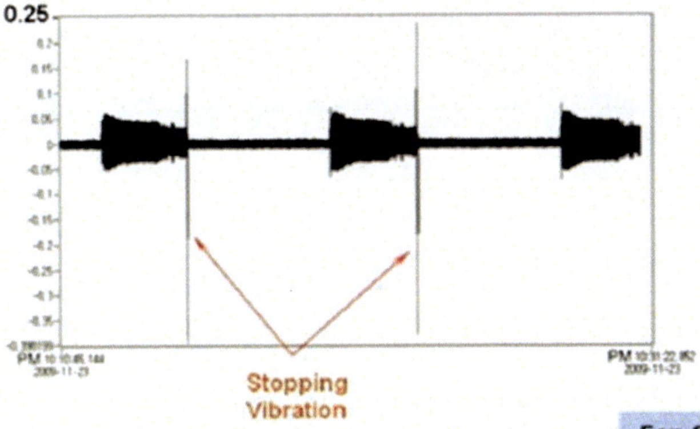

Fan Off Condition

	25°C	32°C	43°C (3800RPM)
3min On/3min Off	No	No	Reproduced
3min On/5min Off	No	No	Reproduced
10min On/10min Off	No	No	No

(b) Noise in refrigerator

Fig. 8.59 Unsuccessful products in the 1st ALT

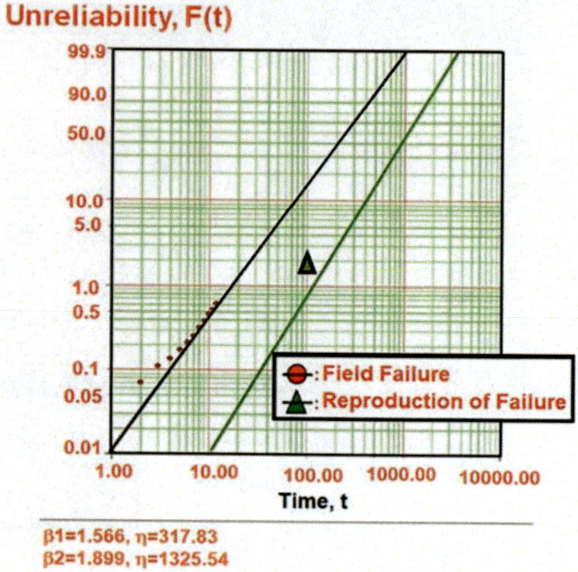

Fig. 8.60 Market data and outcomes of ALT on the Weibull plot

the hitting (or interference) between the stator frame and the upper shell. Therefore, the tests pinpointed the design defects in the compressor (see Fig. 8.61a). For the shape parameter, the approximated value on the plot was 1.9 from the graphical investigation of the ALT outcomes and market data on a Weibull chart. The vital missing parameter in the design phase of the ALT was a gap between the stator frame and the upper shell. These design defects shall create noise when the compressor ceases abruptly. To lessen the noise issues in the frame, the form of the stator frame was redesigned. As the test structure of the compressor assembly was altered to have more than a 6-mm gap, the gap size grew from 2.9 to 7.5 mm (Figs. 8.61b and 8.62).

The design standard of the newly designed samples was more than the lifetime objective—a B1 life 10 years. The affirmed value, β, on the Weibull plot was 1.9. As the 2nd ALT began, the test cycles recomputed in Eq. (7.29) for 100 sample pieces were 21,400. In the 2nd ALT, no issues were discovered with the compressor in 21,400 cycles. We expected that the altered designs would be successful. Table 8.11 supplies an abridge of the ALT outcomes. With the altered designs, the B1 life of the samples in the 2nd ALT extends by more than 10.0 years.

8.8 Refrigerator Compressor Subjected to Repeated Loads

To store food, the refrigerator utilizing a vapor-compression refrigeration cycle has to supply chilled air from the evaporator to the freezer and refrigerator sections. The refrigeration unit is made up to a compressor, a condenser, a capillary tube, and an evaporator. The compressor takes the refrigerant from the evaporator and thus

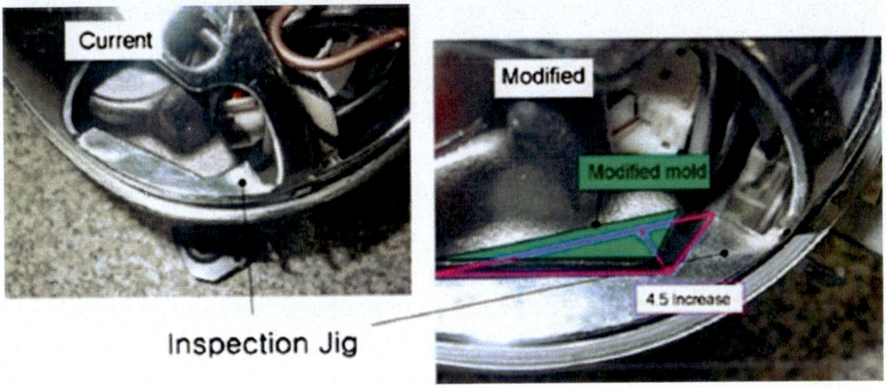

(a) Altered inspection jig

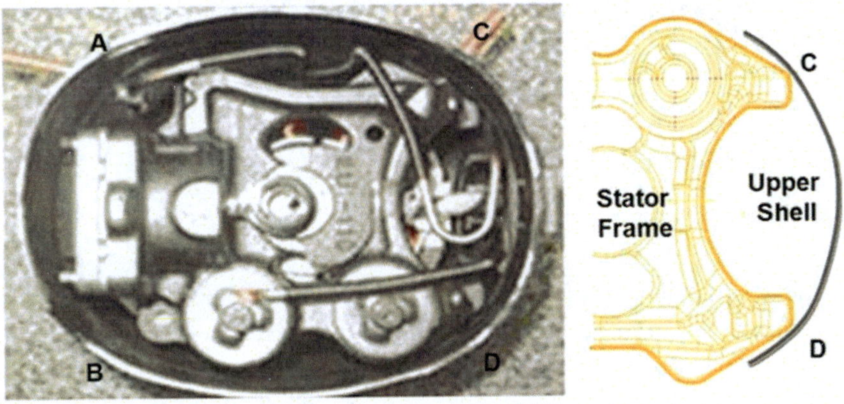

Gap between frame and upper shell (Spec.: 6mm↑)			
A	B	C	D
8.5	9.1	3.2	2.9

(b) Gap between the stator frame and the upper shell

Fig. 8.61 Altered inspection jig and gaps

compresses it, which moves it to the condenser in the VCR cycle. In the process, a reciprocating compressor grows the refrigerant pressure from that in the evaporator to that in the condenser and is subjected to repetitive stresses by the operation of the crankshaft. As consumers demand refrigerators which utilize less energy, it is required to enhance the useful energy efficiency of the refrigerator. One method to enhance efficiency is to alter the compressor in the refrigerator. The main parts in a domestic compressor is made up to the crankshaft, piston assembly, stator and its frame, valve plate, and suction reed valve (Fig. 8.63).

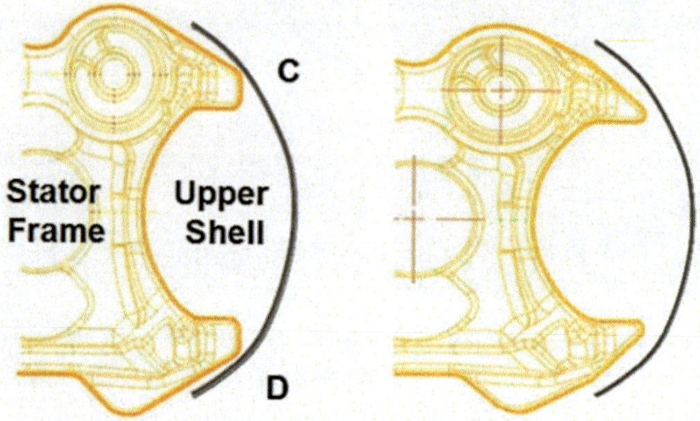

Gap between frame and upper shell (Spec.: 6mm↑)				
	A	B	C	D
Old Part	8.5	9.1	3.2	2.9
New Part	8.3	8.8	7.7	7.5

Fig. 8.62 Modified stator frame in the 2nd ALT

In the field, compressors in the refrigerators were locked due to wear, which was the continuing loss of matter from the exterior of a crankshaft and caused loss of the cooling function. Based on the customer use circumstances, we knew that compressors were subjected to repeated pressure loads during the refrigerator functioning (Fig. 8.64).

In assessing the design of a refrigeration cycle, it was required to decide both the condensing temperature, T_c, and evaporating temperature, T_e. The mass flow rate of refrigerant in a compressor shall be expressed as follows:

$$\dot{m} = PD \times \frac{\eta_v}{v_{\text{suc}}} \tag{8.58}$$

The mass flow rate in the capillary tube shall be defined as

$$\dot{m}_{\text{cap}} = A\left[\frac{-\int_{P_3}^{P_4} \rho \mathrm{d}P}{\frac{2}{D} f_m \Delta L + \ln\left(\frac{\rho_3}{\rho_4}\right)}\right]^{0.5} \tag{8.59}$$

By conservation of mass, the mass flow rate in the capillary tube shall be stabilized with that of a compressor. That is,

$$\dot{m} = \dot{m}_{\text{cap}} \tag{8.60}$$

Table 8.11 Results of ALT

<table>
<tr><td></td><td>1st ALT</td><td>2nd ALT</td></tr>
<tr><td></td><td>Initial design</td><td>Second design</td></tr>
<tr><td>In 21,400 cycles, locking is less than 1</td><td>100 cycles: 2/100 noise
100 cycles: 99/100 OK</td><td>21,400 cycles: 100/100 OK
20,000 cycles: 100/100 OK</td></tr>
<tr><td>Compressor structure</td><td colspan="2">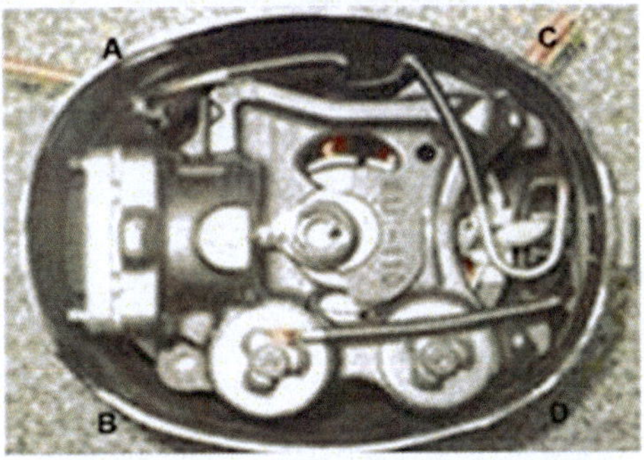

<table>
<tr><td colspan="4">Gap between frame and upper shell (Spec.: 6mm↑)</td></tr>
<tr><td>A</td><td>B</td><td>C</td><td>D</td></tr>
<tr><td>8.5</td><td>9.1</td><td>3.2</td><td>2.9</td></tr>
</table></td></tr>
<tr><td>Material and specification</td><td>C1: Modification of the frame shape</td><td></td></tr>
</table>

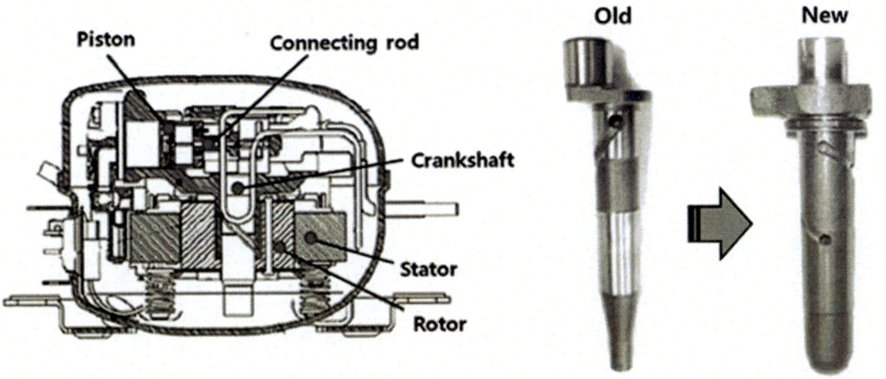

Fig. 8.63 Altered compressor and crankshaft

The energy balance between the energy released by the refrigerant and the heat transfer by the temperature difference in the condenser might be expressed as

$$Q_c = \dot{m}(h_2 - h_3) = (T_c - T_o)/R_c \tag{8.61}$$

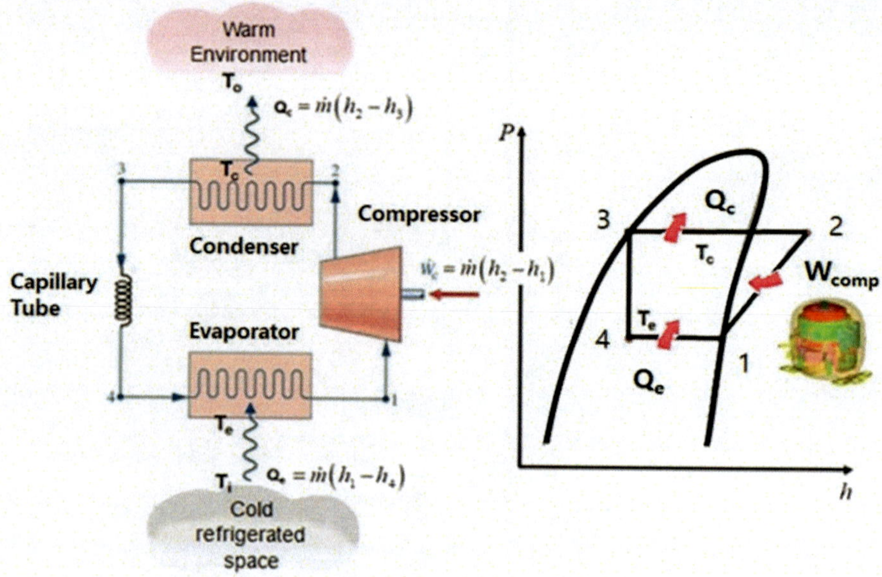

(a) Refrigeration cycle (b) P-h diagram

Fig. 8.64 Functional design idea of the compressor system in the refrigeration cycle

The energy conservation between the energy absorbed by the refrigerant and the heat transfer by the temperature difference in the evaporator could be expressed as

$$Q_c = \dot{m}(h_1 - h_4) = (T_i - T_e)/R_e \tag{8.62}$$

When nonlinear Eqs. (8.60) through (8.62) are resolved, the mass flow rate, $\dot{m}$, evaporator temperature, T_e, and condenser temperature, T_c, shall be attained. Because the saturation pressure, P_{sat}, is a function of temperature, the evaporator pressure, P_e (or condenser pressure P_c), shall be attained as follows:

$$P_e = f(T_e) \quad \text{or} \quad P_c = f(T_c) \tag{8.63}$$

The stress source in a domestic compressor might come from the pressure difference between the discharge pressure, P_{dis}, and suction pressure, P_{suc}. The pressure difference between evaporator pressure P_e and condenser pressure P_c shall be close,

$$\Delta P = P_{\text{dis}} - P_{\text{suc}} \cong P_c - P_e \tag{8.64}$$

For a compressor, the time-to-failure in Eq. (7.19), *TF*, shall be altered as

Table 8.12 ALT circumstances in vapor compression cycles

System circumstances		Worst case	ALT	*AF*
Pressure (MPa)	High side	1.07	1.96	3.36③ (=(① / ②)^2)
	Low side	0.0	0.0	
	ΔP	1.07①	1.96②	
Temp. (°C)	Dome temp.	90	120	1.37④
Total *AF* (= ③ × ④)		–		4.58

$$TF = A(S)^{-n} \exp\left(\frac{E_a}{kT}\right) = A(\Delta P)^{-n} \exp\left(\frac{E_a}{kT}\right) \tag{8.65}$$

The *AF* from Eq. (7.20) shall be defined as

$$AF = \left(\frac{S_1}{S_0}\right)^n \left[\frac{E_a}{k}\left(\frac{1}{T_0} - \frac{1}{T_1}\right)\right] = \left(\frac{\Delta P_1}{\Delta P_0}\right)^{\lambda} \left[\frac{E_a}{k}\left(\frac{1}{T_0} - \frac{1}{T_1}\right)\right] \tag{8.66}$$

The refrigerator compressor operates as follows: 0–43 °C with humidity changing from 0 to 95%, and 0.2–0.24 g of acceleration. The number of compressor operations is influenced by the customer operating states. In the field, the reciprocating compressor occurs approximately 10–24 times each day. For a compressor lifetime of 10 years, 87,600 usage cycles occur.

For this worst occasion, the dome temperature in the compressor was 90 °C, and the pressure was 1.07 MPa. For the ALTs, the compressor dome temperature was 120 °C, and the pressure was 1.96 MPa. Employing a cumulative damage exponent, λ, of 2.0, the AF from Eq. (8.66) was observed to be roughly 4.58 (Table 8.12).

For the life objective—a B1 life 10 years, the mission cycles computed from Eq. (7.29) for one hundred sample units were 19,000 cycles if the shape parameter, β, was presumed to be 2.0. The parametric ALT was assured to have a B1 life 10 years with an approximately 60% confidence level if it fails less than one time in 19,000 cycles. Figure 8.65 manifests the experimental equipment for compressor ALT and duty cycles. By exerting the duty cycles, we shall discover the design failure.

The apparatus utilized in parametric ALT was a straightforward VCR equipment which made up an evaporator, compressor, capillary tube, and condenser. The inlet to condenser was at the top, and the condenser outlet was at the bottom. The condenser inlet had a high-side pressure gauge and was constructed with quick coupling. At the condenser inlet, a refrigerator dryer was set at right angles to a horizontal plane. To operate the condenser fan, a thermal switch was put on the condenser tubing at the head of the condenser. The evaporator inlet was at the bottom. Pressure gauges were put at a location where the evaporator outlet was put to remove refrigerant and charge it to the low side. The condenser outlet was added to the exit of the evaporator with a capillary tube. The compressor on mounting rubber pads was added to the condenser vent and evaporator exit. The temperature of the test room in an insulated

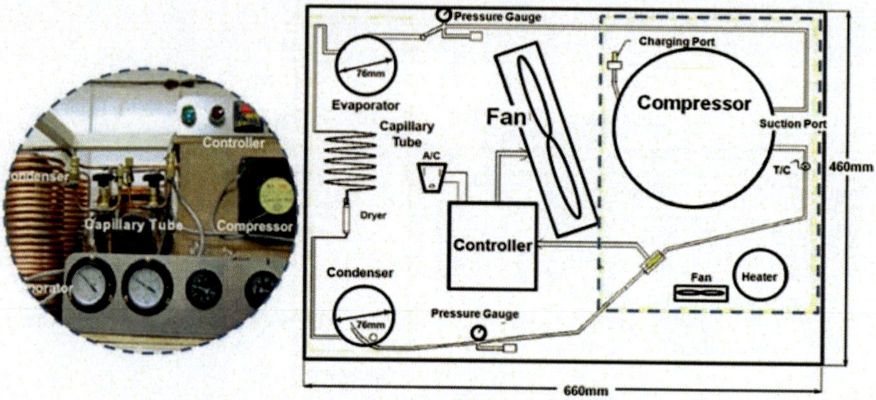

(a) A drawing of the test system

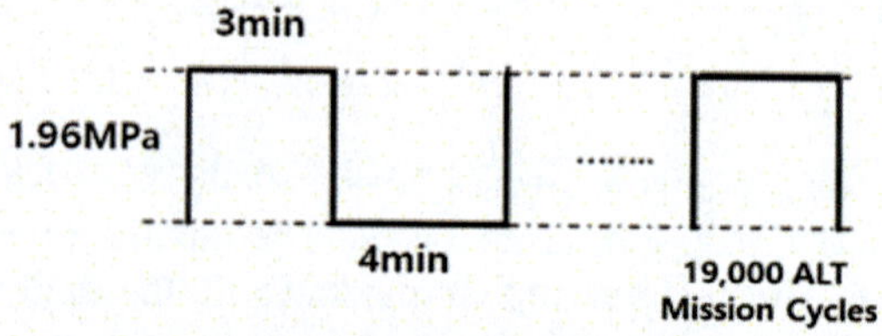

(b) Duty cycle

Fig. 8.65 Apparatus utilized in parametric ALT and duty cycles

box was continued by two 60-watt lamps and a fan. Put on the compressor top, a thermal switch runs a 51 m^3/h fan.

In a French door refrigerator, compressors with the newly designed crankshafts were locked due to lubrication problems and wear. Field data appointed that the failed compressors could have design flaws. In other words, there was the problem of oil lubrication when a crankshaft was repeatedly subjected to the tribological stress on its outside and to relative motion with a solid counter body such as a connecting rod. Thus, we discovered that the seriously wore crankshaft on its top and body when the unsuccessful product was broken down (market and 1st ALT). At initial starting, due to the lubrication problem and direct contact, the repetitive pressure loads in the compressor could generate undue adhesive wear on the outside of the crankshaft. As the wear particles were separated and blocked the refrigerant line in the compressor, the compressor was locked in the end. Engineers should reproduce these design problems and modify them before product launches. Figure 10 displays a picture comparing the troublesome compressor from the field and that from the 1st ALT, separately.

In the 1st ALT, the two compressors were locked at 7500 cycles and 10,000 cycles. As the unsuccessful compressors from the 1st ALT and the market were broken, significant wear was shown in some regions of the crankshaft where there was no lubrication—the motion area between the connecting rod and crankshaft as well as the

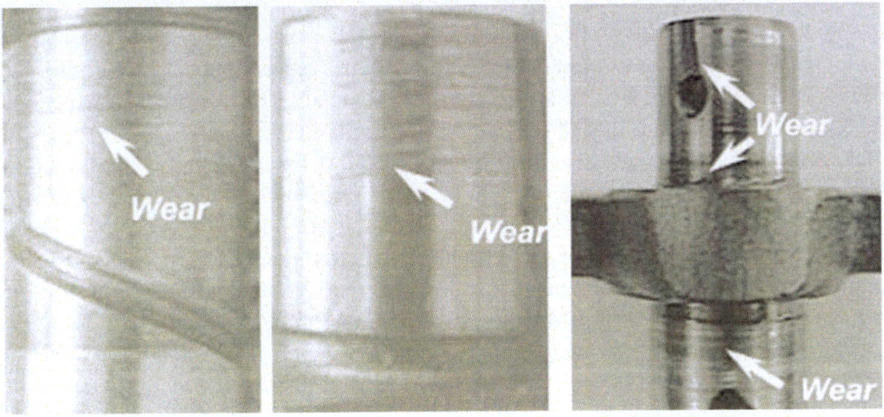

(a) Unsuccessful product in themarket

(b) Unsuccessful sample in the 1st ALT

Fig. 8.66 Failed product in the market and 1st ALT

rotating area between the crankshaft and cylinder block. As manifested in Fig. 8.66b, the tests affirmed that the refrigerator compressor was not acceptably designed to secure correct lubrication. As the compressor was beginning, we recognized that the poor lubrication in combination with the repeated pressure loads might bring troublesome wear on the outside of the crankshaft. The troublesome aspect of the 1st ALT was alike to that of the samples from the field.

In a French door refrigerator, it was discovered that two compressors with the newly designed crankshafts were locked due to wear at 10,504 cycles. Market data designated that the damaged products may have had a design defect. That is, there was the problem of oil lubrication when a crankshaft was repeatedly subjected to the tribological stress on its outside and relative motion with a solid counter body, such as a connecting rod. Thus, we discovered that the crankshaft was seriously worn on its top and body when the unsuccessful product was broken down (Market and first ALT). At the initial phase, due to the lubrication problem and direct contact, the repetitive pressure loads in the compressor might create undue adhesive wear on the outside of the crankshaft. As the wear particles were detached and blocked the refrigerant line in the compressor, the compressor was in the end locked. An engineer should reproduce these design problems and alter them before the product releases. Figure 8.66 manifests photos comparing the unsuccessful product from the market and that from the 1st ALT.

In the 1st ALT, the two compressors were locked at 7500 cycles and 10,000 cycles. As the unsuccessful compressors from the 1st ALT and the market were broken, significant wear was shown in some regions of the crankshaft where there was no lubrication—the motion region between the connecting rod and crankshaft as well as the rotating region between the crankshaft and cylinder block. As manifested in Fig. 8.66b, the tests affirmed that the refrigerator compressor was not acceptably designed to secure correct lubrication. As the compressor was beginning, we recognized that the bad lubrication in combination with the repeated pressure loads might bring troublesome wear on the exterior of the crankshaft. The troublesome aspect of the 1st ALT was alike to that of the samples from the field.

As seen in Fig. 8.67, we know that the basic causes of the crankshaft wear started in the design defects—(1) the deficiency of oil in regions which should have lubricated, (2) weak crankshaft matter (FCD450), and (3) low starting RPM. The design defects might suddenly give rise to compressor lock-up when subjected to repetitive pressure loading. To enhance the lubrication design issues on the outside of the crankshaft, it was altered by growing the RPM from 1650 RPM to 2050 RPM, moving lubrication holes, attaching new grooves, and altering the shaft matter FCD500 (Fig. 8.68).

For the life objective—a B1 life of 10 years, the test cycles recomputed from Eq. (7.29) for one hundred sample units were 19,000 cycles if the shape parameter was supposed to be 2.0. This parametric ALT was designed to ensure a B1 life of 10 years with approximately a 60% level of confidence that it would fail less than once during 19,000 cycles.

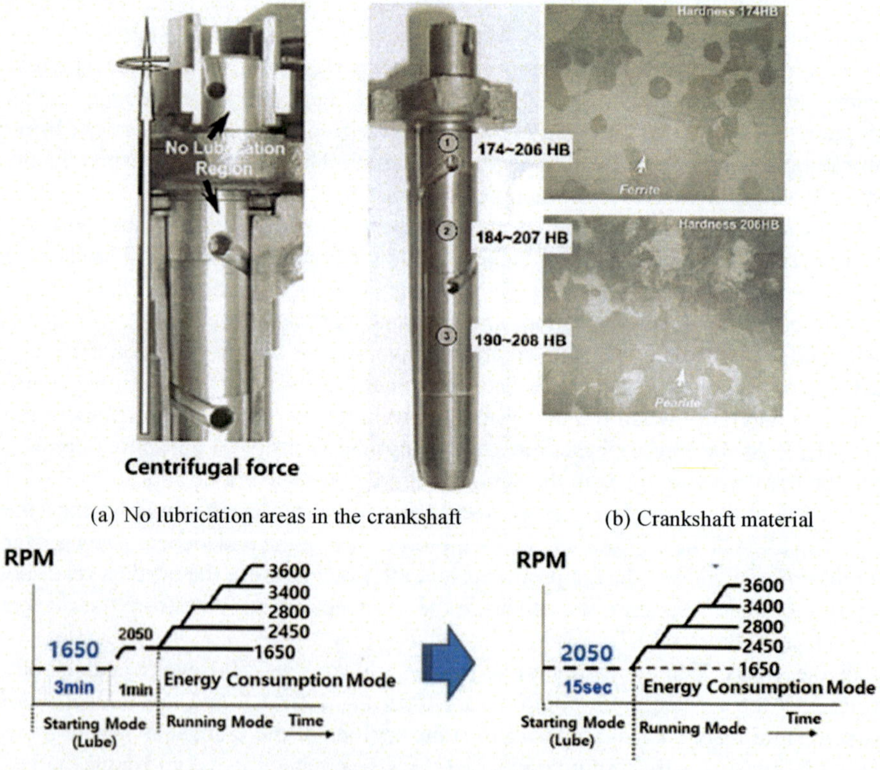

(a) No lubrication areas in the crankshaft (b) Crankshaft material

Fig. 8.67 Basic causes of the failure of 1st ALT

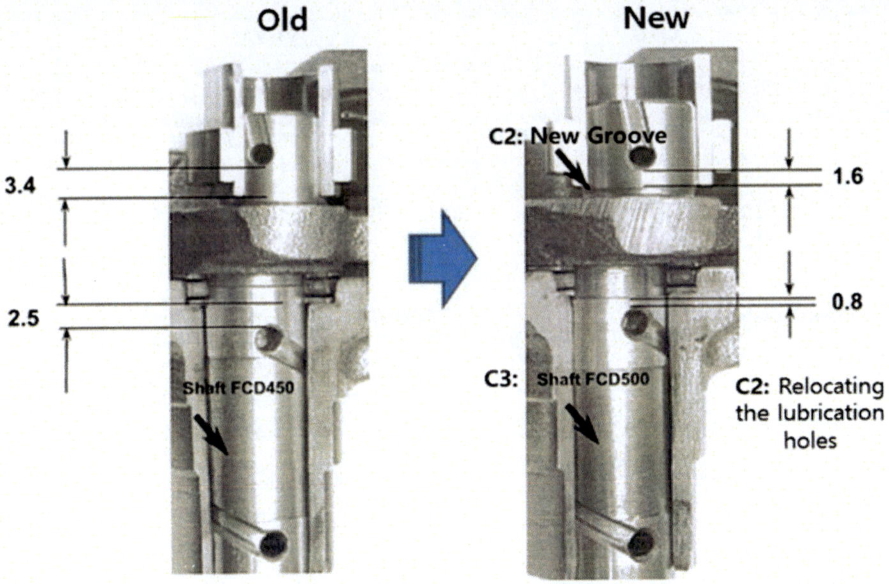

Fig. 8.68 Modified crankshaft in the 1st ALT: relocating the lubrication holes (C2), adding the new groove (C3), and modifying matter (C3: FCD450 → FCD500)

In the 2nd ALT, as the compressors were broken down, the crankshaft wear due to interference between the crankshaft and a thrust washer was discovered at 19,000 cycles (Fig. 8.69). To enhance the design assure in the second ALT, the minimum clearance between the crankshaft and washer was enlarged from 0.141 to 0.480 mm (Fig. 8.70).

(a) Failed products in the 2nd ALT (b) its root cause

Fig. 8.69 Unsuccessful products in the 2nd ALT

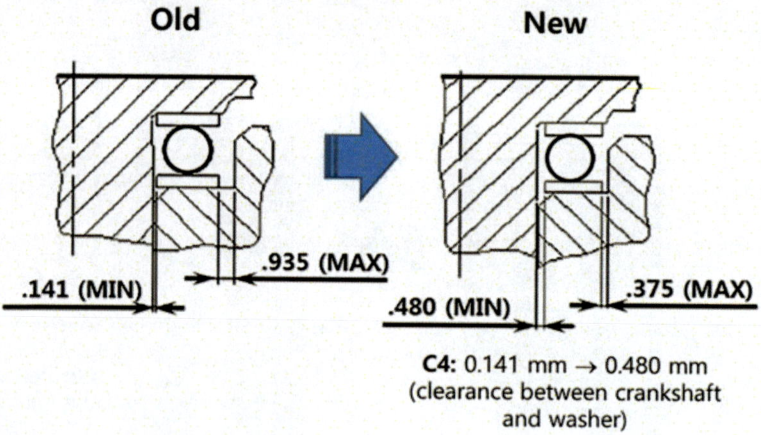

Fig. 8.70 Altered crankshaft in 2nd ALT: altered minimum clearance between crankshaft and washer (0.141 mm → 0.480 mm)

To endure the design issues of crankshaft due to the repetitive pressure loads, the improved design can be abridged as follows: (1) altering the starting RPM, C1, from 1650 RPM to 2050 RPM (Fig. 8.67c); (2) relocating the lubrication holes and attaching the new groove, C2 (Figs. 8.67a and 8.68); (3) altering the matter crankshaft, C3, from FCD450 to FCD500 (Figs. 8.67b and 8.68); and (4) growing the minimum clearance between crankshaft and washer, C4, from 0.141 mm to 0.480 mm (Fig. 8.70).

With the design alternations, the domestic compressor with the newly designed crankshaft might work effortlessly over its lifetime because there were no design defects till 20,000 operation cycles. Table 8.13 is an abridge of parametric ALTs. After the course of three ALTs with corrective actions, the compressor life was enhanced to have a B1 life of 10.0 years. The parametric ALTs might enable engineers to evaluate the design in the mechanical system.

8.9 Lifetime of a Localized Designed Pneumatic Cylinder in an Automatic Assembly Line

Engineers frequently employ pneumatics because they are silent, cleaned, and do not necessitate big spaces for fluid storage. The pneumatic cylinder is a relatively straightforward mechanical structure and shall be utilized in some end–user mechanisms of a mechanical product. As manufacturers use machine tools in an automatic assembly line, they want their staff to comfortably change the tools by exactly pushing them. Pneumatic cylinders shall be employed in equipment to help move or open mechanical devices. A new pneumatic cylinder integrated with a machine tool shall be designed with a specific, anticipated lifetime.

Table 8.13 Results of ALT

	1st ALT	2nd ALT	3rd ALT
	Initial design	Second design	Final Design
In 19,000 mission cycles there is no problem in compressor	10,504 cycles: 2/100 locking 19,000 cycles: 28/100 OK	19,000 cycles: 2/100 wear 19,000 cycles: 28/100 OK	19,000 cycles: 100/100 OK 20,000 cycles: 100/100 OK
	Crackshaft structure		
Photo	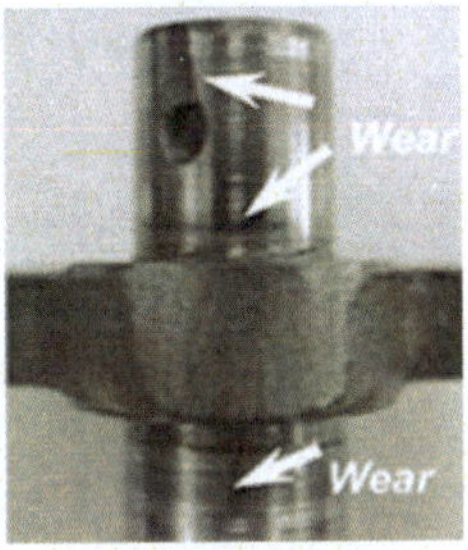		–
Action plans	C1: Starting RPM (1650 → 2050) C2: Relocating hole and one new groove C3: Material (FCD450 → FCD500)	C4: Clearance (0.141 → 0.480 mm) (Modification of washer dimension)	

To convey a load along a linear path to the involved destination, some mechanical parts in the pneumatic cylinder are needed. They consist of a rod cap, piston, head cap, etc. Depending on the expected consumer use conditions, the pneumatic cylinder in machine tools was often subjected to repetitive mechanical pressure loads during normal operations. There were design defects in a newly designed pneumatic cylinder which reduced its anticipated life, but it was not realized at first which components in the cylinder were unsuccessful. Identifying and correcting them necessitated implementation of the structured reliability method (Fig. 8.71).

Based on the market data, as the cylinder was subjected to repetitive impact pressure loads over its life during regular end–user usage, the fractured (or hardened and worn) piston seal brought the cylinder to spill and stop working. The field data from the unsuccessful cylinders due to the wrong-chosen matter rubber or polyurethane were notable for comprehending the use patterns of customers and helping to decide design flaws which needed to be altered in the product. Based on the market data, it

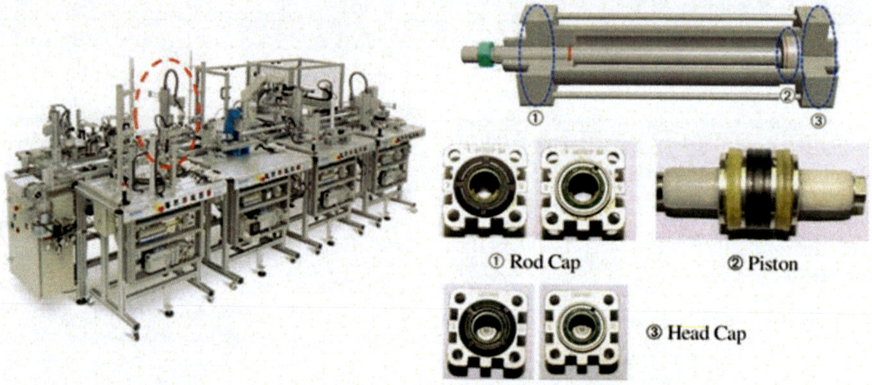

(a) automatic assembly line (b) primary parts.

Fig. 8.71 Pneumatic cylinder in the automatic assembly line

was required to reproduce and pinpoint the basic cause(s) of the failure of the problematic cylinder. If the lifetime objective were set, it would be possible to pinpoint which part(s) in the cylinder should be modified to increase its life by ALT.

As market data were evaluated, the troublesome products manifested leakage around the piston. As the cylinder was subjected to repetitive pressure loading, it encountered the presumed failure: (1) overheating of the piston seal; (2) hardening and wear of the piston seal; and (3) exceeding the minimum operation pressure (MOP) and stroke time (ST) due to leakage.

Because the automatic assembly line, comprising the cylinder, had to be stopped, manufacturers should request the unsuccessful product changed. To address this issue, it was critical to reproduce the failure mode(s) of the cylinder in simulated circumstances in a space or building provided for engineering apparatus. The problematic cylinders which originated from the market had design defects—inappropriate seal (metal) matter (or kind) (Fig. 8.72).

Initially, to supply the endurance of loads of the cylinder, a new metal seal made of nickel-iron alloy (Fe-36Ni Invar Alloy) possessing 36% Ni in the cylinder was utilized because of its fine sealing characteristics. Due to their hardness and comparatively high oxidation resistance at high temperatures, nickel alloys have been utilized for making gas turbines, rocket engines, propellers in boats, desalination, stainless steel cutlery, nickel-cadmium battery, coins, jewelry, and household utensils. It is also uncomplicated to work and shall be drawn into wire. After the fatigue strength of nickel alloys in the current design of a cylinder was evaluated through ALT, design alternatives shall be discovered for attaining the life objective.

The pressure difference of the cylinder due to loading shall be defined as:

$$\Delta P = \Delta F_{int}/A \tag{8.67}$$

where ΔF_{int} = force difference in a pneumatic cylinder due to load.

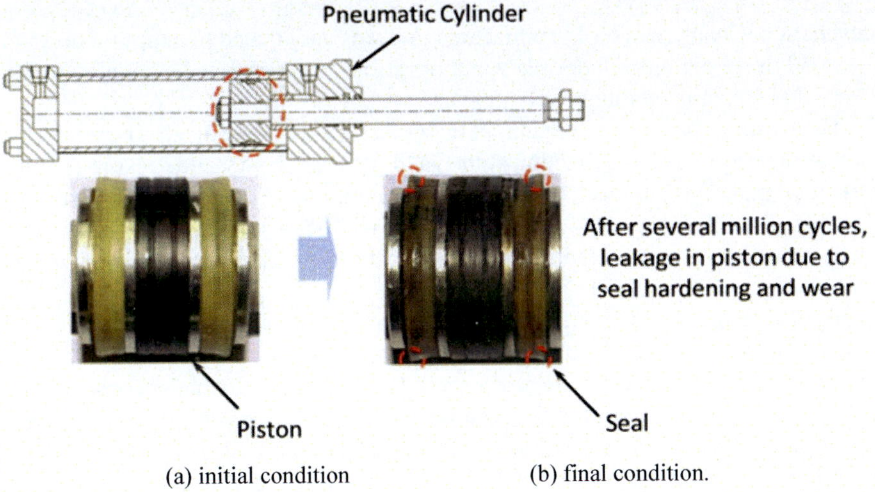

Fig. 8.72 Failed piston seal in the field

Under elevated circumstances, the life-stress model (LS model) in Eq. (7.19) shall be expressed as:

$$TF = A(S)^{-n}\exp\left(\frac{E_a}{kT}\right) = B(\Delta P)^{-\lambda}\exp\left(\frac{E_a}{kT}\right) \tag{8.68}$$

where A and B are constants, k is Boltzman's constant, E_a is the activation energy, T is the absolute temperature, n is the quotient, and λ is the cumulative damage exponent in Palmgren–Miner's rule.

To achieve the acceleration factor (AF), which shall be expressed as the correlation between the accelerated stress quantities and common operation circumstances, it shall be modified as follows:

$$AF = \left(\frac{S_1}{S_0}\right)^n\left[\frac{E_a}{k}\left(\frac{1}{T_0} - \frac{1}{T_1}\right)\right] = \left(\frac{\Delta P_1}{\Delta P_0}\right)^\lambda\left[\frac{E_a}{k}\left(\frac{1}{T_0} - \frac{1}{T_1}\right)\right] \tag{8.69}$$

As the majority of elevated testing is performed at usual (room) temperatures, Eq. (8.70) may be expressed as:

$$AF = \left(\frac{S_1}{S_0}\right)^n = \left(\frac{\Delta P_1}{\Delta P_0}\right)^\lambda \tag{8.70}$$

For a cylinder in a machine center, the usual functioning temperatures varied from 0 to 43 °C with relative humidity varying from 0 to 95%. The vibration circumstances presumed for working the cylinder ranged from 0.2 to 0.24 g of acceleration. As the cylinder worked, the piston moved a load along a linear path to the necessitated destination. Changing a tool in a machine tool was anticipated to happen a mean of 10 to 30 times per day. Under the worst occasions, the cylinder functioning for 10 years shall experience roughly 109,500 use cycles.

The cylinder is made of carbon, cast, stainless, alloy, etc., steel. The allowable stresses are expressed as a function of the yield stress (F_y) or tensile stress (F_u) of the structural material. For steel, the range of yield strength, F_y, and ultimate or tensile strength, F_u, utilized are 248–345 MPa and 400–483 MPa, respectively. The cylinder necessitated to convey a load along a linear path to the needed destination, which was worked at 0.63 MPa, 23 °C.

To decide the stress level for ALT, based on the allowable usage range of the cylinder in bench-marked data which were attained from other crucial companies, we employed the step-stress life test, which shall assess the life under constant used conditions for some accelerated loads (or load masses). As the dissimilar stress levels due to loads were altered, we could notice the failure times of the cylinder at specific stress levels, such as 0.8, 1.2, and 1.4 MPa, which depended on load masses in a variety of implementations. For a parametric ALT, the pressure difference was eventually raised to 0.8 MPa with a cumulative damage exponent, λ, of 2. The whole AF calculated from Eq. (8.70) was 1.6 (Table 8.14).

The test cycles of the ALTs computed from Eq. (7.29) were 200,000 cycles for 10 samples if the shape parameter, β, was supposed to be 2.0. The parametric ALT was designed to assure a life objective—B1 life of 10 years—with an approximate 60% level of confidence that it would be unsuccessful less than once during 200,000 cycles at the necessitated pressure difference.

As manifested in Fig. 8.73, to evaluate the design of the cylinder, reliability testing was carried out for 10 samples. The test specification for cylinders are 160 mm (stroke) and 32 mm (bore size). To apply the pressure loading, we regulated the lumped mass hung at the end. It takes one second to complete a cycle in which the cylinder slides to and from stroke (160 mm). The control console was used to operate the testing equipment and control the number of tests, the testing time, and the starting or stopping of the apparatus. As the start button was pressed on the controller panel, the pneumatic cylinder in the test apparatus pushed the adapted lumped mass and was subjected to a pressure loading of 0.8 MPa.

To put the stress level by the step-stress life test, we checked the failure time at the following stress level—0.8 MPa, 1.2 MPa, and 1.4 MPa which shall be attained from the load mass. For 0.8 MPa, the cylinder made sound at 6000 cycles and 10,000 cycles. For 1.2 MPa, the cylinder made noise near 4000 cycles and 8000 cycles. On the other hand, for 1.4 MPa, the cylinder made sound at 1000 cycles. Therefore, we determined the stress level to be 0.8 MPa for ALT because it has a relatively fine data-linearity compared to other stress levels.

Table 8.14 Pneumatic cylinder ALT circumstances

System states	Worst case	ALT	AF
Pressure (MPa)	0.63	0.8	1.6

(a) Test apparatus (b) ALT for cylinders installed on load masses.

Fig. 8.73 Equipment for the ALTs

At first, when 0.8 MPa as the accelerated pressure in a cylinder was loaded, the cylinder (n = 10) made noise at 6000 cycles, 10,500 cycles, and 11,000 cycles which have a 70 dB (maximum) and bring moderate-to-severe hearing loss. As disassembling three troublesome samples, one was a partly cracked sample, and two were chipped due to the repetitive pressure loading. The fractured surfaces were investigated utilizing scanning electron microscopy (SEM). The position of the multiplicative crack initiation and the microscopic growth at the right beginning of the crack due to repetitive loading were recognized. We therefore identified that loading was perhaps the most usual cause of failures for the partly cracked seal in cylinders. That is, as pressure loading was frequently exerted to the seal along the axial orientation of the cylinder, it might bring to circumstances which resulted in such a failure–chipped and partially cracked sample. It was decided that this failure originated from an appropriate seal matter, such as Fe-36Ni Invar Alloy. Though this metal (Fe-36Ni Invar Alloy) had fine sealing characteristics, it in the end failed under repetitive loading. The seal matter in the cylinder was altered from Fe-36Ni Invar Alloy to a silicone rubber, made up to silicone—itself a polymer—holding silicon together with carbon, hydrogen, and oxygen (C1) (Fig. 8.74).

Some cylinders in the 2nd ALT (n = 10) was unsuccessful at 50,000 cycles (one sample), 100,000 cycles (two samples), 110,000 cycles (one sample), and 115,000 cycles (one sample). The failure mode of the samples was over the minimum working pressure and piston stroke time. As the troublesome cylinders were broke down, the hardening and wear of piston seals were similar to samples returned from the field. The failure mechanism occurred as follows: (1) repeated operating stress, (2) friction heat in seal, (3) lube vaporization in seal, (4) elevating seal hardening, and (5) elevating seal wear. The forms and locations of failure in the samples attained from the 2nd ALT and the market (or usual circumstance) were identical (Fig. 8.75).

The 2nd ALT failure and field failure data manifested an alike pattern on a Weibull chart (Fig. 8.76). As the data for the 2nd ALT and field had alike gradients on the chart, each loading circumstance of the 2nd ALT and that from the market over the product life were alike. It might be expected that the test samples in the laboratory would break similarly to those in the market. For the shape parameter, β, the last shape

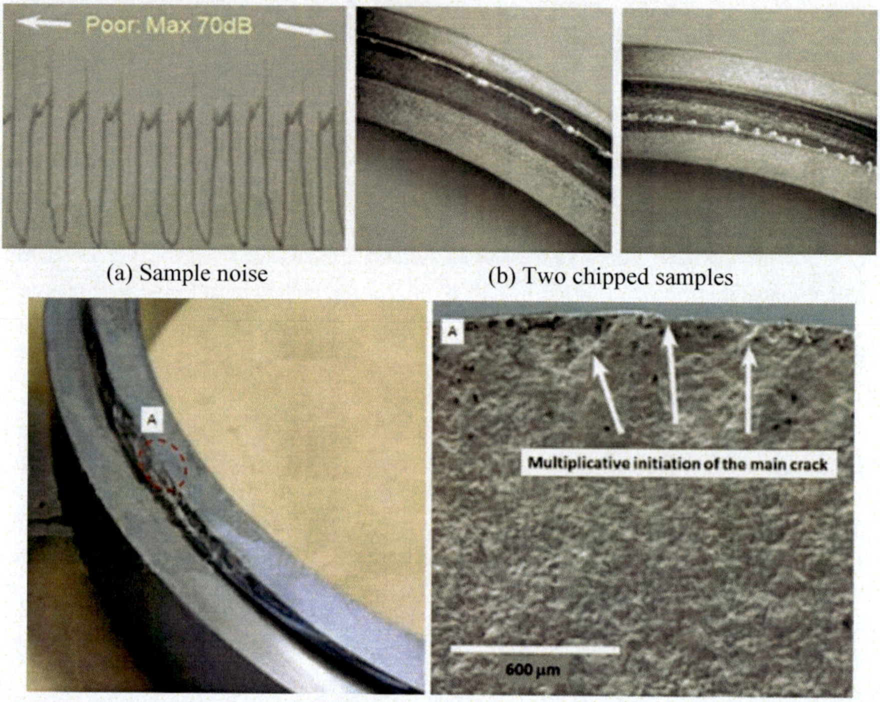

(a) Sample noise (b) Two chipped samples

(c) Partially cracked sample

Fig. 8.74 Design problems of the cylinder in the 1st ALT

parameter from the plot was affirmed to be 1.97, compared with the approximated value of 2.0. Based on both test results in the Weibull chart, the parametric ALT was successful because it recognized the design flaws which were accountable for the field failures. Because of the resemblance in both photos in Fig. 8.75 and similar gradient in the Weibull chart in Fig. 8.76, these ALTs were functional in helping recognize the design defects which were judged to be the failures of the cylinders returned from the field. These failures also decided the product life.

To discover the failures of the cylinder under an accelerated circumstance of 1.2 MPa and 23 °C, reliability testing was performed. As a consequence, the cylinder was unsuccessful in 5 out of 6 samples as follows: roughly 30,000 cycles (one sample), 70,000 cycles (two samples), and 80,000 cycles (two samples). The failure mode of the samples was over the minimum operation pressure and stroke time piston. As the unsuccessful pneumatic cylinders were disassembled, hardening and wear of the piston seal were found, and the rod cap was blocked by a slurry. The repetitive pressure loading of the cylinder brought the wear of the piston seal, which produced the slurry and its flow into the port. The slurry was thus gathered at the port and blocked (Fig. 8.77).

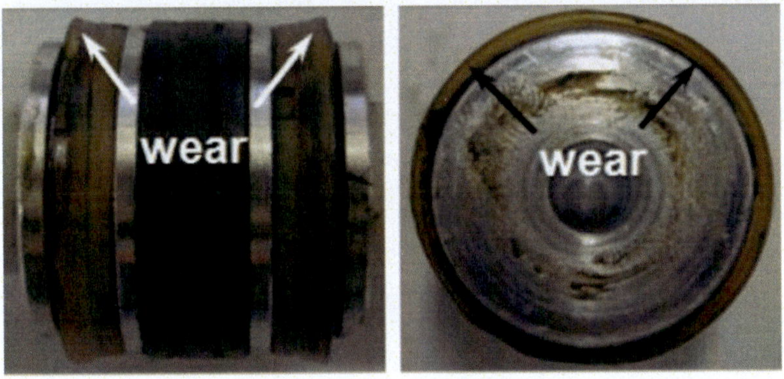

(a) unsuccessful products in thetesting data
at the field (orusual circumstance)

(b) unsuccessful products after ALT

Fig. 8.75 Unsuccessful products in the market and ALT

Fig. 8.76 Field statistics and results of ALT on the Weibull plot

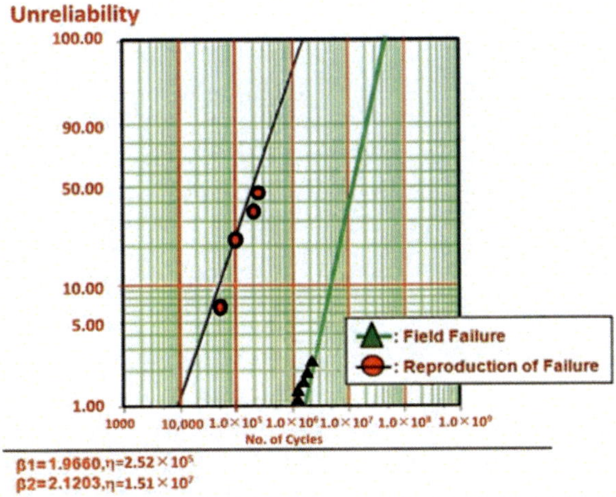

Fig. 8.77 Blocked rod cap discovered by parametric ALT

As closely investigating the product failure in the 2nd ALT, it was decided that the hardening and wear of the piston seal in the cylinder originated from utilizing an inappropriate matter. For this occasion, it was a silicone rubber. Though this matter had fine sealing characteristics, it was still unsafe to breakdown under repetitive pressure loading. The seal material, C2, was altered from silicone rubber to a (thermoset) polyurethane with excellent abrasion and wear resistance, which supplied improved reliability versus a normal rubber-based seal.

In the 3rd ALT, there were no design issues in the cylinder till the ALT extended 220,000 cycles. It was thus deduced that the design alternations obtained from the 1st and 2nd ALTs were functional in fulfilling the wanted life of the product. Figure 8.78 and Table 8.15 abridge the parametric ALT outcomes. With the altered designs, the cylinder samples were assured to get the life objective—B1 life 10 years with approximately a 60% confidence level.

8.10 Drawer System in a French Refrigerator

One favored kind of refrigerator is the French door refrigerator. Figure 8.79 manifests one with a newly designed drawer system. It consists of a box, two guide rails, and a support in the center between the two drawers. Food is stocked in the drawers. The drawer system might be designed to withstand the working conditions subjected to it by the customer. In the United States, the typical consumer opens drawers in the refrigerator to stock food five to ten times per day. Storing food in the French door refrigerator requires repetition as follows: (1) the drawer is opened, (2) food is put in it, and thus (3) it is closed. The drawers have dissimilar quantities of food stocked in them when the consumer employs it.

The drawer system returned from the field had been fracturing, bringing consumers to request for substitution. As the drawers failed to be subjected to repetitive stresses from openings/closings, it was understandable that there was a design issue with the drawer system. Unsuccessful drawers from the market manifested that the drawer system had critical design flaws, which included stress risers—thin

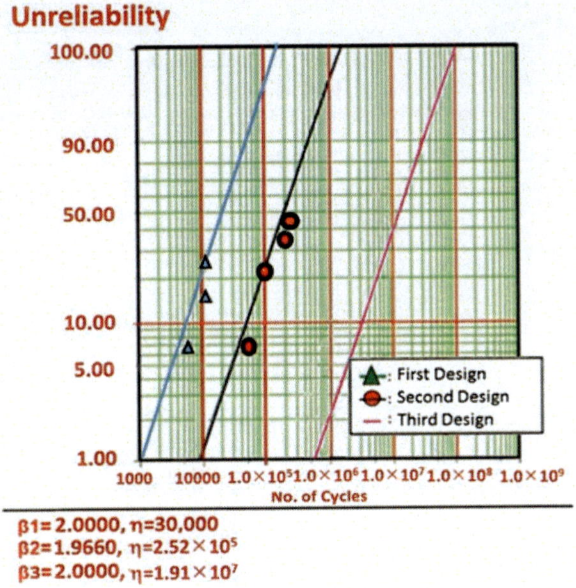

Fig. 8.78 Results of ALT on the Weibull chart

Table 8.15 Pneumatic cylinder ALT outcomes

Parametric ALT	First ALT	Second ALT	Third ALT
	Original design	Design	Final design
In 220,000 cycles, there are no issues in the pneumatic cylinder	6000 cycles: 1/10 noise (partially cracked sample) 10,500 cycles: 1/10 noise 11,000 cycles: 1/10 noise (two chipped samples)	50,000 cycles: 1/10 fail 100,000 cycles: 1/10 fail 110,000 cycles: 1/10 fail 115,000 cycles: 1/10 fail (hardening and wear of piston seal)	220,000 cycles: 10/10 OK
Structure			
Action plans	C1: material: Fe-36Ni Invar Alloy → silicone rubber	C2: material: Silicone rubber → (thermoset) polyurethane	

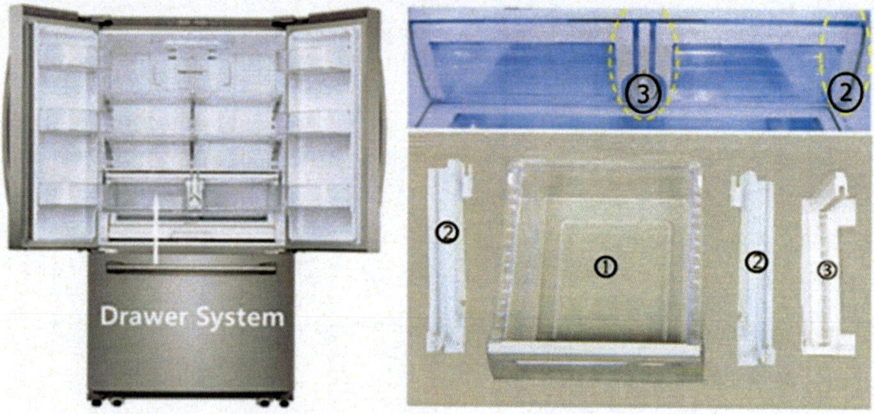

(a) French refrigerator (b) Mechanical components of drawer: 1) a box, 2) left/right of the guide rail, and 3) a support center

Fig. 8.79 French refrigerator with a new drawer system

ribs and sharp corner angles. These flaws produced failures (cracks) in the drawers once they were subjected to repetitive opening and closing with food loads in the drawers although the drawer was designed to withstand repetitive food loading under expected consumer functioning circumstances (Fig. 8.80).

As consumers opened the drawer, they normally took food out or put food into it. Depending on the end-user working circumstances, the drawer system encountered repetitive loading as food was loaded/unloaded and the drawer was opened and closed. To exactly work the drawer system, many mechanical structural components in the drawer assembly should be designed to handle the anticipated loading from the customers. Because the concentrated stresses in a mechanical system happen at stress raisers such as sharp corner angles, it is critical to decide these design defects experimentally and thus alter them.

From the drawer system and its free-body diagram (Fig. 8.80b), we recognized that the drawing force originated from the food weight. The applied force in the drawer shall be defined as

$$F_{\text{box}} = \mu W_{\text{load}} \tag{8.71}$$

As the stress in a drawer system depends on the applied food load, the LS prototype from Eq. (7.19) shall be expressed as

$$TF = A(S)^{-n} = A(F_{box})^{-\lambda} = A(\mu W)^{-\lambda} = B(W)^{-\lambda} \tag{8.72}$$

where A and B are constants.

The *AF* shall be defined as

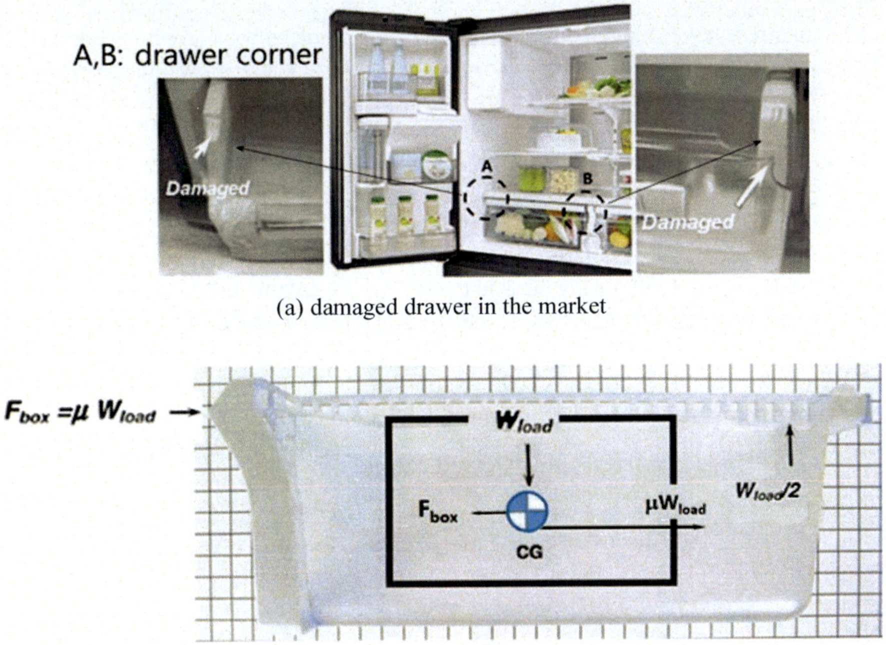

(a) damaged drawer in the market

(b) Free-body diagram

Fig. 8.80 Damaged drawer after use

$$AF = \left(\frac{S_1}{S_0}\right)^n = \left(\frac{F_1}{F_0}\right)^\lambda = \left(\frac{\mu W_1}{\mu W_0}\right)^\lambda = \left(\frac{W_1}{W_0}\right)^\lambda \quad (8.73)$$

For the French-door refrigerator, including the drawer system, the environmental (or working) consumer circumstances are roughly 0–43 °C with 0.2–0.24 g of acceleration and a relative humidity varying from 0 to 95%. As formerly mentioned, the drawer cycles per day were between 5 and 10 times. With the design criterion of a product life for 10 years, L_B^*, the drawer was subjected to 36,500 use cycles for the worst case.

Under a life objective—B1 life 10 years—if the number of life cycles L_B^* and AF are calculated for the assigned sample size, the actual mission cycles, h_a, could be decided from Eq. (7.29). ALT apparatus can then be constructed and operated in accordance with the anticipated customer usage circumstances of the drawer system. Through parameter ALTs, we might achieve the design defects for a new mechanical product.

The greatest force applied by the consumer in stocking food, W_1, was 0.059 kN (6 kg_f). To determine the stress quantity for ALT, we utilized the step-stress life test, which shall evaluate the lifetime under constant used conditions for various accelerated food weights, such as 0.088 kN (9 kg_f), 0.117 kN (12 kg_f), and 0.137 kN (14 kg_f). As some stress levels are altered, the failure cycles of the drawer system at a particular stress level shall be observed. That is, the crack at first increases slowly, but the growth accelerates (i.e., *da*/*dN* increases) as the crack size glows due to elevated load. Thus, we shall identify the failure time, in which it fulfills a critical size and failure occurs at the design weak points.

For ALT, the applied force, W_2, doubled to 0.117 kN (12 kgf). With a cumulative damage factor, λ, of 2, the AF was 4.0 from Eq. (8.73). To attain the design defects of a new drawer, the lifetime objective could be put to be more than B1 life 10 years. First, we presumed that the shape parameter β was 2.0, and the real test cycles calculated from Eq. (7.29) were 37,000 cycles for six sample units. If this ALT fails less than once during 37,000 cycles, the life for the drawer will be reassured to be B1 life 10 years.

To indicate the number of test cycles, beginning, and ending of the equipment, etc., a testing equipment with a control console was employed to function the samples. As the start knob on the controller panel gave the starting signal, the simple hand-shaped arms pushed and pulled the drawer. The greatest mechanical food force due to elevated load (0.117 kN) was applied on the drawer (Fig. 8.81).

In the 1st ALT, we discovered the failure time of the next stress levels: 0.088 kN (9 kgf), 0.117 kN (12 kgf), and 0.137 kN (14 kgf). For 0.088 kN (9 kgf), the cover of the drawer fractured at 14,000 cycles, 19,000 cycles, and 21,000 cycles. For 0.117 kN (12 kgf), it fractured at 3800 cycles and 4800 cycles. For 0.137 kN (14 kgf), it fractured at 550 cycles, 650 cycles, and 800 cycles. To look into the fracture surfaces, they were detected by SEM. We discovered voids produced because of falling-out particles (Fig. 8.82).

Eventually, we decided the stress level to be 0.117 kN (12 kgf) for the parametric ALT because it had comparatively satisfactory data linearity compared with the other stress levels. It can also be seen that increasing the repetitive food weight has the effect of shifting the left of failure time as the stress range (or intensity range ΔK) increased and the crack growth rate moved up, but it did not affect the slope of the growth rate curve and shape parameter β (Fig. 8.83).

At first, when 1.17 kN (12 kgf), as the elevated weight in the drawer was loaded, the left/right rollers on the rail were broken away, and the center support rail was detorted so that the drawer system did not glide any more. Because of inadequate strength due to design defects, the draw was altered by enlarging the roller support to 7 mm (C2) on the guide rail, as well as adding up strengthened ribs on the center support rail (C1) (Fig. 8.84 and Table 8.16).

As deliberately noticing the returned drawers from the field and the 1st ALT, failure places were discovered to be in the junction areas of the drawer cover and its body structure as a consequence of high repeated stress and anticipated high da/dN. Figure 8.85 also supplies a Weibull chart of the ALT consequences compared to the data from the field. As the two patterns had alike gradients on its chart, it was identified that each loading of the 1st ALT and the field was alike for the operation circumstances. For the shape parameter, β, the final shape parameter from the plot was confirmed to be 4.2, compared with the approximated value—2.0. Based on both the test outcomes and the Weibull chart, this ALT was successful because it recognized the design defects which were responsible for the market failures. Based on the

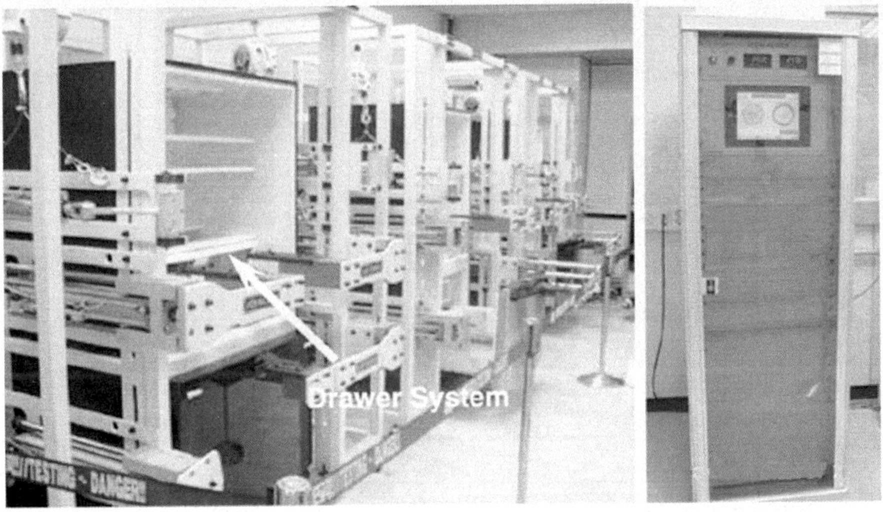

(a) Apparatus (b) Controller

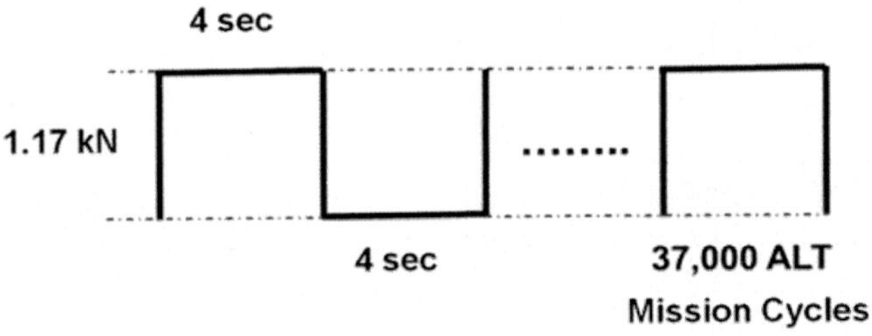

(c) Duty cycles of repetitive food load F

Fig. 8.81 Equipment utilized in ALT and duty cycles

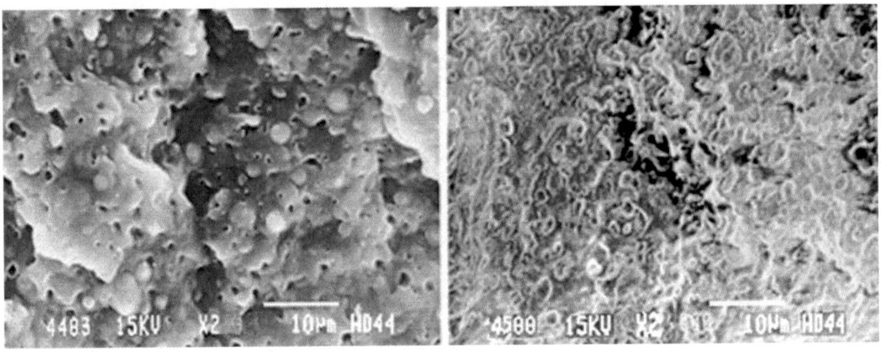

Fig. 8.82 SEM micrographs of fracture surfaces

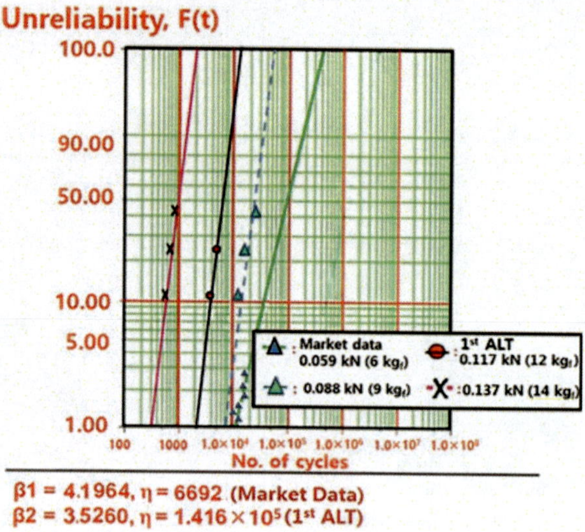

Fig. 8.83 Outcome of the step-stress life test on a Weibull chart

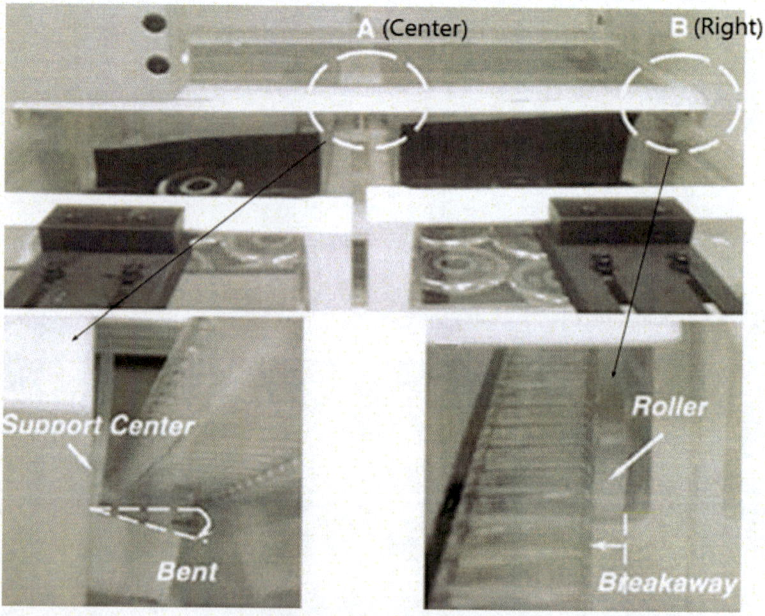

Fig. 8.84 Design problems of the drawer in (static) loading

Table 8.16 Modified designs of center support and rail in initial loading

Center support	Guide rail
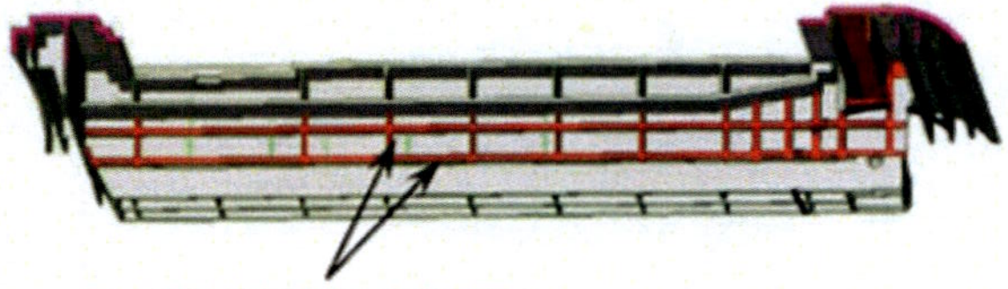	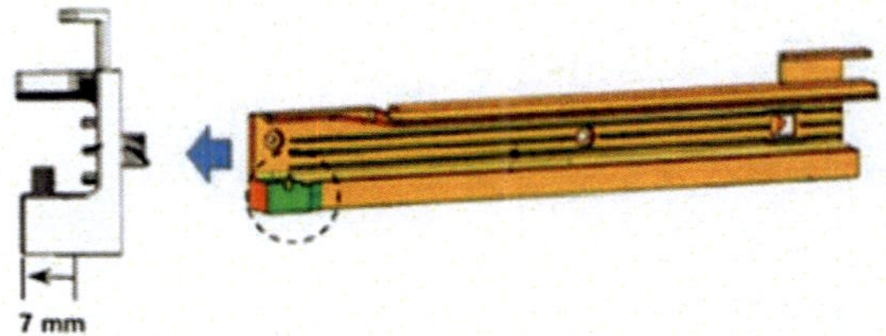

photos from the ALT & the market and Weibull chart, the tests helped in recognizing the issues in the design accountable for field failures and the poor product life.

Due to design flaws such as no corners in the high stress areas of intersection (A), the repetitive loading of the drawers in conjunction with these structural flaws may have been fracturing the drawer cover. These design flaws shall be modified by (1) making thicker reinforced ribs, Rib1, C3, from T2.0 mm to T3.0 mm and (2) exerting the fillets, Fillet1, C4, from R0.0 mm to R1.0 mm (Fig. 8.85).

With the affirmed shape parameter, β, being 4.2, the actual mission cycles decided from Eq. (18) were 19,000 cycles for six samples. If the drawer was unsuccessful at less than once for 13,000 cycles, its life might be ensured to be B1 life 10 years. In the 2nd ALT, the fractured guide rail and sunken roller in the center support rail was unsuccessful at 6000 cycles. As deliberately examining the product failure in the 2nd ALT, the guide rail in the drawer had no enforced rib and inadequate corner rounding to withstand the repetitive loading of the drawer. To enhance the design of the guide rail, it was modified by (1) attaching enforced ribs, C5, and (2) enlarging the corner rounding, C6, from R3 mm to R4 mm. The center support rail in the drawer was

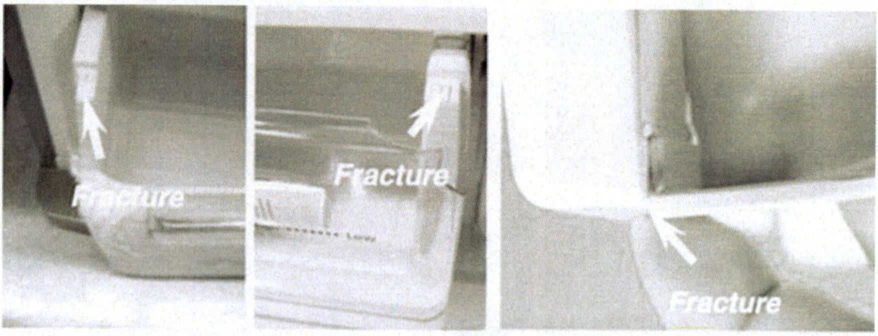

(a) Troublesome product after 1st ALT (b) Product with crack in field

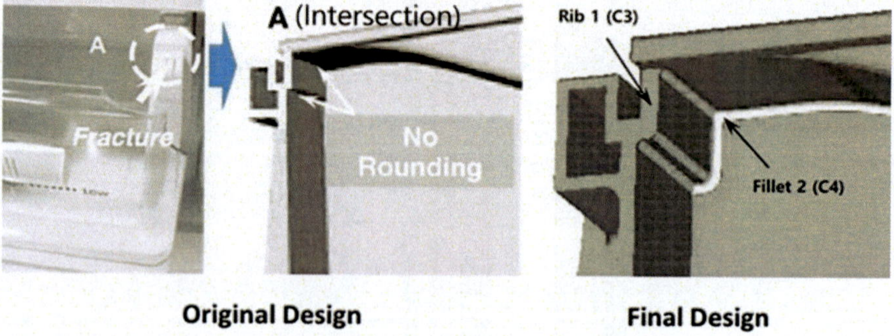

(c) Design alternation of drawer components.

Fig. 8.85 Design alternation of drawer components in the 1st ALT

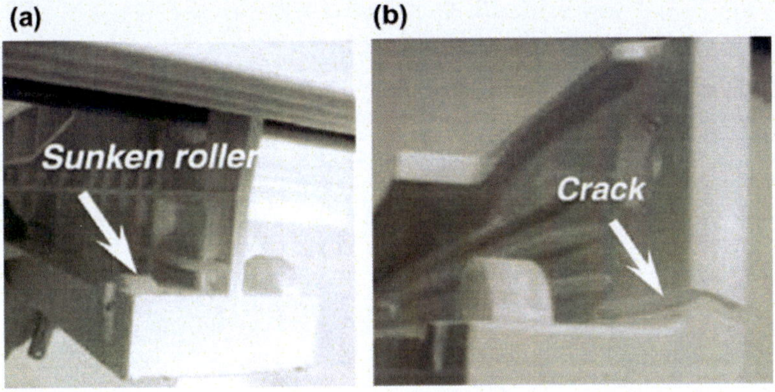

Fig. 8.86 Problems of drawer components in the 2nd ALT. **a** Sunken roller on support rail. **b** Cracked guide rail

modified by (1) enlarging the roller rib, C7, from L0.0 mm to L2.0 mm (Fig. 8.86 and Table 4).

As the drawer design was improved, the lifetime of the new samples was anticipated to be more than the B1 life 10 years. To affirm the lifetime target of the design of the drawer, the 3rd ALT was carried out. Because the affirmed value, β, on the Weibull chart was 4.2, for the lifetime objective—B1 life 10 years—the actual mission cycles in Eq. (18) were 19,000 for a sample size of six. In the 3rd ALT, there were no problems in the drawer till the ALT fulfilled 22,000 cycles. Therefore, the modified designs attained from the 1st and 2nd ALT procedures were accomplishing the lifetime target.

Table 8.17 is an abridge of the redesigned center support and (left/right) rail. With the altered designs, the drawer was ensured to fulfil the life target—B1 life 10 years.

8.11 Improving the Lifetime of a Hinge Kit System (HKS) in a Refrigerator

As a customer works a refrigerator door, they want to comfortably close the door. A new HKS was designed for the refrigerator (Fig. 8.87) to increase the comfort of opening and closing the door for the customer. As opening/closing the door, the HKS was subjected to repetitive impact loads over the life of the domestic refrigerator. To withstand the loads of the HKS, new metals—standard austenitic ductile iron (18 wt% Ni)—for the torsional shaft were a key metal component used. Due to their low cost and outstanding workability, ductile cast irons have been employed for some mechanical components. They have good monotonic strength and high ductility compared to malleable cast irons and gray cast irons. The fatigue strength

Table 8.17 Abridge for the altered center support and (left/right) rail

Center support	Rail
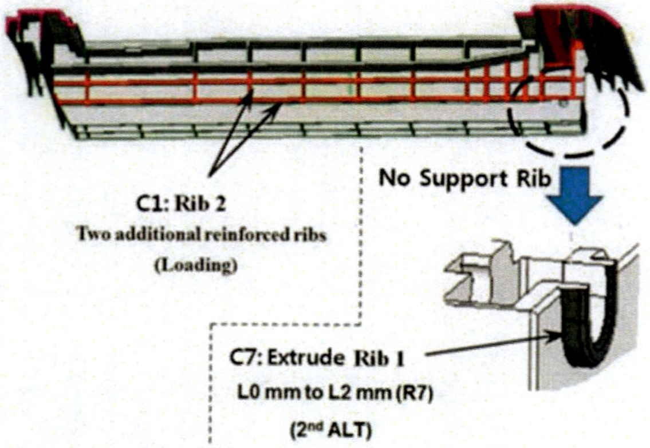	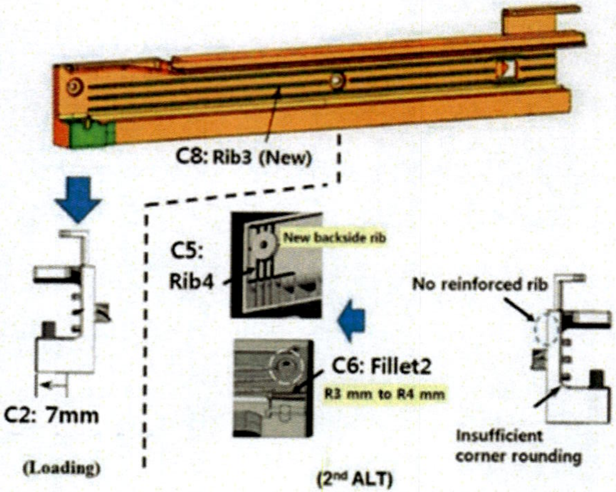

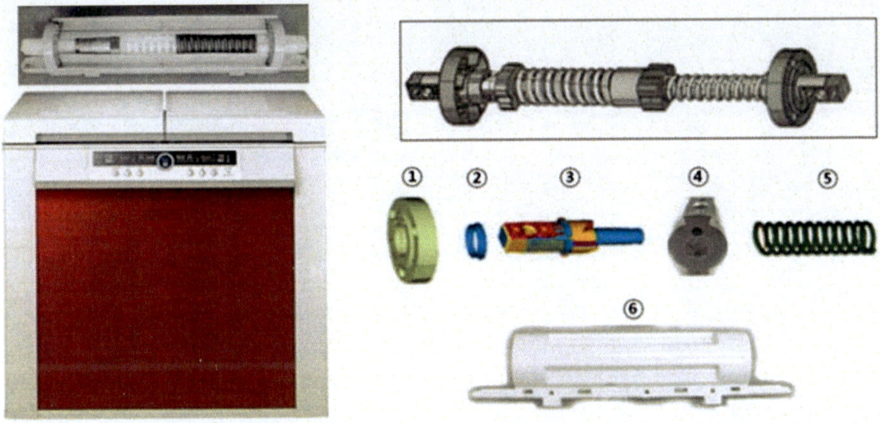

(a) Commercial Kimchi refrigerator

(b) HKS parts: a kit cover ①, torsional shaft (ductile iron) ④, spring ⑤, and kit housing ⑥

Fig. 8.87 Commercial Kimchi refrigerator and its HKS

of ductile cast irons is relatively lower than those of steels and alloys with identical quantities of monotonic strength because of their distinguishing microstructure holding graphite particles and casting defects. The fatigue strength of ductile cast iron in the contemporary HKS design was assessed through parametric ALT.

The HKS depicted in Fig. 8.87b consisted of a kit cover, torsional shaft (ductile iron), spring, and kit housing. To appropriately do its function for a product life, the HKS shall be designed to withstand the working conditions subjected to it by the consumers who employ the refrigerator. In the Korean domestic market, the typical consumer opened and closed the refrigerator door from three to ten times per day. Storing food in the refrigerator had some repetitive working process: (1) Open the door of refrigerator, (2) place the food into it, and thus (3) close it. The HKS had dissimilar mechanical impact loadings when the consumer employed it.

As seen in Fig. 8.88, the HKS in the field had been fracturing, bringing consumers to demand that the refrigerator be substituted. As subject to repetitive impact stresses in utilizing the refrigerator door, it was decided that the troublesome HKS started from some design flaws. Field data also designated that the returned products had critical design issues on the structure, including stress risers—sharp corner angles and thin ribs. These design flaws prevented the HKS from enduring the repeated impact loads during the openings/closings and resulted in a crack that propagated to its end. The HKS was at first designed to withstand repetitive impact loading under customer usage circumstances.

As customers operated the refrigerator door, they could take out and put in food. Depending on the end-user working circumstances, the HKS encountered repetitive impact loading in the process. To accurately work with the HKS, many mechanical structural components in the HKS assembly need to be designed robustly. As the

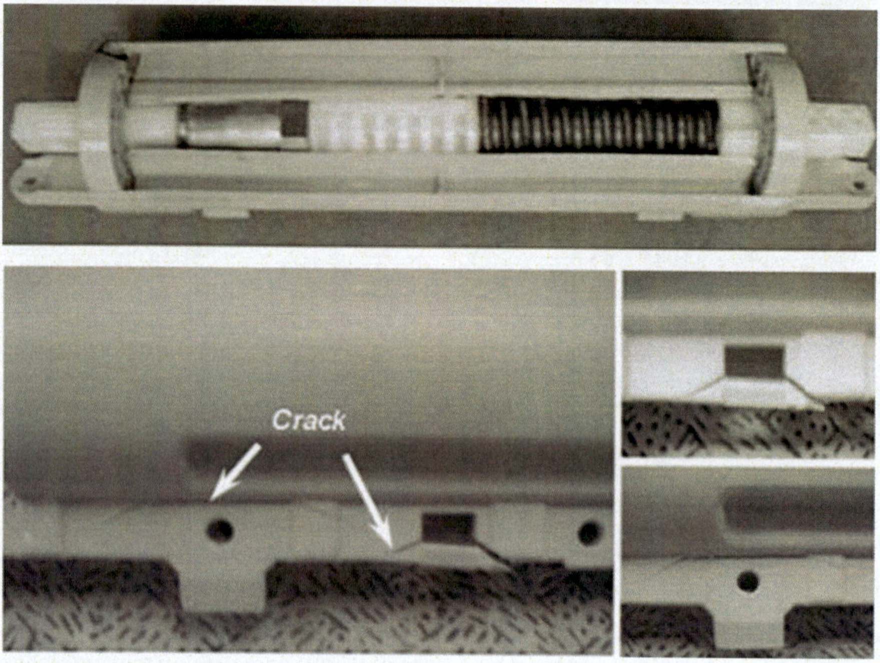

Fig. 8.88 A damaged HKS after a time of usage

concentrated stress in the mechanical system was revealed at stress raisers such as sharp corner angles, it was critical to reveal these design defects experimentally. As a consequence, engineers shall alter the design.

As shown in Fig. 8.89, from the functional design concepts of a mechanical HKS, we recognized that the impact force on the HKS originated from the door weight. That is, the moment balance around the HKS shall be expressed as

$$M_0 = W_{door} \times b \tag{8.74}$$

$$(8.74) = T_0 = F_0 \times R \tag{8.75}$$

where b is the distance from the HKS to the center of gravity (CG) of the door.

To escalate the impact on the HKS, added elevated weight was attached. The moment balance around the HKS with an elevated weight shall be expressed as

$$M_1 = M_0 + M_A = W_{door} \times b + W_A \times a \tag{8.76}$$

$$(8.76) = T_1 = F_1 \times R \tag{8.77}$$

where a is the distance from the HKS to the accelerated weight.

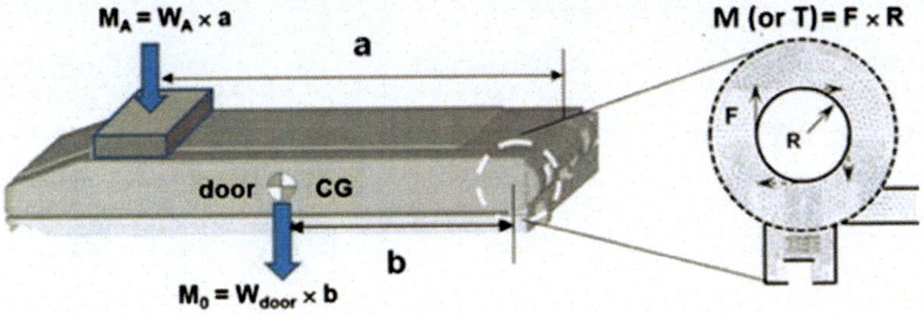

Fig. 8.89 Robust design schematic of the HKS

As the time to failure relied on the impact force due to moment, it was controlled during the ALT. Under the same working circumstances, the life-stress model in Eq. (7.19) shall be defined as

$$TF = A(S)^{-n} = AT^{-\lambda} = A(F \times R)^{-\lambda} = B(F)^{-\lambda} \tag{8.78}$$

The AF from Eq. (7.21) shall be defined as

$$AF = \left(\frac{S_1}{S_0}\right)^n = \left(\frac{T_1}{T_0}\right)^n = \left(\frac{F_1 \times R}{F_0 \times R}\right)^n = \left(\frac{F_1}{F_0}\right)^n \tag{8.79}$$

For a refrigerator including the HKS, the environmental (or working) consumer circumstances were approximately 0–43 °C with a relative humidity ranging from 0 to 95% and 0.2–0.24 g of acceleration. As formerly stated, the number of openings/closings of the HKS per day changed from 3 to 10 times. With a design standard of a product life for 10 years, L_B^*, the HKS has 36,500 use cycles in the worst occasion.

Under a life objective—B1 life 10 years—if the number of lifetime cycles L_B^* and AF are calculated for the assigned sample size, the real mission cycles, h_a, would be attained from Eq. (7.29). Thus, the ALT apparatus shall be established and carried out in accordance with the working course of the HKS. Through parameter ALTs, the design defects for the new mechanical system shall be recognized.

The maximum impact force due to the door weight applied by the consumer in employing the refrigerator, F_1, was 1.1 kN. To decide the stress quantity for ALT, we utilized the step-stress life test, which shall evaluate the lifetime under constant used conditions for various elevated weights. As the stress level changed, the failure times of the HKS at a specific stress level were noticed. Finally, for an ALT with an elevated weight, we decided that the applied impact force, F_2, was 2.76 kN. With an accumulative damage exponent, λ, of 2, the AF was 6.3 from Eq. (8.79). To attain the design defects of a newly designed HKS, a lifetime objective should be more than B1 life 10 years. If the shape parameter β was 2.0, the number of test cycles calculated from Eq. (7.29) would be 23,000 cycles for 6 sample units. If this ALT failed less than once for 23,000 cycles, the life for the HKS would be ensured to be B1 life 10 years (Fig. 8.90).

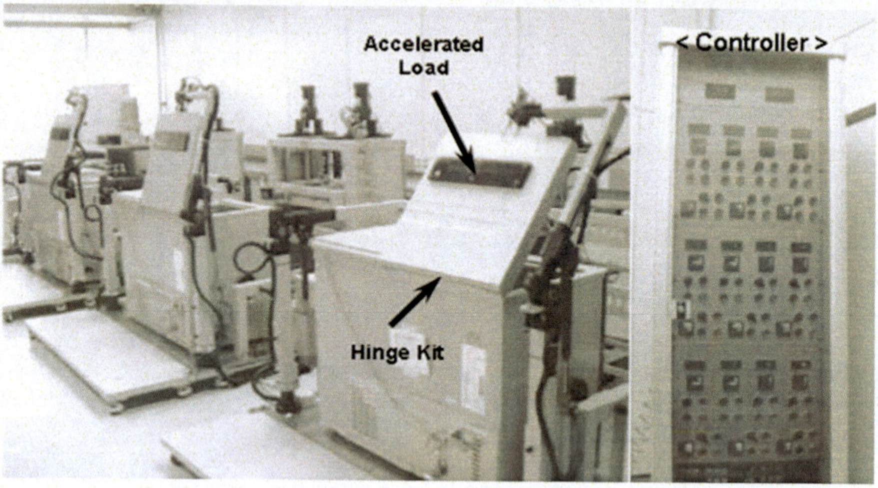

(a) ALT equipment

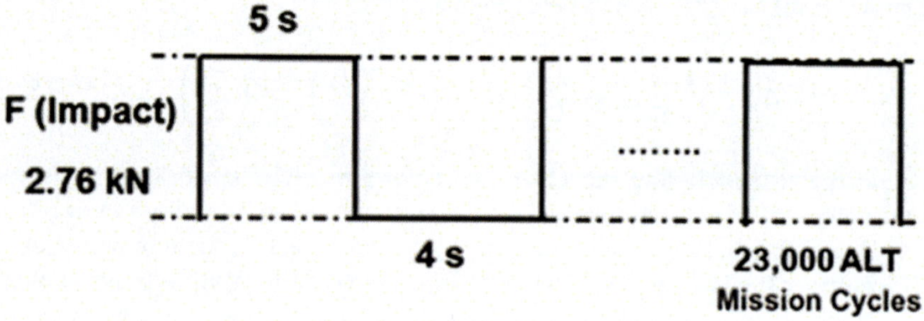

(b) Duty cycles of repeated impact load F.

Fig. 8.90 Apparatus utilized in ALT and controller

The control panel was employed to operate the testing equipment—the number of test cycles, beginning or ending the equipment, etc. When the start knob on the controller console gave the starting signal, the straight hand-shaped arms clasped and raised the refrigerator door. As the door was shut, it was applied on the HKS with the maximum mechanical impact force due to the elevated load (2.76 kN).

In the 1st ALT, the housing of the HKS was unsuccessful at 3000 cycles. Figure 8.96 manifests the unsuccessful product from the field and the 1st ALT. Upon deliberately noticing the failure places from the field and the 1st ALT, it was discovered that the failures were around the housing and its support in the HKS structure as a consequence of high impact stress (Fig. 8.91).

Figure 8.92 supplies a graphical presentation of the 1st ALT outcomes and the failure data from the market manifested on the Weibull chart. As the two patterns had alike gradients on the chart, each loading state of the 1st ALT and the market

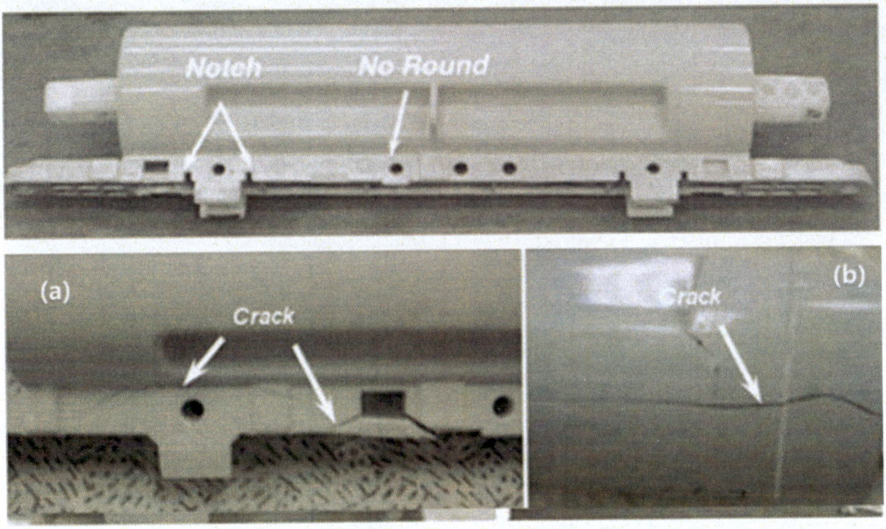

(a) Unsuccessful products in the market (b) crack after 1st ALT.

Fig. 8.91 Unsuccessful products in the market and crack after the 1st ALT

over the product life were similar under the operational circumstances of customers. Therefore, it might be anticipated that the test samples shall be unsuccessful like those in the market. For the shape parameter, β, the final shape parameter from the plot was confirmed to be 2.0, compared with the estimated value—2.0. Based on both test results in the Weibull chart, the parametric ALT was successful because it recognized the design defects which were responsible for the market failures. In other words, as proved by two things—the visual representation in the pictures and similar gradients in the Weibull chart—these structured ways were valid in recognizing the troublesome designs which explained the failures from the market. These failures determine the refrigerator life.

Due to the design flaw of no support in the high-stress regions, the repetitive impact loading in conjunction with this structural imperfection shall have generated fracturing of the HKS housing. This design flaw shall be modified by attaching support ribs C1 (Fig. 8.93). Stress analysis, which shall be integrated with fatigue analysis and parametric ALT, was carried out by utilizing a finite element analysis (FEA). As the HKS was fastened against the wall (or outside) as the boundary circumstances, simple impact loads, as seen in Fig. 8.89, were exerted. Utilizing materials and processing conditions alike to those of the finished HKS, the constitutive properties of the matters such as HIPS (HKS housing) were decided. The greatest stresses for the old and new designs were assessed, individually. Based on these outcomes, the suitableness of the current designs for the HKS housing was assessed. After altering the new designs to enhance the design against fatigue, the approximated stress concentrations in the HKS housing lessened from 21.2 to 15.0 MPa using FEM investigation. It was anticipated that this new design would be successful in

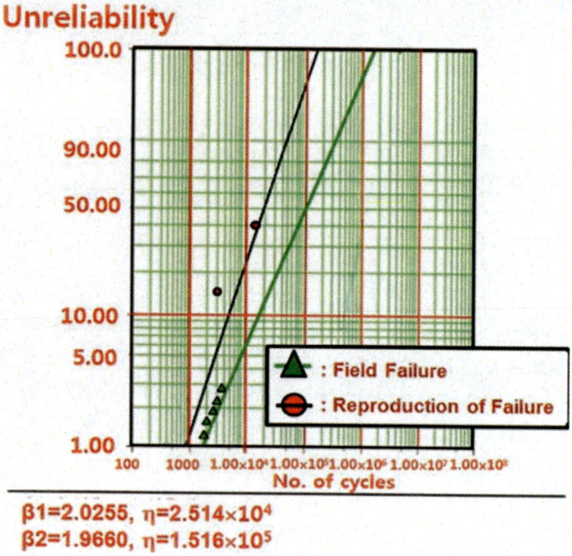

Fig. 8.92 Field data and 1st ALT on the Weibull chart

lessening fatigue failure of the HKS housing when subjected to repetitive loads under customer use circumstances.

With the affirmed shape parameter β of 2.0, the actual mission cycles calculated from Eq. (7.29) were 23,000 cycles for the six sample units. If the HKS was unsuccessful at less than once for 23,000 cycles, its life might be ensured to be B1 life 10 years. As manifested in Fig. 8.94, in the 2nd ALT, from the surface corner, the torsional shaft in the HKS which was made of ductile iron failed at 12,000 cycles. Such ductile cast iron explains a major family of metals which are exclusively employed for gears, automobile crankshafts, dies, and numerous machine components because of its fine machinability, fatigue strength, and high modulus of elasticity. They have a mass fraction (%) as follows: carbon (3.0–3.7), silicon (1.2–2.3), manganese (0.25), magnesium (0.07), and phosphorus (0.03).

As deliberately inspecting the product failure in the 2nd ALT, the torsional shaft in the HKS had inadequate strength to withstand the repetitive impact loading of the opening/closing of the door. As subjected to repetitive impact loads, the stress amplification of mechanical parts such as the torsional shaft in the HKS not only happened at minute flaws or cracks on a microscopic level of matter but could also occur in stress concentrations such as in sharp corners, fillets, holes, and notches on the macroscopic range which are usually described as stress raisers. For instance, the stress concentration at the sharp-edged corners relied on the fillet radius. To enhance the HKS design, the torsional shaft was modified by taking it more rounding from R0.5 mm to R2.0 mm, C2 (Fig. 8.94).

For the HKS upgrade, the design basis of new samples was decided to be more than the life objective—B1 life 10 years. To affirm the design of the HKS, a 3rd ALT was carried out. As the confirmed value, β, on the Weibull chart was 2.0, for the life

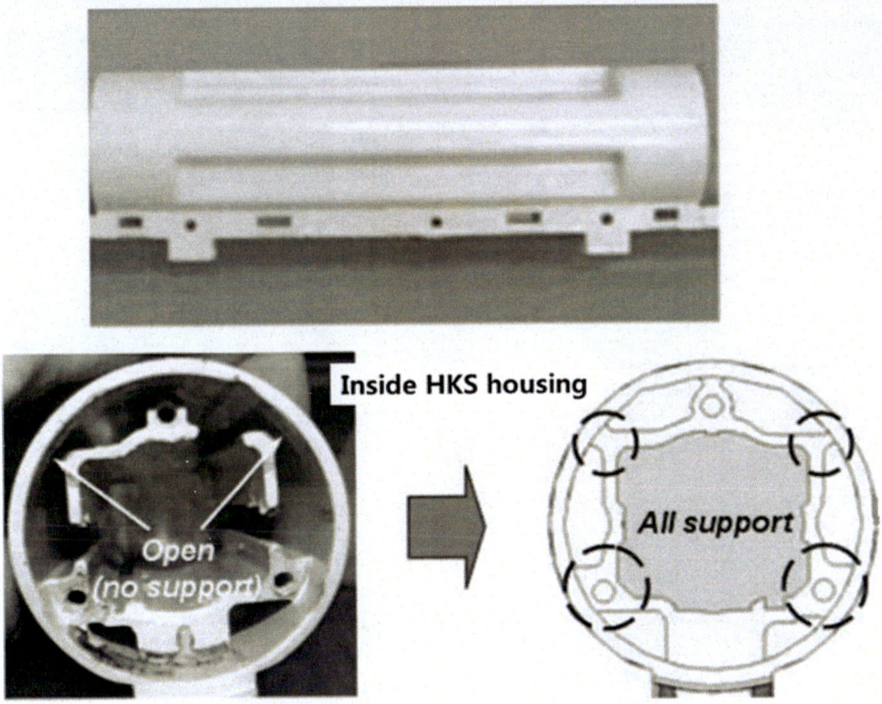

(a) Its basic cause (b) Design modifications

Fig. 8.93 Structure of unsuccessful HKS in the 1st ALT

objective—B1 life 10 years—the actual mission cycles in Eq. (20) were 23,000 for the six-sample size. In the 3rd ALT, there were no design problems in the HKS till the experiment reached 23,000 cycles. It was thus deduced that the modified designs attained from the 1st and 2nd ALTs were efficient.

Table 8.18 supplies an abridge of the ALT outcomes. With the design alternations, the HKS was ensured to have a life objective—B1 life 10 years. That is, we recognized that the product would have 99% reliability (or 1% unreliability) for 10 years with a yearly failure rate of 0.1%.

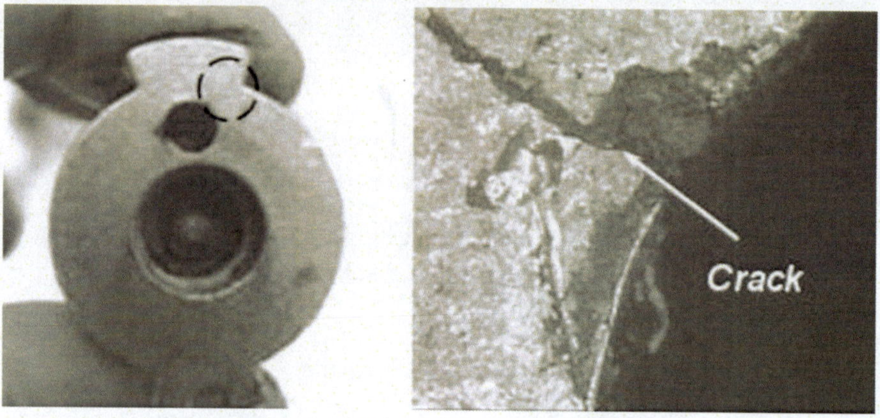

(a) Root cause

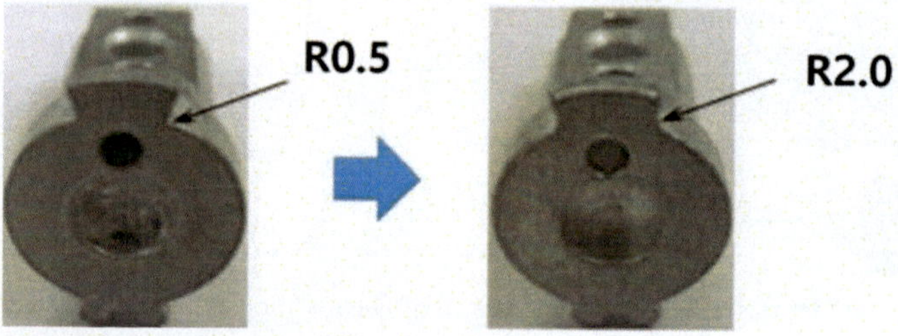

(b) Design modification

Fig. 8.94 Redesigned HKS structure

Table 8.18 Outcomes of ALT

Parametric ALT	1st ALT	2nd ALT	3rd ALT
	Initial design	Second design	Last design
In 23,000 cycles, there are no problems in the HKS	3000 cycles: 2/6 fracture (HKS housing)	12,000 cycles: 4/6 crack (torsional shaft)	23,000 cycles: 6/6 OK 41,000 cycles: 6/6 OK
HKS structure	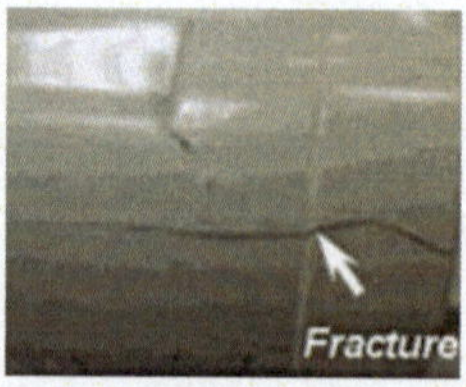		–
Action plans	C1: No → 2 support ribs	C2: R0.5 mm → R2.0 mm Roundness corner of torsional shaft	–